LINEAR PROGRAMMING

LINEAR PROGRAMMING

JAMES P. IGNIZIO
University of Virginia

TOM M. CAVALIER
The Pennsylvania State University

 PRENTICE HALL, Englewood Cliffs, New Jersey 07632

Library of Congress Cataloging-in-Publication Data

Ignizio, James P.
   Introduction to linear programming / James P. Ignizio, Tom M.
Cavalier.
      p.   cm.
   Includes bibliographical references and index.
   ISBN 0-13-183757-5
   1. Linear programming.   I. Cavalier, Tom M.   II. Title.
T57.74.I355 1994
   519.7'2--dc20   93-22179
                                                        CIP

Acquisitions Editor: Marcia Horton
Production Editor: Bayani Mendoza de Leon
Copy Editor: Peter Zurita
Production Coordinator: Dave Dickey
Editorial Assistant: Dolores Mars
Series Logo Design: Judith Winthrop

 © 1994 by Prentice-Hall, Inc.
A Paramount Communications Company
Englewood Cliffs, New Jersey 07632

Printed in the United States of America

10 9 8 7 6 5 4 3 2 1

ISBN 0-13-183757-5

Prentice-Hall International (UK) Limited, *London*
Prentice-Hall of Australia Pty. Limited, *Sydney*
Prentice-Hall Canada Inc., *Toronto*
Prentice-Hall Hispanoamericana, S.A., *Mexico*
Prentice-Hall of India Private Limited, *New Delhi*
Prentice-Hall of Japan, Inc., *Tokyo*
Simon & Schuster Asia Pte. Ltd., *Singapore*
Editora Prentice-Hall do Brasil, Ltda., *Rio de Janeiro*

Jim dedicates this book to his wife, Cynthia.

Tom dedicates this book to his children, Ian and Sierra.

CONTENTS

* Starred sections may be omitted in an introductory course with no appreciable loss of continuity.

14   MULTIOBJECTIVE MODELS   **541**

PREFACE

In this introductory level text on linear programming, a thorough, up-to-date, and comprehensive summary of the philosophies and procedures used in the modeling, solution, and analysis of so-called linear programming problems is provided. The text is for the senior-level college student or for the first-year graduate student having some previous exposure to linear algebra—and earlier versions (or selected portions) have been used in such courses over the past decade. An associated solution manual is provided for all exercises and is available to those adopting the text for classroom use.

The text is divided into three parts. Part 1, consisting of Chapters 2 through 8, addresses linear programming in general—emphasizing the development, presentation, and illustration (via numerous examples) of the fundamentals necessary to model, solve, and analyze linear programs. Although the coverage is relatively traditional, some very unique aspects should be noted. First, coverage is provided in Chapter 7 of recent results with regard to alternative methods to the simplex algorithm, in particular the affine scaling variants of the Karmarkar algorithm. Second, Chapter 8 deals with the use of linear programming in information technology—particularly as a means to analyze large amounts of data. Topics covered include prediction/forecasting, pattern classification/pattern recognition, clustering analysis, input–output analysis, and even the design and training of neural networks—*all achieved by means of linear programming*.

Part 2 addresses *integer* linear programming, including the network simplex method (Chapter 9), transportation and assignment problems (Chapter 10), and general integer programming methods (Chapter 11). In addition, the very important and all too often neglected area of heuristic programming is addressed in

Chapter 12. Further, the coverage of heuristics in that chapter is extended to such currently popular heuristic methods as genetic algorithms, simulated annealing, and various related techniques that are—as of late—often associated with the field of artificial intelligence.

Finally, in Part 3, the topic of multiple objective optimization is addressed. This is accomplished by an original, unified approach to both modeling and solution—the *multiplex* concept. Topics covered include various multiobjective philosophies, their models, and their solution and analysis via a single algorithm. Discussions and demonstrations are given to how such problems may be solved via conventional linear programming algorithms and software.

The text may be employed in a variety of ways, depending upon the needs/ interests of the reader or the purpose of the associated course. For example, a one-term introductory-level course in linear programming might cover Chapters 1, 2, 3, 4, and 6 (omitting the starred sections) followed by Chapter 10 and selected topics from Chapters 8, 11, and 13. A more advanced course in linear programming might cover Chapter 1 and all of Part 1(that is, Chapters 2 to 8) with additional topics selected from Chapters 11 and 12. A course emphasizing network and integer models might cover Chapters 1, 3, 4, and 6 (omitting the starred sections) followed by Part 2 (that is, Chapters 9 to 12). A course devoted solely to multiobjective models and methods might cover Chapter 1 and Part 3 (that is, Chapters 13 to 17). Finally, we would advocate the use of group projects as a part of any course, and the use of various existing softwares for the solution of a variety of LP models (and the choice of software is left to the reader or course instructor). From experience, a combination of lectures, examples/exercises, projects, and computer implementation has—we believe—best served to reinforce the understanding and appreciation of both the power and limitations of the linear programming method.

In this text, the emphasis is to provide a basis for the understanding and appreciation of the truly remarkable power of the linear programming method. As such, emphasis is not placed on the computer implementation of the tools associated with the overall methodology. However, there now exist numerous, inexpensive, commercial linear programming packages that the instructor and/or student may wish to use to accompany the text. Although we certainly advocate such computational support, we would hope that the main emphasis is, and will be, that of the understanding of the fundamental concepts, theory, and solution methods that serve to comprise the linear programming method.

We would like to acknowledge the reviewers selected by Prentice Hall, Professors Wonjang Baek of Mississippi State University and Romesh Saigal of the University of Michigan for taking the time to evaluate the manuscript.

James P. Ignizio
Charlottesville, Virginia

Tom M. Cavalier
University Park, Pennsylvania

INTRODUCTION AND OVERVIEW

THE LINEAR DECISION MODEL

Despite claims of "seat-of-the pants" decision making, divine revelations, and intuition, the human mind is simply not equipped to perform a thorough, objective, and systematic analysis of the vast majority of the decisions so often encountered when dealing with complex, real-world problems. Consequently, the majority of the credible, and successful, approaches to decision making employ an aid: a *model* of the problem under investigation. Such models may include mathematical formulas, decision trees, knowledge bases, neural networks, production rules, and flow charts. However, such models do not, as some managers fear, *make* decisions. Instead, they are—or should be—used to *complement* the decision process, to clarify the situation, to gain insight and understanding into the problem, and thus to provide for a measurable improvement in the decisions and policies ultimately set forth. It would then seem obvious that the better the model, the better should be the resulting decision.

In this text, our attention is restricted to a single, yet exceptionally useful and extremely important type of quantitative model. This is the linear model: a mathematical model consisting solely of linear functions. A linear function, in turn, is one in which all terms consist of a single *continuous-valued* variable and in which each variable is raised to the power of 1. Thus, functions (1.1) and (1.2) are linear functions, whereas functions (1.3) and (1.4) are not.

$$f(x_1, x_2) = x_1 + x_2 \tag{1.1}$$

$$f(x_1, x_2, x_3) = 3x_1 + 7x_2 - 8x_3 \tag{1.2}$$

$$f(x_1, x_2) = 2x_1^2 + 3x_2^{3.7} \tag{1.3}$$

$$f(x_1, x_2) = 4x_1 - 3x_1x_2 + 2x_2^2 \tag{1.4}$$

Function (1.3) is not linear because both x_1 and x_2 are raised to powers other than 1. Function (1.4) is not linear because it contains a term $(-3x_1x_2)$ that consists of more than a single variable—as well as a term $(2x_2^2)$ having a variable raised to a power other than 1.

Although the type of problem encompassed by the linear model, at this point, may appear severely restricted, this is most definitely not the case in actual practice. As we shall see, numerous real-world problems either have a linear form or may be reasonably approximated by such a form. Further, with some relatively minor extensions of the solution method (or methods) employed, one may extend the basic linear programming approach to models containing discrete and/or integer-valued variables. Moreover, it is even possible to further extend the solution approach to encompass nonlinear functions via the solution of a sequence of linear models (each being a linear approximation, within some prescribed small neighborhood of solutions, of the actual model)—and, in fact, the concepts described herein serve, at the least, to form the foundation for many nonlinear-solution methods.

WHAT IS LINEAR PROGRAMMING? A PREVIEW

The general approach to the modeling and solution of linear mathematical models, and more specifically those models that seek to optimize a (linear) measure of performance, is denoted as *linear programming* or less often (although possibly more appropriately) as linear optimization. Realize that the word *program* denotes a *solution* to a problem (for example, the policies to be implemented or the values of the decision variables). Thus, the term *linear programming* is used to indicate a solution method to linear mathematical models—or to simply describe the practice of modeling and solving linear mathematical models. As such, linear programming is not and should not be confused with computer programming.

Although based on the ideas, notions, and work of numerous earlier investigators, the first truly comprehensive and effective approach to both the modeling and solving of linear programming problems was developed by George Dantzig and his colleagues in 1947. The solution approach itself is termed the simplex algorithm. At about the inception of this development, large-scale electronic (digital) computers became practical realities—thus permitting the application of the simplex algorithm to problems approaching real-world sizes.

Originally, the linear mathematical models of interest had but a single measure of performance to be optimized, subject to the satisfaction of a set of linear equations or inequalities. For example, a linear programming (LP) model might appear as

$$\text{maximize profit} = 7x_1 + 10x_2 \tag{1.5}$$

subject to

$$x_1 + x_2 \le 10 \tag{1.6}$$

$$3x_1 - x_2 \ge 4 \tag{1.7}$$

$$x_1 \ge 0 \tag{1.8}$$

$$x_2 \ge 0 \tag{1.9}$$

Here, the measure of performance to be optimized is profit, as represented by function (1.5). Such a function is termed the *objective function*. The rest of the model, functions (1.6) through (1.9), comprises the *constraints*. More recently, attention has been directed to linear programming models having more than a single objective function—a topic designated as linear *multiobjective* programming. Although the majority of this text deals with the traditional (single-objective) LP model, multiobjective optimization is covered in Part 3.

Until the early 1980s, Dantzig's simplex algorithm was considered the only truly efficient approach to the solution of large-scale linear programs. However, in 1984, Karmarkar developed a quite different approach. Whereas the simplex method visits the boundary points of the constraint set (known as the feasible region), Karmarkar's method is an interior-point algorithm, that is, the iterative process constructs a trajectory through the interior of the feasible region. Karmarkar's method and variants thereof have proven to be worthwhile competitors to the simplex algorithm and an affine scaling variant of Karmarkar's method is presented in Chapter 7.

As we shall discover, if one can place a problem into the form of a linear programming model (that is, a linear program), then there are some extremely powerful methods for not only solving such models, but also for the analysis of the sensitivity of the solution to changes in model parameters and for assistance in answering a variety of "what if" questions. And this is true whether dealing with single or multiobjective models. However, and this is a point that all too often tends to be forgotten as one becomes immersed in the details of solution methods, it should always be realized that *the key ingredient in linear programming (or any solution methodology) is the* **model**. And no matter how sophisticated the solution approach, it only plays a supporting role in problem solving.

PURPOSE

The obvious purpose of this text is to teach the reader how to

1. Develop models of linear decision problems.
2. Solve these models and relate their solution to the actual problem.

However, an additional purpose is to provide a text that is useful to a wide audience. Most important, this is a text for those who not only wish to thoroughly understand linear programming, but to implement the procedure in the real world. Care has been taken to relate the solution procedures developed to the underlying theory at a level commensurate with the intended audience. Perhaps the more mathematically inclined may find that the text lacks some of the rigor and sophistication with which they feel comfortable. Those wishing to delve into the more rigorous details of the theory and the more esoteric aspects of LP are directed to the references. The mathematics is limited to only that required for a serious understanding of the methods and procedures involved. In addition, rather than concentrating solely on problem solution, the *interpretation* and *implementation* of the results obtained are clarified and stressed. Although some of the theory surrounding linear programming can be mathematically intense, the technique itself is not particularly complex. Thus, a straightforward, rather than mathematically elegant, presentation of the modeling, solution processes, and extensions of linear programming is given.

APPLICATIONS OF LINEAR PROGRAMMING: A PREVIEW

In the following chapters, the application of LP to a wide variety of problem types is discussed. Here, the diversity of the approach is indicated simply by listing just some of these applications.

Traditional Applications

Almost any existing text on LP will either mention or describe the method's application to such problems as

- *blending* (for example, the blending of petroleum products to produce different grades of gasoline, the blending of meat by-products to form sausage and lunch meat, and the blending of ingredients to form a particular chemical product)
- *mixes* (for example, the determination of how many of each type of product to produce during a given time period, the determination of the weapons mix to be used against a given threat, media planning and selection, academic resource allocation, investment mixes, program selection and capital budgeting, and the selection and planting of crops)
- *diet problem* (for example, the determination of menus so as to satisfy one's daily need for calories, vitamins, and minerals—while, it is hoped, limiting fat and cholesterol intake)
- *production scheduling* (for example, the determination of how many items to produce each time period so as to satisfy customer demand, production capacity, and storage limitations)

- *assignment* (for example, the assignment of workers to tasks so as to minimize costs or time, the assignment of missiles to targets, and the deployment of antennas within phased arrays)
- *transportation/dispatching* (for example, the determination of the shipment scheme necessary to satisfy customer demand while minimizing transport costs and a plan for solid-waste collection and routing)

Applications in Artificial Intelligence and Information Technology

Although most texts focus on the more traditional applications just listed, linear programming can be used as an adjunct to or in place of methods that are now associated with the fields of artificial intelligence (AI) and information technology. In fact, as we shall see, there are certain very significant advantages in the use of LP in these areas. Such applications are discussed in Chapter 8 and include

- prediction/forecasting (for example, the prediction of cost and system performance, and the forecast of demand)
- pattern classification (also known as pattern recognition, discriminant analysis, and classification—the assignment of objects to predetermined classes)
- cluster analysis (also known as grouping, unsupervised learning—the determination of groups according to some measure, or measures, of similarity of those objects within each group)

DEALING WITH PROBLEMS OF SIZE AND COMPLEXITY BEYOND THE CAPABILITIES OF LINEAR PROGRAMMING

Linear programming is a powerful and remarkably versatile tool, but it is certainly not a panacea and admittedly there is a host of problems for which it is simply not appropriate. Further, there are problems of such size and complexity that, although of "proper form," they exceed the capabilities of the methodology even when implemented on the most powerful computers. Such problems are most typically dealt with by means of heuristic methods—or heuristic programming. Although not guaranteeing optimal solutions, such procedures do generate (if properly constructed) *acceptable* results. While this is not a text on heuristic programming (or its various adjuncts such as expert systems, genetic algorithms, tabu search, and simulated annealing), the reader needs and deserves some exposure to this approach. As such, a full chapter (Chapter 12) examines this topic. It is hoped that the brief introduction provided in that chapter will engender further interest on the part of the reader.

INTENDED AUDIENCE AND PREREQUISITES

The intended audience is virtually anyone with a desire to learn about linear programming and the solution of a remarkably diverse class of decision models. The techniques of linear programming, for the most part, have been developed over the past four to five decades (although linear algebra, which forms the foundation of linear programming, is of course considerably older). Initially, the methodology was of interest only to those in applied mathematics, operations research, and economics. However, as word of the power and success of the method spread and as the number of actual implementations increased, interest developed in many other disciplines. For example, we have encountered, in previous classes on linear programming, students from such fields as

- business administration
- marketing
- logistics
- economics
- virtually all fields of engineering
- operations research/management science
- geography
- forestry
- computer science
- statistics
- accounting
- food services and hotel management
- agriculture
- environmental science
- earth and mineral science
- meteorology
- anthropology
- biology

The prerequisites for the study of this text are, primarily, a knowledge of elementary linear algebra and, most important, a desire to learn. The actual mathematical operations involved in LP are surprisingly few and simple, being limited to addition, subtraction, multiplication, division, and elementary matrix operations. The text is designed for use either as an upper-level (i.e., junior or senior) undergraduate course or for a first-year graduate-level course.

WHAT IS DIFFERENT ABOUT THIS TEXT?

All authors believe that their book is unique—and significantly so. In fact, it would be difficult to justify the labor involved in the preparation of a textbook if one did not feel this way. The authors of this text are no exception. Those aspects that set this book apart from others on LP are described in what follows.

This text provides a comprehensive coverage and treatment of both the classical simplex algorithm for LP models as well as the more recent interior-point methods (such as those introduced by Karmarkar). This blend provides the reader with a complete and up-to-date picture of linear programming algorithms as used by today's analysts. Coverage of both continuous and integer models is also provided—and this includes the network simplex method as well as heuristic methods for general linear integer models. Included among the latter methods are methods of heuristic programming—including such techniques as genetic "algorithms."

Further, some rather nonconventional applications of LP are addressed. Specifically, the use of LP in the development of models for prediction, pattern classification, and clustering are covered. Here, LP is shown to provide an effective, robust, and completely nonparametric approach to this very important set of problem types.

In addition, the topic of linear models having multiple objectives as well as hard and soft constraints is addressed. This treatment is enhanced by an original, *unified* approach to the modeling and solution process, an approach designated as *multiplex*. By means of the multiplex concept, both the traditional linear programming model as well as a variety of multiobjective problem types can be modeled and solved.

Finally, we are convinced that learning is enhanced by means of illustrative examples and thus the material presented is accompanied by both numerical examples and specially selected exercises. A solution manual for the exercises is available for those adopting the text for classroom use.

OVERVIEW OF THE MATERIAL TO FOLLOW

The text is divided into three parts. The first, which includes Chapters 2 through 8, is focused for the most part on single-objective, continuous linear programming models and solution methods. In Chapter 2, we describe certain basic notions with regard to linear programming and then provide examples of some of the more common and better known applications. Chapters 3 through 6 then focus on the traditional (simplex) solution process used for such models, as well as the tableau representations, duality, economic analysis, dual simplex, revised simplex, and sensitivity analysis. In Chapter 7, we then discuss the issue of problem complexity as well as alternatives to the simplex method. More specifically, we discuss the new interior-point method of Karmarkar and an affine scaling variant. Finally, in Chapter 8, we discuss some rather unconventional applications of the linear programming process. Specifically, we describe how LP may be used to develop decision models in the support of the solution of problems involving prediction, classification, and clustering—topics that are oftentimes associated with the areas of artificial intelligence and information technology. And we even demonstrate how linear programming may be used to both design and train neural networks.

Part 2 addresses network and linear integer models. In particular, network flow problems and the network simplex method are presented in Chapter 9. This is followed by other, related network models—including the linear transportation and linear assignment problems. We then introduce some general integer programming algorithms including branch and bound, implicit enumeration, and cutting-plane methods. Finally, in Chapter 12, we provide an introduction to heuristic methods (that is, heuristic programming, genetic "algorithms," simulated annealing, tabu search, expert systems, and neural networks). In the event that an exact method cannot solve a problem—an event oftentimes encountered in problems of integer programming—a (well-designed) heuristic programming procedure is often the only rational and effective alternative.

In Part 3, the multiobjective model and its unified solution approach (multiplex) are presented. Chapters 13 and 14 deal with modeling multiobjective systems and Chapter 15 focuses on a unified approach for problem solution. Chapter 16 deals with duality in multiobjective models and implementation of methods of sensitivity analysis. Chapter 17 provides a brief summary of the extensions of such concepts to integer and nonlinear multiobjective problems.

Finally, the appendix provides a brief review of the elements of linear algebra.

COURSE-STRUCTURE RECOMMENDATIONS

The material in the text may be employed in a variety of ways, depending upon the needs/interests of the reader or the purpose of the associated course. For example, a one-term introductory-level course in linear programming might cover Chapters 1, 2, 3, 4, and 6 (omitting the starred sections) followed by Chapter 10 and selected topics from Chapters 8, 11, and 13. A more advanced course in linear programming might cover Chapter 1 and all of Part 1 (that is, Chapters 2 to 8) with additional topics selected from Chapters 11 and 12. A course emphasizing network and integer models might cover Chapters 1, 3, 4, and 6 (omitting the starred sections) followed by Part 2 (that is, Chapters 9 to 12). A course devoted solely to multiobjective models and methods might cover Chapter 1 and Part 3 (that is, Chapters 13 to 17). Finally, we would advocate the use of group projects as a part of any course, and the use of the various existing software packages for the solution of a variety of LP models (and we leave the choice of software to the reader or course instructor). From experience, a combination of lectures, examples/exercises, projects, and computer implementation has best served to reinforce the understanding and appreciation of both the power and limitations of the linear programming method.

CHAPTER **2**

THE (CONVENTIONAL) LINEAR PROGRAMMING MODEL

CHAPTER OVERVIEW

This chapter begins with a discussion of model types and a general philosophy for modeling. This general discussion is followed by the specific form and assumptions of the linear programming model that is to be addressed throughout this text. Several examples are then presented that outline some of the more common modeling techniques in linear programming. The chapter concludes with a discussion of model validation. In real-life applications, much more time is spent deriving an acceptable model than actually optimizing the model, so the time spent studying the ideas presented in this chapter is time well spent.

MODELS AND MODEL TYPES

Most decision problems of interest, and virtually all of those of importance in the real world, can be thought of as occurring in large and complex systems. One could hardly hope to evaluate alternative courses of action by actually trying these in the system. Instead, a model that represents the system is normally used for such experimentation. Consequently, models perform a valuable, if not essential, role in the decision-making process.

Quite often, however, a reader becomes so involved with the model and *its* solution he or she forgets that it is indeed the model, not the actual system, that is being analyzed. If the model is a "good" model, its solution should represent a good approximation to the solution of the actual system. But it is rarely *the* solution to the actual system. This does not negate the importance and usefulness of the model, but it is a factor that must always be kept in mind.

Models may take on a wide variety of forms, including among the best known:

- Scale model
- Pictorial model
- Flow chart (or network)
- Matrix
- Mathematical model

A typical scale model could be characterized by the aerodynamic model of an airplane. Such a model "looks" almost exactly like the actual aircraft except for its smaller size. By employing the model in a wind tunnel, the aerodynamic stability of the model, and thus of the actual aircraft, may be observed. As such, the scale model allows the investigator to observe, economically, the performance of the model under certain conditions. Such a model, however, does not generally lend itself to a systematic *optimization* of the system being modeled. That is, it serves as an *evaluative* tool rather than a means for system optimization.

The pictorial model is usually a two-dimensional photograph or sketch of a system. An example might be an aerial photograph of a region. Such a photo then could be used by an electrical power company to decide where to construct its power lines. Again, although such a model aids the decision process, it does not actually lend itself to an optimizing procedure.

The flow chart is a special type of pictorial model. However, rather than simply depicting spatial relationships, the flow chart illustrates the *interrelationships* among the components. Such charts are useful in a wide variety of areas, including

Depicting the inputs and outputs of system components and the flows between these components

Timing activities, and their sequence, in the scheduling of a project

Providing steps and logic flow of a procedure that is to be simulated

In some instances, the flow chart is used only as a decision aid, but with the help of the methodology of network analysis, one can actually perform an optimization of the system (actually, of the model of the system).

The matrix is also a common model for decision analysis. Like the flow chart, one may also sometimes employ optimizing procedures on the matrix model. In many instances, the flow-chart model and the matrix model may be interchangeable. Consider, for example, the flow chart (or network) of Figure

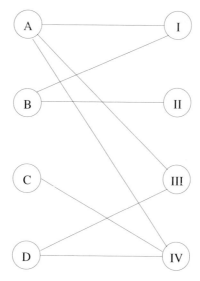

Figure 2.1 Flow graph or network model.

2.1. The nodes (circles) on the left represent workers (designated A, B, C, and D) and the nodes on the right represent jobs (designated I, II, III, and IV). If a worker can perform a given job, a link exists between the worker and the job node. Otherwise there is no link. For example, in Figure 2.1, worker A can perform jobs I, III, and IV, but not job II.

The same network could also be represented by the matrix of Table 2.1. Here the workers head the rows of the matrix and the jobs head the columns. If a worker can perform a given job, there is a 1 in the cell at the intersection of the worker row and job column. Otherwise a 0 is placed in the cell.

TABLE 2.1. EQUIVALENT MATRIX MODEL

Worker	I	II	III	IV
A	1	0	1	1
B	1	1	0	0
C	0	0	0	1
D	0	0	1	1

The primary model of interest in this text is the mathematical model (that is, a set of mathematical functions that represent the problem under consideration). For example, the statement that profit equals sales minus costs may be written, in some cases, mathematically as

$$P = sx_1 - c(x_1 + x_2)$$

or   (2.1)

$$P = sx_1 - cx_1 - cx_2$$

where

P = profit in dollars

s = sales price of each unit of product

c = total cost of each unit of product

x_1 = total amount of products sold

x_2 = total amount of products not sold

Thus, $x_1 + x_2$ is the total amount of products produced. Equation (2.1) is a linear function because each term contains only a single variable and, in turn, each variable is raised to the power of 1.

The mathematical model is the most abstract of all the models discussed. That is, whereas the scale model of a building or airplane is immediately recognizable, one cannot so readily identify the problem that is being mathematically modeled. However, the mathematical model more than compensates for this shortcoming by being (generally speaking) the easiest model to manipulate, analyze, and optimize. Further, it can be used to represent an extremely wide variety of actual problems.

GENERAL GUIDELINES IN MODEL BUILDING

Before discussing the construction of the linear programming model, it is important to take note of several rules and/or guidelines in model development. These rules form a philosophy of model construction that, if followed, should result in a model that is more credible, more useful, and less costly than if a more-or-less brute-force approach were taken.

One should first consider the primary *purpose* of the model. Is it to be used to simulate a system? To evaluate a system? To optimize a system? Or to simply describe a system? What are the results going to be used for? How much accuracy is required? Over what time scale is the model to be used? Over what range of inputs must it respond? What are the budget and time restrictions on model development? Who will use the final model and/or its results?

Usually, in answering such questions as these, a general class of model, to best suit all requirements, may be identified. One must next determine, more specifically, the actual *type* of model to be used. That is, if one wishes to optimize a system, should a network model, matrix model, or mathematical model be employed? Although the mathematical model is usually the most powerful and flexible, this power and flexibility may not be needed for the specific problem under consideration. When isolating the specific type of model, one should try to avoid one of the most common and costly of all mistakes encountered in model construction. That is to *force* the problem to fit a particular model type. All too often, one becomes so enthralled with one type of model (or perhaps it is the only type the analyst is familiar with) that he or she tries to place every problem into that particular format. Rather than fitting the problem to a model, one should fit the most appropriate model to the problem. That is, the model that seems to most *naturally* fit the problem should be employed.

The *level* of the model is an extremely important, often abused, factor in model building. "Level" refers to the amount of "detail" presented by the model, such as the number of variables considered, the number of relationships and interrelationships presented, and so forth. It is an unfortunate misconception that the more "detailed" (or, in other words, the larger, more complex, and "sophisticated") the model, the "better" the model. Often, just the reverse is true. These highly "detailed" models may be a cluttered mass of assumptions and inaccuracies whose results are meaningless.

The proper way to determine the level of a model is to begin with a model that is as simplified as deemed reasonable. That is, try to absolutely minimize the number of variables (and other factors) considered. In general, even in very large and complex systems, only relatively few variables really have a significant impact on the system's output. Once this preliminary model is constructed, it is evaluated. Add to the model *only* that detail believed absolutely necessary to refine the model (and its results) to the level necessary for actual use.

Those who attempt to build highly detailed models usually find (after considerable time and expense) that the model is too cumbersome to use or too computationally burdensome to employ even on the largest computer. They then attempt to "strip" the level of detail until they obtain a "workable" model. On the other hand, when following the foregoing guidelines, a decrease in time and cost is achieved and one should also obtain a more credible and efficient model.

Another consideration in model development, closely related to model level, is the definition of the *system*, its limits, its inputs and outputs, and its components, or subsystems. A particularly useful aid here is the use of a flowchart or network model—regardless of the final model to be used to represent the system. This concept is illustrated in one of the forthcoming examples.

DEFINITIONS

Before describing the development of the basic linear programming model, consider the following definitions of terms commonly employed in model construction.

> VARIABLE A variable, usually denoted as $x_j (j = 1, \ldots, n)$, is a factor subject to change within the problem. That is, its value may change, or at least change within certain limits.

DECISION (CONTROL or STRUCTURAL) VARIABLE   A variable that is both under the control of the decision maker and could have an impact on the solution to the problem of interest is a decision, control, or structural variable.

CONTINUOUS VARIABLE   A variable that may take on *any* values between an upper and lower limit is continuous.

DISCRETE VARIABLE   A variable that may take on only certain prescribed values is discrete. For example, if x_1 can take on only the values 0, 1, 5/2, and 10.32, then x_1 is discrete. A special subclass of discrete variables consists of those that can take on only integer values (such as $-5, 0, 2, 3, 8$), and these are often called *integer* variables.

LINEAR FUNCTION   A linear function contains terms each of which is composed of only a single, continuous variable raised to (and only to) the power of 1. No functions such as cos x, log x, or exp x may be involved.

NONLINEAR FUNCTION   A nonlinear function is basically the complement of a linear function. That is, more than a single variable may appear in a single term and the variables may be raised to any power. Strictly speaking, even if a function satisfies the conditions listed in the previous definition, it is not considered linear if any of the variables involved is discrete. However, it is common practice to call such a function a *linear discrete (or integer) function*.

MATHEMATICAL MODEL   A mathematical model consists of a set of related mathematical functions whose purpose is to simulate the response of the system being modeled. A *linear mathematical model* consists of solely linear functions; a *nonlinear mathematical model* involves one or more nonlinear functions.

EQUATION   A mathematical equation, represented as $f(\mathbf{x}) = b$ (that is, some function of the variables $\mathbf{x} = (x_1, x_2, \ldots, x_n)^t$ is *equal* to a constant right-hand-side value, b), expresses the equivalence between the function on the left and the function on the right (which, in this text, is usually a constant). (Here, the notation $\mathbf{x}^t$ denotes the transpose of the vector $\mathbf{x}$.)

INEQUALITY   Consider the previous definition. If a function on the left can *equal or exceed* the function or constant on the right, this is called an inequality and is represented mathematically as

$$f(\mathbf{x}) \geq b$$

Similarly, if the function on the left can be *less than or equal* to the function or constant on the right, we have

$$f(\mathbf{x}) \leq b$$

OBJECTIVE   An objective is represented by a mathematical function of the decision variables. Such a function usually represents the desires of the decision maker, such as to maximize profit or minimize cost. Objective functions may also be linear or nonlinear in form, although this text concentrates on those of a linear form. It is important to note that the right-hand side of an objective function (that is, its value) is left unspecified. That is, the two most typical forms of objective functions are

$$\text{maximize } f(\mathbf{x}) \text{ or minimize } f(\mathbf{x})$$

CONSTRAINT A constraint is a mathematical equation or inequality that represents a restriction due to a resource or technological limitation. The mathematical form of a constraint is

$$f(\mathbf{x}) \leq b$$

or

$$f(\mathbf{x}) \geq b$$

or

$$f(\mathbf{x}) = b$$

depending upon the situation.

BASIC STEPS IN THE LINEAR PROGRAMMING MODEL FORMULATION

The three basic steps in the linear programming model formulation are

1. Determine the decision (or control or structural) variables.
2. Formulate the objective function.
3. Formulate the constraints.

These three basic steps lead to what is generally termed the (single-objective) linear programming model.

Determination of the Decision Variables

The decision variables within a problem are those over which one actually has *control*. Consequently, they are often referred to as control variables. A set of decision variables is generally denoted as $\mathbf{x}$, or the solution vector, or the program. The optimal set or program is termed $\mathbf{x}^*$. The main thrust (at least initially) in linear programming is to determine the values of $\mathbf{x}^*$.

Consider, for example, the problem involved in insulating your house so as to reduce utility costs. Numerous variables exist in such a problem, including the following:

1. Amount of attic insulation to be installed.
2. Amount of side-wall insulation to be installed.
3. Amount of caulking to be done.
4. Number (and perhaps type) of storm windows.
5. Number (and perhaps type) of insulating draperies or curtains used.
6. Amount of insulation to be installed around the hot-water tank.
7. Temperatures experienced.
8. Wind velocity and direction.
9. Amount of sunshine incident on the house.
10. Number of individuals within the house.
11. Number of times per day that a door or garage door is opened.
12. Cost of utilities furnished.

Now, of these variables, only the first six are directly controllable. Thus, although variables 7 through 12 will certainly combine to determine the resultant cost of your utilities, they are not within your control. Consequently, only the first six variables should appear in your model as decision variables.

It is important that you both identify and *define* your decision variables. The definition of the decision variables should appear directly prior to the actual mathematical formulation. For example, in the home-insulation example, the variables could be defined as follows:

x_1 = amount, in feet, of 6-inch attic insulation installed

x_2 = amount, in pounds, of side-wall insulation installed

x_3 = amount, in the number of tubes, of caulking used

x_4 = number of storm windows (this may have to be divided further into types and styles)

x_5 = number of yards of drapery material used (again, this may very well have to be divided further)

x_6 = amount, in feet, of hot-water-tank insulation used

Formulation of the Objective

The next step in model formulation is the specification of the objective. The objective is generally the result of the desire of the decision maker and may typically be one of the following:

- Maximize profit
- Minimize costs
- Minimize overtime

- Maximize resource utilization (personnel, machinery, or processes)
- Minimize labor turnover
- Minimize machine downtime
- Minimize risk (to the firm, to the environment, etc.)
- Maximize the probability that a given process remains within certain control limits
- Minimize the deviation from a standard

Formulation of the Constraints

The final step in the mathematical formulation is to determine the constraint set. Generally, a constraint (or restriction) is the result of a resource or technological limitation, such as

- Limited raw material
- Limited budget
- Limited time
- Limited personnel
- Limited ability or skills

The overall guideline here always should be to try to identify a minimal number of constraints necessary to adequately represent the problem.

THE GENERAL FORM OF THE LINEAR PROGRAMMING MODEL

The general form of the (single-objective) linear programming model can now be stated mathematically.

Find $\mathbf{x} = (x_1, x_2, \ldots , x_n)^t$ so as to optimize (either maximize or minimize) the objective function subject to the specified constraints.

$$\text{optimize } z = c_1 x_1 + c_2 x_2 + \cdots + c_n x_n \tag{2.2}$$

subject to

$$a_{1,1} x_1 + a_{1,2} x_2 + \cdots + a_{1,n} x_n \{\leq, =, \geq\} b_1 \tag{2.3}$$

$$a_{2,1} x_1 + a_{2,2} x_2 + \cdots + a_{2,n} x_n \{\leq, =, \geq\} b_2 \tag{2.4}$$

$$\cdot$$
$$\cdot$$
$$\cdot$$

$$a_{m,1} x_1 + a_{m,2} x_2 + \cdots + a_{m,n} x_n \{\leq, =, \geq\} b_m \tag{2.5}$$

$$x_1, x_2, \ldots , x_n \geq 0 \tag{2.6}$$

Note that each constraint (2.3–2.5) may be either a type I inequality ($\leq$), a type II inequality ($\geq$), or an equality ($=$). Also, in some instances, the nonnegativity restrictions (2.6) on the decision variables may not be appropriate. That is, there may be cases in which the decision variables may take on negative values. However, it is always possible, through the use of simple linear transformations, to convert any linear programming model into the foregoing form.

Finally, realize that the purpose of the linear programming model is to reflect, as closely as possible, the problem as perceived by the decision maker(s). The basic modeling procedure will now be illustrated with a series of examples. First, however, we describe the assumptions inherent to the linear programming approach.

ASSUMPTIONS OF THE LINEAR PROGRAMMING MODEL

Every mathematical modeling technique operates under certain basic assumptions, and linear programming is no exception. There are four basic assumptions that ensure that a real situation can be represented as a linear programming problem.

CERTAINTY   The problem data (c_j, b_i, and $a_{i,j}$; $i = 1, \ldots, m; j = 1, \ldots, n$) are assumed to be known with certainty. That is, there is no stochastic element to the data.

PROPORTIONALITY   The contribution of a decision variable, x_j, to the objective function is $c_j x_j$ and its contribution to the ith constraint is $a_{i,j} x_j$. That is, the contribution of x_j is always directly proportional to the level of the variable x_j. This simply means, for example, that if the value of x_j doubles, then its contribution to the objective function also doubles. There are no setup costs, discounts, or economies of scale.

ADDITIVITY   This assumption ensures, for example, that the total cost is the sum of the cost contributions of each individual variable. That is, the contributions from individual variables combine linearly in both the objective function and the constraints. There are no interactions that could reduce or increase the level of the combined contributions.

DIVISIBILITY   Divisibility implies that the decision variables are continuous variables, that is, they can be divided into fractional parts.

Despite what appears to be somewhat restrictive assumptions, linear programming remains one of the most widely used modeling techniques.

EXAMPLES OF LINEAR PROGRAMMING MODEL FORMULATION

In lieu of real-world experience, the best way to learn how to establish mathematical formulations is through illustrative examples. In this section, a number of such illustrations is presented using the basic steps previously defined.

Example 2.1: A Product-Mix Problem

The product-mix problem is typical of linear programming problems in which there is competition for limited resources among several products. The problem is to allocate the resources to the production of the various products so as to maximize profit (or minimize cost).

HiTech, Inc., a small manufacturing firm produces two microwave switches, Switch A and Switch B. The return per unit of Switch A is $20, whereas the return per unit of Switch B is $30. Because of contractual commitments, HiTech must manufacture at least 25 units of Switch A per week, and based on the present demand for its products, it can sell all that it can manufacture. However, it wishes to maximize profit while determining the production sizes to satisfy various limits resulting from a small production crew. These include

> Assembly hours: 240 hours available per week
>
> Testing hours: 140 hours available per week

Switch A requires 4 hours of assembly and 1 hour of testing, and Switch B requires 3 hours and 2 hours, respectively.

Determination of the decision variables.   The problem is obviously to determine the optimal number of each type of switch to manufacture based on the limited resources available. The variables directly under HiTech's control are

$$x_1 = \text{amount of Switch A manufactured per week}$$

$$x_2 = \text{amount of Switch B manufactured per week}$$

Formulation of the objective.   The overall objective is to maximize weekly profit and because the unit returns for switches A and B are $20 and $30, respectively, the objective can be written as follows:

$$\text{maximize } z = 20x_1 + 30x_2 \quad \text{(profit per week)}$$

Formulation of the constraints.   Based on the consumption rates of the two switches and the limited resources available, the production constraints can be formulated as follows:

$$4x_1 + 3x_2 \leq 240 \quad \text{(assembly hours per week)}$$

$$x_1 + 2x_2 \leq 140 \quad \text{(testing hours per week)}$$

Also, the minimum requirement for Switch A is given simply by

$$x_1 \geq 25 \quad \text{(Switch A demand per week)}$$

and the nonnegative restrictions on each variable are written as

$$x_1 \geq 0$$

$$x_2 \geq 0$$

As a result, the mathematical model for this problem may be summarized as follows:

$$\text{maximize } z = 20x_1 + 30x_2 \tag{2.7}$$

subject to

$$4x_1 + 3x_2 \leq 240 \tag{2.8}$$

$$x_1 + 2x_2 \leq 140 \tag{2.9}$$

$$x_1 \geq 25 \tag{2.10}$$

$$x_1, x_2 \geq 0 \tag{2.11}$$

Due to the assumptions of linear programming, this model assumes that the decision variables are continuous, and, consequently, it is possible that the optimal solution may require a fractional number of switches to be manufactured. This may or may not create a problem for the analyst. For example, if the batch sizes are very large, a fractional unit may not have a significant impact on the solution. However, if the batch sizes are small, a fractional unit may indeed have a major impact on the solution. In any case, rounding off the solution is never guaranteed to yield the optimal, or even a good, solution and it may be necessary to add the additional restrictions that the decision variables are integer. The model then becomes a linear integer program. Solution techniques for linear integer programs are addressed in Chapter 11.

Example 2.2: An Investment-Planning Problem

An investment-planning problem is another example of the optimal allocation of limited resources, in this case, investment capital.

An investor has decided to invest a total of $50,000 among three investment opportunities: savings certificates, municipal bonds, and stocks. The annual return on each investment is estimated to be 7%, 9%, and 14%, respectively. The investor does not intend to invest his annual interest returns (that is, he plans to use the interest to finance his desire to travel). He would like to maximize his yearly return while investing a minimum of $10,000 in bonds. Also, the investment in stocks should not exceed the combined total investment in bonds and savings certificates. And, finally, he should invest between $5,000 and $15,000 in savings certificates.

Determination of the decision variables. The problem is to determine the proper allocation of the resources ($50,000) among the three investment opportunities. Thus, the decision variables are

x_1 = dollars invested in savings certificates

x_2 = dollars invested in municipal bonds

x_3 = dollars invested in stocks

Formulation of the objective.   The objective is to maximize yearly return, and based on the estimated annual returns of the three investments, the objective can be written:

$$\text{maximize } z = 0.07x_1 + 0.09x_2 + 0.14x_3 \quad \text{(yearly return)}$$

Formulation of the constraints.   The goal to invest a minimum of $10,000 in bonds is written as

$$x_2 \geq 10,000 \quad \text{(investment in bonds)}$$

The restriction that the investment in stocks (x_3) should not exceed the combined total investment in bonds and savings certificates ($x_1 + x_2$) may be formulated simply as

$$x_3 \leq x_1 + x_2 \quad \text{(stock restriction)}$$

or

$$x_3 - x_1 - x_2 \leq 0$$

The investment limits of $5,000 and $15,000 on savings certificates yield the constraint

$$5,000 \leq x_1 \leq 15,000 \quad \text{(savings certificates)}$$

However, this last restriction is best formulated as two constraints:

$$x_1 \geq 5,000$$

$$x_1 \leq 15,000$$

Finally, the restriction on the total investment is given simply as

$$x_1 + x_2 + x_3 \leq 50,000 \quad \text{(total investment)}$$

and the nonnegativity restrictions on each variable are written as

$$x_1 \geq 0$$

$$x_2 \geq 0$$

$$x_3 \geq 0$$

The linear programming model for this problem may then be summarized as follows:

$$\text{maximize } z = 0.07x_1 + 0.09x_2 + 0.14x_3 \tag{2.12}$$

subject to

$$x_2 \geq 10,000 \tag{2.13}$$

$$-x_1 - x_2 + x_3 \leq 0 \tag{2.14}$$

$$x_1 \geq 5,000 \tag{2.15}$$

$$x_1 \leq 15,000 \tag{2.16}$$

$$x_1 + x_2 + x_3 \leq 50,000 \tag{2.17}$$

$$x_1, x_2, x_3 \geq 0 \tag{2.18}$$

Example 2.3: A Product-Blending Problem

There is a wide variety of problems in which certain basic components of raw materials are combined, or blended, to produce a product that satisfies certain specifications. Typical of such problems are the blending of gasolines, the blending of feeds for animals, and the mixture of meats to produce sausage or lunch meats. The specifications of the blend (i.e., the recipe) together with restrictions (such as government requirements) are given, and the task is to produce a blend that minimizes total cost (or maximizes total profit) while satisfying these restrictions.

The Sierra Refining Company produces two grades of unleaded gasoline, Grade 1 and Grade 2, which it supplies to its chain of service stations for $48 and $53 per barrel, respectively. Both grades of gasoline are blended from Sierra's inventory of gasoline components and must meet the specifications in Table 2.2. The characteristics of the components in inventory are found in Table 2.3.

TABLE 2.2   GASOLINE SPECIFICATIONS

Gasoline	Minimum octane rating	Maximum demand (barrels/wk)	Minimum deliveries (barrels/wk)
Grade 1	87	80,000	60,000
Grade 2	93	40,000	15,000

TABLE 2.3   COMPONENTS CHARACTERISTICS

Gasoline component	Octane rating	Inventory (barrels)	Cost ($/barrel)
1	86	70,000	33
2	96	60,000	37

What quantities of the two components should be blended into the two gasolines in order to maximize weekly profit?

Determination of the decision variables.   The control variables in this problem are the quantities of each of the two components that should be blended into the two gasolines:

$x_{i,j}$ = barrels of Component i blended into Grade j gasoline per week;

$i = 1, 2; j = 1, 2$

Formulation of the objective.   Assuming that gasoline components combine linearly, the total amount of Grade 1 gasoline is given by $(x_{1,1} + x_{2,1})$ and the total amount of Grade 2 gasoline is given by $(x_{1,2} + x_{2,2})$. Similarly, the total amounts of Component 1 and Component 2 used are given by $(x_{1,1} + x_{1,2})$ and $(x_{2,1} + x_{2,2})$, respectively. The objective function can now be formulated by noting that profit = sales − cost.

$$\text{maximize } z = 48(x_{1,1} + x_{2,1}) + 53(x_{1,2} + x_{2,2}) - 33(x_{1,1} + x_{1,2})$$
$$- 37(x_{2,1} + x_{2,2}) \quad \text{(profit per week)}$$

which simplifies to

$$\text{maximize } z = 15x_{1,1} + 20x_{1,2} + 11x_{2,1} + 16x_{2,2}$$

Formulation of the constraints. In order to model the minimum octane-rating requirement, one has to be able to determine the octane level when varying quantities of the two refined components are combined. Again, it is necessary to assume that octanes blend linearly. The octane level in a mixture of components is obtained by computing the weighted average of the octane in the mixture. This is done by dividing the total octane in the mixture by the number of barrels in the mixture. Thus, the minimum octane constraint for Grade 1 gasoline can be written mathematically as

$$\frac{86x_{1,1} + 96x_{2,1}}{x_{1,1} + x_{2,1}} \geq 87 \quad \text{(minimum octane rating for Grade 1)}$$

Note that this constraint is nonlinear; however, multiplying both sides by $(x_{1,1} + x_{2,1})$ and simplifying yield the equivalent linear constraint:

$$-x_{1,1} + 9x_{2,1} \geq 0$$

Similarly, the minimum octane restriction for Grade 2 gasoline is given by

$$\frac{86x_{1,2} + 96x_{2,2}}{x_{1,2} + x_{2,2}} \geq 93 \quad \text{(minimum octane rating for Grade 2)}$$

or

$$-7x_{1,2} + 3x_{2,2} \geq 0$$

The minimum and maximum distribution requirements for the two grades of gasoline may be formulated as follows:

$$x_{1,1} + x_{2,1} \geq 60,000 \quad \text{(minimum deliveries of Grade 1)}$$
$$x_{1,2} + x_{2,2} \geq 15,000 \quad \text{(minimum deliveries of Grade 2)}$$
$$x_{1,1} + x_{2,1} \leq 80,000 \quad \text{(maximum demand for Grade 1)}$$
$$x_{1,2} + x_{2,2} \leq 40,000 \quad \text{(maximum demand for Grade 2)}$$

Similarly, the supply constraints are given by

$$x_{1,1} + x_{1,2} \leq 70,000 \quad \text{(inventory of Component 1)}$$
$$x_{2,1} + x_{2,2} \leq 60,000 \quad \text{(inventory of Component 2)}$$

and the nonnegativity restrictions are

$$x_{1,1}, x_{1,2}, x_{2,1}, x_{2,2} \geq 0$$

Finally, the complete linear programming model can be written as

$$\text{maximize } z = 15x_{1,1} + 20x_{1,2} + 11x_{2,1} + 16x_{2,2} \tag{2.19}$$

subject to

$$-x_{1,1} + 9x_{2,1} \geq 0 \tag{2.20}$$

$$-7x_{1,2} + 3x_{2,2} \geq 0 \tag{2.21}$$

$$x_{1,1} + x_{2,1} \geq 60{,}000 \tag{2.22}$$

$$x_{1,2} + x_{2,2} \geq 15{,}000 \tag{2.23}$$

$$x_{1,1} + x_{2,1} \leq 80{,}000 \tag{2.24}$$

$$x_{1,2} + x_{2,2} \leq 40{,}000 \tag{2.25}$$

$$x_{1,1} + x_{1,2} \leq 70{,}000 \tag{2.26}$$

$$x_{2,1} + x_{2,2} \leq 60{,}000 \tag{2.27}$$

$$x_{1,1}, x_{1,2}, x_{2,1}, x_{2,2} \geq 0 \tag{2.28}$$

Under the required assumptions of linearity, this mathematical representation accurately models the blending requirements. However, note that information regarding the original objective coefficients is lost in the final formulation. That is, the original data concerning the purchase and sales prices do not appear in the final objective (2.19). This may be of no concern. However, if this pricing information is important to the analyst (for example, in postoptimality or sensitivity analysis), it can be preserved at the expense of introducing additional variables. This will slightly increase the overall complexity of the model, but, in return, the model will provide additional valuable information to the analyst. There is a definite trade-off, and it is up to the analyst to decide based on the output requirements of the model. To illustrate this process, let g_1 and g_2 represent the total barrels of Grade 1 and Grade 2, respectively. Then

$$g_1 = x_{1,1} + x_{2,1} \quad \text{(total barrels of Grade 1)}$$

or, equivalently,

$$x_{1,1} + x_{2,1} - g_1 = 0$$

Similarly,

$$x_{1,2} + x_{2,2} - g_2 = 0 \quad \text{(total barrels of Grade 2)}$$

Also, in the same manner, the total barrels of Component 1 and Component 2, h_1 and h_2, respectively, can be expressed as follows:

$$x_{1,1} + x_{1,2} - h_1 = 0 \quad \text{(total barrels of Component 1)}$$

$$x_{2,1} + x_{2,2} - h_2 = 0 \quad \text{(total barrels of Component 2)}$$

By utilizing these new relationships, the preceding model can be reformulated as follows using simple substitutions:

$$\text{maximize profit } z = 48g_1 + 53g_2 - 33h_1 - 37h_2 \tag{2.29}$$

subject to

$$x_{1,1} + x_{2,1} - g_1 = 0 \qquad (2.30)$$

$$x_{1,2} + x_{2,2} - g_2 = 0 \qquad (2.31)$$

$$x_{1,1} + x_{1,2} - h_1 = 0 \qquad (2.32)$$

$$x_{2,1} + x_{2,2} - h_2 = 0 \qquad (2.33)$$

$$-x_{1,1} + 9x_{2,1} \geq 0 \qquad (2.34)$$

$$-7x_{1,2} + 3x_{2,2} \geq 0 \qquad (2.35)$$

$$g_1 \geq 60,000 \qquad (2.36)$$

$$g_2 \geq 15,000 \qquad (2.37)$$

$$g_1 \leq 80,000 \qquad (2.38)$$

$$g_2 \leq 40,000 \qquad (2.39)$$

$$h_1 \leq 70,000 \qquad (2.40)$$

$$h_2 \leq 60,000 \qquad (2.41)$$

$$x_{1,1}, x_{1,2}, x_{2,1}, x_{2,2}, g_1, g_2, h_1, h_2 \geq 0 \qquad (2.42)$$

Note that constraints (2.30–2.33) have been added to the model due to the definitions of the new variables. Also, note that the objective function now reflects the original cost coefficients, and constraints (2.36–2.41) have been simplified by substitution. Although this model is slightly more complex than the previous model, it has the advantage that the analyst can extract additional information from this latter model.

Example 2.4: A Transportation Problem

In many applications, it is necessary to determine a shipping schedule for distributing goods from several warehouses (or production centers) to several retail outlets (or customers). Due to proximity and mode of transportation, the cost of shipping a unit between each warehouse and retail outlet may vary from location to location. In addition, supplies available for shipping from the warehouses and units demanded at the retail outlets may also vary. The task then is to determine the number of units to ship from each warehouse to each retail outlet while minimizing total shipping costs.

A manufacturer has three warehouses that supply finished product to four retail outlets. The warehouses have 6000, 9000, and 4000 units available, and the demands at the retail outlets are projected to be 3900, 5200, 2700, and 6400 units. The per unit costs (in dollars) of shipping from each warehouse to each retail outlet are given in Table 2.4. The manufacturer needs to determine the minimum-cost shipping schedule that satisfies all demands.

Determination of the decision variables.   The decision variables are the quantities shipped from each warehouse to each retail outlet:

$x_{i,j}$ = quantity shipped from Warehouse i to Retail Outlet j; $i = 1, 2, 3$; $j = 1, 2, 3, 4$

TABLE 2.4 UNIT SHIPPING COSTS

Warehouse	Retail outlets			
	1	2	3	4
1	7	3	8	4
2	9	5	6	3
3	4	6	9	6

Formulation of the objective. The objective function is simply the sum of all the units shipped between each warehouse and each retail outlet multiplied by the unit shipping costs. Thus, the objective can be written as

$$\text{minimize } z = 7x_{1,1} + 3x_{1,2} + 8x_{1,3} + 4x_{1,4} + 9x_{2,1} + 5x_{2,2} + 6x_{2,3} + 3x_{2,4}$$
$$+ 4x_{3,1} + 6x_{3,2} + 9x_{3,3} + 6x_{3,4} \quad \text{(total shipping cost)}$$

Formulation of the constraints. Note that the total number of units being shipped *from* Warehouse i is given by $(x_{i,1} + x_{i,2} + x_{i,3} + x_{i,4})$. Similarly, the total number of units being shipped *to* Retail Outlet j is $(x_{1,j} + x_{2,j} + x_{3,j})$. Thus, because limited supplies are available at each warehouse, we can write the following supply constraints:

$$x_{1,1} + x_{1,2} + x_{1,3} + x_{1,4} \leq 6000 \quad \text{(supply at Warehouse 1)}$$

$$x_{2,1} + x_{2,2} + x_{2,3} + x_{2,4} \leq 9000 \quad \text{(supply at Warehouse 2)}$$

$$x_{3,1} + x_{3,2} + x_{3,3} + x_{3,4} \leq 4000 \quad \text{(supply at Warehouse 3)}$$

The demand constraints for the retail outlets follow in a similar manner:

$$x_{1,1} + x_{2,1} + x_{3,1} = 3900 \quad \text{(demand at Retail Outlet 1)}$$

$$x_{1,2} + x_{2,2} + x_{3,2} = 5200 \quad \text{(demand at Retail Outlet 2)}$$

$$x_{1,3} + x_{2,3} + x_{3,3} = 2700 \quad \text{(demand at Retail Outlet 3)}$$

$$x_{1,4} + x_{2,4} + x_{3,4} = 6400 \quad \text{(demand at Retail Outlet 4)}$$

Noting that the nonnegativity restrictions are

$$x_{1,1}, x_{1,2}, x_{1,3}, x_{1,4}, x_{2,1}, x_{2,2}, x_{2,3}, x_{2,4}, x_{3,1}, x_{3,2}, x_{3,3}, x_{3,4} \geq 0$$

the complete model may be written as follows:

$$\text{minimize } z = 7x_{1,1} + 3x_{1,2} + 8x_{1,3} + 3x_{1,4} + 9x_{2,1} + 5x_{2,2} + 6x_{2,3} + 3x_{2,4}$$
$$+ 4x_{3,1} + 6x_{3,2} + 9x_{3,3} + 6x_{3,4} \tag{2.43}$$

subject to

$$x_{1,1} + x_{1,2} + x_{1,3} + x_{1,4} \leq 6000 \tag{2.44}$$

$$x_{2,1} + x_{2,2} + x_{2,3} + x_{2,4} \leq 9000 \tag{2.45}$$

$$x_{3,1} + x_{3,2} + x_{3,3} + x_{3,4} \leq 4000 \qquad (2.46)$$

$$x_{1,1} + x_{2,1} + x_{3,1} = 3900 \qquad (2.47)$$

$$x_{1,2} + x_{2,2} + x_{3,2} = 5200 \qquad (2.48)$$

$$x_{1,3} + x_{2,3} + x_{3,3} = 2700 \qquad (2.49)$$

$$x_{1,4} + x_{2,4} + x_{3,4} = 6400 \qquad (2.50)$$

$$x_{1,1}, x_{1,2}, x_{1,3}, x_{1,4}, x_{2,1}, x_{2,2}, x_{2,3}, x_{2,4}, x_{3,1}, x_{3,2}, x_{3,3}, x_{3,4} \geq 0 \qquad (2.51)$$

Observe that all the coefficients of the variables in constraints (2.44–2.50) are either one or zero. This is one of the characteristics of the transportation problem that allows it to be solved more efficiently than standard linear programming problems. The transportation problem is discussed further in Chapter 10.

Example 2.5: A Production-Scheduling and Inventory-Control Problem

Many firms produce products that are highly subject to seasonal sales fluctuations. When they try to follow such fluctuations, with fluctuating production rates, their production costs tend to increase. However, with uniform production, they usually discover a buildup in inventory, resulting in large storage costs. Consequently, in production scheduling and inventory control, the predicted demand and predicted production costs per unit for each period and the production capacity for each period are given. The problem is to minimize the combined production and inventory costs while satisfying the demand and not exceeding production limits.

IMC, Inc., needs to schedule the monthly production of a certain item for the next 4 months. The unit production cost is estimated to be $12 for the first 2 months and $14 for the last two months. The monthly demands are 400, 750, 950, and 900 units. IMC can produce a maximum of 800 units each month. In addition, the company can employ overtime during the second and third months, which increases monthly production by an additional 200 units. However, the cost of production increases by $4 per unit. Excess production can be stored at a cost of $3 per unit per month, but a maximum of 50 units can be stored in any month. Assuming that beginning and ending inventory levels are zero, how should the production be scheduled so as to minimize the total costs?

Determination of the decision variables.   It is clear from the problem statement that IMC must decide on the level of production in each month as well as the amount of inventory to carry. Thus, the decision variables in the problem are

x_i = amount produced during regular time in month i; $i = 1, \ldots, 4$

y_i = amount produced during overtime in month i; $i = 2, 3$

w_i = amount in inventory at the end of month i; $i = 1, 2, 3$

Formulation of the objective.   The objective is to minimize the combined cost of regular time production, overtime production, and carrying inventory. This can be stated mathematically as

minimize $z = 12x_1 + 12x_2 + 14x_3 + 14x_4 + 16y_2 + 18y_3 + 3w_1 + 3w_2 + 3w_3$ (cost)

Formulation of the constraints.   From the flow diagram in Figure 2.2, it can be seen that the total product flow into Month 1 is x_1, whereas the total product flow out of Month 1 is $(400 + w_1)$. Thus, using the basic idea of conservation of flow, the production constraint for Month 1 is

$$x_1 = 400 + w_1 \quad \text{(Month 1)}$$

or

$$x_1 - w_1 = 400$$

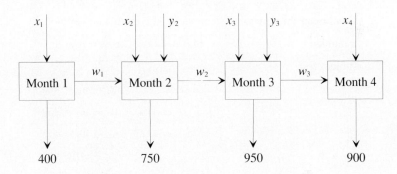

Figure 2.2   Production and inventory model.

Again, using the idea of flow conservation, the constraint for Month 2 can be written as

$$x_2 + y_2 + w_1 = 750 + w_2 \quad \text{(Month 2)}$$

or

$$x_2 + y_2 + w_1 - w_2 = 750$$

The constraints for Months 3 and 4 follow in a similar manner:

$$x_3 + y_3 + w_2 - w_3 = 950 \quad \text{(Month 3)}$$

$$x_4 + w_3 = 900 \quad \text{(Month 4)}$$

The regular-time production-capacity constraints may be written simply as

$$x_1 \leq 800 \quad \text{(regular-time production capacity in Month 1)}$$

$$x_2 \leq 800 \quad \text{(regular-time production capacity in Month 2)}$$

$$x_3 \leq 800 \quad \text{(regular-time production capacity in Month 3)}$$

$$x_4 \leq 800 \quad \text{(regular-time production capacity in Month 4)}$$

and the overtime constraints are given by

$$y_2 \leq 200 \quad \text{(overtime production capacity in Month 2)}$$

$$y_3 \leq 200 \quad \text{(overtime production capacity in Month 3)}$$

Finally, the inventory capacity restrictions are written as

$$w_1 \leq 50 \quad \text{(inventory capacity in Month 1)}$$

$$w_2 \leq 50 \quad \text{(inventory capacity in Month 2)}$$

$$w_3 \leq 50 \quad \text{(inventory capacity in Month 3)}$$

with nonnegativity restrictions

$$x_1, x_2, x_3, x_4, y_2, y_3, w_1, w_2, w_3 \geq 0$$

Combining the foregoing steps yields the mathematical model:

minimize $z = 12x_1 + 12x_2 + 14x_3 + 14x_4 + 16y_2 + 16y_3 + 3w_1 + 3w_2 + 3w_3$ (2.52)

subject to

$$x_1 - w_1 = 400 \tag{2.53}$$

$$x_2 + y_2 + w_1 - w_2 = 750 \tag{2.54}$$

$$x_3 + y_3 + w_2 - w_3 = 950 \tag{2.55}$$

$$x_4 + w_3 = 900 \tag{2.56}$$

$$x_1 \leq 800 \tag{2.57}$$

$$x_2 \leq 800 \tag{2.58}$$

$$x_3 \leq 800 \tag{2.59}$$

$$x_4 \leq 800 \tag{2.60}$$

$$y_2 \leq 200 \tag{2.61}$$

$$y_3 \leq 200 \tag{2.62}$$

$$w_1 \leq 50 \tag{2.63}$$

$$w_2 \leq 50 \tag{2.64}$$

$$w_3 \leq 50 \tag{2.65}$$

$$x_1, x_2, x_3, x_4, y_2, y_3, w_1, w_2, w_3 \geq 0 \tag{2.66}$$

MODEL VALIDITY

Some of the questions that will, or at least should, enter the reader's mind are: How "good" is the model that has been developed? Does it really represent the actual system and respond to alternative policies as would the actual system? As has been previously warned, the results obtained through the solution of a model are only as good as the model itself. Thus far, the discussion has been confined to the approach that should be taken to *develop* a model. It is believed that if these steps were followed, the resulting model would be better than one developed without consideration given to such a systematic approach. However, even then, one may wonder about the validity of the model. Consequently, this section presents some ideas on how to be satisfied with the validity of the model.

First, the reader must face the fact that the mathematical model of a real problem is never perfect (unless, perhaps, the problem is of a very trivial nature). Second, an accurate, absolute *measure* of validity simply does not exist. That is, one cannot speak with certainty of one model as having a validity of, say, 75 units while another has a validity of, say, 90 units. Validity is a multidimensional and highly subjective concept that does not lend itself to such simple approaches. Consequently, when one speaks of model validity, it should be recognized that this validity is an imperfect measure based very much on faith, a concept somewhat unnerving to those with an analytical bent and used to rigorous proofs and absolute definitions.

This difficulty in the measurement of model validity usually leads one to establish a *relative ranking* of models according to their perceived validity. That is, although one may be unable to assign an accurate measure of validity to a given model, one is generally comfortable with a comparison of the perceived validity between two models. If one believes that there is a distinguishable difference in validity between models, one model may be labeled as having a higher degree of validity than the other. The next question is: On what basis may the validity of models be compared? A validation procedure consisting of four phases is suggested:

1. An evaluation of model structure.
2. An evaluation of model logic.
3. An evaluation of the design and/or input data.
4. An evaluation of model response.

An Evaluation of Model Structure

As previously discussed, there are two basic approaches commonly used in model development. The first is to start with a model that contains all aspects and variables of the system. One then attempts to simplify this model (which is generally far too cumbersome for analysis) step by step to the point at which any further simplification would so distort the model as to be unacceptable. The second method, and by far the more preferable, is to begin with a preliminary, simplified model of the system. Such a model represents only the most basic factors and operations of the system. Detail is added to this model until one is satisfied that the responses of interest are accurately represented (obviously, a subjective phase in which faith in the model becomes of central importance). The first approach is both wasteful and time-consuming, whereas the second approach is more systematic, logical, and much more likely to lead to a model that is structurally correct. That is, the elements of the system and their place within the system are accurately represented.

An Evaluation of Model Logic

The first evaluation examined the representation of those elements within the system that are believed to have a (significant) impact on the system's responses. The accuracy of the representation of the interrelationships and interactions between those elements is now determined. If the model logic truly reflects the system logic, the model will react to a stimulus (for example, a change in policy) in the same manner as would the actual system. A common procedure for the evaluation of model logic is to stimulate the model with a representative range of inputs and observe the resultant model output.

Generally, it is not essential that the output response values of the model be of the same value as the actual system. Rather, it is the *relative difference* in the outputs that is of importance. For example, for a given policy, do profits rise or fall?

An Evaluation of the Design and/or Input Data

All too often, a great deal of work is performed in the development of a model, only to be negated by the use of poor or incorrect data. The data used in model development may be roughly classified into two types: (1) the design data, or that information used to actually construct the model, and (2) the input data, or data used to stimulate the system. Inattention to either may seriously degrade the validity of a model.

Data collection and verification may well be the most overlooked portion of model construction. Unfortunately, virtually all textbooks contribute to this problem, at least indirectly, by presenting the reader with a misleading impression of the data-collection process. Textbook presentations are, and must be, simplified. One common simplification is that the data needed for that model development and/or solution are usually presented directly to the reader as was done in the foregoing examples, and thus the reader seldom, if ever, faces the problems involved in the actual collection of data. Unfortunately, not only is the validity of the model dependent on these data, but, also, the process of data collection often consumes the major amount of time and resources when dealing with actual problems. In fact, it can require months to compile *valid* data for a mathematical model.

All too often data are collected *before* one has decided on the basic form of the model to be used. As a result, many of the data are useless in that they do not fit the requirements of the model finally developed. The amount of wasted effort may be considerable. What one should do *first* is decide on the basic form of the model, construct a preliminary model, and *then* identify its specific data needs. One may then collect only those data that are actually needed to support the model.

The sources of data vary extensively based upon the particular situation. These sources may include historical records (if the system has been in existence), theoretical data (which are generally projected data for systems not yet in existence), and ongoing records (for systems in existence). If historical records are available, one must compare the system in its present (or proposed state) versus the state of the system at the time over which the historical records were collected. If no significant differences exist, confidence in the validity of the data is increased.

Although the use of ongoing records is usually a particularly attractive source of data, one often does not have the time to wait for the collection of these data.

An Evaluation of Model Response

The true validation of a model is often said to be reflected solely in its ability to predict the behavior of the system that has been modeled. Such a premise can be carried to the absurd. For example, some observers have noted a correlation between the state of the economy and the hemlines on women's dresses. High hemlines have occurred simultaneously with a good state of the economy, whereas low hemlines have occurred in times of depression or recession. The fact that such a "model" has happened to be an accurate one does not mean that it is valid.

Even if one accepts the premise of future verification, it is not always possible to wait for such results. Validation of model response must then be accomplished through the input of estimated, historical, and ongoing data. Again, however, the fact that the model performs "reasonably" with such inputs is not, by any means, an absolute guarantee of its validity or reaction to future data.

Although consideration of the four evaluation areas is by no means perfect nor completely objective, it does provide a practical means to consider validation on a relative basis and to compare the validity of models. If one can establish some degree of confidence in the structure, logic data, and model response, one can establish some faith in the model.

SUMMARY

In this chapter, we presented an introduction to the modeling process. The basic concepts introduced included several working definitions as well as general modeling guidelines. Some basic modeling strategies for traditional (single-objective) linear programming problems were illustrated through examples. The reader should recognize that modeling is a process that is learned more from practice and experience than anything else.

EXERCISES

2.1. Two recent graduates have decided to enter the field of microcomputers. They intend to manufacture two types of microcomputers, Comp386 and Comp486. Because of the interest in microcomputers, they can (presently) sell all that they could possibly produce. However, they wish to size the production rate so as to satisfy various estimated limits with a small production crew. These include

Assembly hours: 150 hours per week

Test hours: 70 hours per week

The Comp386 requires 4 hours of assembly and 3 hours of testing, and the Comp486 consumes 6 hours and 3.5 hours, respectively. Profit for the Comp386 is estimated at $300 per unit; that of Comp486 is $450 per unit. Develop a linear programming model that maximizes weekly profit.

2.2. An automotive firm produces three types of cars: a large luxury car, a midsized car, and a compact car. The gasoline mileage figures, predicted sales, and profit figures for each type of car are given in Table 2.5. Government regulations state that the average gasoline mileage for the company's entire line of cars should equal or exceed 30 mile per gallon (mpg). The firm wishes to maximize its profits. Formulate a linear programming model for this problem.

TABLE 2.5.

Car	Mileage (mpg)	Profit/car	Demand
Luxury	18	$600	600,000
Mid-size	29	$460	800,000
Compact	38	$320	700,000

2.3. Greentree Farms owns 500 acres of tillable farmland that is used to grow corn, wheat, soybeans, and oats. On average, each acre of corn, wheat, soybeans, and oats yields 110, 35, 32, and 55 bushels, respectively. In order to receive federal subsidies, no more than 120 acres of soybeans can be planted. At least 10,000 bushels of corn are required due to a contractual agreement with a dairy farm. In addition, the total acreage of wheat should equal or exceed the total combined acreage of oats and soybeans. If a bushel of corn, wheat, soybeans, and oats sells for $0.36, $0.90, $0.82, $0.98, respectively, formulate a linear programming model to determine the optimal acreage of each crop to plant.

2.4. A manufacturing firm needs to schedule the monthly production of two seasonal items for the next 6 months. The unit production cost of Item A is estimated to be $15 for the first 2 months, $16 for the third and fourth months, and $18 for the last two months. The unit production cost for Item B is estimated to be $8 for the first 3 months and $10 for the last 3 months. The monthly demands for Item A are 200, 250, 400, 650, 700, 450 units, and the monthly demands for Item B are 160, 180, 370, 500, 420, 350 units. The firm can produce a maximum of 800 units per month. Excess production can be stored from one month to the next at a cost of $2 per unit, but a maximum of 200 total units can be stored in any given month. Assuming that beginning inventory levels are zero, how should the production be scheduled so as to minimize the total costs?

2.5. A refinery produces three grades (A, B, C) of gasolines from three different sources of crude oil (I, II, III). Any crude oil can be used to produce any of the gasolines as long as the specifications in Table 2.6 are met.

TABLE 2.6

Grade of gasoline	Specifications	Selling price/gallon
A	Not less than 50% crude I, not more than 30% crude II	$1.39
B	Not less than 35% crude I, not more than 45% crude II	$1.24
C	Not more than 20% crude III	$1.18

The maximum amount of crude oil available per period and their costs are

Crude I: 10,000 gallons, $1.10 cost/gallon
Crude II: 9,000 gallons, $0.84 cost/gallon
Crude III: 3,000 gallons, $0.90 cost/gallon

The oil refinery naturally wants to maximize profit. Formulate a linear programming model.

2.6. Three different investment options are available at the beginning of each year during the next 6-year period. The durations of the investments are 1 year, 3 years, and 5 years. The 1-year investment yields a total return of 5.1%, the 3-year investment yields a total return of 16.2%, and the 5-year investment yields a total return of 28.5%. If an initial investment of $10,000 is made and all available funds are invested at the beginning of each year, formulate a linear programming model to determine the investment pattern that results in the maximum available cash at the end of the sixth year.

2.7. PDQ Manufacturing Company produces two products, widgets and gadgets. Each widget and gadget requires several basic machining operations to produce. PDQ has five different machining centers, and some of the required machining operations can be performed at more than one of the centers. Consequently, there are several alternative ways of producing each widget and each gadget. Table 2.7 summarizes the unit production times required for the various operations.

TABLE 2.7

		Unit production times (hours)				
Product	Method	Center 1	Center 2	Center 3	Center 4	Center 5
Widget	1	0.25	0.13	—	0.20	—
	2	—	0.34	0.15	—	0.28
	3	0.25	—	0.42	—	—
Gadget	1	—	0.18	—	—	0.32
	2	0.26	—	0.22	—	0.14
	3	0.20	—	—	0.30	—
	4	—	0.12	0.18	0.20	—

The unit cost of widgets produced using Methods 1, 2, and 3 are \$1.35, \$1.28, and \$1.47, respectively. Similarly, the unit cost for gadgets produced using Methods 1, 2, 3, and 4 are \$1.14, \$1.19, \$1.26, and \$1.16, respectively. The weekly demand for widgets and gadgets are 320 and 250, respectively, and each machining center is available for 80 hours per week. Formulate a linear programming problem to find the least-cost production schedule.

2.8. A firm manufactures chicken feed by mixing three different ingredients. Each ingredient contains four key nutrients: protein, fat, vitamin s, and mineral t. The amount of each nutrient contained in 1 kilogram of the three basic ingredients is summarized in Table 2.8.

TABLE 2.8

Ingredient	Protein (grams)	Fat (grams)	Vitamin s (units)	Mineral t (grams)
1	25	11	235	12
2	45	10	160	6
3	32	7	190	10

The costs per kilogram of Ingredients 1, 2, and 3 are \$0.55, \$0.42, and \$0.38, respectively. Each kilogram of the feed must contain at least 35 grams of protein, a minimum of 8 grams of fat and a maximum of 10 grams of fat, at least 200 units of vitamin s, and at least 10 units of mineral t. Formulate a linear programming model for finding the feed mix that has the minimum cost per kilogram.

2.9. C, S, & C, Inc., must decide how much to invest in a number of alternative investment opportunities, which are summarized in Table 2.9.

TABLE 2.9

Investment opportunity	Country	Expected return (%)	Maximum investment ($ million)
1	United States	9	2
2	Japan	10	12
3	Canada	8	8
4	United States	7	6
5	Kuwait	7	10
6	United States	6	4
7	United States	11	9

Company policy requires that the total amount invested inside the United States should be at least as much as invested outside the United States. In addition, of the total amount invested in North America, at most 20% should be invested in Canada. C, S, & C has $40 million to invest and obviously wants to maximize its total expected return on investment. Formulate a linear programming model for solving this problem.

2.10. A small foundry needs to schedule the production of four different castings during the next week. The production requirements of each casting are summarized in Table 2.10.

TABLE 2.10

Product	Unit production times (minutes)				
	Pouring	Cleaning	Grinding	Inspection	Packing
A	3	8	10	1	3
B	1	12	6	1	5
C	2	6	9	1	3
D	1	7	7	1	2

The unit profit for Products A, B, C, and D are $18, $15, $13, and $14, respectively. Current demands indicate that all castings that are made can be sold; however contracts dictate that at least 200 units of Product A and 300 units of Product D be produced. The estimated time available for each of the operations during the next week are

Pouring: 40 hours

Cleaning: 80 hours

Grinding: 80 hours

Inspection: 20 hours

Packing: 40 hours

2.11. Acme Fuel, Inc., has two refineries where fuel oil is produced. Refinery A has the capacity to produce a maximum of 275,000 gallons per week, and the corresponding figure for Refinery B is 350,000 gallons. Acme has four regional distribution centers that receive fuel oil directly from the refineries. The shipping cost per gallon are summarized in Table 2.11.

TABLE 2.11

Refinery	Center 1	Center 2	Center 3	Center 4
A	$0.12	$0.07	$0.09	$0.11
B	$0.08	$0.10	$0.09	$0.10

The projected demands at Distribution Centers 1, 2, 3, and 4 are 120,000, 70,000, 185,000, and 200,000 gallons, respectively. To help maintain a uniform work load, it is management policy that the ratio of scheduled production to a refinery's capacity must be the same for the two refineries. Formulate a linear programming model to find the minimum-cost shipping pattern.

2.12. Zoltar, Inc., a manufacturer of automobile radios, has received an order for 10,000 standard AM/FM radios and an order for 6,000 digital AM/FM radios with built-in cassette decks. Because of other contractual commitments, Zoltar may not be able to produce the total order on its own. That is, it may be necessary to subcontract the production of some of the radios. Another manufacturer, Positron, Inc., has agreed to supply Zoltar with standard radios at a cost of $65 per unit and digital radios at $115 per unit. Zoltar must decide how many of each type of radio to produce in its own production facility and how many to purchase from Positron. Zoltar's production data is given in Table 2.12.

TABLE 2.12

Product	Assembly	Inspection	Packaging	Unit Production cost
	Unit production times (hours)			
Standard	4.2	0.4	0.2	$59
Digital	5.1	0.7	0.2	$103

Zoltar estimates that its production hours available for assembly, inspection, and packaging will be 40,000, 6,000, and 3,000, respectively. Zoltar will be receiving $70.20 for each standard radio and $126.50 for each digital radio. Formulate a linear programming model to determine the most profitable way for Zoltar to fulfill this order.

2.13. Alpha-Beta, Inc., needs to schedule the weekly production of widgets and gadgets for the next 4 weeks. The production of widgets and gadgets requires machining a single raw material. Each widget requires 2 pounds of raw material and 24 minutes of machine time; each gadget requires 2.4 pounds of raw material and 18 minutes of machine time. Each pound of raw material currently costs $2.10, but the price is expected to rise $0.20 per pound each week. There is no storage area available for raw material and Alpha-Beta must purchase the raw material used in a given week during that week. The total machine time available during each week is 120 hours at a cost of $18 per hour. In addition, 40 hours of overtime are available each week at a cost of $30 per hour. However, the cost of a machine-hour is expected to rise 10% during the last week. The projected weekly demands for widgets are 100, 120, 200, 150 units, and the weekly demands for Gadgets are 80, 90, 115, 160 units. A maximum of 40 units of excess production can be stored from one week to the next at a cost of $3 per unit. Assuming that beginning inventory levels are zero, develop a linear programming model to determine the optimal production and inventory schedules.

2.14. The preliminary layout of a communications network has been determined and is illustrated in the accompanying network diagram. The links of this network are depicted as branches, or arcs between each node pair, and the cost of transmitting a single message unit is indicated beside each link. The nodes of the network (with the exception of nodes 1 and 5, which are simply transmission and reception terminals) actually represent repeater stations (whose purposes are to receive, amplify, error check, and transmit any messages received, per second, at the station). Further,

TABLE 2.13

Link	Maximum capacity (messages/second)	Transition cost/message
1,2	300,000	$3
1,3	400,000	6
2,4	400,000	4
3,4	100,000	3
3,5	300,000	7
4,5	300,000	3

TABLE 2.14

Repeater Node	Cost per messages/second
2	$2.50
3	$2.00
4	$3.00

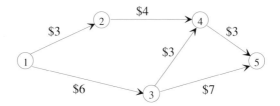

there are limits to the messages per second that may be transmitted via each link. We wish to determine the minimal-cost design of this network if the maximum (that is, worst-case) message load between terminals (nodes) 1 and 5 is to be 500,000 messages per second. The accompanying network diagram and Tables 2.13 and 2.14 provide a summary of the data to be used in constructing a linear programming model for this problem.

3

FOUNDATIONS OF THE SIMPLEX METHOD

CHAPTER OVERVIEW

In this chapter, we develop the theoretical foundation of the simplex method from both a geometric and an algebraic viewpoint. In the process, we establish important notation and terminology that will be used throughout the remainder of the text. The geometry of the problem is discussed first, and the reader will learn to solve simple linear programming problems through a straightforward graphical technique. By utilizing the intuition gained from this graphical procedure, the underlying theoretical aspects of the geometry of the problem are discussed. In essence, this enables us to conclude that in order to determine the optimal solution of a linear programming problem, we only need to consider a *finite* number of candidate solution points. However, to actually find these candidate points for linear programs of nontrivial size, it is necessary to examine the linear program from an algebraic viewpoint. This algebraic viewpoint relies heavily on linear algebra, and, in particular, on the concept of a basic solution to a linear system. The chapter concludes with theoretical results that tie the algebraic and geometric concepts together. Thus, at the conclusion of this chapter, the reader should have a thorough understanding of the foundations of the simplex method, and will be ready to discuss actual implementation issues, which are presented in Chapter 4.

CONVERTING A LINEAR PROGRAM INTO STANDARD FORM

Recall from Chapter 2 that a linear program may be written in the general form:

$$\text{optimize } z = c_1 x_1 + c_2 x_2 + \cdots + c_n x_n \tag{3.1}$$

subject to

$$a_{1,1} x_1 + a_{1,2} x_2 + \cdots + a_{1,n} x_n \{\leq, =, \text{ or } \geq\} b_1 \tag{3.2}$$

$$a_{2,1} x_1 + a_{2,2} x_2 + \cdots + a_{2,n} x_n \{\leq, =, \text{ or } \geq\} b_2 \tag{3.3}$$

$$\cdot$$
$$\cdot$$
$$\cdot$$

$$a_{m,1} x_1 + a_{m,2} x_2 + \cdots + a_{m,n} x_n \{\leq, =, \text{ or } \geq\} b_m \tag{3.4}$$

$$x_1, x_2, \ldots, x_n \geq 0 \tag{3.5}$$

or written more compactly, in summation notation,

$$\text{optimize } z = \sum_{j=1}^{n} c_j x_j \tag{3.6}$$

subject to

$$\sum_{j=1}^{n} a_{i,j} x_j \{\leq, =, \text{ or } \geq\} b_i; \, i = 1, \ldots, m \tag{3.7}$$

$$x_j \geq 0; \, j = 1, \ldots, n \tag{3.8}$$

In general, it is far easier to deal with equations than with inequalities. In this section, we provide simple techniques that allow one to convert any inequality into an equation by means of introducing some additional variables into the formulation.

For simplicity in procedure, we shall assume that regardless of the form of the inequality, the right-hand side is nonnegative (that is, $b_i \geq 0$). Thus, if the constraint initially has a negative b_i, we multiply the entire constraint by -1 and reverse the direction of the inequality.

Constraint Conversion

First, consider an inequality, say, constraint r, of the following form:

$$\sum_{j=1}^{n} a_{r,j} x_j \leq b_r \tag{3.9}$$

We introduce a new variable, $s_r \geq 0$, called the *slack* variable, so that

$$\sum_{j=1}^{n} a_{r,j} x_j + s_r = b_r \tag{3.10}$$

That is,

$$s_r = b_r - \sum_{j=1}^{n} a_{r,j} x_j \tag{3.11}$$

In words, s_r is the (nonnegative) difference between the right-hand-side constant and the original left-hand side and, thus, it "takes up the slack." Physically, it often represents the amount of resource r (that is, b_r) that is unused or idle. From a mathematical view, it allows us to express an inequality of the form (3.9) in a more convenient equality format.

Next, consider an inequality, say, constraint t, of the form

$$\sum_{j=1}^{n} a_{t,j} x_j \geq b_t \tag{3.12}$$

In this case, we introduce a new variable, $s_t \geq 0$, so that

$$\sum_{j=1}^{n} a_{t,j} x_j = b_t + s_t \tag{3.13}$$

Now, rearranging (3.13), we obtain

$$\sum_{j=1}^{n} a_{t,j} x_j - s_t = b_t \tag{3.14}$$

Note that in this case,

$$s_t = \sum_{j=1}^{n} a_{t,j} x_j - b_t \tag{3.15}$$

That is, it represents the (nonnegative) difference between the left-hand side and the right-hand side of the inequality in (3.12). Such a variable is termed a *surplus* variable. Physically, the surplus variable represents the amount by which we *exceed* the right-hand side, and thus "surplus" seems to be a fairly appropriate term.

The Objective Function

The choice of decision variables (that is, the x_j's) directly affects the value of the objective function. This holds true as well for the slack and surplus variables. As a result, each variable introduced in the constraint conversion process should also be introduced, with a proper coefficient, into the objective function. To make this process considerably easier, we shall consider the standard form of the objective function to be *maximization*. This in no way eliminates the consideration of minimization-type objectives because if a function z is to be minimized, we can use the simple equivalence:

$$\text{minimize } z \equiv -\text{maximize } (-z) \tag{3.16}$$

Thus, given a maximization objective z, and p surplus and slack variables, the modified objective is

$$\text{maximize } z = \sum_{j=1}^{n} c_j x_j + \sum_{k=1}^{p} c_k s_k \qquad (3.17)$$

The first term, $\sum_{j=1}^{n} c_j x_j$, is simply the original objective function, and the second term corresponds to the impact of the slack and surplus variables. The question that still remains is: What are the values of the c_k's? If there is a cost or profit associated with idle resources or a surplus, the values of the c_k's accordingly should reflect these values. However, in most textbooks, it is usually assumed that these costs or profits are zero and, as a result, one normally sees each c_k given a value of zero. This is convenient and speeds the formulation process, but it is a dangerous habit to fall into because, in actual practice, there is sometimes a nonzero value associated with a slack or surplus variable.

Example 3.1: Converting a Linear Program into Standard Form

Given the following linear programming model, convert both the constraints and objective function into standard form.

$$\text{minimize } z = 7x_1 - 3x_2 + 5x_3 \quad \text{(cost in dollars)} \qquad (3.18)$$

subject to

$$x_1 + x_2 + x_3 \geq 9 \qquad (3.19)$$

$$3x_1 + 2x_2 + x_3 \leq 12 \qquad (3.20)$$

$$x_1, x_2, x_3 \geq 0 \qquad (3.21)$$

Let us assume that the cost of a surplus unit in the first constraint is zero, and the cost of slack in the second constraint is \$1.50 per unit. The resultant model is

$$\text{maximize } z' = -7x_1 + 3x_2 - 5x_3 + 0s_1 - 1.5s_2 \qquad (3.22)$$

subject to

$$x_1 + x_2 + x_3 - s_1 = 9 \qquad (3.23)$$

$$3x_1 + 2x_2 + x_3 + s_2 = 12 \qquad (3.24)$$

$$x_1, x_2, x_3, s_1, s_2 \geq 0 \qquad (3.25)$$

Notice that the original objective, z, was to minimize *cost*. Thus, the objective $z' = -z$ must be to maximize the negative of cost, which is, of course, profit. Also, because the cost of a slack resource unit is \$1.50, it is a *negative* contribution to the profit objective.

Quite often, textbooks do not distinguish between a decision variable (x_j) or a slack or surplus variable and the foregoing model would be written:

$$\text{maximize } z' = -7x_1 + 3x_2 - 5x_3 + 0x_4 - 1.5x_5 \qquad (3.26)$$

subject to

$$x_1 + x_2 + x_3 - x_4 = 9 \tag{3.27}$$

$$3x_1 + 2x_2 + x_3 + x_5 = 12 \tag{3.28}$$

$$x_j \geq 0; j = 1, \ldots, 5 \tag{3.29}$$

This latter equivalent formulation will be easier to deal with in subsequent sections. In general, this formulation will be referred to as the *standard form* of a linear program and can be summarized as follows:

$$(\text{LP}) \text{ maximize } z = \sum_{j=1}^{n} c_j x_j \tag{3.30}$$

subject to

$$\sum_{j=1}^{n} a_{i,j} x_j = b_i; i = 1, \ldots, m \tag{3.31}$$

$$x_j \geq 0; j = 1, \ldots, n \tag{3.32}$$

Notation and Definitions

The material that follows will be made easier if we introduce some vector and matrix notation. First, we generally let the vector **x** represent *all* the variables in the converted model (that is, the model in which the appropriate slack and surplus variables have been introduced). Thus, **x** may include some slack and surplus variables. It is a vector of order n.

We may then write the standard linear programming model as

$$(\text{LP}) \text{ maximize } z = \mathbf{cx} \tag{3.33}$$

subject to

$$\mathbf{Ax} = \mathbf{b}$$

$$\mathbf{x} \geq \mathbf{0}$$

where the data are given by

$\mathbf{A} = m \times n$ matrix of the coefficients of the constraints, that is,

$$\mathbf{A} = \begin{pmatrix} a_{1,1} & a_{1,2} & \cdots & a_{1,n} \\ a_{2,1} & a_{2,2} & \cdots & a_{2,n} \\ \vdots & \vdots & \ddots & \vdots \\ a_{m,1} & a_{m,2} & \cdots & a_{m,n} \end{pmatrix} = (\mathbf{a}_1, \mathbf{a}_2, \ldots, \mathbf{a}_n)$$

$$\mathbf{b} = m\text{-vector of right-hand sides, that is, } \mathbf{b} = \begin{pmatrix} b_1 \\ b_2 \\ \vdots \\ b_m \end{pmatrix}$$

$$\mathbf{c} = n\text{-vector of objective coefficients, that is, } \mathbf{c} = (c_1, c_2, \ldots, c_n)$$

and the variables are given by the n-vector:

$$\mathbf{x} = \begin{pmatrix} x_1 \\ x_2 \\ \vdots \\ x_n \end{pmatrix} \in E^n \quad (\text{Euclidean } n\text{-space})$$

Example 3.2: Using Matrix Notation

Consider the following linear programming model, in which x_4 and x_5 are slack variables.

$$\text{maximize } z = 5x_1 + 7x_2 + x_3 + 0x_4 + 0x_5 \tag{3.34}$$

subject to

$$x_1 + 3x_2 - x_3 + x_4 = 12 \tag{3.35}$$

$$5x_1 + 6x_2 + x_5 = 24 \tag{3.36}$$

$$x_1, x_2, x_3, x_4, x_5 \geq 0 \tag{3.37}$$

Then this problem may be written as

$$\text{maximize } z = \mathbf{cx} \tag{3.38}$$

subject to

$$\mathbf{Ax} = \mathbf{b}$$

$$\mathbf{x} \geq \mathbf{0}$$

where

$$\mathbf{A} = \begin{pmatrix} 1 & 3 & -1 & 1 & 0 \\ 5 & 6 & 0 & 0 & 1 \end{pmatrix}$$

$$\mathbf{b} = \begin{pmatrix} 12 \\ 24 \end{pmatrix}$$

$$\mathbf{c} = (5 \quad 7 \quad 1 \quad 0 \quad 0)$$

$$\mathbf{x} = \begin{pmatrix} x_1 \\ x_2 \\ \vdots \\ x_5 \end{pmatrix}$$

Before concluding this section, we introduce some terminology concerning the solution space of a linear programming problem.

FEASIBLE SOLUTION A solution is feasible if it satisfies all the constraints of the linear program (including nonnegativity), for example, $\mathbf{x}$ is a feasible solution of problem (LP) if $\mathbf{Ax} = \mathbf{b}$ and $\mathbf{x} \geq \mathbf{0}$. The set of all feasible solutions is called the *feasible region*.

INFEASIBLE SOLUTION Any point that does *not* satisfy all the constraints and the nonnegativity conditions is infeasible.

OPTIMAL SOLUTION A point $\mathbf{x}^*$ is an optimal solution to a maximization linear program if $\mathbf{x}^*$ is a feasible solution and $\mathbf{cx}^* \geq \mathbf{cx}$ for all feasible solutions $\mathbf{x}$.

GRAPHICAL SOLUTION OF TWO-DIMENSIONAL LINEAR PROGRAMS

Prior to presenting the geometrical concepts that form the foundation of the simplex method, we present a graphical method for solving simple problems involving only two variables. It would be unusual to discover a problem in the real world that consisted of no more than two variables. Consequently, it should not be inferred that graphical analysis is a practical approach to linear programming. It is, however, a superb teaching tool and visual aid in linear programming.

The mechanics of the graphical procedure are brief and straightforward. We first plot all constraints on a rectangular (Cartesian) coordinate system in which each axis represents one decision variable. If a constraint is an *equality*, it will plot as a *straight line* (in general, this is called a *hyperplane*, and will be defined formally later). However, if the constraint is an *inequality*, it defines a *region* (this region is formally called a *halfplane* or *halfspace* and will be discussed in detail later) that is bounded by the straight line obtained when the constraint is considered an equality. To clarify this, suppose we wish to graph the set of points (x_1, x_2) satisfying the inequality constraint

$$3x_1 + 2x_2 \leq 12 \qquad (3.39)$$

To identify the region defined by this constraint, we begin by graphing the corresponding linear equation,

$$3x_1 + 2x_2 = 12 \qquad (3.40)$$

This straight line divides the Cartesian plane into two regions (halfplanes). It is quite easy to determine which side of the line defined in (3.40) corresponds to the region defined by the inequality in (3.39). Simply choose any point, P, that is not on the line and check if P satisfies the inequality. If so, then all points that lie on the same side as P satisfy the inequality. Otherwise all points on the opposite side as P satisfy the inequality. For example, choosing $P = (0, 0)$, we see that $3(0) + 2(0) = 0 < 6$; therefore, all points on the same side as $(0,0)$ satisfy the inequality. The region corresponding to (3.39) is plotted in Figure 3.1.

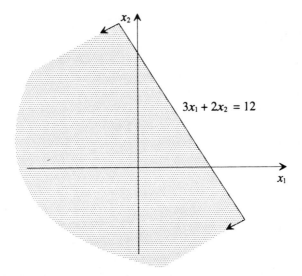

Figure 3.1 Region (halfplane) defined by an inequality constraint.

We now use the following example to illustrate how to graphically solve a linear program with two decision variables.

Example 3.3: A Maximization Problem

Find x_1 and x_2 so as to

$$\text{maximize } z = 2x_1 + 3x_2 \tag{3.41}$$

subject to

$$x_1 - 2x_2 \leq 4 \tag{3.42}$$

$$2x_1 + x_2 \leq 18 \tag{3.43}$$

$$x_2 \leq 10 \tag{3.44}$$

$$x_1, x_2 \geq 0 \tag{3.45}$$

First, we must identify the feasible region of the model. Labeling one axis x_1 and the other x_2, we establish our coordinate system, as shown in Figure 3.2(a). Note that the nonnegativity restrictions, $x_1, x_2 \geq 0$, require that we only consider points, (x_1, x_2), in the first quadrant. Next, the region identified by each constraint is plotted. Considering the first constraint ($x_1 - 2x_2 \leq 4$) initially, we graph the corresponding linear equation ($x_1 - 2x_2 = 4$) and identify the region defined by this constraint in the first quadrant. This region is depicted in Figure 3.2(b). We then repeat this process with the second and third constraints, which results in Figures

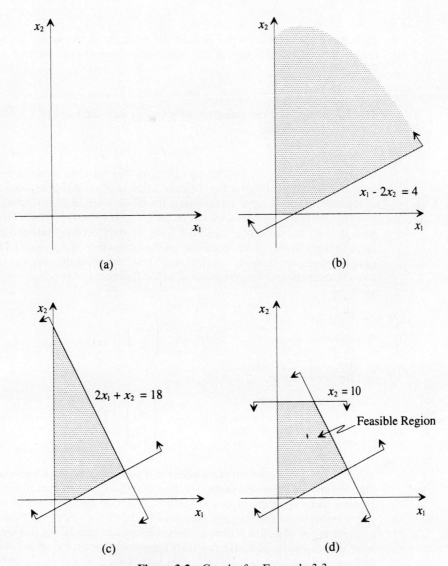

Figure 3.2 Graphs for Example 3.3.

3.2(c) and 3.2(d). The final graph in Figure 3.2(d) represents the feasible region of the problem, that is, the set of points that satisfy all of the constraints.

The final step is to determine the point(s) in the feasible region that yield the maximum value of the objective function:

$$z = 2x_1 + 3x_2 \tag{3.46}$$

To better understand how this is done, let us begin by examining the level curves (isoprofit lines, isocost lines) of the objective function. For example, $z = 13$ defines the line

$$2x_1 + 3x_2 = 13 \tag{3.47}$$

That is, any point on this line gives an objective function value of $z = 13$. Similarly, $z = 30$ defines the line

$$2x_1 + 3x_2 = 30 \tag{3.48}$$

Clearly, these represent parallel lines because they have the same slope. Thus, the level curves of the objective function are a family of parallel lines. We simply need to identify that level curve that contacts the feasible region (that is, contains at least one feasible point) and corresponds to the greatest objective value. Thus, once we have defined the slope of the parallel lines, we only need to slide this line of fixed slope through the set of feasible points in the direction of improving z. The direction of improving z can be quite easily identified by examining the *gradient* of the objective function. Recall that the gradient of the function $z = f(x_1, x_2) = c_1x_1 + c_2x_2$ is given by

$$\nabla f(x_1, x_2) = \begin{pmatrix} \dfrac{\partial z}{\partial x_1} \\ \dfrac{\partial z}{\partial x_2} \end{pmatrix} = \begin{pmatrix} c_1 \\ c_2 \end{pmatrix} \tag{3.49}$$

and for our example,

$$\nabla f(x_1, x_2) = \begin{pmatrix} \dfrac{\partial z}{\partial x_1} \\ \dfrac{\partial z}{\partial x_2} \end{pmatrix} = \begin{pmatrix} 2 \\ 3 \end{pmatrix} \tag{3.50}$$

Recall also that the gradient of a function at a point is *normal* to the level curve of the function and always points in the *direction of steepest ascent*, that is, the direction of greatest increase of the objective function. Thus, to find the optimal solution of a two-variable linear program, we only need to sketch the vector corresponding to the gradient of the objective function. This is illustrated graphically in Figure 3.3. The level curves of the objective are then normal to this vector. For a maximization problem, we would slide the level curves in the direction of the gradient (direction of increasing z) until they reach the boundary of the solution space. Similarly, for a minimization problem, we would slide the level curves in the direction opposite the gradient (direction of decreasing z) until they reach the boundary of the solution space.

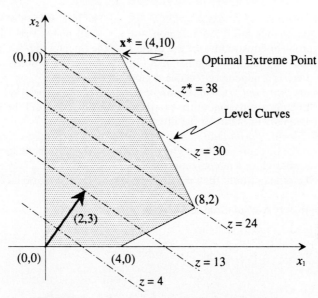

Figure 3.3 Level curves and optimal solution.

By using the foregoing technique, the optimal solution to Example 3.3 is determined to be $(x_1^*, x_2^*) = (4,10)$, as illustrated in Figure 3.3. The corresponding optimal objective value is computed as $z^* = 2(4) + 3(10) = 38$.

Note that the optimal point $\mathbf{x}^*$ lies directly on the constraints $x_2 \leq 10$ and $2x_1 + x_2 \leq 18$; that is, these constraints are satisfied as equalities by $\mathbf{x}^*$ (that is, $x_2^* = 10$; $2x_1^* + x_2^* = 18$). In general, those constraints that are satisfied as equalities by a given point, say, $\bar{\mathbf{x}}$, are said to be *binding* (or *tight* or *active*) at $\bar{\mathbf{x}}$. Those constraints that are not satisfied as equalities at $\bar{\mathbf{x}}$ are said to be *nonbinding*. For example, in the previous problem, the constraint $x_1 - 2x_2 \leq 4$ is nonbinding at $\mathbf{x}^*$ because $x_1^* - 2x_2^* < 4$.

Based on this discussion, we are now ready to state the steps involved in the graphical solution of a two-dimensional linear programming problem:

Graphical solution procedure

STEP 1. *Define the coordinate system.* Sketch the axes of the coordinate system and associate, with each axis, a specific decision variable.

STEP 2. *Plot the constraints.* Establish the line (in the case of an equality) or region (in the case of an inequality) associated with each constraint.

STEP 3. *Identify the resultant solution space.* The *intersection* of all the regions in Step 2 determines the *feasible* region, the set of all points that simultaneously satisfy all problem constraints. In the event that the intersection is empty, no solution exists that will satisfy *all* constraints. In this case, the problem is termed *infeasible*. If there is a nonempty feasible region, go to Step 4.

STEP 4. *Identify the gradient of the objective function.* The level curves of the objective function are normal to the gradient of the objective and the vector corresponding to the gradient points in the direction of increasing z.

STEP 5. *Identify the optimal solution(s).* For a maximization problem, slide the level curves in the direction of the gradient until they reach the boundary of the feasible region. Similarly, for a minimization problem, slide the level curves in the direction opposite the gradient until they reach the boundary of the feasible region.

Example 3.4: A Minimization Problem

Find x_1 and x_2 so as to

$$\text{minimize } z = x_1 + 3x_2 \tag{3.51}$$

subject to

$$x_1 - 2x_2 \leq 4 \tag{3.52}$$

$$-x_1 + x_2 \leq 3 \tag{3.53}$$

$$x_2 \geq 2 \tag{3.54}$$

$$x_1, x_2 \geq 0 \tag{3.55}$$

The feasible region defined by the constraints (3.52–3.55) is plotted in Figure 3.4. Notice that this feasible region is *unbounded*, whereas the feasible region associated with Example 3.3 is *bounded*. Now, as before, consider the level curves associated with the objective function. The set of parallel lines corresponding to these level curves is plotted in Figure 3.4 and we again note that they are normal to the objective gradient, $(c_1, c_2) = (1, 3)$. Because the objective is to minimize z, we slide the level curves in the direction opposite the gradient of the objective to determine that the optimal solution is $(x_1^*, x_2^*) = (0, 2)$ and $z^* = 6$.

Now, suppose that the objective in Example 3.4 is to *maximize* $z = x_1 + 3x_2$ instead of to minimize z. Observe from Figure 3.4 that we can slide the level curves in the direction of the gradient and never reach the boundary of the feasible region. That is, the value of z can be made arbitrarily large. When this occurs, we say that the linear programming problem has an *unbounded objective value* or that the linear programming problem has *no finite optimum*.

Example 3.5: Alternative Optimal Solutions

Find x_1 and x_2 so as to

$$\text{maximize } z = 6x_1 + 3x_2 \tag{3.56}$$

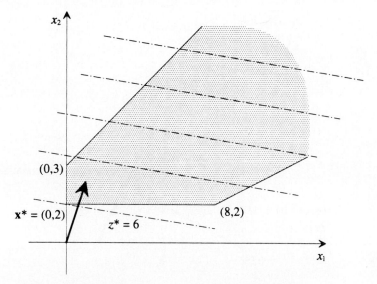

Figure 3.4   Optimal solution of minimization problem.

subject to

$$2x_1 + x_2 \leq 16 \tag{3.57}$$

$$x_1 + x_2 \leq 10 \tag{3.58}$$

$$x_1, x_2 \geq 0 \tag{3.59}$$

The feasible region for this example is plotted in Figure 3.5. In this case, we see that the feasible region is nonempty and bounded. The level curves associated with the objective function are also plotted in Figure 3.5. In this case, however, sliding the level curves in the direction of the gradient does not identify a single extreme point as the optimal solution. Instead, the entire line segment connecting the extreme points (6, 4) and (8, 0) coincides with the level curve, $z^* = 48$, which corresponds to the optimal value of z. When this occurs, we say that we have *alternative optimal solutions* and any point on the line segment joining (6, 4) and (8, 0) is an optimal solution.

The reader should note that in those instances in which a finite optimal solution existed, an optimal solution occurred at a corner point of the feasible region. This a very intuitive notion that is formalized in the following sections. These "corner" points are more precisely referred to as *extreme points*, and as we will show, if a linear program has a finite optimal solution, then, indeed, it has an extreme-point optimal solution.

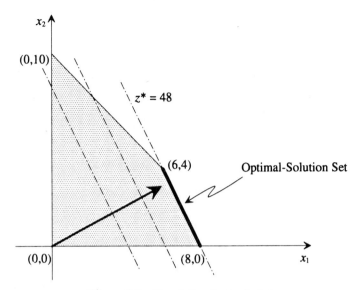

Figure 3.5   Graph for Example 3.5.

CONVEX SETS AND POLYHEDRAL SETS

In the preceding section, we introduced a graphical procedure for solving simple linear programming problems. This was done to motivate the geometrical aspects of the simplex method. We begin the discussion of the geometry of the problem by presenting several definitions that form the foundation of the development that is to follow.

> **HYPERPLANE**   A hyperplane (line in two dimensions, plane in three dimensions) is the set of points $\mathbf{x} = (x_1, \ldots, x_n)^t \in E^n$ that satisfy $\mathbf{ax} = b$, where $\mathbf{a} = (a_1, \ldots, a_n) \in E^n$, $\mathbf{a} \neq \mathbf{0}$, and $b \in E^1$ (i.e., b is a scalar).

Utilizing this definition, note that the linear system defined by $\mathbf{Ax} = \mathbf{b}$ is simply a collection of m hyperplanes.

> **HALFSPACE**   A closed halfspace corresponding to the hyperplane $\mathbf{ax} = b$ is either of the sets $H^+ = \{\mathbf{x} : \mathbf{ax} \geq b\}$ or $H^- = \{\mathbf{x} : \mathbf{ax} \leq b\}$. When these halfspaces are defined as $\{\mathbf{x} : \mathbf{ax} > b\}$ or $\{\mathbf{x} : \mathbf{ax} < b\}$, they are called *open halfspaces*.

It is easily seen that vector $\mathbf{a}$ is the *gradient* of linear function $f(\mathbf{x}) = \mathbf{ax}$, and thus is *normal* to the hyperplane and points in the direction of increasing $\mathbf{ax}$, as depicted in Figure 3.6.

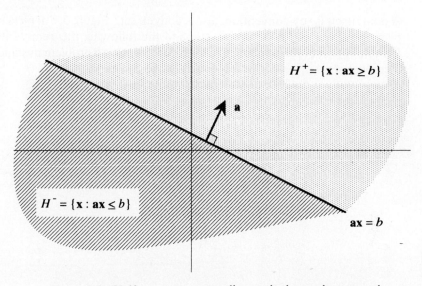

Figure 3.6 Halfspaces corresponding to the hyperplane $\mathbf{ax} = b$.

POLYHEDRAL SET A polyhedral set is the intersection of a finite number of halfspaces. Thus, the constraint set $S = \{\mathbf{x} : \mathbf{Ax} \leq \mathbf{b}, \mathbf{x} \geq \mathbf{0}\}$ is a polyhedral set because it is the intersection of m halfspaces corresponding to $\mathbf{Ax} \leq \mathbf{b}$ and n halfspaces corresponding to $\mathbf{x} \geq \mathbf{0}$.

CONVEX SET A set S is convex if, for *any* two points, say, $\mathbf{x}_1, \mathbf{x}_2 \in S$, then the line segment joining these two points lies entirely within S. Mathematically, this means that if $\mathbf{x}_1, \mathbf{x}_2 \in S$, then $\alpha\mathbf{x}_1 + (1 - \alpha)\mathbf{x}_2 \in S$ for all $\alpha \in [0, 1]$.

The expression $\mathbf{x} = \alpha\mathbf{x}_1 + (1 - \alpha)\mathbf{x}_2 \in S$, $\alpha \in [0, 1]$ parametrically defines the line segment joining $\mathbf{x}_1$ and $\mathbf{x}_2$ and is called the *convex combination* of $\mathbf{x}_1$ and $\mathbf{x}_2$. If $\alpha \in (0, 1)$, then it is called a *strict convex combination*. The concept of convex combination can be generalized to any finite number of points as follows:

$$\mathbf{x} = \sum_{i=1}^{p} \alpha_i \mathbf{x}_i \qquad (3.60)$$

where

$$\sum_{i=1}^{p} \alpha_i = 1 \qquad (3.61)$$

$$\alpha_i \geq 0; \; i = 1, \ldots, p \qquad (3.62)$$

A point itself is, by convention, also a convex set. Figure 3.7 depicts some examples of convex and nonconvex sets, and the following theorem verifies that the feasible region corresponding to a linear programming problem in standard form is a convex set.

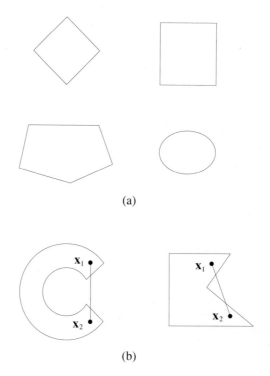

(a)

(b)

Figure 3.7   (a) Convex and (b) nonconvex sets.

Theorem 3.1

The set $S = \{x : \mathbf{Ax} = \mathbf{b}, \mathbf{x} \geq \mathbf{0}\}$ is a convex set.

Proof

Let $\mathbf{x}_1, \mathbf{x}_2 \in S$ and let $\alpha \in [0, 1]$. To complete the proof, it is sufficient to show that $\bar{\mathbf{x}} = \alpha \mathbf{x}_1 + (1 - \alpha)\mathbf{x}_2 \in S$.

Because $\mathbf{x}_1 \in S$, it follows from the definition of S that $\mathbf{Ax}_1 = \mathbf{b}$ and $\mathbf{x}_1 \geq \mathbf{0}$. Similarly, $\mathbf{Ax}_2 = \mathbf{b}$ and $\mathbf{x}_2 \geq \mathbf{0}$. Also $\alpha \in [0, 1]$ implies that $\alpha \geq 0$ and $(1 - \alpha) \geq 0$.

Now, combining these results yields

$$\alpha \mathbf{A} \mathbf{x}_1 = \alpha \mathbf{b} \tag{3.63}$$

$$\alpha \mathbf{x}_1 \geq \mathbf{0} \tag{3.64}$$

$$(1 - \alpha) \mathbf{A} \mathbf{x}_2 = (1 - \alpha) \mathbf{b} \tag{3.65}$$

$$(1 - \alpha) \mathbf{x}_2 \geq \mathbf{0} \tag{3.66}$$

Summing the expressions in (3.63) and (3.65) yields

$$\alpha \mathbf{A} \mathbf{x}_1 + (1 - \alpha) \mathbf{A} \mathbf{x}_2 = \alpha \mathbf{b} + (1 - \alpha) \mathbf{b} \tag{3.67}$$

Similarly, summing (3.64) and (3.66), we obtain

$$\alpha \mathbf{x}_1 + (1 - \alpha) \mathbf{x}_2 \geq \mathbf{0} \tag{3.68}$$

Now, rearranging, (3.67) and (3.68) yield, respectively,

$$\mathbf{A}[\alpha \mathbf{x}_1 + (1 - \alpha) \mathbf{x}_2] = [\alpha + (1 - \alpha)] \mathbf{b} = \mathbf{b} \tag{3.69}$$

and

$$\alpha \mathbf{x}_1 + (1 - \alpha) \mathbf{x}_2 \geq \mathbf{0} \tag{3.70}$$

From (3.69) and (3.70), it is clear that $\mathbf{A} \bar{\mathbf{x}} = \mathbf{b}$ and $\bar{\mathbf{x}} \geq \mathbf{0}$, and thus $\bar{\mathbf{x}} \in S$. ☐

> **EXTREME POINT** A point $\mathbf{x}$ is an extreme point of a given convex set S if it cannot be written as a *strict* convex combination of two other distinct points of S. Geometrically, this means that $\mathbf{x}$ is an extreme point of S if it does not lie on the interior of the line segment joining two other distinct points of S. Mathematically, there does not exist $\mathbf{x}_1, \mathbf{x}_2 \in S$, $\mathbf{x}_1 \neq \mathbf{x}_2$, and $\alpha \in (0, 1)$ such that $\mathbf{x} = \alpha \mathbf{x}_1 + (1 - \alpha) \mathbf{x}_2$. Or, equivalently, if $\mathbf{x}_1, \mathbf{x}_2 \in S$, and $\alpha \in (0, 1)$ and $\mathbf{x} = \alpha \mathbf{x}_1 + (1 - \alpha) \mathbf{x}_2$, then $\mathbf{x} = \mathbf{x}_1 = \mathbf{x}_2$.

In polyhedral sets, these extreme points occur only at the intersection of the hyperplanes that form the boundaries of the polyhedral set. In contrast, all points that lie on the boundary of a closed circle are extreme points.

> **ADJACENT EXTREME POINTS** Two distinct extreme points, say, $\mathbf{x}_1$ and $\mathbf{x}_2$, are adjacent if the line segment joining them is an edge of the convex set.

As we noticed when graphically solving two-dimensional linear programming problems, if an optimal solution exists, then at least one extreme-point (corner-point) optimal solution exists. As such, the concept of the extreme point is very important and, as we will see, the algebraic technique for solving linear programming problems involves determining these extreme points and systematically moving between adjacent extreme points. Adjacent extreme points and edges in E^2 and E^3 are illustrated in Figure 3.8.

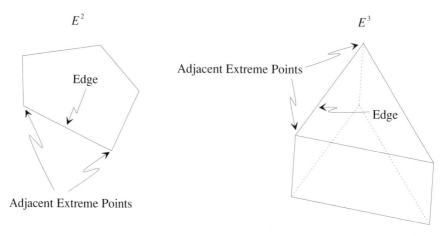

Figure 3.8   Adjacent extreme points and edges.

*EXTREME POINTS, EXTREME DIRECTIONS, AND OPTIMALITY

As previously noted, the concept of extreme point plays a very important role in the study of linear programming. Let us now look at an alternative characterization of an extreme point. Whereas the definition of an extreme point given in the previous section was applicable to any convex set, this alternative characterization is specific to polyhedral sets. In particular, we will consider the polyhedral set S defined by $S = \{\mathbf{x} : \mathbf{Ax} = \mathbf{b}, \mathbf{x} \geq \mathbf{0}\}$.

Theorem 3.2

Let $S = \{\mathbf{x} : \mathbf{Ax} = \mathbf{b}, \mathbf{x} \geq \mathbf{0}\}$, where $\mathbf{A}$ is $m \times n$ and rank$(\mathbf{A}) = m < n$. $\bar{\mathbf{x}}$ is an extreme point of S if and only if $\bar{\mathbf{x}}$ is the intersection of n linearly independent hyperplanes.

Proof

($\rightarrow$) Let $\bar{\mathbf{x}}$ be an extreme point of S. We need to show that $\bar{\mathbf{x}}$ is the intersection of n linearly independent hyperplanes.

Note from the definition of S that $\bar{\mathbf{x}}$ lies on the m linearly independent hyperplanes forming the constraint set $\mathbf{Ax} = \mathbf{b}$. Thus, to have n linearly independent hyperplanes, at least $n - m$ of the hyperplanes forming the constraints $\mathbf{x} \geq \mathbf{0}$ would have to be binding at $\bar{\mathbf{x}}$. Suppose that fewer than $n - m$ of these hyperplanes are binding at $\bar{\mathbf{x}}$, thus giving us fewer than n linearly independent hyper-

---

* Starred sections may be omitted in an introductory course with no appreciable loss of continuity.

planes defining the point $\bar{\mathbf{x}}$. Without loss of generality, assume that

$$\bar{x}_i = 0, \qquad \text{for } i = 1, \ldots, p \tag{3.71}$$

and

$$\bar{x}_i > 0, \qquad \text{for } i = p + 1, \ldots, n \tag{3.72}$$

where $p < n - m$. Then, $\bar{\mathbf{x}}$ is a solution of the $m + p$ linearly independent hyperplanes:

$$\mathbf{Ax} = \mathbf{b} \tag{3.73}$$

$$x_i = 0, \qquad \text{for } i = 1, \ldots, p \tag{3.74}$$

For notational convenience, let us represent the system in (3.73–3.74) by $\mathbf{Qx} = \mathbf{h}$. Then, $\mathbf{Q\bar{x}} = \mathbf{h}$. Also, note that the $\mathbf{Qx} = \mathbf{h}$ is a system of $m + p$ linearly independent equations in n unknowns, where $m + p < n$. Thus, the columns of $\mathbf{Q}$ are linearly dependent, and, consequently, there exists a vector $\mathbf{y} \in E^n$ such that

$$\mathbf{Qy} = \mathbf{0} \tag{3.75}$$

Now, consider the points defined by

$$\tilde{\mathbf{x}} = \bar{\mathbf{x}} + \lambda \mathbf{y} \tag{3.76}$$

and

$$\hat{\mathbf{x}} = \bar{\mathbf{x}} - \lambda \mathbf{y} \tag{3.77}$$

where $\lambda > 0$. Then,

$$\mathbf{Q\tilde{x}} = \mathbf{Q}(\bar{\mathbf{x}} + \lambda \mathbf{y}) = \mathbf{Q\bar{x}} + \lambda \mathbf{Qy} = \mathbf{h} + \lambda \mathbf{0} = \mathbf{h} \tag{3.78}$$

and

$$\mathbf{Q\hat{x}} = \mathbf{Q}(\bar{\mathbf{x}} + \lambda \mathbf{y}) = \mathbf{Q\bar{x}} + \lambda \mathbf{Qy} = \mathbf{h} + \lambda \mathbf{0} = \mathbf{h} \tag{3.79}$$

But, from the definition of the system $\mathbf{Qx} = \mathbf{h}$ in (3.73–3.74), it follows from (3.78) and (3.79) that

$$\mathbf{A\tilde{x}} = \mathbf{b} \tag{3.80}$$

$$\tilde{x}_i = 0, \qquad \text{for } i = 1, \ldots, p \tag{3.81}$$

and

$$\mathbf{A\hat{x}} = \mathbf{b} \tag{3.82}$$

$$\hat{x}_i = 0, \qquad \text{for } i = 1, \ldots, p \tag{3.83}$$

Also, because $\bar{x}_j > 0$, for all $j = p + 1, \ldots, n$, there exists $\lambda > 0$ such that

$$\tilde{x}_j = \bar{x}_j + \lambda y_j > 0, \qquad \text{for } j = p + 1, \ldots, n \tag{3.84}$$

and

$$\hat{x}_j = \bar{x}_j - \lambda y_j > 0, \qquad \text{for } j = p + 1, \ldots, n \tag{3.85}$$

It then follows from (3.80), (3.81), and (3.84) that $\tilde{\mathbf{x}} \in S$. Similarly, (3.82), (3.83), and (3.85) imply that $\hat{\mathbf{x}} \in S$.

Finally, note from (3.76) and (3.77) that $\bar{\mathbf{x}} = (1/2)\tilde{\mathbf{x}} + (1/2)\hat{\mathbf{x}}$. Thus, $\bar{\mathbf{x}}$ can be written as a strict convex combination of two distinct points of S and hence is not an extreme point of S. This is a contradiction and thus $\bar{\mathbf{x}}$ is the intersection of n linearly independent hyperplanes.

($\leftarrow$) Let $\bar{\mathbf{x}}$ be the intersection of n linearly independent hyperplanes. Show that $\bar{\mathbf{x}}$ is an extreme point of S.

Without loss of generality, let us denote the n linearly independent hyperplanes defining $\bar{\mathbf{x}}$ by

$$\mathbf{Ax} = \mathbf{b} \tag{3.86}$$

$$x_i = 0, \qquad \text{for } i = 1, \ldots, n - m \tag{3.87}$$

Now, let $\alpha \in (0, 1)$ and let $\tilde{\mathbf{x}}, \hat{\mathbf{x}} \in S$. Suppose that

$$\bar{\mathbf{x}} = \alpha\tilde{\mathbf{x}} + (1 - \alpha)\hat{\mathbf{x}} \tag{3.88}$$

To show that $\bar{\mathbf{x}}$ is an extreme point, we need to show that $\bar{\mathbf{x}} = \tilde{\mathbf{x}} = \hat{\mathbf{x}}$. Combining (3.87) and (3.88), we see that

$$\bar{x}_i = \alpha\tilde{x}_i + (1 - \alpha)\hat{x}_i = 0, \qquad \text{for } i = 1, \ldots, n - m \tag{3.89}$$

Because $\alpha, (1 - \alpha) > 0$, and $\tilde{x}_i, \hat{x}_i \geq 0$, we obtain, from (3.89), that

$$\bar{x}_i = \tilde{x}_i = \hat{x}_i = 0, \qquad \text{for } i = 1, \ldots, n - m \tag{3.90}$$

Let $\mathbf{A} = (\mathbf{a}_1, \mathbf{a}_2, \ldots, \mathbf{a}_n)$. Because $\mathbf{A\bar{x}} = \mathbf{A\tilde{x}} = \mathbf{A\hat{x}} = \mathbf{b}$, it follows from (3.90) that

$$\sum_{j=n-m+1}^{n} \bar{x}_j\mathbf{a}_j = \sum_{j=n-m+1}^{n} \tilde{x}_j\mathbf{a}_j = \sum_{j=n-m+1}^{n} \hat{x}_j\mathbf{a}_j = \mathbf{b} \tag{3.91}$$

But $\bar{\mathbf{x}}$ was the unique solution to (3.86–3.87). Thus, the columns $\mathbf{a}_{n-m+1}, \ldots, \mathbf{a}_n$ are linearly independent, and it follows from (3.91) that

$$\bar{x}_i = \tilde{x}_i = \hat{x}_i, \qquad \text{for } i = n - m + 1, \ldots, n \tag{3.92}$$

Combining (3.90) and (3.92), we have $\bar{\mathbf{x}} = \tilde{\mathbf{x}} = \hat{\mathbf{x}}$, and it follows that $\bar{\mathbf{x}}$ is an extreme point of S. $\square$

We have now established an equivalent characterization of the extreme points of a polyhedral set. That is, an extreme point of a polyhedral set is the intersection of n linearly independent hyperplanes defining the set. A simple example of this can be observed in Figure 3.8. Note that in E^2, each extreme point is the intersection of two linearly independent hyperplanes (lines), and in E^3, each extreme point is the intersection of three linearly independent hyperplanes (planes). This characterization will be essential in establishing the extreme-point optimality results in the following section. Of equal importance, however, in establishing these optimality results are the concepts of directions and extreme directions of a polyhedral set.

> **DIRECTION** A nonzero vector $\mathbf{d} = (d_1, d_2, \ldots, d_n)^t \in E^n$ is a direction of a convex set S if $\mathbf{x} + \lambda\mathbf{d} \in S$ for all $\mathbf{x} \in S$ and $\lambda \geq 0$.

Note that the set $\mathbf{x} + \lambda\mathbf{d}$, $\lambda \geq 0$, defines a *ray* with vertex $\mathbf{x}$ in the direction $\mathbf{d}$.

> **EXTREME DIRECTION** A direction $\mathbf{d}$ of a set S is an extreme direction if it cannot be written as a *positive* combination of two distinct directions of S. That is, there does not exist directions $\mathbf{d}_1$, $\mathbf{d}_2$ of S, $\mathbf{d}_1 \neq \mathbf{d}_2$, and $\alpha_1, \alpha_2 > 0$ such that $\mathbf{d} = \alpha_1\mathbf{d}_1 + \alpha_2\mathbf{d}_2$.

It should be clear from the definition of direction that a convex set must be unbounded to have a nonempty direction set. Directions and extreme directions play an important role in characterizing the optimal solution of a linear programming problem. The following theorem characterizes the direction set of an arbitrary polyhedral set.

Theorem 3.3

Let $S = \{\mathbf{x} : \mathbf{A}\mathbf{x} = \mathbf{b}, \mathbf{x} \geq \mathbf{0}\}$. Then, $\mathbf{d}$ is a direction of S if and only if $\mathbf{d} \in D = \{\mathbf{d} : \mathbf{A}\mathbf{d} = \mathbf{0}, \mathbf{d} \geq \mathbf{0}, \mathbf{d} \neq \mathbf{0}\}$. (Note that D is simply the set of nonnegative, nonzero solutions satisfying the homogeneous system corresponding to $\mathbf{A}\mathbf{x} = \mathbf{b}$.)

Proof
($\rightarrow$) Let $\mathbf{d}$ be a direction of S. Then, by definition, $\mathbf{d} \neq \mathbf{0}$ and $\mathbf{x} + \lambda\mathbf{d} \in S$ for all $\mathbf{x} \in S$ and $\lambda \geq 0$. Therefore, for all $\lambda \geq 0$,

$$\mathbf{A}(\mathbf{x} + \lambda\mathbf{d}) = \mathbf{A}\mathbf{x} + \lambda\mathbf{A}\mathbf{d} = \mathbf{b} + \lambda\mathbf{A}\mathbf{d} = \mathbf{b} \tag{3.93}$$

and

$$\mathbf{x} + \lambda\mathbf{d} \geq \mathbf{0} \tag{3.94}$$

Setting $\lambda = 1$ in (3.93) clearly yields $\mathbf{A}\mathbf{d} = \mathbf{0}$. Also, (3.94) specifies that $\mathbf{d} \geq \mathbf{0}$, otherwise one can choose λ to be an arbitrarily large positive number and the inequality in (3.94) will be violated.

($\leftarrow$) Let $\mathbf{d} \in D = \{\mathbf{d} : \mathbf{A}\mathbf{d} = \mathbf{0}, \mathbf{d} \geq \mathbf{0}, \mathbf{d} \neq \mathbf{0}\}$ and let $\mathbf{x} \in S$. By the definition of D, it follows that $\mathbf{d} \neq \mathbf{0}$. Thus, all that remains to be shown is that $\mathbf{x} + \lambda\mathbf{d} \in S$ for all $\lambda \geq 0$.

$$\mathbf{A}(\mathbf{x} + \lambda\mathbf{d}) = \mathbf{A}\mathbf{x} + \lambda\mathbf{A}\mathbf{d} = \mathbf{b} + \mathbf{0} = \mathbf{b} \tag{3.95}$$

and because $\mathbf{x}, \mathbf{d} \geq \mathbf{0}$, $\lambda \geq 0$, it follows immediately that $\mathbf{x} + \lambda\mathbf{d} \geq \mathbf{0}$. Thus, $\mathbf{x} + \lambda\mathbf{d} \in S$ for all $\lambda \geq 0$. □

In an analogous manner, it can be shown that the direction set of the convex set $S = \{\mathbf{x} : \mathbf{A}\mathbf{x} \leq \mathbf{b}, \mathbf{x} \geq \mathbf{0}\}$ is given by the set $D = \{\mathbf{d} : \mathbf{A}\mathbf{d} \leq \mathbf{0}, \mathbf{d} \geq \mathbf{0}, \mathbf{d} \neq \mathbf{0}\}$. Thus, in general, the direction set of a convex set can be thought of as the set of nonzero solutions of the corresponding homogeneous system. These concepts are illustrated graphically in the following example.

Example 3.6: Directions and Extreme Directions

Consider the set

$$S = \{\mathbf{x} = (x_1, x_2)^t : -x_1 + x_2 \leq 2, x_1 + x_2 \geq 1, x_1 - 2x_2 \leq 2, x_1, x_2 \geq 0\}$$

Then the direction set D of S is given by

$$D = \{\mathbf{d} =$$
$$(d_1, d_2)^t : -d_1 + d_2 \leq 0, d_1 + d_2 \geq 0, d_1 - 2d_2 \leq 0, d_1, d_2 \geq 0, (d_1, d_2) \neq (0, 0)\}$$

These two sets are illustrated graphically in Figure 3.9. Notice that, geometrically, set D is determined by sliding the constraints of S until they pass through the origin, thus specifying the associated homogeneous system. The extreme directions of S are easily found to be $\mathbf{d}_1 = (1, 1)^t$ and $\mathbf{d}_2 = (2, 1)^t$.

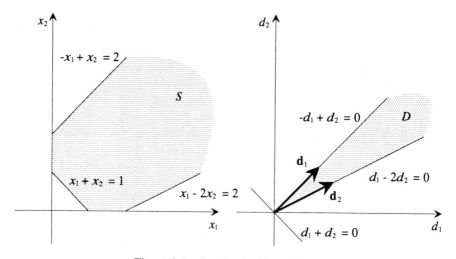

Figure 3.9 Graphs for Example 3.6.

The concepts of extreme points and extreme directions play a very important role in establishing the fundamental optimality results for linear programming. Before stating the theorem that will enable us to proceed with this development, let us consider a graphical interpretation of the usefulness of extreme points

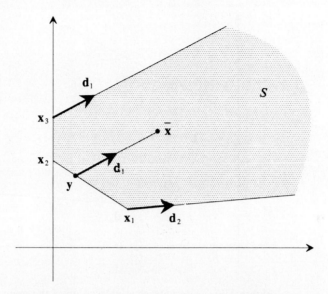

Figure 3.10   Extreme points and extreme directions of S.

and extreme directions. Consider the unbounded polyhedral set S in Figure 3.10. Clearly, set S has three extreme points, $\mathbf{x}_1$, $\mathbf{x}_2$, and $\mathbf{x}_3$, and two extreme directions, $\mathbf{d}_1$ and $\mathbf{d}_2$. Now, consider the point $\bar{\mathbf{x}}$ depicted in Figure 3.10.

Observe that $\bar{\mathbf{x}}$ can be written in the form

$$\bar{\mathbf{x}} = \mathbf{y} + \lambda \mathbf{d}_1 \qquad (3.96)$$

for some $\lambda > 0$. That is, $\bar{\mathbf{x}}$ lies on the *ray* with vertex $\mathbf{y}$ and direction $\mathbf{d}_1$. Also, because $\mathbf{y}$ lies on the edge connecting $\mathbf{x}_1$ and $\mathbf{x}_2$, then $\mathbf{y}$ can be written as a convex combination of $\mathbf{x}_1$ and $\mathbf{x}_2$. That is,

$$\mathbf{y} = \alpha \mathbf{x}_1 + (1 - \alpha)\mathbf{x}_2 \qquad (3.97)$$

for some $\alpha \in (0, 1)$. Now substituting (3.97) into (3.96) yields

$$\bar{\mathbf{x}} = \alpha \mathbf{x}_1 + (1 - \alpha)\mathbf{x}_2 + \lambda \mathbf{d}_1, \qquad \alpha \in (0, 1), \qquad \lambda > 0 \qquad (3.98)$$

or written more completely

$$\bar{\mathbf{x}} = \alpha \mathbf{x}_1 + (1 - \alpha)\mathbf{x}_1 + 0\mathbf{x}_3 + \lambda \mathbf{d}_1 + 0\mathbf{d}_2, \qquad \alpha \in (0, 1), \qquad \lambda > 0 \qquad (3.99)$$

Thus, the arbitrary point $\bar{\mathbf{x}}$ can be written as a convex combination of the extreme points, $\mathbf{x}_1$, $\mathbf{x}_2$, and $\mathbf{x}_3$, and a nonnegative combination of the extreme directions, $\mathbf{d}_1$ and $\mathbf{d}_2$. Observe that this representation is not necessarily unique, as indicated in

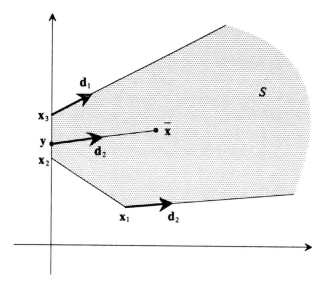

Figure 3.11   Alternate characterization of the point $\bar{\mathbf{x}}$.

Figure 3.11. This idea is generalized in the following theorem, called the Representation Theorem, which, because of its intuitive geometric nature, is provided without proof. The interested reader is referred to Bazaraa, Jarvis, and Sherali (1990) and Murty (1983).

Theorem 3.4

Let $S = \{\mathbf{x} : \mathbf{Ax} = \mathbf{b}, \mathbf{x} \geq \mathbf{0}\}$ and let E be the set of extreme points of S and let D be the set of extreme directions of S. Then:

(1) S has at least one extreme point and at most a finite number of extreme points, that is, the set E is nonempty and finite, $E = \{\mathbf{x}_1, \mathbf{x}_2, \ldots, \mathbf{x}_p\} \neq \phi$

(2) S is unbounded if and only if S has at least one extreme direction, that is, S is unbounded if and only if D is nonempty

(3) if S is unbounded, then S has a finite number of extreme directions, that is, $D = \{\mathbf{d}_1, \mathbf{d}_2, \ldots, \mathbf{d}_q\} \neq \phi$

(4) if $\mathbf{x} \in S$, then $\mathbf{x}$ can be written as a convex combination of the extreme points plus a nonnegative combination of the extreme directions, that is,

$$\mathbf{x} = \sum_{i=1}^{p} \alpha_i \mathbf{x}_i + \sum_{j=1}^{q} \lambda_j \mathbf{d}_j \tag{3.100}$$

where

$$\sum_{i=1}^{p} \alpha_i = 1 \qquad (3.101)$$

$$\alpha_i \geq 0, \qquad \text{for all } i = 1, \ldots, p \qquad (3.102)$$

$$\lambda_j \geq 0, \qquad \text{for all } j = 1, \ldots, q \quad \square \qquad (3.103)$$

Although the importance of this result may not be immediately obvious, this theorem provides the machinery necessary for establishing the foundation of the simplex algorithm. This fundamental result specifies that if an optimal solution exists, we only need to examine the extreme points of the feasible region in order to find an optimal solution. This is a very important concept, because, in general, the search for an optimal solution is reduced from the infinitely many points comprising the feasible region to a finite set of candidate (extreme) points. These results are established formally through the following theorems.

Theorem 3.5

Let $S = \{x : Ax = b, x \geq 0\}$ and consider the following linear program:

$$(LP) \quad \text{maximize } z = cx$$

$$\text{subject to } x \in S.$$

Suppose S is an unbounded set with extreme points $E = \{x_1, x_2, \ldots, x_p\} \neq \phi$ and extreme directions $D = \{d_1, d_2, \ldots, d_q\} \neq \phi$. Let z^* represent the optimal objective value of (LP). Then z^* is finite (that is, $z^* < \infty$), if and only if $cd_j \leq 0$ for all $d_j \in D$. And, furthermore, if a finite optimal solution exists, then an extreme-point optimal solution exists.

Proof

We begin the proof by rewriting (LP) using the results of Theorem 3.4. Let $x \in S$. Then by part (4) of Theorem 3.4, we can write x as in (3.100–3.103). Now, substituting this into (LP) yields the following equivalent linear program:

$$(LP') \quad \text{maximize } z = c \left(\sum_{i=1}^{p} \alpha_i x_i + \sum_{j=1}^{q} \lambda_j d_j \right) = \sum_{i=1}^{p} (cx_i)\alpha_i + \sum_{j=1}^{q} (cd_j)\lambda_j \qquad (3.104)$$

subject to

$$\sum_{i=1}^{p} \alpha_i = 1 \qquad (3.105)$$

$$\alpha_i \geq 0, \qquad \text{for all } i = 1, \ldots, p \qquad (3.106)$$

$$\lambda_j \geq 0, \qquad \text{for all } j = 1, \ldots, q \qquad (3.107)$$

Observe from (3.104–3.107) that the variables in (LP′) are $\alpha_i \geq 0$, $i = 1, \ldots, p$, and $\lambda_j \geq 0$, $j = 1, \ldots, q$. Thus, the total number of variables is $p + q$, the total number of extreme points and extreme directions. In general, this is a very *impractical* formulation because the number of variables may be astronomical. However, this formulation is of theoretical value, as we shall see in the remainder of the proof.

($\rightarrow$) By contradiction, suppose that $\mathbf{cd}_j > 0$ for some $\mathbf{d}_j \in D$. Then, clearly, because $\mathbf{cd}_j$ is the coefficient of λ_j in (3.104), we can make z arbitrarily large by making λ_j arbitrarily large. This is possible because λ_j is only constrained by $\lambda_j \geq 0$. This contradicts the fact that z^* is finite. Thus, $\mathbf{cd}_j \leq 0$ for all $\mathbf{d}_j \in D$.

($\leftarrow$) Conversely, suppose that $\mathbf{cd}_j \leq 0$ for all $\mathbf{d}_j \in D$. Then, because (LP′) is a maximization problem, it follows that $\lambda_j = 0$, for all $j = 1, \ldots, q$. This results in the simplified problem:

$$\text{(LP′)} \quad \text{maximize } z = \sum_{i=1}^{p} (\mathbf{cx}_i)\alpha_i \tag{3.108}$$

subject to

$$\sum_{i=1}^{p} \alpha_i = 1 \tag{3.109}$$

$$\alpha_i \geq 0, \qquad \text{for all } i = 1, \ldots, p \tag{3.110}$$

Now choose extreme point $\mathbf{x}_k$ so that $\mathbf{cx}_k \geq \mathbf{cx}_i$, for all $i = 1, \ldots, p$. Clearly, this is possible because there are finitely many extreme points. Then

$$z = \sum_{i=1}^{p} (\mathbf{cx}_i)\alpha_i \leq \sum_{i=1}^{p} (\mathbf{cx}_k)\alpha_i = (\mathbf{cx}_k) \sum_{i=1}^{p} \alpha_i = (\mathbf{cx}_k)(1) = \mathbf{cx}_k \tag{3.111}$$

Therefore, $\mathbf{cx}_k$ provides an upper bound for the objective value. Thus, z is finite, and in fact $z^* = \mathbf{cx}_k$ and $\mathbf{x}_k$ is an optimal extreme point. $\square$

Corollary 3.6

Suppose S is a nonempty bounded set with extreme points $E = \{\mathbf{x}_1, \mathbf{x}_2, \ldots, \mathbf{x}_p\} \neq \phi$. Let z^* represent the optimal objective value of (LP). Then z^* is finite (that is, $z^* < \infty$), and, furthermore, an extreme-point optimal solution exists.

Recall that $\mathbf{cd}_j = \|\mathbf{c}\| \|\mathbf{d}_j\| \cos \theta$, where $\|\mathbf{v}\|$ represents the Euclidean norm of the vector $\mathbf{v}$, and θ is the angle between the vectors. Thus, the foregoing theorem states that (LP) has a finite optimal solution if and only if the angle between the objective gradient $\mathbf{c}$ and each extreme direction $\mathbf{d}_j$ is greater than or equal to 90°. Or, equivalently, if some extreme direction $\mathbf{d}_j$ makes a strict acute angle with $\mathbf{c}$, then (LP) has an unbounded objective value.

We now conclude this discussion with an example that serves to illustrate the basic concepts established in the preceding theorems. The reader should take special care in understanding these concepts because they play a vital role in the overall development of the simplex method.

Example 3.7: Using Theorem 3.5 to Characterize the Optimum of a Linear Program

Consider the following simple linear program:

$$\text{maximize } z = -3x_1 - 2x_2 \tag{3.112}$$

subject to

$$-2x_1 + x_2 \leq 2 \tag{3.113}$$

$$x_1 + x_2 \geq 1 \tag{3.114}$$

$$x_1 - 2x_1 \leq 1 \tag{3.115}$$

$$x_1, x_2 \geq 0 \tag{3.116}$$

The feasible region of this linear program is depicted in Figure 3.12. Note that the feasible region has three extreme points, $\mathbf{x}_1 = (0, 2)^t$, $\mathbf{x}_2 = (0, 1)^t$, and $\mathbf{x}_3 = (1, 0)^t$, and two extreme directions, $\mathbf{d}_1 = (1, 2)^t$ and $\mathbf{d}_2 = (2, 1)^t$. (These representations of $\mathbf{d}_1$ and $\mathbf{d}_2$ are not unique.) From (3.112), the objective gradient is $\mathbf{c} = (-3, -2)$, which yields

$$\mathbf{cx}_1 = (-3 \ -2)\begin{pmatrix} 0 \\ 2 \end{pmatrix} = -4$$

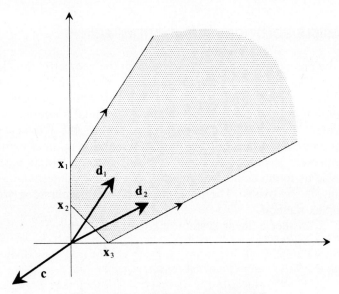

Figure 3.12 Graph for Example 3.7.

Similarly,

$$\mathbf{cx}_2 = -2$$

$$\mathbf{cx}_3 = -3$$

$$\mathbf{cd}_1 = -7$$

$$\mathbf{cd}_2 = -8$$

Because $\mathbf{cd}_1 < 0$ and $\mathbf{cd}_2 < 0$, the linear program has a finite optimal solution. And because it has a finite optimal solution, it has an extreme-point optimal solution. Finally, because $\mathbf{cx}_2 > \mathbf{cx}_1$ and $\mathbf{cx}_2 > \mathbf{cx}_3$, then clearly the optimal extreme point is $\mathbf{x}_2 = (0, 1)^t$ with $z^* = \mathbf{cx}_2 = -2$.

Alternatively, one may look at the reformulation provided by (3.104–3.107):

$$\text{maximize } z = -4\alpha_1 - 2\alpha_2 - 3\alpha_3 - 5\lambda_1 - 8\lambda_2 \tag{3.117}$$

subject to

$$\alpha_1 + \alpha_2 + \alpha_3 = 1 \tag{3.118}$$

$$\alpha_1, \alpha_2, \alpha_3 \geq 0 \tag{3.119}$$

$$\lambda_1, \lambda_2 \geq 0 \tag{3.120}$$

Obviously, the optimal solution of this problem is $\alpha_2 = 1$, $\alpha_1 = \alpha_3 = \lambda_1 = \lambda_2 = 0$, which corresponds precisely to extreme point $\mathbf{x}_2$.

BASIC FEASIBLE SOLUTIONS AND EXTREME POINTS

In the preceding sections, we developed the geometric foundation of the simplex method and established the importance of extreme points in the search for an optimal solution. That is, if a linear programming problem has a finite optimal solution, then it has an extreme-point optimal solution. However, unless the problem is very small (at most three variables), the graphical and geometric techniques are not sufficient to actually determine an optimal solution. In this section, we present a method for characterizing extreme points algebraically. This will enable us to develop a systematic procedure for determining an optimal extreme point, if one exists. The resulting systematic algebraic approach is, in fact, the simplex method.

Consider a linear system of equations given by

$$\mathbf{Ax} = \mathbf{b} \tag{3.121}$$

where

$\mathbf{A}$ is a given $m \times n$ matrix,

$$\mathbf{b} \text{ is a given } m\text{-vector, i.e., } \mathbf{b} = \begin{pmatrix} b_1 \\ b_2 \\ \vdots \\ b_m \end{pmatrix}$$

$$\mathbf{x} = \begin{pmatrix} x_1 \\ x_2 \\ \vdots \\ x_n \end{pmatrix} \in E^n$$

Assume that the rank($\mathbf{A}$) = $m \leq n$. That is, assume that $\mathbf{A}$ has full row rank, or, equivalently, the rows of $\mathbf{A}$ are linearly independent. Also assume that the columns of $\mathbf{A}$ can be reordered so that $\mathbf{A}$ can be written in partitioned form as

$$\mathbf{A} = (\mathbf{B} : \mathbf{N}) \tag{3.122}$$

where

$\mathbf{B} = m \times m$ nonsingular matrix, designated the *basis matrix*

$\mathbf{N} = m \times (n - m)$ matrix (the matrix of nonbasic columns)

Based on this partitioning of matrix $\mathbf{A}$, the linear system given in (3.121) can be recast in the form

$$\mathbf{B}\mathbf{x}_B + \mathbf{N}\mathbf{x}_N = \mathbf{b} \tag{3.123}$$

where vector $\mathbf{x}$ has been partitioned as

$$\mathbf{x} = \begin{pmatrix} \mathbf{x}_B \\ \mathbf{x}_N \end{pmatrix}$$

to correspond precisely to the partitioning of matrix $\mathbf{A}$. Now, because $\mathbf{B}$ is nonsingular, the inverse of $\mathbf{B}$ exists, and we may premultiply both sides of (3.123) by $\mathbf{B}^{-1}$ to obtain

$$\mathbf{B}^{-1}\mathbf{B}\mathbf{x}_B + \mathbf{B}^{-1}\mathbf{N}\mathbf{x}_N = \mathbf{B}^{-1}\mathbf{b} \tag{3.124}$$

This simplifies to

$$\mathbf{x}_B + \mathbf{B}^{-1}\mathbf{N}\mathbf{x}_N = \mathbf{B}^{-1}\mathbf{b} \tag{3.125}$$

and solving for $\mathbf{x}_B$ yields

$$\mathbf{x}_B = \mathbf{B}^{-1}\mathbf{b} - \mathbf{B}^{-1}\mathbf{N}\mathbf{x}_N \tag{3.126}$$

Therefore, we have expressed vector $\mathbf{x}_B$ as the constant vector $\mathbf{B}^{-1}\mathbf{b}$ less the term $\mathbf{B}^{-1}\mathbf{N}\mathbf{x}_N$. Now setting $\mathbf{x}_N = \mathbf{0}$, we see that (3.126) results in $\mathbf{x}_B = \mathbf{B}^{-1}\mathbf{b}$. The solution

$$\mathbf{x} = \begin{pmatrix} \mathbf{x}_B \\ \mathbf{x}_N \end{pmatrix} = \begin{pmatrix} \mathbf{B}^{-1}\mathbf{b} \\ \mathbf{0} \end{pmatrix}$$

is called a *basic solution*, with vector $\mathbf{x}_B$ called the vector of *basic variables*, and $\mathbf{x}_N$ is called the vector of *nonbasic variables*.

If, in addition, $\mathbf{x}_B = \mathbf{B}^{-1}\mathbf{b} \geq \mathbf{0}$, then

$$\mathbf{x} = \begin{pmatrix} \mathbf{B}^{-1}\mathbf{b} \\ \mathbf{0} \end{pmatrix}$$

is called a *basic feasible solution* of the system defined by

$$\mathbf{Ax} = \mathbf{b} \tag{3.127}$$

$$\mathbf{x} \geq \mathbf{0} \tag{3.128}$$

Finally, if $\mathbf{x}_B = \mathbf{B}^{-1}\mathbf{b} > \mathbf{0}$, then

$$\mathbf{x} = \begin{pmatrix} \mathbf{B}^{-1}\mathbf{b} \\ \mathbf{0} \end{pmatrix}$$

is called a *nondegenerate* basic feasible solution. Otherwise if at least one element of $\mathbf{x}_B$ is zero, then $\mathbf{x}$ is called a *degenerate* basic feasible solution.

Note that (3.127) and (3.128) correspond to the standard representation of the constraint set of a linear programming problem. This algebraic characterization of a solution of this linear constraint set will be tied to the geometric concept of an extreme point by Theorem 3.7, but first we review the concept of a basic feasible solution through the following example.

Example 3.8: Basic Solutions

Consider the linear system

$$2x_1 + x_2 + x_3 + x_4 = 15 \tag{3.129}$$

$$x_1 + 3x_2 + x_3 - x_5 = 12 \tag{3.130}$$

This linear system can be written in the form

$$\mathbf{Ax} = \mathbf{b}$$

where

$$\mathbf{A} = \begin{pmatrix} 2 & 1 & 1 & 1 & 0 \\ 1 & 3 & 1 & 0 & -1 \end{pmatrix}$$

$$\mathbf{b} = \begin{pmatrix} 15 \\ 12 \end{pmatrix}$$

$$\mathbf{x} = \begin{pmatrix} x_1 \\ x_2 \\ \vdots \\ x_5 \end{pmatrix}$$

Because **A** is a 2×5 matrix, each basic solution will have $m = 2$ basic variables and $n - m = 3$ nonbasic variables. Now consider the basis matrix **B** formed from the first and third columns (*in that order*) of **A**, that is,

$$\mathbf{B} = (\mathbf{a}_1, \mathbf{a}_3) = \begin{pmatrix} 2 & 1 \\ 1 & 1 \end{pmatrix} \tag{3.131}$$

Because **B** is nonsingular, it is a suitable basis matrix and

$$\mathbf{B}^{-1} = \begin{pmatrix} 1 & -1 \\ -1 & 2 \end{pmatrix} \tag{3.132}$$

Thus, the basic solution corresponding to the basis matrix **B** is

$$\mathbf{x} = \begin{pmatrix} \mathbf{x}_B \\ \mathbf{x}_N \end{pmatrix} \tag{3.133}$$

where

$$\mathbf{x}_N = \begin{pmatrix} x_2 \\ x_4 \\ x_5 \end{pmatrix} = \begin{pmatrix} 0 \\ 0 \\ 0 \end{pmatrix} \tag{3.134}$$

$$\mathbf{x}_B = \begin{pmatrix} x_{B,1} \\ x_{B,2} \end{pmatrix} = \begin{pmatrix} x_1 \\ x_3 \end{pmatrix} = \mathbf{B}^{-1}\mathbf{b} = \begin{pmatrix} 1 & -1 \\ -1 & 2 \end{pmatrix} \begin{pmatrix} 15 \\ 12 \end{pmatrix} = \begin{pmatrix} 3 \\ 9 \end{pmatrix} \tag{3.135}$$

In the context of (3.127–3.128), this solution would be termed a *nondegenerate* basic *feasible* solution because $\mathbf{x}_B > \mathbf{0}$. We are now ready to prove a fundamental result that establishes the relationship between the geometric notion of an extreme point and the algebraic concept of a basic feasible solution. This will form the basis of our solution procedure in Chapter 4.

Theorem 3.7
Let $S = \{\mathbf{x} : \mathbf{Ax} = \mathbf{b}, \mathbf{x} \geq \mathbf{0}\}$, where **A** is $m \times n$, and rank(**A**) $= m < n$. **x** is an extreme point of S if and only if **x** is a basic feasible solution.

Proof
($\rightarrow$) Let **x** be an extreme point of S. We need to show that **x** is a basic feasible solution.

Because **x** is an extreme point, **x** is the intersection of n linearly independent hyperplanes. Note, from the definition of S, that $\mathbf{Ax} = \mathbf{b}$ provides m of these hyperplanes. Thus, the remaining $n - m$ hyperplanes must come from the non-negativity constraints $\mathbf{x} \geq \mathbf{0}$. That is, at least $n - m$ of the constraints $\mathbf{x} \geq \mathbf{0}$ are satisfied as equalities by the extreme point **x**. Let us denote $n - m$ of these binding constraints by $\mathbf{x}_N = \mathbf{0}$. Then the extreme point **x** is the unique solution of the n linearly independent hyperplanes, $\mathbf{Ax} = \mathbf{b}$, $\mathbf{x}_N = \mathbf{0}$. Let $\mathbf{x}_B$ represent the remaining m components of **x** and partition the matrix **A** to correspond to the vectors $\mathbf{x}_B$ and $\mathbf{x}_N$; that is, $\mathbf{A} = (\mathbf{B} : \mathbf{N})$. Then the extreme point **x** is the unique solution of the system, $\mathbf{Bx}_B + \mathbf{Nx}_N = \mathbf{b}$, $\mathbf{x}_N = \mathbf{0}$. It then follows that $\mathbf{x}_B$ is the unique solution of $\mathbf{Bx}_B = \mathbf{b}$ and, thus, $\mathbf{B}$ is invertible and a basis matrix. Therefore,

$$\mathbf{x} = \begin{pmatrix} \mathbf{x}_B \\ \mathbf{x}_N \end{pmatrix}$$

is a basic solution and, clearly, $\mathbf{x}$ is also feasible because $\mathbf{x}$ is an extreme point.

($\leftarrow$) Let $\mathbf{x}$ be a basic feasible solution. Show that $\mathbf{x}$ is an extreme point of S. Because $\mathbf{x}$ is a basic feasible solution, there exists a basis matrix $\mathbf{B}$ such that

$$\mathbf{x} = \begin{pmatrix} \mathbf{x}_B \\ \mathbf{x}_N \end{pmatrix} = \begin{pmatrix} \mathbf{B}^{-1}\mathbf{b} \\ \mathbf{0} \end{pmatrix}$$

But this implies that $\mathbf{x}$ is the unique solution to the system, $\mathbf{Bx}_B + \mathbf{Nx}_N = \mathbf{b}$, $\mathbf{x}_N = \mathbf{0}$, or, equivalently, $\mathbf{Ax} = \mathbf{b}$, $\mathbf{x}_N = \mathbf{0}$. Because $\mathbf{Ax} = \mathbf{b}$ represents m hyperplanes and $\mathbf{x}_N = \mathbf{0}$ is an additional $n - m$ hyperplanes, $\mathbf{x}$ is the intersection of n linearly independent hyperplanes and, hence, is an extreme point. $\square$

Example 3.9: Basic Feasible Solutions and Extreme Points

Consider, again, the two-variable problem of Example 3.3. After the addition of slack variables, we obtain

$$\text{maximize } z = 2x_1 + 3x_2 \tag{3.136}$$

subject to

$$x_1 - 2x_2 + x_3 = 4 \tag{3.137}$$

$$2x_1 + x_2 + x_4 = 18 \tag{3.138}$$

$$x_2 + x_5 = 10 \tag{3.139}$$

$$x_1, x_2, x_3, x_4, x_5 \geq 0 \tag{3.140}$$

Thus, the problem can be rewritten as

$$\text{maximize } z = \mathbf{cx}$$

subject to

$$\mathbf{Ax} = \mathbf{b}$$

$$\mathbf{x} \geq \mathbf{0}$$

where

$$\mathbf{A} = \begin{pmatrix} 1 & -2 & 1 & 0 & 0 \\ 2 & 1 & 0 & 1 & 0 \\ 0 & 1 & 0 & 0 & 1 \end{pmatrix}$$

$$\mathbf{b} = \begin{pmatrix} 4 \\ 18 \\ 10 \end{pmatrix}$$

$$\mathbf{c} = (2 \quad 3 \quad 0 \quad 0 \quad 0)$$

$$\mathbf{x} = \begin{pmatrix} x_1 \\ x_2 \\ \vdots \\ x_5 \end{pmatrix}$$

The feasible region is graphed in the x_1–x_2 plane in Figure 3.13. Notice that the boundary of the feasible region consists of five hyperplanes:

$$x_1 - 2x_2 = 4 \tag{3.141}$$

$$2x_1 + x_2 = 18 \tag{3.142}$$

$$x_2 = 10 \tag{3.143}$$

$$x_1 = 0 \tag{3.144}$$

$$x_2 = 0 \tag{3.145}$$

Observe also that each of these hyperplanes is associated with a particular variable, which has the value zero along the hyperplane. For example, from (3.137), x_3 has the value zero along the hyperplane defined in (3.141). Thus, x_3 can be thought of as being nonbasic on the hyperplane $x_1 - 2x_2 = 4$. The nonbasic variable associated with each bounding hyperplane is noted in Figure 3.13.

The coefficient matrix $\mathbf{A}$ is 3×5, therefore, each basic solution will have $m = 3$ basic variables and $n - m = 2$ nonbasic variables. Consider the basis matrix

$$\mathbf{B} = (\mathbf{a}_2, \mathbf{a}_3, \mathbf{a}_4) = \begin{pmatrix} -2 & 1 & 0 \\ 1 & 0 & 1 \\ 1 & 0 & 0 \end{pmatrix} \tag{3.146}$$

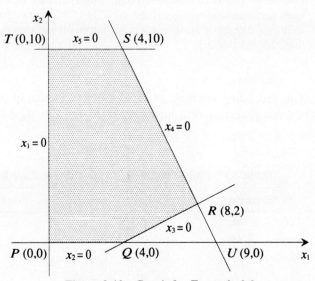

Figure 3.13 Graph for Example 3.9.

Because **B** is nonsingular, it is invertible and forms a suitable basis. The basic solution corresponding to **B** can now be computed as

$$\mathbf{x}_N = \begin{pmatrix} x_1 \\ x_5 \end{pmatrix} = \begin{pmatrix} 0 \\ 0 \end{pmatrix} \tag{3.147}$$

$$\mathbf{x}_B = \begin{pmatrix} x_{B,1} \\ x_{B,2} \\ x_{B,3} \end{pmatrix} = \begin{pmatrix} x_2 \\ x_3 \\ x_4 \end{pmatrix} = \mathbf{B}^{-1}\mathbf{b} = \begin{pmatrix} 10 \\ 24 \\ 8 \end{pmatrix} \tag{3.148}$$

Notice that because x_1 and x_5 are nonbasic variables, this basic solution corresponds to extreme point T in Figure 3.13. This is true because point T is the intersection of the hyperplanes associated with nonbasic variables x_1 and x_5. This also checks out mathematically because the coordinates of this extreme point are $(x_1, x_2) = (0,10)$.

Now consider the basic solution corresponding to the basis matrix

$$\mathbf{B} = (\mathbf{a}_1, \mathbf{a}_3, \mathbf{a}_5) = \begin{pmatrix} 1 & 1 & 0 \\ 2 & 0 & 0 \\ 0 & 0 & 1 \end{pmatrix} \tag{3.149}$$

Following the procedure defined earlier, we obtain

$$\mathbf{x}_N = \begin{pmatrix} x_2 \\ x_4 \end{pmatrix} = \begin{pmatrix} 0 \\ 0 \end{pmatrix} \tag{3.150}$$

$$\mathbf{x}_B = \begin{pmatrix} x_{B,1} \\ x_{B,2} \\ x_{B,3} \end{pmatrix} = \begin{pmatrix} x_1 \\ x_3 \\ x_5 \end{pmatrix} = \mathbf{B}^{-1}\mathbf{b} = \begin{pmatrix} 9 \\ -5 \\ 10 \end{pmatrix} \tag{3.151}$$

Because $x_3 < 0$, this is a basic *infeasible* solution. Also, because x_2 and x_4 are nonbasic, observe that this basic solution corresponds to point U in Figure 3.13. Obviously, this is an infeasible point. Finally, notice that the constraint that is violated corresponds precisely to the variable x_3, which has a negative value in the basic solution. The basic feasible solutions for this linear programming problem are summarized in Table 3.1.

The preceding example, although quite simple, illustrates the relationship between the geometry of a linear program and the algebra of the problem. The

TABLE 3.1   BASIC FEASIBLE SOLUTIONS FOR EXAMPLE 3.9

Extreme point	Basic variables	Nonbasic variables
P	$x_3 = 4,\ x_4 = 18,\ x_5 = 10$	$x_1 = 0,\ x_2 = 0$
Q	$x_1 = 4,\ x_4 = 10,\ x_5 = 10$	$x_2 = 0,\ x_3 = 0$
R	$x_1 = 8,\ x_2 = 2,\ x_5 = 8$	$x_3 = 0,\ x_4 = 0$
S	$x_1 = 4,\ x_2 = 10,\ x_3 = 20$	$x_4 = 0,\ x_5 = 0$
T	$x_2 = 10,\ x_3 = 24,\ x_4 = 8$	$x_1 = 0,\ x_5 = 0$

reader should also observe that adjacent extreme points differ by exactly one basic and one nonbasic variable. Thus, we can move from an extreme point to an adjacent extreme point by simply interchanging one basic and one nonbasic variable. This idea is central to the operation of the simplex method and is discussed in detail in the following chapter.

SUMMARY

A considerable amount of material has been presented in this chapter. The diligent reader may, at this time, conclude that the simplex method, of which we have laid the theoretical foundation, is a very complex algorithm. However, as we will see in Chapter 4, this is not the case. The operations of the simplex method are actually quite straightforward, but, as we mentioned earlier, the theoretical development of this chapter is necessary to lend a firm understanding of the underlying process.

We now briefly summarize the highlights of this chapter.

1. The standard form of a linear programming problem may be expressed in the concise form:

$$(\text{LP}) \quad \text{maximize } z = \mathbf{c}\mathbf{x}$$

$$\text{subject to}$$

$$\mathbf{A}\mathbf{x} = \mathbf{b}$$

$$\mathbf{x} \geq \mathbf{0}$$

where the data is given by

$\mathbf{A} = m \times n$ matrix of the coefficients of the constraints

$\mathbf{b} = m$-vector of right-hand sides

$\mathbf{c} = n$-vector of objective coefficients

and the variables are given by the n-vector, $\mathbf{x}$.

2. Any linear programming problem may be placed in standard form. This may involve changing a minimization objective to a maximization objective by multiplying the objective by -1 and converting inequality constraints to equality constraints by the addition of slack or surplus variables.

3. If a linear programming problem has a finite optimal objective value, then at least one optimal extreme point exists.

4. Each basic feasible solution of the linear system $\mathbf{A}\mathbf{x} = \mathbf{b}$, $\mathbf{x} \geq \mathbf{0}$, corresponds to an extreme point of the set $S = \{\mathbf{x} : \mathbf{A}\mathbf{x} = \mathbf{b}, \mathbf{x} \geq \mathbf{0}\}$.

EXERCISES

3.1. Transform the following linear program into the standard form given by (3.33).

$$\text{minimize } z = 2x_1 - 3x_2 + 5x_3 + x_4$$

subject to

$$-x_1 + 3x_2 - x_3 + 2x_4 \leq -12$$
$$5x_1 + x_2 + 4x_3 - x_4 \geq 10$$
$$3x_1 - 2x_2 + x_3 - x_4 = -8$$
$$x_1, x_2, x_3, x_4 \geq 0$$

3.2. Modern Furniture, Inc., produces two types of wooden chairs. The manufacture of Chair A requires 2 hours of assembly time and 4 hours of finishing time. Chair B requires 3 hours to assemble and 3 hours to finish. Modern estimates that next week 72 hours will be available for assembly operations, and 108 hours will be available in the finishing shop. The unit profits for Chairs A and B are $10 and $9, respectively. If it is estimated that the maximum demand for Chair B is 16, what is the optimal product mix? Formulate a linear programming model and solve the resulting model graphically.

3.3. Solve the following linear program graphically.

$$\text{minimize } z = 4x_1 + 5x_2$$

subject to

$$3x_1 + 2x_2 \leq 24$$
$$x_1 \geq 5$$
$$3x_1 - x_2 \leq 6$$
$$x_1, x_2 \geq 0$$

3.4. Solve the following linear program graphically.

$$\text{minimize } z = x_1 - 4x_2$$

subject to

$$x_1 + x_2 \leq 12$$
$$-2x_1 + x_2 \leq 4$$
$$x_2 \leq 8$$
$$x_1 - 3x_2 \leq 4$$
$$x_1, x_2 \geq 0$$

3.5. Solve the following linear program graphically.

$$\text{maximize } z = 6x_1 + 8x_2$$

subject to

$$x_1 + 4x_2 \leq 16$$

$$3x_1 + 4x_2 \leq 24$$

$$3x_1 - 4x_2 \leq 12$$

$$x_1, x_2 \geq 0$$

3.6. Solve the following linear program graphically.

$$\text{maximize } z = x_1 + 2x_2$$

subject to

$$-2x_1 + x_2 \leq 2$$

$$2x_1 + 5x_2 \geq 10$$

$$x_1 - 4x_2 \leq 2$$

$$x_1, x_2 \geq 0$$

3.7. Show that the halfspace $H^- = \{\mathbf{x} : \mathbf{ax} \leq \alpha\}$ is a convex set.

3.8. Let

$$\mathbf{a}_1 = \begin{pmatrix} 4 \\ 2 \end{pmatrix}, \qquad \mathbf{a}_2 = \begin{pmatrix} -2 \\ 6 \end{pmatrix}, \qquad \mathbf{a}_3 = \begin{pmatrix} 2 \\ 5 \end{pmatrix}$$

Illustrate graphically the following:
(a) The set of all linear combinations of $\mathbf{a}_1$, $\mathbf{a}_2$, and $\mathbf{a}_3$.
(b) The set of all nonnegative linear combinations of $\mathbf{a}_1$, $\mathbf{a}_2$, and $\mathbf{a}_3$.
(c) The set of all convex combinations of $\mathbf{a}_1$, $\mathbf{a}_2$, and $\mathbf{a}_3$.

3.9. Given the polyhedral set $S = \{(x_1, x_2) : x_1 + x_2 \leq 10, -x_1 + x_2 \leq 6, x_1 - 4x_2 \leq 0\}$.
(a) Find all extreme points of S.
(b) Represent the point $\mathbf{x} = (2,4)$ as a convex combination of the extreme points.

3.10. Let S_1 and S_2 be convex sets. Show that the set $S_1 \cap S_2$ is a convex set. Is this also true of $S_1 \cup S_2$?

3.11. Given the following system of linear equations.

$$x_1 + 3x_2 - x_3 + x_4 = 30$$

$$2x_1 + x_2 + 2x_3 + x_4 = 15$$

(a) Find all basic solutions.
(b) For each basic solution, specify the basic and nonbasic variables and the basis matrix $\mathbf{B}$.

3.12. Consider the following linear programming problem.

$$\text{maximize } z = 2x_1 + x_2$$

subject to

$$x_1 + 2x_2 \leq 20$$

$$-3x_1 + 4x_2 \leq 20$$

$$3x_1 + 2x_2 \leq 36$$

$$x_1, x_2 \geq 0$$

(a) Sketch the feasible region.

(b) Write the problem in standard equality form by adding slack variables.

(c) Identify the defining variable for each hyperplane bounding the feasible region, and specify the basic and nonbasic variables for each extreme point.

(d) Graphically determine the optimal extreme point and specify the optimal basis matrix.

3.13. Given the polyhedral set $S = \{(x_1, x_2) : x_1 + x_2 \geq 6, x_1 \geq 2, -2x_1 + x_2 \leq 4, x_1 - x_2 \leq 4, x_1, x_2 \geq 0\}$.

(a) Find all extreme points and extreme directions of S.

(b) Represent the point $\mathbf{x} = (5,8)$ as a convex combination of the extreme points and a nonnegative combination of the extreme directions.

3.14. Consider the following linear programming problem.

$$\text{maximize } z = -7x_1 + 2x_2$$

subject to

$$2x_1 + x_2 \geq 6$$

$$-3x_1 + x_2 \leq 9$$

$$x_2 \geq 4$$

$$2x_1 - 4x_2 \leq 6$$

$$x_1, x_2 \geq 0$$

(a) Sketch the feasible region and identify the optimal solution.

(b) Identify all extreme points and extreme directions.

(c) Reformulate the problem in terms of convex combinations of the extreme points and nonnegative combinations of the extreme directions as in (3.104–3.107). Solve the resulting problem and interpret the solution.

(d) Change the objective function to maximize $z = 4x_1 - x_2$ and repeat part (c).

(e) Discuss the practicality of using the procedure in parts (c) and (d) for large problems.

3.15. Let $S = \{\mathbf{x} : \mathbf{Ax} \leq \mathbf{b}, \mathbf{x} \geq \mathbf{0}\}$ and let z^* be the optimal objective value associated with the linear program: maximize $z = \mathbf{cx}$ subject to $\mathbf{x} \in S$. Suppose that a constraint is added to the problem and results in the new feasible region S' with the corresponding optimal objective value z'. What is the relationship between S and S'? What is the relationship between z^* and z'?

3.16. Let $S = \{\mathbf{x} : \mathbf{Ax} \leq \mathbf{b}, \mathbf{x} \geq \mathbf{0}\}$ and let z^* be the optimal objective value associated with the linear program: maximize $z = \mathbf{cx}$ subject to $\mathbf{x} \in S$. Suppose that a constraint is deleted from the problem and results in the new feasible region S' with the corresponding optimal objective value z'. What is the relationship between S and S'? What is the relationship between z^* and z'?

3.17. Mathematically characterize the set of objective coefficients (c_1, c_2) for which the following linear program has a finite optimal solution.

$$\text{maximize } z = c_1 x_1 + c_2 x_2$$

subject to

$$x_1 - 2x_2 \leq 8$$

$$x_1 \geq 2$$

$$2x_1 + 3x_2 \geq 12$$

$$x_1, x_2 \geq 0$$

3.18. Consider the problem: (LP) maximize $\mathbf{cx}$ subject to $\mathbf{Ax} \leq \mathbf{b}$, $\mathbf{x} \geq \mathbf{0}$. Suppose that $\bar{\mathbf{x}}$ and $\hat{\mathbf{x}}$ are both optimal solutions to (LP). Show that $\alpha\bar{\mathbf{x}} + (1 - \alpha)\hat{\mathbf{x}}$ is also optimal to (LP) for all $\alpha \in [0,1]$.

3.19. Let S be a nonempty polyhedral set defined by $S = \{\mathbf{x} \in E^n : \mathbf{Ax} \leq \mathbf{b}\}$. Assuming that the dimension of S is n, formulate a linear programming problem for determining the largest n-dimensional sphere that can be completely contained in S.

3.20. Let S be a nonempty convex subset of E^n. Show that $\bar{\mathbf{x}}$ is an extreme point of S if and only if the set $S - \{\bar{\mathbf{x}}\}$ is a convex set.

CHAPTER 4

THE SIMPLEX ALGORITHM: TABLEAUX AND COMPUTATION

CHAPTER OVERVIEW

In Chapter 3, we used graphical methods to solve several simple linear programming problems, and to illustrate various linear programming outcomes, such as unique and alternative optimal solutions, and an unbounded objective. We also established theoretical results that show that if a finite optimal solution to a linear programming problem exists, then an extreme-point optimal solution exists. Thus, in the search for an optimal solution, we only need to consider the extreme points of the feasible region. This is a very important result because the solution space has been reduced to a finite number of candidate points. Algebraically, these extreme points correspond to basic feasible solutions of the linear constraint set, $\mathbf{Ax} = \mathbf{b}$, $\mathbf{x} \geq \mathbf{0}$.

In this chapter, we present a systematic method for iteratively moving from one extreme point to an adjacent extreme point in the search for an optimal solution. The method is first discussed algebraically so that the reader can better understand the actual mode of operations. However, it is generally far more convenient to employ one of the many tabular formats that have been developed. These tabular formats allow us to utilize exactly the same concepts as we would algebraically, but they simplify the "bookkeeping" aspects of the simplex method. These tabular formats are usually referred to as *simplex tableaux*.

Following the introduction of the simplex tableau, we discuss the problem of finding an initial basic feasible solution to a linear programming problem. This involves the possible introduction of a number of what are called *artificial variables*. These artificial variables help define a convenient starting basis, and an initial basic solution can then be found by the *two-phase method*.

The chapter concludes with a discussion of the convergence properties of the simplex algorithm, and the problems that may be encountered through *degeneracy* and *cycling*.

ALGEBRA OF THE SIMPLEX METHOD

Consider the standard linear programming problem:

$$\text{(LP)} \quad \text{maximize } z = \mathbf{cx}$$

$$\text{subject to}$$

$$\mathbf{Ax} = \mathbf{b}$$

$$\mathbf{x} \geq \mathbf{0}$$

Recall that a basic feasible solution to this problem corresponds to an extreme point of the feasible region and is characterized mathematically by partitioning matrix $\mathbf{A}$ into a nonsingular basis matrix $\mathbf{B}$ and the matrix of nonbasic columns $\mathbf{N}$. That is,

$$\mathbf{A} = (\mathbf{B}:\mathbf{N}) \tag{4.1}$$

Based on this partitioning, the linear system $\mathbf{Ax} = \mathbf{b}$ can be rewritten to yield

$$\mathbf{Bx}_B + \mathbf{Nx}_N = \mathbf{b} \tag{4.2}$$

This simplifies to

$$\mathbf{x}_B + \mathbf{B}^{-1}\mathbf{Nx}_N = \mathbf{B}^{-1}\mathbf{b} \tag{4.3}$$

and solving for $\mathbf{x}_B$ in terms of $\mathbf{x}_N$ yields

$$\mathbf{x}_B = \mathbf{B}^{-1}\mathbf{b} - \mathbf{B}^{-1}\mathbf{Nx}_N \tag{4.4}$$

Now setting $\mathbf{x}_N = \mathbf{0}$, we see that (4.4) results in $\mathbf{x}_B = \mathbf{B}^{-1}\mathbf{b}$. The solution

$$\mathbf{x} = \begin{pmatrix} \mathbf{x}_B \\ \mathbf{x}_N \end{pmatrix} = \begin{pmatrix} \mathbf{B}^{-1}\mathbf{b} \\ \mathbf{0} \end{pmatrix}$$

is called a *basic solution*, with vector $\mathbf{x}_B$ called the vector of *basic variables*, and $\mathbf{x}_N$ is called the vector of *nonbasic variables*. If, in addition, $\mathbf{x}_B = \mathbf{B}^{-1}\mathbf{b} \geq \mathbf{0}$, then

$$\mathbf{x} = \begin{pmatrix} \mathbf{B}^{-1}\mathbf{b} \\ \mathbf{0} \end{pmatrix}$$

is called a *basic feasible solution*.

Now consider the objective function $z = \mathbf{cx}$. Partitioning the cost vector $\mathbf{c}$

into basic and nonbasic components (i.e., $\mathbf{c} = (\mathbf{c}_B, \mathbf{c}_N)$), the objective function can be recast as

$$z = \mathbf{c}_B \mathbf{x}_B + \mathbf{c}_N \mathbf{x}_N \qquad (4.5)$$

Now, substituting the expression for $\mathbf{x}_B$ defined in (4.5) into (4.4) yields

$$z = \mathbf{c}_B(\mathbf{B}^{-1}\mathbf{b} - \mathbf{B}^{-1}\mathbf{N}\mathbf{x}_N) + \mathbf{c}_N \mathbf{x}_N \qquad (4.6)$$

which can be rewritten as

$$z = \mathbf{c}_B\mathbf{B}^{-1}\mathbf{b} - (\mathbf{c}_B\mathbf{B}^{-1}\mathbf{N} - \mathbf{c}_N)\mathbf{x}_N \qquad (4.7)$$

Thus, we have written z as the constant $\mathbf{c}_B\mathbf{B}^{-1}\mathbf{b}$ less the term $(\mathbf{c}_B\mathbf{B}^{-1}\mathbf{N} - \mathbf{c}_N)\mathbf{x}_N$. And setting $\mathbf{x}_N = \mathbf{0}$, we see that (4.7) results in $z = \mathbf{c}_B\mathbf{B}^{-1}\mathbf{b}$, which is the objective value corresponding to the current basic feasible solution. Therefore, the current extreme-point solution can be represented in *canonical form:*

$$z = \mathbf{c}_B\mathbf{B}^{-1}\mathbf{b} - (\mathbf{c}_B\mathbf{B}^{-1}\mathbf{N} - \mathbf{c}_N)\mathbf{x}_N \qquad (4.8)$$

$$\mathbf{x}_B = \mathbf{B}^{-1}\mathbf{b} - \mathbf{B}^{-1}\mathbf{N}\mathbf{x}_N \qquad (4.9)$$

with the current basic feasible solution given as

$$z = \mathbf{c}_B\mathbf{B}^{-1}\mathbf{b} \qquad (4.10)$$

$$\mathbf{x} = \begin{pmatrix} \mathbf{x}_B \\ \mathbf{x}_N \end{pmatrix} = \begin{pmatrix} \mathbf{B}^{-1}\mathbf{b} \\ \mathbf{0} \end{pmatrix} \geq \mathbf{0} \qquad (4.11)$$

The canonical representation (4.8–4.9) forms the foundation upon which the simplex method is built. Now letting J denote the index set of the nonbasic variables, observe that (4.8–4.9) can be rewritten as follows:

$$z = \mathbf{c}_B\mathbf{B}^{-1}\mathbf{b} - \sum_{j \in J} (\mathbf{c}_B\mathbf{B}^{-1}\mathbf{a}_j - c_j)x_j \qquad (4.12)$$

$$\mathbf{x}_B = \mathbf{B}^{-1}\mathbf{b} - \sum_{j \in J} (\mathbf{B}^{-1}\mathbf{a}_j)x_j \qquad (4.13)$$

Recall from Chapter 3 that adjacent extreme points differ by exactly one basic and nonbasic variable. The central idea behind the simplex method is to move from an extreme point to an improving adjacent extreme point by interchanging a column of $\mathbf{B}$ and $\mathbf{N}$.

Checking for Optimality

The first question that should enter the reader's mind is: When will such an exchange improve the objective function? This can be answered in a rather straightforward manner by examining the canonical representation of z in (4.12). In the current basic feasible solution, $\mathbf{x}_N = \mathbf{0}$, that is, the nonbasic variables are at their *lower bound* and can only be increased from their current value of zero.

Observe that the coefficient $-(c_B B^{-1} a_j - c_j)$ of x_j in (4.12) represents the *rate of change* of z with respect to the nonbasic variable x_j. That is,

$$\frac{\partial z}{\partial x_j} = -(c_B B^{-1} a_j - c_j) \tag{4.14}$$

Thus, if $\partial z / \partial x_j > 0$, then increasing the nonbasic variable x_j will increase z. The quantity $(c_B B^{-1} a_j - c_j)$ is sometimes referred to as the *reduced cost* and for convenience is usually denoted by $(z_j - c_j)$. We can thus state the optimality conditions for a maximization linear programming problem.

Optimality conditions (maximization problem). The basic feasible solution represented by (4.11) will be optimal to (LP) if

$$\frac{\partial z}{\partial x_j} = -(z_j - c_j) = -(c_B B^{-1} a_j - c_j) \le 0, \qquad \text{for all } j \in J$$

or, equivalently, if

$$z_j - c_j = c_B B^{-1} a_j - c_j \ge 0, \qquad \text{for all } j \in J$$

Note that because $z_j - c_j = 0$ for all basic variables, then the optimality conditions could also be stated simply as $z_j - c_j \ge 0$, for all $j = 1, \ldots, n$.

If $z_j - c_j > 0$, for all $j \in J$, then the current basic solution will be the *unique* optimal solution because increasing any nonbasic variable results in a *strict* decrease in the objective function. However, if some nonbasic variable x_k has $z_k - c_k = 0$, then increasing x_k does not change the objective value, and in the absence of degeneracy, entering x_k will lead to an alternative extreme point with the same objective value. When this occurs, we say that there are *alternative optimal solutions*.

Determining the Entering Variable

Suppose there exists some nonbasic variable x_k with a reduced cost $z_k - c_k < 0$. Then $\partial z / \partial x_k > 0$ and the objective function can be improved (increased) by increasing x_k from its current value of zero. Typically, we choose to increase that nonbasic variable that forces the greatest rate of change of the objective, that is, the nonbasic variable with the most negative $z_j - c_j$. (This process of choosing that variable that results in the greatest rate of change of the objective is generally referred to as the *steepest-ascent rule*.) The selected variable x_k is called the *entering variable*. That is, x_k is going to enter the basic vector. To maintain a basic vector with m components, we must exchange it for some variable that is currently basic; this variable is called the *departing variable*. Mathematically, we need to form a new basis by exchanging a_k with some column of the current basis matrix B. The following theorems establish under what conditions this exchange is valid. That is, the theorems provide the conditions under which the resulting matrix will still form a valid basis.

Theorem 4.1

Let $\mathbf{B} = (\mathbf{b}_1, \mathbf{b}_2, \ldots, \mathbf{b}_m)$ be a basis for E^m, and let $\mathbf{a} \in E^m$, $\mathbf{a} \neq \mathbf{0}$. Then $\mathbf{a}$ can be written *uniquely* as a linear combination of $\mathbf{b}_1, \mathbf{b}_2, \ldots, \mathbf{b}_m$.

Proof

Because $\mathbf{B} = (\mathbf{b}_1, \mathbf{b}_2, \ldots, \mathbf{b}_m)$ is a basis for E^m, then, clearly, it is possible to write $\mathbf{a}$ as a linear combination of $\mathbf{b}_1, \mathbf{b}_2, \ldots, \mathbf{b}_m$. We need to show that this representation is unique. Suppose it is not unique. That is, suppose that $\mathbf{a}$ can be written as

$$\mathbf{a} = \sum_{j=1}^{m} \lambda_j \mathbf{b}_j, \text{ where } \lambda_j \in E^1, \text{ for all } j = 1, \ldots, m \tag{4.15}$$

and

$$\mathbf{a} = \sum_{j=1}^{m} \mu_j \mathbf{b}_j, \text{ where } \mu_j \in E^1, \text{ for all } j = 1, \ldots, m \tag{4.16}$$

Subtracting (4.16) from (4.15) yields

$$\mathbf{0} = \sum_{j=1}^{m} (\lambda_j - \mu_j) \mathbf{b}_j \tag{4.17}$$

But because $\mathbf{b}_1, \mathbf{b}_2, \ldots, \mathbf{b}_m$ form a basis, they are linearly independent. Thus, from (4.17), we see that

$$(\lambda_j - \mu_j) = 0, \qquad \text{for all } j = 1, \ldots, m \tag{4.18}$$

It then follows that $\lambda_j = \mu_j$, for all $j = 1, \ldots, m$, and uniqueness is established.

□

Therefore, given a basis for E^m, any vector $\mathbf{a}$ in E^m can be represented as a linear combination of the basis vectors in a unique manner. The following theorem establishes under what conditions a new basis may be formed by exchanging $\mathbf{a}$ for one of the basis vectors.

Theorem 4.2

Let $\mathbf{B} = (\mathbf{b}_1, \mathbf{b}_2, \ldots, \mathbf{b}_m)$ be a basis for E^m, and let $\mathbf{a} \in E^m$, $\mathbf{a} \neq \mathbf{0}$ be represented by $\mathbf{a} = \sum_{j=1}^{m} \lambda_j \mathbf{b}_j$. Without loss of generality, suppose $\lambda_m \neq 0$. Then $\mathbf{b}_1, \mathbf{b}_2, \ldots, \mathbf{b}_{m-1}, \mathbf{a}$ form a basis for E^m.

Proof

To complete the proof, we simply need to show that $\mathbf{b}_1, \mathbf{b}_2, \ldots, \mathbf{b}_{m-1}, \mathbf{a}$ are linearly independent. By contradiction, suppose they are not. Then there exists $\gamma_1, \gamma_2, \ldots, \gamma_{m-1}, \delta \in E^1$, which are not all zero such that

$$\sum_{j=1}^{m-1} \gamma_j \mathbf{b}_j + \delta \mathbf{a} = \mathbf{0} \tag{4.19}$$

Clearly, $\delta \neq 0$, otherwise $\sum_{j=1}^{m-1} \gamma_j \mathbf{b}_j = \mathbf{0}$, which contradicts the fact that $\mathbf{b}_1$, $\mathbf{b}_2$, . . . , $\mathbf{b}_{m-1}$ are linearly independent. But

$$\mathbf{a} = \sum_{j=1}^{m} \lambda_j \mathbf{b}_j \qquad (4.20)$$

Substituting (4.20) into (4.19) yields

$$\sum_{j=1}^{m-1} \gamma_j \mathbf{b}_j + \delta \sum_{j=1}^{m} \lambda_j \mathbf{b}_j = \sum_{j=1}^{m-1} (\gamma_j + \delta\lambda_j)\mathbf{b}_j + \delta\lambda_m \mathbf{b}_m = \mathbf{0} \qquad (4.21)$$

Because $\delta \neq 0$ and $\lambda_m \neq 0$, it follows that $\delta\lambda_m \neq 0$ and (4.21) contradicts the fact that $\mathbf{b}_1$, $\mathbf{b}_2$, . . . , $\mathbf{b}_m$ are linearly independent. Thus, $\mathbf{b}_1$, $\mathbf{b}_2$, . . . , $\mathbf{b}_{m-1}$, $\mathbf{a}$ are linearly independent and form a basis for E^m. $\square$

Determining the Departing Variable

Let us now investigate the consequences of the preceding results. Consider the vector of coefficients of the nonbasic variable x_k in (4.13) and let

$$\boldsymbol{\alpha}_k = \mathbf{B}^{-1}\mathbf{a}_k \qquad (4.22)$$

The elements of vector $\boldsymbol{\alpha}_k$ can be interpreted in two different ways. Both of these interpretations provide important information that is useful in understanding the mechanics of the simplex method.

First, multiplying both sides of (4.22) by $\mathbf{B}^{-1}$ yields

$$\mathbf{a}_k = \mathbf{B}\boldsymbol{\alpha}_k = (\mathbf{b}_1, \mathbf{b}_2, \ldots, \mathbf{b}_m) \begin{pmatrix} \alpha_{1,k} \\ \alpha_{2,k} \\ \vdots \\ \alpha_{m,k} \end{pmatrix} = \sum_{j=1}^{m} \alpha_{j,k}\mathbf{b}_j \qquad (4.23)$$

Recall from Theorem 4.1 that $\mathbf{a}_k$ can be written uniquely as a linear combination of the columns of the basis matrix $\mathbf{B}$. Thus, (4.23) identifies that, in fact, the elements of $\boldsymbol{\alpha}_k$ specify how to write $\mathbf{a}_k$ as a linear combination of the columns of $\mathbf{B}$. Therefore, $\mathbf{a}_k$ may be exchanged with any column $\mathbf{b}_j$ of $\mathbf{B}$ for which $\alpha_{j,k} \neq 0$.

Second, from (4.13), note that the rate of change of the basic variables with respect to the nonbasic variable x_k is given by

$$\frac{\partial \mathbf{x}_B}{\partial x_k} = -\mathbf{B}^{-1}\mathbf{a}_k = -\boldsymbol{\alpha}_k \qquad (4.24)$$

That is, if the nonbasic variable x_k is increased from its current value of zero while holding all other nonbasic variables at zero, then the basic variables will change according to the relationship

$$\mathbf{x}_B = \mathbf{B}^{-1}\mathbf{b} + x_k(-\mathbf{B}^{-1}\mathbf{a}_k) = \mathbf{B}^{-1}\mathbf{b} - x_k\boldsymbol{\alpha}_k \qquad (4.25)$$

And because all variables must remain nonnegative, it follows that

$$\mathbf{x}_B = \mathbf{B}^{-1}\mathbf{b} - x_k\boldsymbol{\alpha}_k \geq \mathbf{0} \tag{4.26}$$

Now let

$$\mathbf{B}^{-1}\mathbf{b} = \boldsymbol{\beta} = \begin{pmatrix} \beta_1 \\ \beta_2 \\ \vdots \\ \beta_m \end{pmatrix} \tag{4.27}$$

Then, from (4.26–4.27),

$$\begin{pmatrix} \beta_1 \\ \beta_2 \\ \vdots \\ \beta_m \end{pmatrix} - x_k \begin{pmatrix} \alpha_{1,k} \\ \alpha_{2,k} \\ \vdots \\ \alpha_{m,k} \end{pmatrix} \geq \mathbf{0} \tag{4.28}$$

and an upper bound on x_k can be found quite easily as

$$x_k \leq \text{minimum} \left\{ \frac{\beta_i}{\alpha_{i,k}} : \alpha_{i,k} > 0 \right\} \tag{4.29}$$

This process is termed the *minimum ratio test* and provides a very simple method for determining the maximum value of the entering variable. Essentially, we are finding the smallest value of the entering variable x_k, which results in a basic variable assuming the value zero. The basic variable that is forced to zero as a result of this increase in x_k is called the *departing variable*. (That is, it departs or leaves the basic vector.)

Optimality Conditions and Directions*

Let the current basic feasible solution be given by

$$\mathbf{x} = \begin{pmatrix} \mathbf{x}_B \\ \mathbf{x}_N \end{pmatrix} = \begin{pmatrix} \mathbf{B}^{-1}\mathbf{b} \\ \mathbf{0} \end{pmatrix} \geq \mathbf{0} \tag{4.30}$$

and suppose that the nonbasic variable x_k is the entering variable. Then, as discussed previously, as x_k is increased, the values of the basic variables change according to

$$\mathbf{x}_B = \mathbf{B}^{-1}\mathbf{b} + x_k(-\mathbf{B}^{-1}\mathbf{a}_k) = \mathbf{B}^{-1}\mathbf{b} + x_k(-\boldsymbol{\alpha}_k) \tag{4.31}$$

Because all the nonbasic variables except for x_k are held at zero, then the nonbasic

---

* Starred sections may be omitted in an introductory course with no appreciable loss of continuity.

variables change according to

$$\mathbf{x}_N = \mathbf{0} + x_k \mathbf{e}_k = \mathbf{0} + x_k \begin{pmatrix} 0 \\ 0 \\ \vdots \\ 1 \\ \vdots \\ 0 \end{pmatrix} \tag{4.32}$$

where the 1 in $\mathbf{e}_k$ appears in the kth position.

Thus, as x_k is increased, the solution

$$\mathbf{x} = \begin{pmatrix} \mathbf{x}_B \\ \mathbf{x}_N \end{pmatrix}$$

is moving in the direction

$$\mathbf{d} = \begin{pmatrix} -\boldsymbol{\alpha}_k \\ \mathbf{e}_k \end{pmatrix} \tag{4.33}$$

Now consider the product $\mathbf{cd}$, where $\mathbf{c}$ is the gradient of the objective function.

$$\mathbf{cd} = (\mathbf{c}_B, \mathbf{c}_N) \begin{pmatrix} -\boldsymbol{\alpha}_k \\ \mathbf{e}_k \end{pmatrix}$$

$$= -\mathbf{c}_B \boldsymbol{\alpha}_k + \mathbf{c}_N \mathbf{e}_k = -\mathbf{c}_B \mathbf{B}^{-1} \mathbf{a}_k + c_k = -z_k + c_k = -(z_k - c_k) \tag{4.34}$$

Recall that if $z_k - c_k < 0$, then entering x_k will improve the objective. Thus, from (4.34), we see that moving in direction $\mathbf{d}$ will improve the objective function if $\mathbf{cd} > 0$ (that is, $z_k - c_k < 0$). But $\mathbf{cd} > 0$ implies that $\mathbf{c}$ makes an *acute* angle with $\mathbf{d}$. Similarly, moving in direciton $\mathbf{d}$ will not improve the objective function if $\mathbf{cd} \leq 0$. These results are interpreted graphically in Figure 4.1.

Suppose that we are currently at extreme point P (the origin). Note that at extreme point P, both possible directions, $\mathbf{d}_{P1}$ and $\mathbf{d}_{P2}$, make acute angles with $\mathbf{c}$. Therefore, both are improving directions. Moving in the direction $\mathbf{d}_{P1}$, we arrive at extreme point Q. Of the two directions at Q, only $\mathbf{d}_{Q2}$ is an improving direction because $\mathbf{d}_{Q1}$ makes an *obtuse* angle with $\mathbf{c}$. Now moving in the improving direction $\mathbf{d}_{Q2}$, we reach extreme point R. Clearly, R is the optimal extreme point because both directions at R make *obtuse* angles with $\mathbf{c}$ and, consequently, there does not exist an improving direction. This is precisely a graphical interpretation of the simplex algorithm.

Checking for an Unbounded Objective

Suppose that x_k is chosen as the entering variable (that is, $z_k - c_k < 0$). However, when examining vector $\boldsymbol{\alpha}_k$ for the minimum ratio test, we find that $\alpha_{i,k} \leq 0$, for all i. Then, from (4.28), x_k can be increased without bound, that is, as x_k is increased from its current value of zero, no basic variable decreases in value. Thus, if $z_k - c_k < 0$ and $\boldsymbol{\alpha}_k \leq \mathbf{0}$, then the objective function can be increased indefinitely and no

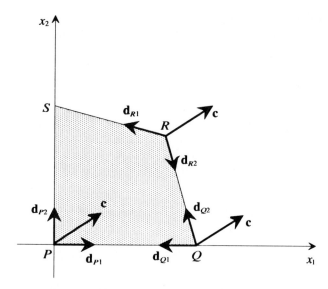

Figure 4.1   Optimality conditions and directions.

finite optimal solution exists. In fact, the objective function can be increased indefinitely by moving along the ray defined by

$$\begin{pmatrix} \mathbf{B}^{-1}\mathbf{b} \\ \mathbf{0} \end{pmatrix} + x_k \begin{pmatrix} -\alpha_k \\ \mathbf{e}_k \end{pmatrix} \tag{4.35}$$

where again

$$\mathbf{e}_k = \begin{pmatrix} 0 \\ 0 \\ \vdots \\ 1 \\ \vdots \\ 0 \end{pmatrix} \tag{4.36}$$

and the 1 appears in the kth position. Note that the vertex of this ray is the current basic feasible solution that is given by

$$\begin{pmatrix} \mathbf{x}_B \\ \mathbf{x}_N \end{pmatrix} = \begin{pmatrix} \mathbf{B}^{-1}\mathbf{b} \\ \mathbf{0} \end{pmatrix} \tag{4.37}$$

and the direction of the ray is given by

$$\mathbf{d} = \begin{pmatrix} -\alpha_k \\ \mathbf{e}_k \end{pmatrix} = \begin{pmatrix} -\mathbf{B}^{-1}\mathbf{a}_k \\ \mathbf{e}_k \end{pmatrix} \tag{4.38}$$

Before presenting the simplex method in tableau form, it is important to see the actual algebraic operations that are being performed by the algorithm. This will be demonstrated using the following simple example.

Example 4.1: Algebra of the Simplex Method

To lend some continuity to the development of the simplex method, let us again consider Example 3.9.

$$\text{maximize } z = 2x_1 + 3x_2 \qquad (4.39)$$

subject to

$$x_1 - 2x_2 + x_3 = 4 \qquad (4.40)$$

$$2x_1 + x_2 + x_4 = 18 \qquad (4.41)$$

$$x_2 + x_5 = 10 \qquad (4.42)$$

$$x_1, x_2, x_3, x_4, x_5 \geq 0 \qquad (4.43)$$

As before, the data for this problem can be summarized as follows:

$$\mathbf{A} = \begin{pmatrix} 1 & -2 & 1 & 0 & 0 \\ 2 & 1 & 0 & 1 & 0 \\ 0 & 1 & 0 & 0 & 1 \end{pmatrix} \qquad (4.44)$$

$$\mathbf{b} = \begin{pmatrix} 4 \\ 18 \\ 10 \end{pmatrix} \qquad (4.45)$$

$$\mathbf{c} = (2 \quad 3 \quad 0 \quad 0 \quad 0) \qquad (4.46)$$

Also, the feasible region in the x_1–x_2 plane is depicted in Figure 4.2.

We begin the solution process by choosing a convenient starting basis matrix **B**. We do not want to choose an arbitrary matrix **B**, but, instead, because the solution will be determined by $\mathbf{B}^{-1}$, we will *always* choose the starting basis matrix **B** = **I**. (As we will see in subsequent sections, this may involve introducing additional variables, called artificial variables.) Observe from (4.44) that this results in

$$\mathbf{B} = (\mathbf{a}_3, \mathbf{a}_4, \mathbf{a}_5) = \begin{pmatrix} 1 & 0 & 0 \\ 0 & 1 & 0 \\ 0 & 0 & 1 \end{pmatrix} = \mathbf{I}$$

$$\mathbf{x}_B = \begin{pmatrix} x_{B,1} \\ x_{B,2} \\ x_{B,3} \end{pmatrix} = \begin{pmatrix} x_3 \\ x_4 \\ x_5 \end{pmatrix}$$

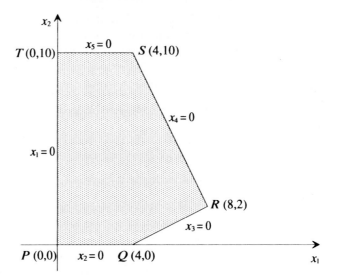

Figure 4.2   Graph for Example 4.1.

Now form the canonical representation as in (4.8–4.9) by solving for z and $\mathbf{x}_B$ in terms of $\mathbf{x}_N$. Because $\mathbf{B} = \mathbf{I}$, this is a trivial process and yields

$$z = 2x_1 + 3x_2 \tag{4.47}$$

$$x_3 = 4 - x_1 + 2x_2 \tag{4.48}$$

$$x_4 = 18 - 2x_1 - x_2 \tag{4.49}$$

$$x_5 = 10 - x_2 \tag{4.50}$$

The starting solution, which is obtained by setting the nonbasic variables equal to zero, can be summarized as follows:

$$z = 0$$

$$\mathbf{x}_B = \begin{pmatrix} x_{B,1} \\ x_{B,2} \\ x_{B,3} \end{pmatrix} = \begin{pmatrix} x_3 \\ x_4 \\ x_5 \end{pmatrix} = \begin{pmatrix} 4 \\ 18 \\ 10 \end{pmatrix}$$

$$\mathbf{x}_N = \begin{pmatrix} x_1 \\ x_2 \end{pmatrix} = \begin{pmatrix} 0 \\ 0 \end{pmatrix}$$

$$\mathbf{B} = (\mathbf{a}_3, \mathbf{a}_4, \mathbf{a}_5) = \begin{pmatrix} 1 & 0 & 0 \\ 0 & 1 & 0 \\ 0 & 0 & 1 \end{pmatrix}$$

Observe from Figure 4.2, that this starting solution corresponds to extreme point P, the origin. This can be seen quite readily by noting that x_1 and x_2 are the nonbasic variables.

From (4.47), we see that $\partial z/\partial x_1 = -(z_1 - c_1) = 2 > 0$ and $\partial z/\partial x_2 = -(z_2 - c_2) = 3 > 0$, that is, $z_1 - c_1 = -2 < 0$ and $z_2 - c_2 = -3 < 0$. Thus, increasing either of the nonbasic variables x_1 or x_2 will increase the value of z and the current solution is not optimal. Because $\partial z/\partial x_2 > \partial z/\partial x_1$, let us choose to increase variable x_2 (that is, x_2 is the *entering* variable). The next step is to find the departing variable using the minimum ratio test. Note that the nonnegativity restrictions are not included in the canonical representation and must be enforced implicitly. As x_2 is increased, we must ensure that x_3, x_4, and x_5 as defined in (4.48–4.50) remain nonnegative. Because nonbasic variable x_1 is being held at zero, we see from (4.48–4.50) that the values of the basic variables are given by

$$\mathbf{x}_B = \begin{pmatrix} x_3 \\ x_4 \\ x_5 \end{pmatrix} = \boldsymbol{\beta} - x_2\boldsymbol{\alpha}_2 = \begin{pmatrix} 4 \\ 18 \\ 10 \end{pmatrix} - x_2 \begin{pmatrix} -2 \\ 1 \\ 1 \end{pmatrix} \geq \mathbf{0} \tag{4.51}$$

From (4.51), $\partial x_3/\partial x_2 = 2$, and thus x_3 increases 2 units for each unit increase in x_2. Therefore, x_3 will always remain positive as x_2 is increased. However, this is not true of x_4 and x_5 because $\partial x_4/\partial x_2 = -1$ and $\partial x_5/\partial x_2 = -1$. Notice from (4.51) that x_4 will remain nonnegative as long as $x_2 \leq 18/1$, and, similarly, x_5 will remain nonnegative as long as $x_2 \leq 10/1$. Thus, by the minimum ratio test, the maximum value of x_2 is equal to minimum $\{18, 10\} = 10$. Equation (4.50) is called the *blocking equation* and x_5 is the blocking variable, or *departing variable*. A new canonical representation is now derived by solving for z and the new set of basic variables in terms of the new set of nonbasic variables. This can be done by solving for $x_2 = 10 - x_5$ in the blocking equation and using this representation of x_2 to eliminate x_2 from the remaining equations. This process is called a *pivot* and results in the new canonical representation:

$$z = 2x_1 + 3(10 - x_5) = 30 + 2x_1 - 3x_5 \tag{4.52}$$

$$x_3 = 4 - x_1 + 2(10 - x_5) = 24 - x_1 - 2x_5 \tag{4.53}$$

$$x_4 = 18 - 2x_1 - (10 - x_5) = 8 - 2x_1 + x_5 \tag{4.54}$$

$$x_2 = 10 - x_5 \tag{4.55}$$

The current solution and basis matrix can be summarized as follows:

$$z = 30$$

$$\mathbf{x}_B = \begin{pmatrix} x_{B,1} \\ x_{B,2} \\ x_{B,3} \end{pmatrix} = \begin{pmatrix} x_3 \\ x_4 \\ x_2 \end{pmatrix} = \begin{pmatrix} 24 \\ 8 \\ 10 \end{pmatrix}$$

$$\mathbf{x}_N = \begin{pmatrix} x_1 \\ x_5 \end{pmatrix} = \begin{pmatrix} 0 \\ 0 \end{pmatrix}$$

$$\mathbf{B} = (\mathbf{a}_3, \mathbf{a}_4, \mathbf{a}_2) = \begin{pmatrix} 1 & 0 & -2 \\ 0 & 1 & 1 \\ 0 & 0 & 1 \end{pmatrix}$$

Note that $\mathbf{a}_2$ has replaced $\mathbf{a}_5$ in the basis matrix $\mathbf{B}$. Also observe, from Figure 4.2, that graphically, we have moved from extreme point P to extreme point T.

This solution is not yet optimal because, from (4.52), we see that $z_1 - c_1 = -2 < 0$ (that is, $\partial z/\partial x_1 > 0$). Thus, x_1 is chosen as the entering variable. As before, the basic variables can be written as

$$\mathbf{x}_B = \begin{pmatrix} x_3 \\ x_4 \\ x_2 \end{pmatrix} = \boldsymbol{\beta} - x_1\boldsymbol{\alpha}_1 = \begin{pmatrix} 24 \\ 8 \\ 10 \end{pmatrix} - x_1 \begin{pmatrix} 1 \\ 2 \\ 0 \end{pmatrix} \geq \mathbf{0} \tag{4.56}$$

From (4.56), x_3 will remain nonnegative as long as $x_2 \leq 24/1 = 24$, and, similarly, x_4 and x_2 will remain nonnegative as long as $x_2 \leq 8/2 = 4$ and $x_2 \leq \infty$, respectively. Therefore, the minimum ratio test yields the minimum $\{24, 4\} = 4$ and x_4 is the departing variable. The pivot operation results in

$$z = 30 + 2[4 - (\tfrac{1}{2})x_4 + (\tfrac{1}{2})x_5] - 3x_5 = 38 - x_4 - 2x_5 \tag{4.57}$$

$$x_3 = 24 - [4 - (\tfrac{1}{2})x_4 + (\tfrac{1}{2})x_5] - 2x_5 = 20 + (\tfrac{1}{2})x_4 - (\tfrac{5}{2})x_5 \tag{4.58}$$

$$x_1 = 4 - (\tfrac{1}{2})x_4 + (\tfrac{1}{2})x_5 \tag{4.59}$$

$$x_2 = 10 - x_5 \tag{4.60}$$

Graphically, we have moved from extreme point T to extreme point S. Clearly, this solution is optimal because, from (4.57), $z_4 - c_4 = 1 > 0$ and $z_5 - c_5 = 2 > 0$. The optimal solution can be summarized as follows:

$$z^* = 38$$

$$\mathbf{x}_B^* = \begin{pmatrix} x_{B,1}^* \\ x_{B,2}^* \\ x_{B,3}^* \end{pmatrix} = \begin{pmatrix} x_3^* \\ x_1^* \\ x_2^* \end{pmatrix} = \begin{pmatrix} 20 \\ 4 \\ 10 \end{pmatrix}$$

$$\mathbf{x}_N^* = \begin{pmatrix} x_4^* \\ x_5^* \end{pmatrix} = \begin{pmatrix} 0 \\ 0 \end{pmatrix}$$

$$\mathbf{B} = (\mathbf{a}_3, \mathbf{a}_1, \mathbf{a}_2) = \begin{pmatrix} 1 & 1 & -2 \\ 0 & 2 & 1 \\ 0 & 0 & 1 \end{pmatrix}$$

Note again that the basis matrix has changed by one column.

To simplify the bookkeeping aspects of the simplex method, the algebraic process just described is usually summarized in tabular form. As mentioned previously, these tabular formats are referred to as the *simplex tableaux*. Although some of the intuitive nature of the algebraic operations is lost when using the tableau method, the computational aspects of the method are enhanced. In the following section, we describe the basic form of a simplex tableau and the steps of the simplex algorithm as applied to the tableau.

THE SIMPLEX METHOD IN TABLEAU FORM

Consider again the canonical form represented in (4.8–4.9):

$$z = \mathbf{c}_B \mathbf{B}^{-1} \mathbf{b} - (\mathbf{c}_B \mathbf{B}^{-1} \mathbf{N} - \mathbf{c}_N)\mathbf{x}_N \qquad (4.61)$$

$$\mathbf{x}_B = \mathbf{B}^{-1}\mathbf{b} - \mathbf{B}^{-1}\mathbf{N}\mathbf{x}_N \qquad (4.62)$$

Now, rearranging terms so that all the variables are on the left-hand side of the equation, with the constants on the right-hand side, we have

$$z + (\mathbf{c}_B \mathbf{B}^{-1} \mathbf{N} - \mathbf{c}_N)\mathbf{x}_N = \mathbf{c}_B \mathbf{B}^{-1}\mathbf{b} \qquad (4.63)$$

$$\mathbf{x}_B + \mathbf{B}^{-1}\mathbf{N}\mathbf{x}_N = \mathbf{B}^{-1}\mathbf{b} \qquad (4.64)$$

The simplex tableau is simply a table used to store the coefficients of the algebraic representation in (4.63–4.64). The top row (row 0) of the tableau consists of the coefficients in the objective equation (4.63), and the body of the tableau (rows 1 to m) records the coefficients of the constraint equations (4.64). The general form is as shown in Table 4.1. Or written more compactly, as shown in Table 4.2.

TABLE 4.1   THE SIMPLEX TABLEAU

	z	$\mathbf{x}_B$	$\mathbf{x}_N$	RHS	
z	1	0	$\mathbf{c}_B \mathbf{B}^{-1}\mathbf{N} - \mathbf{c}_N$	$\mathbf{c}_B \mathbf{B}^{-1}\mathbf{b}$	(row 0)
$\mathbf{x}_B$	0	I	$\mathbf{B}^{-1}\mathbf{N}$	$\mathbf{B}^{-1}\mathbf{b}$	(rows 1–m)

TABLE 4.2

	z	x	RHS	
z	1	$\mathbf{c}_B \mathbf{B}^{-1}\mathbf{A} - \mathbf{c}$	$\mathbf{c}_B \mathbf{B}^{-1}\mathbf{b}$	(row 0)
$\mathbf{x}_B$	0	$\mathbf{B}^{-1}\mathbf{A}$	$\mathbf{B}^{-1}\mathbf{b}$	(rows 1–m)

Identifying $\mathbf{B}^{-1}$ from the Simplex Tableau

Observe that each x_j column in the tableau of Table 4.2 is of the form

$$\overset{x_j}{\begin{pmatrix} z_j - c_j \\ \alpha_j \end{pmatrix}} = \overset{x_j}{\begin{pmatrix} \mathbf{c}_B \mathbf{B}^{-1}\mathbf{a}_j - c_j \\ \mathbf{B}^{-1}\mathbf{a}_j \end{pmatrix}}$$

That is, each original $\mathbf{a}_j$ column is updated by multiplying by $\mathbf{B}^{-1}$ to get $\mathbf{B}^{-1}\mathbf{a}_j$, and $z_j - c_j$ is computed by multiplying the updated column $\mathbf{B}^{-1}\mathbf{a}_j$ by $\mathbf{c}_B$ and then subtracting c_j. This is an important observation for two reasons. First, it identifies the key elements in constructing any simplex tableau as $\mathbf{B}^{-1}$ and $\mathbf{c}_B$ along with the original data columns; this will be especially important in discussing the revised simplex method in Chapter 5. Second, because all the columns are updated by multiplying by $\mathbf{B}^{-1}$, if the original $\mathbf{A}$ matrix contains the identity matrix $\mathbf{I}$, then $\mathbf{B}^{-1}$ will occupy the position in the updated tableau that was occupied by $\mathbf{I}$ in the original tableau. For example, suppose that the original identity $\mathbf{I} = (\mathbf{a}_i, \mathbf{a}_j, \mathbf{a}_k)$. Then $\mathbf{B}^{-1} = \mathbf{B}^{-1}\mathbf{I} = \mathbf{B}^{-1}(\mathbf{a}_i, \mathbf{a}_j, \mathbf{a}_k) = (\mathbf{B}^{-1}\mathbf{a}_i, \mathbf{B}^{-1}\mathbf{a}_j, \mathbf{B}^{-1}\mathbf{a}_k) = (\boldsymbol{\alpha}_i, \boldsymbol{\alpha}_j, \boldsymbol{\alpha}_k)$. Thus, $\mathbf{B}^{-1}$ can be found by a suitable rearrangement of a subset of the $\boldsymbol{\alpha}_j$ columns. This provides a method for conveniently identifying $\mathbf{B}^{-1}$ for a given tableau. The tableau can also be written in terms of its component elements, as in Table 4.3.

TABLE 4.3

	z	x_1	x_2		x_n	RHS
z	1	$z_1 - c_1$	$z_2 - c_2$	$\cdots$	$z_n - c_n$	$\mathbf{c}_B\boldsymbol{\beta}$
$x_{B,1}$	0	$\alpha_{1,1}$	$\alpha_{1,2}$	$\cdots$	$\alpha_{1,n}$	β_1
$x_{B,2}$	0	$\alpha_{2,1}$	$\alpha_{2,2}$	$\cdots$	$\alpha_{2,n}$	β_2
$\vdots$	$\vdots$	$\vdots$	$\vdots$	$\ddots$	$\vdots$	$\vdots$
$x_{B,m}$	0	$\alpha_{m,1}$	$\alpha_{m,2}$	$\cdots$	$\alpha_{m,n}$	β_m

We are now ready to summarize the steps of the simplex algorithm as applied to the simplex tableau. The algorithmic steps follow directly from the preceding algebraic analysis and the specific form of the foregoing tableau provided.

The simplex algorithm (maximization problem)

STEP 1. *Check for possible improvement.* Examine the $z_j - c_j$ values in the top row (row 0) of the simplex tableau. If these are all nonnegative, then the current basic feasible solution is optimal; stop. If, however, any $z_j - c_j$ is negative, go to Step 2.

STEP 2. *Check for unboundedness.* If, for any $z_j - c_j < 0$, there is no positive element in the associated $\boldsymbol{\alpha}_j$ vector (i.e., $\boldsymbol{\alpha}_j \leq \mathbf{0}$), then the problem has an unbounded objective value. Otherwise finite improvement in the objective is possible and we go to Step 3.

STEP 3. *Determine the entering variable.* Select as the entering variable, the nonbasic variable with the most negative $z_j - c_j$. Designate this variable as x_k. Ties in the selection of x_k may be broken arbitrarily. The column associated with x_k is called the *pivot column.* Go to Step 4.

STEP 4. *Determine the departing variable.* Use the minimum ratio test to determine the departing basic variable. That is, let

$$\frac{\beta_r}{\alpha_{r,k}} = \text{minimum} \left\{ \frac{\beta_i}{\alpha_{i,k}} : \alpha_{i,k} > 0 \right\}$$

TABLE 4.4 THE SIMPLEX TABLEAU BEFORE PIVOTING

	z	$x_{B,1}$	$\cdots$	$x_{B,r}$	$\cdots$	$x_{B,m}$	$\cdots$	x_k	$\cdots$	x_j	$\cdots$	RHS
								$z_k - c_k$	$\cdots$	$z_j - c_j$	$\cdots$	$\mathbf{c}_B\boldsymbol{\beta}$
z	1	0	$\cdots$	0	$\cdots$	0	$\cdots$					
$x_{B,1}$	0	1	$\cdots$	0	$\cdots$	0	$\cdots$	$\alpha_{1,k}$	$\cdots$	$\alpha_{1,j}$	$\cdots$	β_1
$\cdots$												
$x_{B,r}$	0	0	$\cdots$	1	$\cdots$	0	$\cdots$	$(\alpha_{r,k})$	$\cdots$	$\alpha_{r,j}$	$\cdots$	β_r
$\cdots$												
$x_{B,m}$	0	0	$\cdots$	0	$\cdots$	1	$\cdots$	$\alpha_{m,k}$	$\cdots$	$\alpha_{m,j}$	$\cdots$	β_m

TABLE 4.5 THE SIMPLEX TABLEAU AFTER PIVOTING ON $\alpha_{r,k}$

	z	$x_{B,1}$	$\cdots$	$x_{B,r}$	$\cdots$	$x_{B,m}$	$\cdots$	x_k	$\cdots$	x_j	$\cdots$	RHS
z	1	0	$\cdots$	$\dfrac{-(z_k - c_k)}{\alpha_{r,k}}$	$\cdots$	0	$\cdots$	0	$\cdots$	$z_j - c_j - (z_k - c_k)\dfrac{\alpha_{r,j}}{\alpha_{r,k}}$	$\cdots$	$\mathbf{c}_B\boldsymbol{\beta} - (z_k - c_k)\dfrac{\beta_r}{\alpha_{r,k}}$
$x_{B,1}$	0	1	$\cdots$	$\dfrac{-\alpha_{1,k}}{\alpha_{r,k}}$	$\cdots$	0	$\cdots$	0	$\cdots$	$\alpha_{1,j} - \alpha_{1,k}\dfrac{\alpha_{r,j}}{\alpha_{r,k}}$	$\cdots$	$\beta_1 - \alpha_{1,k}\dfrac{\beta_r}{\alpha_{r,k}}$
$\cdots$	$\cdots$	$\cdots$	$\cdots$	$\cdots$	$\cdots$	$\cdots$	$\cdots$	$\cdots$	$\cdots$	$\cdots$	$\cdots$	$\cdots$
x_k	0	0	$\cdots$	$\dfrac{1}{\alpha_{r,k}}$	$\cdots$	0	$\cdots$	1	$\cdots$	$\dfrac{\alpha_{r,j}}{\alpha_{r,k}}$	$\cdots$	$\dfrac{\beta_r}{\alpha_{r,k}}$
$\cdots$	$\cdots$	$\cdots$	$\cdots$	$\cdots$	$\cdots$	$\cdots$	$\cdots$	$\cdots$	$\cdots$	$\cdots$	$\cdots$	$\cdots$
$x_{B,m}$	0	0	$\cdots$	$\dfrac{-\alpha_{m,k}}{\alpha_{r,k}}$	$\cdots$	1	$\cdots$	0	$\cdots$	$\alpha_{m,j} - \alpha_{m,k}\dfrac{\alpha_{r,j}}{\alpha_{r,k}}$	$\cdots$	$\beta_m - \alpha_{m,k}\dfrac{\beta_k}{\alpha_{r,k}}$

Row r is called the *pivot row*, $\alpha_{r,k}$ is called the *pivot element*, and the basic variable, $x_{B,r}$, associated with row r is the departing variable. Go to Step 5.

STEP 5. *Pivot and establish a new tableau.*

(a) The entering variable x_k is the new basic variable in row r.

(b) Use elementary row operations on the old tableau so that the column associated with x_k in the new tableau consists of all zero elements except for a 1 at the pivot position $\alpha_{r,k}$. (See Tables 4.4 and 4.5.)

(c) Return to Step 1.

The understanding of any algorithm is enhanced through the use of examples and, particularly, by practice. To illustrate this algorithm, we solve the following example problems. To provide a basis for comparison, we will first resolve the problem of Example 4.1. The reader should take special care in comparing the steps of the tableau method with those of the algebraic method used in Example 4.1. Although, with practice, the tableau method is easier from an implementation standpoint, many readers find the algebraic approach more intuitive.

Example 4.2: The Simplex Tableau

The initial tableau for the problem of Example 4.1 is given in Table 4.6. The steps of the algorithm, as applied to this problem, are listed in what follows. Note that the slack variables, x_3, x_4, x_5, are associated with the identity submatrix of $\mathbf{A}$, and thus are the initial basic variables. The nonbasic variables are x_1 and x_2.

TABLE 4.6

	z	x_1	x_2	x_3	x_4	x_5	RHS
z	1	-2	-3	0	0	0	0
x_3	0	1	-2	1	0	0	4
x_4	0	2	1	0	1	0	18
x_5	0	0	①	0	0	1	10

STEP 1. The initial tableau appears in Table 4.6. Because there are negative $z_j - c_j$ (both $z_1 - c_1$ and $z_2 - c_2$), we go to Step 2.

STEP 2. There is no $\alpha_j \leq 0$ associated with a $z_j - c_j < 0$. Thus, finite improvement in the objective is possible and we go to Step 3.

STEP 3. The most negative $z_j - c_j$ is $z_2 - c_2 = -3$. Thus, $k = 2$ and x_2 is the entering variable. Go to Step 4.

STEP 4. We now examine the ratios $\beta_i/\alpha_{i,2}$ where $\alpha_{i,2} > 0$:

$$\frac{\beta_2}{\alpha_{2,2}} = \frac{18}{1} = 18$$

$$\frac{\beta_3}{\alpha_{3,2}} = \frac{10}{1} = 10$$

Thus, $r = 3$, and the departing variable is in row 3. That is, $x_{B,3} = x_5$ is the departing variable.

STEP 5. (a) Because x_2 is the entering variable and x_5 is the departing variable, x_2 replaces x_5 in x_B as the basic variable in row 3.

(b) Row $r = 3$ of the new tableau is obtained by dividing row r of the preceding tableau by $\alpha_{r,k} = 1$ (the pivot element at the intersection of the entering variable column and departing variable row). The remainder of the tableau is updated using elementary row operations. That is, the new objective row is obtained by multiplying pivot row 3 by 3 and adding it to the old objective row. The new row 1 is obtained by multiplying the pivot row by 2 and adding it to old row 1. Finally, the new row 2 is obtained by multiplying the pivot row by -1 and adding to old row 2. The completed second tableau is shown in Table 4.7. (The reader should compare this tableau solution with the algebraic solution in (4.52–4.55).)

Recall from the previous section that $\mathbf{B}^{-1}$ occupies that portion of the tableau associated with the *original identity*. Note that $\mathbf{I} = (\mathbf{a}_3, \mathbf{a}_4, \mathbf{a}_5)$. Thus, $\mathbf{B}^{-1} = \mathbf{B}^{-1}\mathbf{I} = \mathbf{B}^{-1}(\mathbf{a}_3, \mathbf{a}_4, \mathbf{a}_5) = (\mathbf{B}^{-1}\mathbf{a}_3, \mathbf{B}^{-1}\mathbf{a}_4, \mathbf{B}^{-1}\mathbf{a}_5) = (\alpha_3, \alpha_4, \alpha_5)$. That is, $\mathbf{B}^{-1}$ is located beneath slack variables x_3, x_4, and x_5. For the tableau shown in Table 4.7, this corresponds to

$$\mathbf{B}^{-1} = \begin{pmatrix} 1 & 0 & 2 \\ 0 & 1 & -1 \\ 0 & 0 & 1 \end{pmatrix}$$

TABLE 4.7

	z	x_1	x_2	x_3	x_4	x_5	RHS
z	1	-2	0	0	0	3	30
x_3	0	1	0	1	0	2	24
x_4	0	②	0	0	1	-1	8
x_2	0	0	1	0	0	1	10

(c) Return to Step 1.

STEP 1. Because $z_1 - c_1 < 0$, we go to Step 2.

STEP 2. There are positive elements in α_1. Thus, finite improvement in the objective is possible and we go to Step 3.

STEP 3. The most negative (and only negative) $z_j - c_j$ is $z_1 - c_1 = -2$. Thus, $k = 1$ and x_1 is the entering variable. Go to Step 4.

STEP 4. The ratios $\beta_i/\alpha_{i,1}$, where $\alpha_{i,1} > 0$, are

$$\frac{\beta_1}{\alpha_{1,1}} = \frac{24}{1} = 24$$

$$\frac{\beta_2}{\alpha_{2,1}} = \frac{8}{2} = 4$$

Thus, $r = 2$, and the departing variable is $x_{B,2} = x_4$.

STEP 5. (a) Because x_1 is the entering variable and x_4 is the departing variable, x_1 replaces x_4 in x_B as the basic variable in row 2.

(b) Row $r = 2$ of the new tableau is obtained by dividing row r of the preceding tableau by $\alpha_{r,k} = 2$. The new objective row is obtained by multiplying the pivot row 2 by 1 and adding it to the old objective row. New row 1 is obtained by multiplying the pivot row by $-\frac{1}{2}$ and adding it to old row 1. Because row 3 already has a zero in the pivot column, no updating is necessary. The completed third tableau is shown in Table 4.8. (The reader should again compare this solution with the algebraic representation in (4.57–4.60).)

As before, the basis inverse can be identified as

$$\mathbf{B}^{-1} = \begin{pmatrix} 1 & -\frac{1}{2} & \frac{5}{2} \\ 0 & \frac{1}{2} & -\frac{1}{2} \\ 0 & 0 & 1 \end{pmatrix}$$

TABLE 4.8

	z	x_1	x_2	x_3	x_4	x_5	RHS
z	1	0	0	0	1	2	38
x_3	0	0	0	1	$-\frac{1}{2}$	$\frac{5}{2}$	20
x_1	0	1	0	0	$\frac{1}{2}$	$-\frac{1}{2}$	4
x_2	0	0	1	0	0	1	10

(c) Return to Step 1.

STEP 1. Because all $z_j - c_j \geq 0$, the solution given in Table 4.8 is optimal. In fact, because $z_j - c_j > 0$ for the nonbasic variables x_4 and x_5, then this tableau represents the *unique* optimal solution. Thus, the optimal solution is

$$z^* = 38$$
$$x_1^* = 4$$
$$x_2^* = 10$$
$$x_3^* = 20$$
$$x_4^* = 0$$
$$x_5^* = 0$$

Notice that the values of z^*, x_1^*, x_2^*, and x_3^* are read from the RHS column in Table 4.8. The values of x_4 and x_5 must be zero because they are nonbasic. The fact that x_4 and x_5 are zero implies that there is no slack in the last two constraints. That is, the last two resources are totally consumed. Resources that are totally consumed are sometimes referred to as *scarce resources*.

It actually takes considerably longer to *describe* the process of the simplex algorithm than it does to actually *apply* it. Example 4.2 illustrates this, as it should take the average reader no more than a few minutes to solve a problem of this size after some practice.

Example 4.3: Unbounded Objective

$$\text{maximize } z = 5x_1 + 3x_2 \tag{4.65}$$

subject to

$$-x_1 + x_2 \leq 4 \tag{4.66}$$

$$x_1 - 2x_2 \leq 6 \tag{4.67}$$

$$x_1, x_2 \geq 0 \tag{4.68}$$

Adding slack variables results in

$$\text{maximize } z = 5x_1 + 3x_2 \tag{4.69}$$

subject to

$$-x_1 + x_2 + x_3 = 4 \tag{4.70}$$

$$x_1 - 2x_2 + x_4 = 6 \tag{4.71}$$

$$x_j \geq 0, \qquad \text{for all } j \tag{4.72}$$

Tables 4.9 and 4.10 present the first two tableaux when applying the simplex algorithm to the preprocessed model. The objective function can be designated as (mathematically) unbounded in Table 4.10 because $z_2 - c_2 < 0$ and $\alpha_2 = \mathbf{B}^{-1}\mathbf{a}_2 \leq \mathbf{0}$. In fact, the objective function can be made arbitrarily large by moving along the ray defined by

$$\begin{pmatrix} 6 \\ 0 \\ 10 \\ 0 \end{pmatrix} + x_2 \begin{pmatrix} 2 \\ 1 \\ 1 \\ 0 \end{pmatrix}$$

For a real-life problem, the next step would be to review the model. Has an error been made in the constraint formulation, or (more likely) have we overlooked a limited resource or other restriction that when included in the formulation will bound the objective value?

TABLE 4.9

	z	x_1	x_2	x_3	x_4	RHS
z	1	-5	-3	0	0	0
x_3	0	-1	1	1	0	4
x_4	0	①	-2	0	1	6

TABLE 4.10

	z	x_1	x_2	x_3	x_4	RHS
z	1	0	−13	0	5	30
x_3	0	0	−1	1	1	10
x_1	0	1	−2	0	1	6

Example 4.4: Alternative Optimal Solutions

$$\text{maximize } z = 2x_1 + x_2 \qquad (4.73)$$

subject to

$$2x_1 + x_2 \leq 8 \qquad (4.74)$$

$$x_1 + x_2 \leq 5 \qquad (4.75)$$

$$x_1, x_2 \geq 0 \qquad (4.76)$$

Preprocessing, we have

$$\text{maximize } z = 2x_1 + x_2 \qquad (4.77)$$

subject to

$$2x_1 + x_2 + x_3 = 8 \qquad (4.78)$$

$$x_1 + x_2 + x_4 = 5 \qquad (4.79)$$

$$x_j \geq 0, \qquad \text{for all } j \qquad (4.80)$$

The tableaux for this problem are given in Tables 4.11 and 4.12.

TABLE 4.11

	z	x_1	x_2	x_3	x_4	RHS
z	1	−2	−1	0	0	0
x_3	0	②	1	1	0	8
x_4	0	1	1	0	1	5

TABLE 4.12

	z	x_1	x_2	x_3	x_4	RHS
z	1	0	0	1	0	8
x_1	0	1	$\frac{1}{2}$	$\frac{1}{2}$	0	4
x_4	0	0	$\frac{1}{2}$	$-\frac{1}{2}$	1	1

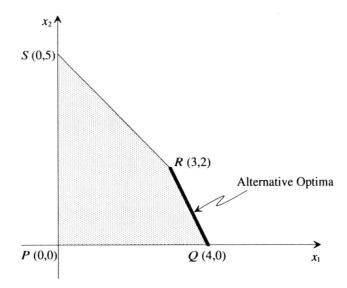

Figure 4.3   Alternative optimal solutions.

Note that the basic feasible solution represented by Table 4.12 is an optimal solution because $z_j - c_j \geq 0$, for all j. However, note also that $z_j - c_j = 0$ for the nonbasic variable x_2. Thus, increasing x_2 from its current value of zero will not change the objective value. Entering x_2 into the basis results in the tableau shown in Table 4.13. Observe that we have moved from the optimal extreme point Q to the optimal extreme point R in Figure 4.3.

TABLE 4.13

	z	x_1	x_2	x_3	x_4	RHS
z	1	0	0	1	0	8
x_1	0	1	0	1	-1	3
x_2	0	0	1	-1	2	2

FINDING AN INITIAL BASIC FEASIBLE SOLUTION

In each of the previous examples, a starting basic feasible solution was quite apparent. For example, if we look at the initial tableau of Example 4.2 (Table 4.6), we see that there is an imbedded $m \times m$ identity matrix $\mathbf{I}$, and the starting basic variables are readily identified by letting $\mathbf{B} = \mathbf{I}$. And because the right-hand-side vector $\mathbf{b}$ is nonnegative, the resulting solution is clearly feasible because $\mathbf{x}_B = \mathbf{B}^{-1}\mathbf{b} = \mathbf{Ib} = \mathbf{b} \geq \mathbf{0}$. However, such a straightforward starting basic feasible solution is not always available. For example, consider the following problem.

Example 4.5

$$\text{maximize } z = 8x_1 + 10x_2 \tag{4.81}$$

subject to

$$x_1 - x_2 = 1 \tag{4.82}$$

$$x_1 + x_2 \leq 9 \tag{4.83}$$

$$x_1 + (\tfrac{1}{2})x_2 \geq 4 \tag{4.84}$$

$$x_1, x_2 \geq 0 \tag{4.85}$$

Now, converting the problem to standard equality form by adding the appropriate slack/surplus variables yields

$$\text{maximize } z = 8x_1 + 10x_2 \tag{4.86}$$

subject to

$$x_1 - x_2 = 1 \tag{4.87}$$

$$x_1 + x_2 + x_3 = 9 \tag{4.88}$$

$$x_1 + (\tfrac{1}{2})x_2 - x_4 = 4 \tag{4.89}$$

$$x_1, x_2, x_3, x_4 \geq 0 \tag{4.90}$$

Therefore, the coefficient matrix is given by

$$\mathbf{A} = \begin{pmatrix} 1 & -1 & 0 & 0 \\ 1 & 1 & 1 & 0 \\ 1 & \tfrac{1}{2} & 0 & -1 \end{pmatrix}$$

Observe that matrix $\mathbf{A}$ does not contain the identity as a submatrix. In fact, $\mathbf{A}$ contains only the second column of the identity matrix. Thus, in its present state, we cannot use $\mathbf{B} = \mathbf{I}$ as a convenient starting basis. Artificial-variable techniques were developed to find a starting basis feasible solution in this all-too-common situation when a nice starting basis is not available. In the following section, we present one of the most common artificial-variable techniques, the two-phase method.

The Two-Phase Method

The general approach of the two-phase method can be described as follows. First, we create an identity submatrix by adding the necessary artificial variables to the original constraints. For Example 4.5, it would be necessary to add two artificial variables, say, x_5 and x_6, to constraints (4.87) and (4.89), respectively. This would result in the following system of constraints.

$$x_1 - x_2 + x_5 = 1 \tag{4.91}$$

$$x_1 + x_2 + x_3 = 9 \tag{4.92}$$

$$x_1 + (\tfrac{1}{2})x_2 - x_4 + x_6 = 4 \tag{4.93}$$

$$x_j \geq 0, \qquad \text{for all } j \tag{4.94}$$

Thus, the coefficient matrix becomes

$$(\mathbf{a}_1, \mathbf{a}_2, \mathbf{a}_3, \mathbf{a}_4, \mathbf{a}_5, \mathbf{a}_6) = \begin{pmatrix} 1 & -1 & 0 & 0 & 1 & 0 \\ 1 & 1 & 1 & 0 & 0 & 0 \\ 1 & \tfrac{1}{2} & 0 & -1 & 0 & 1 \end{pmatrix}$$

Clearly, the identity submatrix is now available with $\mathbf{I} = (\mathbf{a}_5, \mathbf{a}_3, \mathbf{a}_6)$. Note that it was not necessary to add an artificial variable to the second constraint because x_3 appears only in the second constraint with a coefficient of 1. However, by adding these variables, we have changed the problem, and in order to have a solution to the original problem, the artificial variables must be zero. Thus, in phase I, an *artificial objective function* is used and an attempt is made to drive all artificial variables to zero. This artificial objective is to minimize the sum of the artificial variables (or, equivalently, to maximize the negative sum of the artificial variables). If all the artificial variables cannot be driven to zero, then at least one constraint of the original problem is violated, and, consequently, the original problem is infeasible.

Phase II consists of replacing the artificial objective function by the original objective function and using the basic feasible solution found in phase I as a starting point. If no artificial variables were left in the basis at the end of phase I, we simply perform the simplex algorithm until an optimum is reached. If, however, an artificial variable was in the basis (at a zero value) at the conclusion of phase I, we slightly modify the *departing-variable rule*. The specific steps of the two-phase algorithm follow and are illustrated via some examples.

The two-phase algorithm

STEP 1. Establish the problem formulation in a suitable form for the implementation of the simplex algorithm (i.e., convert the objective function to a maximization form and convert all the constraints by adding slack, surplus, and/or artificial variables that are required).

STEP 2. The artificial objective function of phase I is to maximize the negative sum of the artificial variables (that is, equivalently minimize the sum of the artificial variables in an attempt to drive them all to zero).

STEP 3. *Phase I:* Employ the simplex algorithm of the previous section on the problem constructed in Steps 1 and 2. If, at optimality, there are no artificial variables in the basis at a positive value, go to Step 4 (phase II). Otherwise the problem is (mathematically) infeasible and we stop.

STEP 4. *Phase II:* Assign the actual objective function coefficient (the original c_j's) to each variable except for the artificial variables. Any artificial variables in the basis at a zero level are given a c_j value of zero in phase II. Any artificial variables that are not in the basis may be dropped from consideration by striking out their entire associated column in the tableau.

STEP 5. The first tableau of phase II is the final tableau of phase I except for the objective row. Update the objective row using the relationships

$$z_j - c_j = \mathbf{c}_B \mathbf{B}^{-1} \mathbf{a}_j - c_j = \mathbf{c}_B \boldsymbol{\alpha}_j - c_j \tag{4.95}$$

$$z = \mathbf{c}_B \mathbf{B}^{-1} \mathbf{b} = \mathbf{c}_B \boldsymbol{\beta} \tag{4.96}$$

STEP 6. If no artificial variables were in the basis (at zero values) at the end of phase I, we now simply use the simplex algorithm and proceed as usual. If, however, there are artificial variables in the basis, go to Step 7.

STEP 7. We must make sure that the artificial variables in the basis do not ever become positive in phase II. This is accomplished by modifying the departing variable rule of the simplex algorithm as follows:

(a) Determine the entering variable x_k in the usual manner.

(b) Examine the entering variable column $\mathbf{a}_k = \mathbf{B}^{-1}\mathbf{a}_k$. If the $\alpha_{i,k}$ values for any of the artificial variables left in the basis are negative, then choose an artificial variable with a negative $\alpha_{i,k}$ as the departing variable. Otherwise employ the usual departing variable rule.

Example 4.6: The Two-Phase Algorithm

Returning to Example 4.5, we see that the phase I problem is as follows:

$$\text{maximize } Z = -x_5 - x_6 \equiv \text{minimize } x_5 + x_6 \tag{4.97}$$

subject to

$$x_1 - x_2 + x_5 = 1 \tag{4.98}$$

$$x_1 + x_2 + x_3 = 9 \tag{4.99}$$

$$x_1 + (\tfrac{1}{2})x_2 - x_4 + x_6 = 4 \tag{4.100}$$

$$x_j \geq 0, \qquad \text{for all } j \tag{4.101}$$

Note that x_5 and x_6 are artificial variables, whereas x_3 and x_4 are, respectively, slack and surplus variables in the original formulation. Notice also that the objective function of the phase I problem includes only the artificial variables. We now proceed to phase I and attempt to drive all artificial variables to zero (and thus drive Z to zero). It is important to realize that the objective row of the initial tableau must be updated using the relationships $z_j - c_j = \mathbf{c}_B\mathbf{B}^{-1}\mathbf{a}_j - c_j$ and $Z = \mathbf{c}_B\mathbf{B}^{-1}\mathbf{b}$ because x_5 and x_6 are basic and have nonzero objective coefficients. (This update procedure is equivalent to using $Z + x_5 + x_6 = 0$ as the top row and then applying elementary row operations to zero out the costs of the basic variables x_5 and x_6.)

$$z_1 - c_1 = \mathbf{c}_B\mathbf{B}^{-1}\mathbf{a}_1 - c_1 = (-1 \quad 0 \quad -1)\begin{pmatrix} 1 \\ 1 \\ 1 \end{pmatrix} - 0 = -2$$

$$z_2 - c_2 = \mathbf{c}_B\mathbf{B}^{-1}\mathbf{a}_2 - c_2 = (-1 \quad 0 \quad -1)\begin{pmatrix} -1 \\ 1 \\ \tfrac{1}{2} \end{pmatrix} - 0 = \tfrac{1}{2}$$

$$z_3 - c_3 = \mathbf{c}_B\mathbf{B}^{-1}\mathbf{a}_3 - c_3 = (-1 \quad 0 \quad -1)\begin{pmatrix} 0 \\ 1 \\ 0 \end{pmatrix} - 0 = 0$$

$$z_4 - c_4 = \mathbf{c}_B\mathbf{B}^{-1}\mathbf{a}_4 - c_4 = (-1 \quad 0 \quad -1)\begin{pmatrix} 0 \\ 0 \\ -1 \end{pmatrix} - 0 = 1$$

$$z_5 - c_5 = \mathbf{c}_B\mathbf{B}^{-1}\mathbf{a}_5 - c_5 = (-1 \quad 0 \quad -1)\begin{pmatrix} 1 \\ 0 \\ 0 \end{pmatrix} - (-1) = 0$$

$$z_6 - c_6 = \mathbf{c}_B\mathbf{B}^{-1}\mathbf{a}_6 - c_6 = (-1 \quad 0 \quad -1)\begin{pmatrix} 0 \\ 0 \\ 1 \end{pmatrix} - (-1) = 0$$

$$Z = \mathbf{c}_B\mathbf{B}^{-1}\mathbf{b} = (-1 \quad 0 \quad -1)\begin{pmatrix} 1 \\ 9 \\ 4 \end{pmatrix} = -5$$

Tables 4.14 through 4.16 summarize the results of phase I.

TABLE 4.14

	Z	x_1	x_2	x_3	x_4	x_5	x_6	RHS
Z	1	-2	$\frac{1}{2}$	0	1	0	0	-5
x_5	0	①	-1	0	0	1	0	1
x_3	0	2	1	1	0	0	0	9
x_6	0	0	$\frac{1}{2}$	0	-1	0	1	4

TABLE 4.15

	Z	x_1	x_2	x_3	x_4	x_5	x_6	RHS
Z	1	0	$-\frac{3}{2}$	0	1	2	0	-3
x_1	0	1	-1	0	0	1	0	1
x_3	0	0	2	1	0	-1	0	8
x_6	0	0	③	0	-1	-1	1	3

TABLE 4.16

	Z	x_1	x_2	x_3	x_4	x_5	x_6	RHS
Z	0	0	0	0	0	1	0	0
x_1	0	1	0	0	$-\frac{2}{3}$	$\frac{1}{3}$	$\frac{2}{3}$	3
x_3	0	0	0	1	$\frac{4}{3}$	$\frac{1}{3}$	$-\frac{4}{3}$	4
x_2	0	0	1	0	$-\frac{2}{3}$	$-\frac{2}{3}$	$\frac{2}{3}$	2

Table 4.16 represents the optimal solution to phase I because $z_j - c_j \geq 0$, for all j. Because $Z = 0$, then all the artificial variables have been driven to zero. Also, because the artificial variables, x_5 and x_6, have been driven from the basis, their columns may be dropped from the phase II tableau. Updating the objective row using the original objective coefficients, we begin phase II with the tableau of Table 4.17. Note that we only need to compute $z_4 - c_4$ and z because x_1, x_2, and x_3 are basic variables.

$$z_4 - c_4 = \mathbf{c}_B \mathbf{B}^{-1} \mathbf{a}_4 - c_4 = (8 \quad 0 \quad 10) \begin{pmatrix} -\frac{2}{3} \\ \frac{4}{3} \\ -\frac{2}{3} \end{pmatrix} - 0 = -12$$

$$z = \mathbf{c}_B \mathbf{B}^{-1} \mathbf{b} = (8 \quad 0 \quad 10) \begin{pmatrix} 3 \\ 4 \\ 2 \end{pmatrix} = 44$$

Table 4.17 is the same tableau as shown in Table 4.16, except for the objective row. Phase II is summarized in Tables 4.17 and 4.18.

TABLE 4.17

	z	x_1	x_2	x_3	x_4	RHS
z	1	0	0	0	-12	44
x_1	0	1	0	0	$-\frac{2}{3}$	3
x_3	0	0	0	1	$\frac{4}{3}$	4
x_2	0	0	1	0	$-\frac{2}{3}$	2

TABLE 4.18

	z	x_1	x_2	x_3	x_4	RHS
z	1	0	0	9	0	80
x_1	0	1	0	$\frac{1}{2}$	0	5
x_4	0	0	0	$\frac{3}{4}$	1	3
x_2	0	0	1	$\frac{1}{2}$	0	4

Because $z_j - c_j \geq 0$, for all j, in Table 4.18, we can stop as we have found the optimal solution:

$$z^* = 80$$
$$x_1^* = 5$$
$$x_2^* = 4$$
$$x_3^* = 0$$
$$x_4^* = 3$$

Example 4.7: An Infeasible Problem

$$\text{maximize } z = 5x_1 + x_2 \tag{4.102}$$

subject to

$$2x_1 + x_2 \geq 5 \tag{4.103}$$

$$x_2 \geq 1 \tag{4.104}$$

$$2x_1 + 3x_2 \leq 6 \tag{4.105}$$

$$x_1, x_2 \geq 0 \tag{4.106}$$

Let x_3 and x_4 be the surplus variables in constraints (4.103) and (4.104), respectively, and let x_5 be the slack variable constraint (4.105). Then in order to have an identity submatrix, it is necessary to introduce artificial variables x_6 and x_7 into the first and second constraints, respectively. The resulting phase I problem is given as

$$\text{maximize } Z = -x_6 - x_7 \tag{4.107}$$

subject to

$$2x_1 + x_2 - x_3 + x_6 = 5 \tag{4.108}$$

$$x_2 - x_4 + x_7 = 1 \tag{4.109}$$

$$3x_1 + 2x_2 + x_5 = 6 \tag{4.110}$$

$$x_j \geq 0, \qquad \text{for all } j \tag{4.111}$$

Phase I of the two-phase method is presented in Tables 4.19 through 4.21. The final tableau (Table 4.21) indicates optimality (all $z_j - c_j \geq 0$), but Z is *not* zero (because artificial variable x_6 is in the basis at a positive value). Consequently, this

TABLE 4.19

	Z	x_1	x_2	x_3	x_4	x_5	x_6	x_7	RHS
Z	1	-2	-2	1	1	0	0	0	-6
x_6	0	②	1	-1	0	0	1	0	5
x_7	0	0	1	0	-1	0	0	1	1
x_5	0	2	3	0	0	1	0	0	6

TABLE 4.20

	Z	x_1	x_2	x_3	x_4	x_5	x_6	x_7	RHS
Z	1	0	-1	0	1	0	1	0	-1
x_1	0	1	$\frac{1}{2}$	0	0	0	$\frac{1}{2}$	0	$\frac{5}{2}$
x_7	0	0	1	0	-1	0	0	1	1
x_5	0	0	②	1	0	1	-1	0	1

TABLE 4.21

	Z	x_1	x_2	x_3	x_4	x_5	x_6	x_7	RHS
Z	1	0	0	$\frac{1}{2}$	1	$\frac{1}{2}$	$\frac{1}{2}$	0	$-\frac{1}{2}$
x_1	0	1	0	$-\frac{3}{4}$	0	$-\frac{1}{4}$	1	0	$\frac{9}{4}$
x_6	0	0	0	$-\frac{1}{2}$	-1	$-\frac{1}{2}$	$\frac{1}{2}$	1	$\frac{1}{2}$
x_2	0	0	1	$\frac{1}{2}$	0	$\frac{1}{2}$	$-\frac{1}{2}$	0	$\frac{1}{2}$

problem, as modeled, is mathematically infeasible and we may terminate the solution procedure (and try to decide what can be done to find a workable solution to the actual problem). The feasible region of the original problem is graphed in Figure 4.4, and clearly the feasible region is empty.

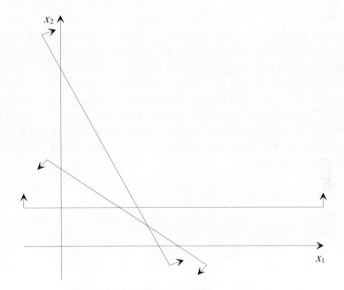

Figure 4.4   Graph for Example 4.7.

Example 4.8: Artificial Variables Left in the Basis

Our final illustration of the two-phase method involves a problem in which an artificial variable remains in the basis, at zero value, at the end of phase I. This is often caused by redundant constraints in the model formulation.

Table 4.22 is the first tableau, for a particular problem, of phase II. Variables x_6 and x_7 are *artificial variables* and are both in the basis at zero values. Under normal simplex rules, variable x_2 would enter the basis and x_4 would leave, resulting in both x_6 and x_7 going positive. This would result in an infeasible solution. However, using the modified departing-variable rule of the two-phase method, we would select either x_6 or x_7 as the departing variable. The reader is invited to complete the operations of phase II.

TABLE 4.22

	z	x_1	x_2	x_3	x_4	x_5	x_6	x_7	RHS
z	1	0	-3	3	0	2	0	0	6
x_1	0	1	-1	3	0	1	0	0	3
x_6	0	0	-2	1	0	3	1	0	0
x_4	0	0	3	-1	1	-1	0	0	9
x_7	0	0	-1	0	0	2	0	1	0

The Big-*M* Method

In the two-phase method, phase I addresses the problem of finding an initial basic feasible solution by seeking to drive all the artificial variables to zero. Recall that this is done by minimizing the sum of the artificial variables while ignoring the original objective function. The phase II problem then uses the basic feasible solution found in phase I along with the original objective function to find the optimal solution. The Big-*M* method is a technique that essentially combines the phase I and phase II problems into a single problem. This is done by including the artificial variables in the original objective function with cost coefficients that implicitly try to drive the artificial variables to zero. To illustrate, consider the following example.

Example 4.9: The Big-*M* Method

$$\text{minimize } z = 3x_1 + x_2 + 4x_3 \tag{4.112}$$

subject to

$$x_1 + x_2 + x_3 \geq 12 \tag{4.113}$$

$$4x_1 - x_2 + x_3 \geq 6 \tag{4.114}$$

$$x_1, x_2, x_3 \geq 0 \tag{4.115}$$

Converting the problem to a maximization problem and preprocessing the constraints yields

$$\text{maximize } z' = -3x_1 - x_2 - 4x_3 \tag{4.116}$$

subject to

$$x_1 + x_2 + x_3 - x_4 = 12 \tag{4.117}$$

$$4x_1 - x_2 + x_3 - x_5 = 6 \tag{4.118}$$

$$x_j \geq 0, \qquad \text{for all } j \tag{4.119}$$

Because neither column of the identity is available, we supplement the problem with two artificial variables, x_6 and x_7, and form the *Big-M problem:*

$$\text{maximize } z' = -3x_1 - x_2 - 4x_3 - Mx_6 - Mx_7 \qquad (4.120)$$

subject to

$$x_1 + x_2 + x_3 - x_4 + x_6 = 12 \qquad (4.121)$$

$$4x_1 - x_2 + x_3 - x_5 + x_7 = 6 \qquad (4.122)$$

$$x_j \geq 0, \qquad \text{for all } j \qquad (4.123)$$

We now solve this problem by the standard simple procedure, while assuming that M is a large positive number. Tables 4.23 and 4.24 present the initial and final tableaux. The reader is encouraged to fill in the details. Because all $z_j - c_j \geq 0$ and the artificial variables have value zero, then Table 4.24 represents the optimal solution to the original problem. Thus,

$$z^* = -z' = -\left(-\tfrac{106}{5}\right) = \tfrac{106}{5}$$

$$x_1^* = \tfrac{18}{5}$$

$$x_2^* = \tfrac{42}{5}$$

$$x_3^* = 0$$

$$x_4^* = 0$$

$$x_5^* = 0$$

The use of $-M$ as the objective coefficient for artificial variables serves its purpose (that is, to drive these variables out of the basis, if possible), but only at the cost of tedious hand calculations. It also presents problems when using the

TABLE 4.23

	z'	x_1	x_2	x_3	x_4	x_5	x_6	x_7	RHS
z'	1	$-5M + 3$	1	$-2M + 4$	M	M	0	0	$-18M$
x_6	0	1	1	1	-1	0	1	0	12
x_7	0	4	-1	1	0	-1	0	1	6

TABLE 4.24

	z'	x_1	x_2	x_3	x_4	x_5	x_6	x_7	RHS
z'	1	0	0	$\tfrac{11}{5}$	$\tfrac{1}{5}$	$\tfrac{2}{5}$	$M - \tfrac{7}{5}$	$M - \tfrac{2}{5}$	$-\tfrac{106}{5}$
x_2	0	0	1	$\tfrac{3}{5}$	$-\tfrac{4}{5}$	$\tfrac{1}{5}$	$\tfrac{4}{5}$	$-\tfrac{1}{5}$	$\tfrac{42}{5}$
x_1	0	1	0	$\tfrac{2}{5}$	$-\tfrac{1}{5}$	$-\tfrac{1}{5}$	$-\tfrac{1}{5}$	$\tfrac{1}{5}$	$\tfrac{18}{5}$

computer. On the computer, M must be assigned some numerical value that must be considerably larger than any of the other objective coefficients. If M is too small, we might obtain a solution with an artificial variable in the basis at a positive value (signifying an infeasible problem) when actually the problem is feasible. However, if M is too large, it may tend to dominate the $z_j - c_j$ values. Roundoff errors (inherent in any digital computer) could well result and impact the final solution. For these reasons, the Big-M method is seldom used in practice. The two-phase method was developed to avoid, or at least alleviate, these difficulties.

UNRESTRICTED VARIABLES AND VARIABLES WITH NEGATIVE LOWER BOUNDS

All of the material presented so far has stressed that all variables must be restricted to nonnegative values when employing the simplex algorithm. This restriction is one that may easily be circumvented by simple substitutions. For example, if x_k is an unrestricted variable, we may simply let

$$x_k = x_k^+ - x_k^-$$

where

$$-\infty \leq x_k \leq +\infty$$
$$x_k^+, x_k^- \geq 0$$

Wherever x_k appears in the problem, we substitute $x_k^+ - x_k^-$. This, of course, increases the size of the problem we have to solve. However, because the columns associated with x_k^+ and x_k^- are linearly dependent (one column is simply the negative of the other), at most one can appear in the basis at a positive value. If x_k^+ is in the basis, then $x_k \geq 0$, whereas if x_k^- is in the basis, then $x_k \leq 0$.

In a similar way, we can also handle variables with negative lower bounds. For example, suppose $x_l \geq -5$. Then, we simply let

$$x_l' = x_l + 5 \geq 0$$

Whenever x_l appears in the problem, we substitute $x_l' - 5$. Both of these techniques are illustrated via the following example.

Example 4.10: Unrestricted Variables and Variables with Negative Lower Bounds

$$\text{maximize } z = -3x_1 + x_2 \tag{4.124}$$

subject to

$$2x_1 - 3x_2 \leq 4 \tag{4.125}$$

$$2x_1 + x_2 \leq 8 \tag{4.126}$$

$$4x_1 - x_2 \leq 16 \tag{4.127}$$

$$x_1 \text{ unrestricted} \tag{4.128}$$

$$x_2 \geq -4 \tag{4.129}$$

Let $x_2' = x_2 + 4 \geq 0$. Also let $x_1 = x_1^+ - x_1^-$, where $x_1^+, x_1^- \geq 0$. Now transform the original problem by substituting $x_2 = x_2' - 4$ and $x_1 = x_1^+ - x_1^-$. This results in

maximize $z = -3(x_1^+ - x_1^-) + (x_2' - 4) = -3x_1^+ + 3x_1^- + x_2' - 4$ (4.130)

subject to

$$-2(x_1^+ - x_1^-) + 3(x_2' - 4) \leq 4 \tag{4.131}$$

$$2(x_1^+ - x_1^-) + (x_2' - 4) \leq 8 \tag{4.132}$$

$$4(x_1^+ - x_1^-) - (x_2' - 4) \leq 16 \tag{4.133}$$

$$x_1^+, x_1^-, x_2' \geq 0 \tag{4.134}$$

The constant (-4) in the objective has no effect on the optimization process; therefore, let $z' = z + 4$. Simplification and the addition of slack variables yields

maximize $z' = -3x_1^+ + 3x_1^- + x_2'$ (4.135)

subject to

$$-2x_1^+ + 2x_1^- + 3x_2' + x_3 = 16 \tag{4.136}$$

$$2x_1^+ - 2x_1^- + x_2' + x_4 = 12 \tag{4.137}$$

$$4x_1^+ - 4x_1^- - x_2' + x_5 = 12 \tag{4.138}$$

$$x_1^+, x_1^-, x_2', x_3, x_4, x_5 \geq 0 \tag{4.139}$$

Tables 4.25 and 4.26 summarize the solution using the simplex algorithm. Note that Table 4.26 represents the optimal solution. Thus, the optimal solution to the original

TABLE 4.25

	z'	x_1^+	x_1^-	x_2'	x_3	x_4	x_5	RHS
z'	1	3	-3	-1	0	0	0	0
x_3	0	-2	②	3	1	0	0	16
x_4	0	2	-2	1	0	1	0	12
x_5	0	4	-4	-1	0	0	1	12

TABLE 4.26

	z'	x_1^+	x_1^-	x_2'	x_3	x_4	x_5	RHS
z'	1	0	0	$\frac{7}{2}$	$\frac{3}{2}$	0	0	24
x_1^-	0	-1	1	$\frac{3}{2}$	$\frac{1}{2}$	0	0	8
x_4	0	0	0	4	1	1	0	28
x_5	0	0	0	5	2	0	1	44

problem can be recovered as follows:

$$z^* = z' - 4 = 24 - 4 = 20$$
$$x_1^* = x_1^+ - x_1^- = 0 - 8 = -8$$
$$x_2^* = x_2' - 4 = 0 - 4 = -4$$
$$x_3^* = 0$$
$$x_4^* = 28$$
$$x_5^* = 44$$

DEGENERACY AND CYCLING

Recall that a degenerate basic feasible solution is one in which at least one basic variable has a value of zero. This does not mean that anything is wrong with the solution. However, degeneracy could possibly create two related problems:

1. The objective function z may not improve when we move from one basis to another.
2. We might, in fact, cycle forever (repeating a sequence of bases) and not ever reach the optimal solution.

We illustrate the concept of a degenerate extreme point and a degenerate pivot via the following example.

Example 4.11: Degeneracy

$$\text{maximize } z = 4x_1 + 3x_2 \tag{4.140}$$

subject to

$$2x_1 + x_2 \leq 8 \tag{4.141}$$
$$x_1 + x_2 \leq 5 \tag{4.142}$$
$$x_1 - x_2 \leq 4 \tag{4.143}$$
$$x_1 \leq 4 \tag{4.144}$$
$$x_1, x_2 \geq 0 \tag{4.145}$$

Preprocessing, we have

$$\text{maximize } z = 4x_1 + 3x_2 \tag{4.146}$$

subject to

$$2x_1 + x_2 + x_3 = 8 \tag{4.147}$$

$$x_1 + x_2 + x_4 = 5 \tag{4.148}$$

$$x_1 - x_2 + x_5 = 4 \tag{4.149}$$

$$x_1 + x_6 = 4 \tag{4.150}$$

$$x_j \geq 0, \qquad \text{for all } j \tag{4.151}$$

The feasible region for this problem is graphed in the x_1–x_2 plane in Figure 4.5. Obviously, in this case, constraints (4.143) and (4.144) are redundant and have no effect on the feasible region. However, detecting such redundant constraints is not an easy task for problems of nontrivial size. Let us ignore this redundancy so that we can illustrate the concept of a degenerate extreme point. Because $m = 4$ and $n = 6$, each basic feasible solution will be characterized by $n - m = 2$ nonbasic variables. Observe, however, that extreme point Q has $x_2 = x_3 = x_5 = x_6 = 0$. Thus, there are

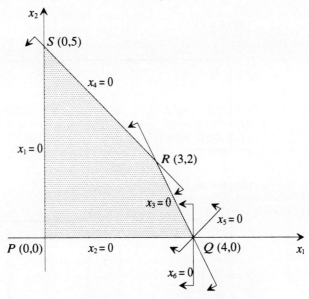

Figure 4.5 Graph for Example 4.11.

$C_2^4 = 4!/[2!(4 - 2)!] = 6$ basic feasible solutions representing Q. Table 4.27 gives the initial simplex tableau. Note that x_1 is chosen as the entering variable, and the minimum ratio test results in a three-way tie. That is, each of x_3, x_5, or x_6 could be chosen as the departing variable. This should be a signal that the next basic feasible solution will be degenerate. Choosing x_5 as the departing variable results in the degenerate solution in Table 4.28, which represents extreme point Q. Note that, as expected, the basic variables x_3 and x_5 are equal to zero. Because $z_2 - c_2 < 0$ in Table 4.28, x_2 is chosen as the entering variable. This results in a minimum ratio of zero with either x_3 or x_6 as the departing variable. Choosing x_6 as the departing variable results in Table 4.29. Note that although the basis has changed, the values of all the variables are the same, and this new solution is also a basic feasible solution representing extreme point Q. Continuing the simplex procedure, we see that after an additional degenerate pivot, in Table 4.30, we eventually end up with the optimal solution in Table 4.31 corresponding to extreme point R.

TABLE 4.27

	z	x_1	x_2	x_3	x_4	x_5	x_6	RHS
z	1	-3	-2	0	0	0	0	0
x_3	0	2	1	1	0	0	0	8
x_4	0	1	1	0	1	0	0	5
x_5	0	①	-1	0	0	1	0	4
x_6	0	1	0	0	0	0	1	4

TABLE 4.28

	z	x_1	x_2	x_3	x_4	x_5	x_6	RHS
z	1	0	-5	0	0	3	0	12
x_3	0	0	3	1	0	-2	0	0
x_4	0	0	2	0	1	-1	0	1
x_1	0	1	-1	0	0	1	0	4
x_6	0	0	①	0	0	-1	1	0

TABLE 4.29

	z	x_1	x_2	x_3	x_4	x_5	x_6	RHS
z	1	0	0	0	0	-2	5	12
x_3	0	0	0	1	0	①	-3	0
x_4	0	0	0	0	1	1	-2	1
x_1	0	1	0	0	0	0	1	4
x_2	0	0	1	0	0	-1	1	0

TABLE 4.30

	z	x_1	x_2	x_3	x_4	x_5	x_6	RHS
z	1	0	0	2	0	0	-1	12
x_5	0	0	0	1	0	1	-3	0
x_4	0	0	0	-1	1	0	①	1
x_1	0	1	0	0	0	0	1	4
x_2	0	0	1	1	0	0	-2	0

TABLE 4.31

	z	x_1	x_2	x_3	x_4	x_5	x_6	RHS
z	1	0	0	1	1	0	0	13
x_5	0	0	0	-2	3	1	0	3
x_6	0	0	0	-1	1	0	1	1
x_1	0	1	0	1	-1	0	0	3
x_2	0	0	1	-1	2	0	0	2

In this example, we saw that in the presence of degeneracy, it is possible to perform pivot operations and remain at the same extreme point. Although, this example did not exhibit the phenomenon of cycling, it is theoretically possible for degeneracy to lead to a computational loop or cycle that is repeated infinitely. In fact, Beale (1955) and Marshall and Suurballe (1969) both provide examples of problems with three constraints and seven variables that exhibit cycling. In practice, however, this has not occurred in *actual* problems and is highly unlikely to present a problem. It is, however, of theoretical interest, and several methods have been developed to prevent the occurrence of cycling. These include the *lexicographic minimum ratio test* (Dantzig, Orden, Wolfe, 1955) and *Bland's cycling prevention rules* (Bland, 1977).

In the absence of degeneracy, the simplex method is guaranteed to stop in a finite number of iterations. This follows directly because there are a finite number of extreme points and each nondegenerate pivot forces the objective function to *strictly increase*. Thus, it is not possible to visit the same extreme point twice. However, in the presence of degeneracy, finite convergence is only guaranteed if a cycling-prevention rule is employed. The basic idea behind cycling-prevention rules is to provide a method for breaking ties in the minimum ratio test in such a way that it is not possible to repeat a basis matrix during a sequence of degenerate pivots. The lexicographic minimum ratio test is one such method and is discussed in the following section.

*The Lexicographic Minimum Ratio Test and Finite Convergence

As previously noted, the simplex algorithm will stop in a finite number of iterations unless cycling occurs as a result of degeneracy. The lexicographic minimum ratio test is designed to prevent cycling from occurring by preventing the simplex method from repeating a basis matrix that has already been used at a prior iteration. Of course, in the absence of degeneracy, each pivot operation results in a strict improvement of the objective function. Thus, bases associated with nondegenerate pivots can never be repeated. Before formally stating the lexicographic minimum ratio test, let us introduce the idea of a lexicographically positive vector.

A vector $\mathbf{v}$ is *lexicographically positive* if $\mathbf{v} \neq \mathbf{0}$, and the first nonzero element of $\mathbf{v}$ is positive. For example, consider the vectors

$$\mathbf{v}_1 = (0 \quad 0 \quad 4 \quad -10)$$

$$\mathbf{v}_2 = (1 \quad -3 \quad 0 \quad -5)$$

$$\mathbf{v}_3 = (0 \quad 3 \quad 0 \quad 1)$$

$$\mathbf{v}_4 = (0 \quad -1 \quad 10 \quad 20)$$

Then $\mathbf{v}_1$, $\mathbf{v}_2$, and $\mathbf{v}_3$ are lexicographically positive, whereas $\mathbf{v}_4$ is *not* lexicographically positive.

Given two vectors, $\mathbf{w}_1, \mathbf{w}_2 \in E^n$, we say that $\mathbf{w}_1$ is *lexicographically greater* than $\mathbf{w}_2$ if the vector $(\mathbf{w}_1 - \mathbf{w}_2)$ is *lexicographically positive*. For example, consider again the vectors $\mathbf{v}_1$, $\mathbf{v}_2$, $\mathbf{v}_3$, and $\mathbf{v}_4$ defined earlier. Then, it is easy to see that $\mathbf{v}_2$ is lexicographically greater than $\mathbf{v}_1$, $\mathbf{v}_3$, and $\mathbf{v}_4$, because $(\mathbf{v}_2 - \mathbf{v}_1)$, $(\mathbf{v}_2 - \mathbf{v}_3)$, $(\mathbf{v}_2 - \mathbf{v}_4)$ are all lexicographically positive. Similarly, $\mathbf{v}_3$ is lexicographically greater than $\mathbf{v}_1$ and $\mathbf{v}_4$, and $\mathbf{v}_1$ is lexicographically greater than $\mathbf{v}_4$. Thus, of the four vectors, we can say that $\mathbf{v}_4$ is the lexicographic minimum. Just as $\mathbf{v}_4$ was the lexicographic minimum among the given vectors, the lexicographic minimum ratio test selects the lexicographic minimum row as the pivot row. In performing this test, however, we are only interested in a specific portion of the simplex tableau.

Suppose that x_k is the entering variable at some iteration of the simplex algorithm. Then recall that the standard minimum ratio test is defined by

$$\frac{\beta_r}{\alpha_{r,k}} = \text{minimum} \left\{ \frac{\beta_i}{\alpha_{i,k}} : \alpha_{i,k} > 0 \right\} \tag{4.152}$$

---

* Starred sections may be omitted in an introductory course with no appreciable loss of continuity.

where

$$\boldsymbol{\beta} = \mathbf{B}^{-1}\mathbf{b} = \begin{pmatrix} \beta_1 \\ \beta_2 \\ \vdots \\ \beta_m \end{pmatrix}$$

$$\boldsymbol{\alpha}_k = \mathbf{B}^{-1}\mathbf{a}_k = \begin{pmatrix} \alpha_{1,k} \\ a_{2,k} \\ \vdots \\ \alpha_{m,k} \end{pmatrix}$$

The variable $x_{B,r}$ is then chosen as the departing variable. In the event of a tie for the minimum ratio, however, the departing variable must be selected from possibly several alternatives. And, as a result of this tie, the next basic feasible solution will be degenerate. The lexicographic minimum ratio test allows us to break these ties in such a way that the minimum ratio test will always result in a unique choice for the departing variable.

For notational convenience, suppose that the columns of the simplex tableau have been reordered so that $\mathbf{B}^{-1}$ occupies the first m columns of the tableau, that is,

$$\mathbf{B}^{-1} = (\boldsymbol{\alpha}_1, \boldsymbol{\alpha}_2, \ldots, \boldsymbol{\alpha}_m) = \begin{pmatrix} \alpha_{1,1} & \alpha_{1,2} & \cdots & \alpha_{1,m} \\ \alpha_{2,1} & \alpha_{2,2} & \cdots & \alpha_{2,m} \\ \vdots & \vdots & \ddots & \vdots \\ \alpha_{m,1} & \alpha_{m,2} & \cdots & \alpha_{m,m} \end{pmatrix} \quad (4.153)$$

Then the lexicographic minimum ratio test will be defined using the matrix

$$(\mathbf{B}^{-1}\mathbf{b}, \mathbf{B}^{-1}) = (\boldsymbol{\beta}, \boldsymbol{\alpha}_1, \boldsymbol{\alpha}_2, \ldots, \boldsymbol{\alpha}_m) = \begin{pmatrix} \beta_1 & \alpha_{1,1} & \alpha_{1,2} & \cdots & \alpha_{1,m} \\ \beta_2 & \alpha_{2,1} & \alpha_{2,2} & \cdots & \alpha_{2,m} \\ \vdots & \vdots & \vdots & \ddots & \vdots \\ \beta_m & \alpha_{m,1} & \alpha_{m,2} & \cdots & \alpha_{m,m} \end{pmatrix} \quad (4.154)$$

Before formally stating the lexicographic minimum ratio test, let us make an observation concerning the matrix defined in (4.154). Recall that the simplex algorithm always starts with $\mathbf{B} = \mathbf{I}$ and $\mathbf{b} \geq \mathbf{0}$. Thus, the initial matrix defined by (4.154) is

$$(\mathbf{B}^{-1}\mathbf{b}, \mathbf{B}^{-1}) = (\mathbf{b}, \mathbf{I}) = \begin{pmatrix} b_1 & 1 & 0 & \cdots & 0 \\ b_2 & 0 & 1 & \cdots & 0 \\ \vdots & \vdots & \vdots & \ddots & \vdots \\ b_m & 0 & 0 & \cdots & 1 \end{pmatrix}$$

Clearly, the rows of this initial matrix are lexicographically positive. In fact, as we will see, the lexicographic minimum ratio test maintains the rows as lexicographically positive throughout the operation of the simplex algorithm. We are now ready to state the lexicographic minimum ratio test.

The lexicographic minimum ratio test

STEP 1. Set $t = 0$ and let $I_t = I_0$ be the set of indices that result in the minimum ratio:

$$\text{minimum} \left\{ \frac{\beta_i}{\alpha_{i,k}} : \alpha_{i,k} > 0 \right\} \tag{4.155}$$

If I_0 is a singleton, that is, if $I_0 = \{r\}$, then $x_{B,r}$ is selected as the departing variable in the usual manner. Otherwise replace t by $t + 1$ and go to Step 2.

STEP 2. I_t be the set of indices that result in the minimum ratio:

$$\text{minimum} \left\{ \frac{\alpha_{i,t}}{\alpha_{i,k}} : i \in I_{t-1} \right\} \tag{4.156}$$

If I_t is a singleton, that is, if $I_t = \{r\}$, then $x_{B,r}$ is selected as the departing variable. Otherwise replace t by $t + 1$ and repeat Step 2.

Note that the lexicographic minimum ratio test will always result in a unique selection because the rows of $\mathbf{B}^{-1}$ are linearly independent. Also, if we divide each row of the matrix $(\boldsymbol{\beta}, \mathbf{B}^{-1})$ by the corresponding element of $\boldsymbol{\alpha}_k$, then the lexicographic minimum ratio rule can be viewed as simply choosing the lexicographic minimum row from the following.

$$\begin{pmatrix} \beta_1/\alpha_{1,k} & \alpha_{1,1}/\alpha_{1,k} & \alpha_{1,2}/\alpha_{1,k} & \cdots & \alpha_{1,m}/\alpha_{1,k} \\ \beta_2/\alpha_{2,k} & \alpha_{2,1}/\alpha_{2,k} & \alpha_{2,2}/\alpha_{2,k} & \cdots & \alpha_{2,m}/\alpha_{2,k} \\ \vdots & \vdots & \vdots & \ddots & \vdots \\ \beta_m/\alpha_{m,k} & \alpha_{m,1}/\alpha_{m,k} & \alpha_{m,2}/\alpha_{m,k} & \cdots & \alpha_{m,m}/\alpha_{m,k} \end{pmatrix}$$

Because we are choosing to pivot in the lexicographic minimum row, it is easily seen that the pivot operation will result in all rows of $(\boldsymbol{\beta}, \mathbf{B}^{-1})$ remaining lexicographically positive. Finally, we are adding a positive multiple of this lexicographically positive row to the objective row. In particular, we multiply row r of $(\boldsymbol{\beta}, \mathbf{B}^{-1})$ by $-(z_k - c_k)/\alpha_{r,k}$ and add it to $(\mathbf{c}_B\boldsymbol{\beta}, \mathbf{c}_B\mathbf{B}^{-1})$. Thus, the vector $(\mathbf{c}_B\boldsymbol{\beta}, \mathbf{c}_B\mathbf{B}^{-1})$ is lexicographically increasing at each iteration and the simplex algorithm can never repeat a basis.

We now illustrate the process of selecting the departing variable based on the lexicographically minimum ratio test via the following example.

Example 4.12: The Lexicographic Minimum Ratio Test

Given the following linear programming problem.

$$\text{maximize } z = 2x_1 + 2x_2 + 3x_3 \tag{4.157}$$

subject to

$$-x_1 + x_2 \le 16 \qquad (4.158)$$

$$2x_1 + x_2 + x_3 \le 24 \qquad (4.159)$$

$$x_1 + x_2 + x_3 \le 24 \qquad (4.160)$$

$$x_1 + x_3 \le 8 \qquad (4.161)$$

$$x_1, x_2, x_3 \ge 0 \qquad (4.162)$$

Let x_4, x_5, x_6, and x_7 denote the slack variables for constraints (4.158–4.161), respectively, and consider the basic feasible solution defined by the tableau in Table 4.32.

TABLE 4.32

	z	x_1	x_2	x_3	x_4	x_5	x_6	x_7	RHS
z	0	0	0	-3	0	2	0	-2	32
x_4	0	0	0	2	1	-1	0	3	16
x_2	0	0	1	-1	0	1	0	-2	8
x_6	0	0	0	1	0	-1	1	1	8
x_1	0	1	0	1	0	0	0	1	8

Observe that choosing x_3 as the entering variable results in a three-way tie for the departing variable because

$$\frac{\beta_1}{\alpha_{1,3}} = \frac{16}{2} = 8$$

$$\frac{\beta_3}{\alpha_{3,3}} = \frac{8}{1} = 8$$

$$\frac{\beta_4}{\alpha_{4,3}} = \frac{8}{1} = 8$$

Let us use the lexicographic minimum ratio test to break the tie and choose the departing variable.

First, note from Step 1 of the lexicographic minimum ratio test that $I_0 = \{1, 3, 4\}$. Because the identity matrix $\mathbf{I} = (\mathbf{a}_4, \mathbf{a}_5, \mathbf{a}_6, \mathbf{a}_7)$, we know that $\mathbf{B}^{-1} = (\boldsymbol{\alpha}_4, \boldsymbol{\alpha}_5, \boldsymbol{\alpha}_6, \boldsymbol{\alpha}_7)$. Therefore, we have

$$(\boldsymbol{\beta}, \mathbf{B}^{-1}) = (\boldsymbol{\beta}, \boldsymbol{\alpha}_4, \boldsymbol{\alpha}_5, \boldsymbol{\alpha}_6, \boldsymbol{\alpha}_7) = \begin{pmatrix} 16 & 1 & -1 & 0 & 3 \\ 8 & 0 & 1 & 0 & -2 \\ 8 & 0 & -1 & 1 & 1 \\ 8 & 0 & 0 & 0 & 1 \end{pmatrix}$$

Note that, as expected, the rows of this matrix are lexicographically positive. Continuing the minimum ratio test with the first column of $\mathbf{B}^{-1}$, we next compute

$$\text{minimum}\left\{\frac{\alpha_{i,4}}{\alpha_{i,3}} : i \in I_0 = \{1, 3, 4\}\right\} = \text{minimum}\left\{\frac{\alpha_{1,4}}{\alpha_{1,3}} = \frac{1}{2}, \frac{\alpha_{3,4}}{\alpha_{3,3}} = \frac{0}{1}, \frac{\alpha_{4,4}}{\alpha_{4,3}} = \frac{0}{1}\right\} = 0$$

We again have a tie for the minimum ratio, which results in $I_1 = \{3, 4\}$. Continuing with the second column of $\mathbf{B}^{-1}$, we have

$$\text{minimum} \left\{ \frac{\alpha_{i,5}}{\alpha_{i,3}} : i \in I_1 = \{3, 4\} \right\} = \text{minimum} \left\{ \frac{\alpha_{3,5}}{\alpha_{3,3}} = \frac{-1}{1}, \frac{\alpha_{4,5}}{\alpha_{4,3}} = \frac{0}{1} \right\} = -1$$

Therefore, $I_2 = \{3\}$ and the lexicographic minimum ratio test selects $x_{B,3} = x_6$ as the departing variable and the pivot element in Table 4.32 is $\alpha_{3,3} = 1$.

SUMMARY

In this chapter, we have presented the simplex algorithm in both an algebraic and tableau format. Whereas the actual methodology of the simplex method is perhaps easier to understand from an algebraic viewpoint, the tableau format provides an easier method for handling the bookkeeping aspects of the simplex method. The reader should strive not only to understand the mechanics of the tableau method, but also the theoretical foundation upon which the mechanical operations are built. In addition, this chapter discussed the two-phase method, a procedure for finding an initial basic feasible solution to a linear program, and the convergence properties of the simplex algorithm. In Chapter 5, we present a method called the revised simplex method, which can be used to reduce the computational burden of the simplex method, especially in large problems.

EXERCISES

4.1. Consider the following linear programming problem.

$$\text{maximize } z = 4x_1 + 3x_2$$

subject to

$$-x_1 + x_2 \leq 6$$
$$2x_1 + x_2 \leq 20$$
$$x_1 + x_2 \leq 12$$
$$x_1, x_2 \geq 0$$

(a) Solve this problem graphically.
(b) Solve this problem by the algebraic simplex method described in Example 4.1, and at each iteration, identify the corresponding extreme-point solution on the graph.
(c) Identify the basic variables, the nonbasic variables, and the basis matrix $\mathbf{B}$ at each iteration.

4.2. Consider the following constraint set.

$$x_2 \leq 9$$

$$x_1 - x_2 \leq 5$$

$$2x_1 + x_2 \leq 22$$

$$x_1, x_2 \geq 0$$

(a) Sketch the feasible region.

(b) Identify the basic variables, the nonbasic variables, and the basis matrix associated with each extreme point of the feasible region.

(c) Suppose that x_2 and the slack variable in the second constraint are the nonbasic variables defining the current extreme-point solution. If x_2 is chosen as the entering variable, what would be the departing variable and which extreme point would correspond to the next basic feasible solution?

4.3. Consider the following linear programming problem.

$$\text{maximize } z = x_1 + 2x_2$$

subject to

$$x_1 + x_2 \leq 16$$

$$-x_1 + x_2 \leq 5$$

$$x_1 \leq 12$$

$$-x_1 + 3x_2 \leq 16$$

$$x_1, x_2 \geq 0$$

(a) Solve this problem graphically.

(b) Solve this problem by the algebraic simplex method described in Example 4.1.

(c) Solve this problem by the simplex algorithm using the simplex tableau.

4.4. Solve the following linear programming problem by the simplex algorithm.

$$\text{minimize } z = 3x_1 - 4x_2 - x_3 - 2x_4 - 3x_5$$

subject to

$$x_1 + x_2 + x_3 - x_4 + 2x_5 \leq 12$$

$$x_1 - 2x_2 - x_3 - x_4 - x_5 \geq -30$$

$$x_1, x_2, x_3, x_4, x_5 \geq 0$$

4.5. Solve the following problem by the simplex algorithm, identifying the basis matrix **B** and the basis inverse **B**$^{-1}$ at each iteration. Is the optimal solution unique? Explain.

$$\text{maximize } z = x_1 + 2x_2 + 5x_3 + x_4$$

subject to

$$x_1 + 2x_2 + x_3 - x_4 \leq 20$$

$$-x_1 + x_2 + x_3 + x_4 \leq 12$$

$$2x_1 + x_2 + x_3 - x_4 \leq 30$$

$$x_1, x_2, x_3, x_4 \geq 0$$

4.6. Consider the following problem.

$$\text{maximize } z = 2x_1 + 2x_2$$

subject to

$$x_2 \leq 10$$

$$x_1 - 3x_2 \leq 2$$

$$x_1 + x_2 \leq 16$$

$$x_1, x_2 \geq 0$$

(a) Verify graphically that this problem has alternative optimal solutions.
(b) Use the simplex algorithm to find all alternative optimal solutions.

4.7. Three products, A, B, and C, are made using two manufacturing processes. The unit production times in hours are given in Table 4.33. The time available for Process 1 is 36 hours, and for Process 2 is 40 hours. Products A, B, and C sell for $9, $6, and $8, respectively. In addition, it is estimated that no more than 6 units of Product C can be sold. Formulate a linear programming problem for determining the optimal product mix, and solve by the simplex method.

TABLE 4.33

	Unit production times (hours)	
Product	Process 1	Process 2
A	2	2
B	1	3
C	2	1

4.8. Consider the basic solution defined by the tableau of Table 4.34. Suppose that the objective function of this problem is given by maximize $z = c_1 x_1 + c_2 x_2$. Give mathematical condition(s) in terms of c_1 and c_2 such that the given basic feasible solution is the unique optimal solution.

TABLE 4.34

	x_1	x_2	x_3	x_4	x_5	RHS
x_3	0	0	1	$-\frac{1}{2}$	$\frac{5}{2}$	20
x_1	1	0	0	$\frac{1}{2}$	$-\frac{1}{2}$	4
x_2	0	1	0	0	1	10

4.9. Consider the following linear programming problem.

$$\text{minimize } z = -5x_1 + 2x_2$$

subject to

$$x_1 \leq 12$$
$$-x_1 + 2x_2 \geq -8$$
$$x_2 \leq 8$$
$$2x_1 + 3x_2 \leq 36$$
$$x_1, x_2 \geq 0$$

(a) Solve this problem graphically.
(b) Solve this problem by the simplex algorithm.

4.10. Consider the optimal tableau shown in Table 4.35. Characterize mathematically the set of all optimal solutions to this problem.

TABLE 4.35

	z	x_1	x_2	x_3	x_4	x_5	x_6	RHS
z	1	0	0	3	0	0	4	8
x_1	0	1	0	0	0	$-\frac{1}{2}$	$\frac{1}{2}$	4
x_4	0	0	0	3	1	1	0	4
x_2	0	0	1	-1	0	$-\frac{1}{2}$	$-\frac{1}{2}$	2

4.11. Consider the following problem.

$$\text{maximize } z = 4x_1 + x_2$$

subject to

$$2x_1 - 3x_2 \leq 12$$
$$-4x_1 + x_2 \leq 8$$
$$x_1, x_2 \geq 0$$

(a) Verify graphically that this problem does not have a finite optimal solution.
(b) Use the simplex method to show that this problem does not have a finite optimal solution.
(c) Use your final tableau to construct an extreme direction of the feasible region, $\mathbf{Ax} = \mathbf{b}, \mathbf{x} \geq \mathbf{0}$.

4.12. Once a variable departs from the basis (during the simplex procedure), can it ever return (i.e., become, once again, basic)? Explain.

4.13. Consider the basic solution defined by the tableau of Table 4.36. Suppose that the objective function of this problem is given by maximize $z = c_1x_1 + c_2x_2$. Give a

TABLE 4.36

	x_1	x_2	x_3	x_4	x_5	RHS
x_3	0	0	1	$-\frac{1}{3}$	$\frac{2}{3}$	10
x_1	1	0	0	$-\frac{2}{3}$	$\frac{1}{3}$	6
x_2	0	1	0	$-\frac{1}{3}$	$-\frac{1}{3}$	2

mathematical condition(s) in terms of c_1 and c_2 such that the given basic feasible solution indicates that no finite optimal solution exists.

4.14. Consider the following problem.

$$\text{maximize } z = 2x_1 + x_2$$

subject to

$$-x_1 + 2x_2 \leq 10$$
$$x_1 \leq 6$$
$$x_1 + x_2 \leq 14$$
$$x_1 - x_2 \leq 6$$
$$x_1, x_2 \geq 0$$

(a) Sketch the feasible region and identify any degenerate extreme points.
(b) For each extreme point, specify all possible basis matrices.
(c) Solve the problem by the simplex algorithm.

4.15. Consider the following problem.

$$\text{maximize } z = 7x_1 + 3x_2 + 2x_3$$

subject to

$$4x_1 + x_2 + x_3 \leq 18$$
$$3x_1 + 2x_2 + x_3 \leq 14$$
$$x_1, x_2, x_3 \geq 0$$

Let the slack variables for the respective constraints be denoted by x_4 and x_5. The tableau of Table 4.37 represents the current basic solution.

TABLE 4.37

	z	x_1	x_2	x_3	x_4	x_5	RHS
z	1				1	1	
x_1	0				1	-1	
x_3	0				-3	4	

(a) Identify the basis inverse corresponding to the given tableau.

(b) Determine the values of the missing entries in the tableau.

(c) Is the tableau optimal? If so, is the optimal solution unique?

4.16. Consider the following problem.

$$\text{maximize } z = 2x_1 - x_4 - 5x_5 + 2x_7$$

subject to

$$x_1 + x_3 - 2x_5 - x_7 = 12$$
$$-x_1 + x_2 + x_4 + 3x_5 - x_7 = 6$$
$$2x_1 - 2x_4 + 6x_5 + x_6 + 4x_7 = 18$$
$$x_j \geq 0, \qquad \text{for all } j$$

The tableau of Table 4.38 represents the current basic solution.

TABLE 4.38

z	x_1	x_2	x_3	x_4	x_5	x_6	x_7	RHS
1	0	1	1	0	a	1	0	e
0	0	3	1	1	b	1	0	f
0	0	1	0	0	c	$\frac{1}{2}$	1	g
0	1	1	1	0	d	$\frac{1}{2}$	0	h

(a) Identify the basic variables and the basis inverse corresponding to the given tableau.

(b) Determine the values of the unknowns in the tableau.

(c) What is the rate of change of z with respect to x_2 (i.e., $\partial z / \partial x_2$)?

(d) What is the rate of change of x_7 with respect to x_6 (i.e., $\partial x_7 / \partial x_6$)?

4.17. Consider the tableau of Table 4.39 corresponding to a maximization problem. Note that the tableau indicates that there is no finite optimal solution. Determine a direction $\mathbf{d} = (d_1, d_2, d_3, d_4, d_5, d_6)^t$ such that $\mathbf{cd} > 0$.

TABLE 4.39

	z	x_1	x_2	x_3	x_4	x_5	x_6	RHS
z	1	0	0	14	0	$\frac{9}{2}$	$-\frac{5}{2}$	26
x_4	0	0	0	15	1	$\frac{7}{2}$	$-\frac{3}{2}$	26
x_1	0	1	0	1	0	$\frac{1}{2}$	$-\frac{1}{2}$	2
x_2	0	0	1	7	0	2	-1	12

4.18. Consider the following problem.

$$\text{maximize } z = 2x_1 + x_2$$

subject to

$$x_1 + x_2 \geq 6$$
$$x_1 + 3x_2 \leq 24$$
$$x_2 \geq 2$$
$$x_1 - x_2 \leq 8$$
$$x_1, x_2 \geq 0$$

(a) Sketch the feasible region and find the optimal solution graphically.
(b) Solve the problem by the two-phase method and track the sequence of solutions on the graph. Note that each solution corresponds to a basic (but not necessarily feasible) solution of the original problem.

4.19. Consider the following problem.

$$\text{maximize } z = 2x_1 + x_2$$

subject to

$$x_1 - x_2 \geq 8$$
$$2x_1 + 3x_2 \leq 24$$
$$2x_1 + x_2 \leq 12$$
$$x_1, x_2 \geq 0$$

(a) Verify graphically that this problem is infeasible.
(b) Use phase 1 of the two-phase method to show that the problem is infeasible.

4.20. Solve the following problem by the two-phase method.

$$\text{minimize } z = 2x_1 - 5x_2 + x_3$$

subject to

$$-x_1 + x_2 + x_3 \geq 5$$
$$-x_1 - x_2 + x_3 = -1$$
$$5x_1 + 3x_2 - x_3 \leq 9$$
$$x_1, x_2, x_3 \geq 0$$

4.21. Consider the following problem.

$$\text{minimize } z = 7x_1 + 3x_2 + 4x_3$$

subject to

$$x_1 + x_2 \geq 10$$
$$2x_1 + 2x_2 + x_3 \leq 40$$

$$-2x_1 + x_2 - x_3 = 22$$

$$x_1, x_2, x_3 \geq 0$$

(a) Solve this problem by the two-phase method.

(b) Solve this problem by the big-M method.

4.22. The tableau of Table 4.40 represents a basic feasible solution of a linear programming problem in which the objective is maximize $z = 10x_1 - 12x_2 + 8x_3$ and x_4, x_5, and x_6 are the slack variables for the respective constraints.

TABLE 4.40

	z	x_1	x_2	x_3	x_4	x_5	x_6	RHS
z	1	0	a	d	2	6	0	88
x_1	0	1	b	1	e	3	0	10
x_2	0	0	c	-2	-1	f	0	g
x_6	0	0	0	3	-1	1	1	8

(a) Determine the values of the unknowns in the tableau.

(b) Identify $\mathbf{B}^{-1}$.

(c) What is the rate of change of z with respect to x_5 (i.e., $\partial z/\partial x_5$)?

(d) What is the rate of change of x_2 with respect to x_3 (i.e., $\partial x_2/\partial x_3$)?

(e) Without explicitly finding vectors $\mathbf{a}_1$, $\mathbf{a}_2$, and $\mathbf{a}_6$, write vector $\mathbf{a}_3$ as a linear combination of $\mathbf{a}_1$, $\mathbf{a}_2$, and $\mathbf{a}_6$.

4.23. Consider the following linear programming problem.

$$\text{maximize } z = 5x_1 + x_2$$

$$\text{subject to}$$

$$x_1 - x_2 \leq 9$$

$$-x_1 - x_2 \leq 2$$

$$2x_1 + x_2 \leq 12$$

$$x_1 \geq 0$$

$$x_2 \text{ unrestricted}$$

(a) Solve this problem graphically.

(b) Transform the problem so that all variables are nonnegative and solve the resulting problem by the simplex algorithm. Derive the solution to the original problem.

4.24. Consider the following problem.

$$\text{maximize } z = x_1 - 2x_2 + 3x_3$$

subject to

$$x_1 - x_2 + x_3 \le 8$$
$$x_1 - x_2 - x_3 \le 12$$
$$x_1 \ge -2$$
$$x_2 \le 0$$

$$x_3 \text{ unrestricted}$$

(a) Transform this problem so that all variables are nonnegative.
(b) Solve the transformed problem by the simplex algorithm.
(c) Use the solution found in part (b) to derive the optimal solution of the original problem.

4.25. Consider the following problem.

$$\text{maximize } z = 2x_1 - 4x_2 + 7x_3$$

subject to

$$x_1 - x_2 + x_3 = 2$$
$$x_1 + x_2 + 2x_3 \le 10$$
$$x_1 + x_2 - x_3 \ge 6$$
$$x_1, x_2, x_3 \ge 0$$

(a) Solve this problem by the two-phase method.
(b) Solve this problem by the Big-M method.

4.26. Solve the following problem by the simplex method.

$$\text{maximize } z = x_1 + 5x_2 + x_3 + 2x_4$$

subject to

$$2x_2 + x_3 + 3x_4 \ge 100$$
$$3x_1 + x_2 + 2x_3 + x_4 \le 900$$
$$-x_1 - x_2 + x_3 - 4x_4 = 300$$
$$x_1, x_2, x_3, x_4 \ge 0$$

4.27. Consider the following linear programming problem and the associated optimal tableau shown in Table 4.41.

$$\text{maximize } z = c_1 x_1 + c_2 x_2 + c_3 x_3$$

subject to

$$a_{1,1} x_1 + a_{1,2} x_2 + a_{1,3} x_3 \le b_1$$

$$a_{2,1}x_1 + a_{2,2}x_2 + a_{2,3}x_3 \le b_2$$

$$x_1, x_2, x_3 \ge 0$$

TABLE 4.41

	z	x_1	x_2	x_3	x_4	x_5	RHS
z	1	0	2	0	1	$\frac{1}{2}$	11
x_3	0	0	−5	1	−2	$\frac{3}{2}$	3
x_1	0	1	3	0	1	$-\frac{1}{2}$	1

Find the values of the $a_{i,j}$, b_i, and c_j.

4.28. Consider the following system of linear equations in the form $\mathbf{Ax} = \mathbf{b}$, where $\mathbf{A}$ is $m \times m$.

$$2x_1 - 2x_2 + x_3 = 2$$

$$-2x_1 + x_2 - 2x_3 = 3$$

$$x_1 + x_2 + x_3 = 6$$

(a) Use phase 1 of the two-phase method to find the unique optimal solution to this system.

(b) Specify the inverse of the coefficient matrix, $\mathbf{A}^{-1}$.

4.29. Solve the following problem by the simplex algorithm using the lexicographic minimum ratio test for cycling prevention.

$$\text{maximize } z = x_1 + 3x_2 - 2x_3$$

$$\text{subject to}$$

$$-2x_1 + x_2 + 3x_3 \le 6$$

$$x_1 + x_2 + 4x_3 \le 10$$

$$x_2 + 2x_3 \le 6$$

$$-2x_1 + 2x_2 + 3x_3 \le 12$$

$$x_1, x_2, x_3, x_4 \ge 0$$

4.30. A company must distribute its product from two warehouse locations to two retail outlets. Warehouse A has a total of 48 units, and Warehouse B has a total of 60 units. Forecasting estimates a demand of at most 36 units for Retail Outlet 1 and 72 units for Retail Outlet 2. The unit shipping costs between each warehouse and retail outlet are given in Table 4.42. The problem is to determine the minimum-cost shipping schedule. Formulate a linear programming model and solve by the simplex method.

TABLE 4.42

Warehouse	Retail Outlet 1	Retail Outlet 2
A	$6	$8
B	$4	$3

4.31. Consider the following system of linear equations.

$$2x_1 + 3x_2 = 6$$

$$2x_1 + 3x_2 = 8$$

$$x_1 + x_2 = 4$$

Clearly, this system of equations has no solution. Given a solution x_1, x_2, define the absolute error in the first equation by $|2x_1 + 3x_2 - 6|$. Similarly, define the error in the second and third equations by $|2x_1 + 3x_2 - 8|$ and $|x_1 + x_2 - 4|$, respectively.

(a) Formulate a linear programming problem for finding x_1 and x_2 that minimizes the sum of the absolute errors.

(b) Formulate a linear programming problem for finding x_1 and x_2 that minimizes the maximum of the absolute errors.

4.32. Consider the polyhedral set defined by the following.

$$-2x_1 + x_2 \le 2$$

$$x_1 + 2x_2 \ge 8$$

$$x_1 - x_2 \le 10$$

$$x_1, x_2 \ge 0$$

(a) Sketch this polyhedral set and graphically determine the extreme directions of this set.

(b) Now consider the following linear programming problem. Solve it by the simplex algorithm and determine all extreme-point optimal solutions. Compare these alternative optimal solutions with the extreme directions found in part (a).

$$\text{maximize } d_1 + d_2$$

subject to

$$-2d_1 + d_2 \le 0$$

$$d_1 + 2d_2 \ge 0$$

$$d_1 - d_2 \le 0$$

$$d_1 + d_2 \le 1$$

$$d_1, d_2 \ge 0$$

(c) Did the method described in part (b) determine the extreme directions of the given polyhedral set? Explain.

4.33. Consider the following linear programming problem. Note that besides the nonnegativity restrictions, there is a only *one* constraint.

$$\text{maximize } z = \sum_{j=1}^{n} c_j x_j$$

subject to

$$\sum_{j=1}^{n} a_j x_j = b$$

$$x_j \geq 0, \qquad \text{for } j = 1, \ldots, n$$

(a) Develop an efficient procedure for checking the feasibility of this problem.
(b) Develop an efficient procedure for checking if the objective is unbounded.
(c) Develop an efficient procedure for determining an optimal solution directly.
(d) Use the method developed in part (c) to solve the following problem.

$$\text{maximize } z = 3x_1 + 10x_2 + 5x_3 + 11x_4 + 7x_5 + 14x_6$$

subject to

$$x_1 + 7x_2 + 3x_3 + 4x_4 + 2x_5 + 5x_6 = 42$$

$$x_1, x_2, x_3, x_4, x_5, x_6 \geq 0$$

4.34. Let $S = \{\mathbf{x} \in E^n : \mathbf{Ax} \leq \mathbf{b}\}$, $T = \{\mathbf{x} \in E^n : \mathbf{Dx} \leq \mathbf{d}\}$. Assuming that S and T are nonempty, develop an efficient procedure for checking whether $S \subset T$.

CHAPTER $\underset{\overline{\overline{}}}{}$ **5**

*SPECIAL SIMPLEX IMPLEMENTATIONS

CHAPTER OVERVIEW

In Chapter 4, we presented the simplex algorithm in considerable detail. The algorithm was addressed from both an algebraic viewpoint as well as in tableau format. However, not much attention was given to computational efficiency. In this chapter, we present an implementation of the simplex algorithm called the revised simplex method. Although the basic elements of the algorithm are precisely the same as those presented in Chapter 4, the algorithm is implemented in such a way that the number of arithmetical operations is kept to a minimum. For very small problems such as those presented as examples in this and other textbooks, the computational savings are not all that significant. In fact, the reader may initially think that the computational burden has increased. However, for large problems, particularly where the number of variables (n) is much greater than the number of constraints (m), the computational savings are indeed significant.

Following the presentation of the revised simplex method, an alternative method for storing the basis inverse is discussed. Recall that successive basis matrices differ by one column. The product form of the inverse allows us to determine the new inverse from the old inverse in an efficient manner. The chapter then concludes with a discussion of the bounded-variables simplex method. This variation of the simplex algorithm is designed to handle simple upper and

---

* Starred sections may be omitted in an introductory course with no appreciable loss of continuity.

lower bounds on the variables in an implicit manner. This is much more efficient than dealing with these bounds as explicit constraints.

THE REVISED SIMPLEX METHOD

Let us again consider the standard linear programming problem

$$(\text{LP}) \qquad \text{maximize } z = \mathbf{cx}$$

$$\text{subject to}$$

$$\mathbf{Ax} = \mathbf{b}$$

$$\mathbf{x} \geq \mathbf{0}$$

Recall from Chapter 4 that the standard simplex tableau can be written in the form shown in Table 5.1. In examining the steps of the simplex method, one sees that in checking for optimality and selecting the entering variable, the essential information consists of the reduced costs of the nonbasic variables. In performing the minimum ratio test and the resulting pivot operation, the necessary information consists of the column $\boldsymbol{\alpha}_k = \mathbf{B}^{-1}\mathbf{a}_k$ associated with the entering variable and the entries in the right-hand side, $\boldsymbol{\beta} = \mathbf{B}^{-1}\mathbf{b}$. All other nonbasic variable information is essentially carried along as excess baggage. That is, by using the standard simplex procedure, each nonbasic variable column is updated at each pivot operation; this is even true of variables that *never* enter the basis throughout the course of the algorithm. This process is very inefficient, especially for computer implementations because it not only results in unnecessary computations, but increases the storage requirements dramatically. Note that any column in the body of the tableau (see Table 5.1) can be readily computed from the original data provided that $\mathbf{B}^{-1}$ is available. Similarly, the top row of the tableau can be generated from the original data if $\mathbf{c}_B\mathbf{B}^{-1}$ is known. The basic idea behind the revised simplex algorithm is to reduce the storage requirements by storing only the information that is essential to perform the steps of the simplex algorithm. This results in not only a reduction in storage requirements, but also results in fewer computations, especially in problems where n is much greater than m.

TABLE 5.1   THE SIMPLEX TABLEAU

		z	x	RHS
z	1	$\mathbf{c}_B\mathbf{B}^{-1}\mathbf{A} - \mathbf{c}$	$\mathbf{c}_B\mathbf{B}^{-1}\mathbf{b}$
$\mathbf{x}_B$	0	$\mathbf{B}^{-1}\mathbf{A}$	$\mathbf{B}^{-1}b$

The information that is stored by the revised simple procedure is depicted in Table 5.2. As we will see in the following section, it is not always the case that $\mathbf{B}^{-1}$ is stored explicitly as in Table 5.2; it can also be stored in *product* or *factored* form. However, for our initial discussion of the revised simplex, we will assume that $\mathbf{B}^{-1}$ is stored explicitly.

TABLE 5.2   THE REVISED SIMPLEX TABLEAU

		RHS
z	$\mathbf{c}_B\mathbf{B}^{-1}$	$\mathbf{c}_B\mathbf{B}^{-1}\mathbf{b}$
$\mathbf{x}_B$	$\mathbf{B}^{-1}$	$\mathbf{B}^{-1}b$

Now, letting $\boldsymbol{\pi} = \mathbf{c}_B\mathbf{B}^{-1}$ and $\boldsymbol{\beta} = \mathbf{B}^{-1}\mathbf{b}$, the revised simplex tableau can be written as shown in Table 5.3. The steps of the revised simplex algorithm can be summarized as follows:

TABLE 5.3   THE REVISED SIMPLEX TABLEAU

		RHS
z	$\boldsymbol{\pi}$	$\mathbf{c}_B\boldsymbol{\beta}$
$\mathbf{x}_B$	$\mathbf{B}^{-1}$	$\boldsymbol{\beta}$

The revised simplex algorithm (maximization problem)

STEP 1. *Check for optimality.* Compute $z_j - c_j = \boldsymbol{\pi}\mathbf{a}_j - c_j$ for each nonbasic variable x_j. If these are all nonnegative then the current basic feasible solution is optimal; stop. Otherwise, select as the entering variable the nonbasic variable with the most negative $z_j - c_j$. Designate this variable as x_k. Ties in the selection of x_k may be broken arbitrarily. Go to Step 2.

STEP 2. *Compute the updated x_k column and check for unboundedness.* Compute $\boldsymbol{\alpha}_k = \mathbf{B}^{-1}\mathbf{a}_k$. If $\boldsymbol{\alpha}_k \leq \mathbf{0}$, then the problem has an unbounded objective value. Otherwise, finite improvement in the objective is possible and we go to Step 3.

STEP 3. *Determine the departing variable.* Append the updated x_k column,

x_k
$z_k - c_k$
$\boldsymbol{\alpha}_k$

to the revised simplex tableau. In the usual manner, determine the departing variable using the minimum ratio test. That is, let

$$\frac{\beta_r}{\alpha_{r,k}} = \text{minimum} \left\{ \frac{\beta_i}{\alpha_{i,k}} : \alpha_{i,k} > 0 \right\}$$

Row r is the pivot row, $\alpha_{r,k}$ is the pivot element, and the basic variable $x_{B,r}$ associated with row r is the departing variable. Go to Step 4.

STEP 4. *Pivot and establish a new tableau.*
(a) The entering variable x_k is the new basic variable in row r.
(b) Pivot on $\alpha_{r,k}$ in the usual manner. This will update $\mathbf{B}^{-1}$, $\boldsymbol{\pi}$, $\boldsymbol{\beta}$, and $\mathbf{c}_B\boldsymbol{\beta}$.
(c) Return to Step 1.

The steps of the algorithm are now illustrated via the following example.

Example 5.1: The Revised Simplex Method

$$\text{maximize } z = 2x_1 + 3x_2 - x_3 + 4x_4 + x_5 - 3x_6 \tag{5.1}$$

subject to

$$x_1 - 2x_2 + x_4 + 4x_5 + (\tfrac{1}{2})x_6 \le 10 \tag{5.2}$$

$$x_1 + x_2 + 3x_3 + 2x_4 + x_5 - x_6 \le 16 \tag{5.3}$$

$$2x_2 + (\tfrac{1}{2})x_2 - x_3 - x_4 + 2x_5 + 5x_6 \le 8 \tag{5.4}$$

$$x_1, x_2, x_3, x_4, x_5, x_6 \ge 0 \tag{5.5}$$

Now, by introducing the slack variables, x_7, x_8, and x_9, the data for the problem can be summarized as follows:

$$\mathbf{A} = \begin{pmatrix} 1 & -2 & 0 & 1 & 4 & \tfrac{1}{2} & 1 & 0 & 0 \\ 1 & 1 & 3 & 2 & 1 & -1 & 0 & 1 & 0 \\ 2 & \tfrac{1}{2} & -1 & -1 & 2 & 5 & 0 & 0 & 1 \end{pmatrix}$$

$$\mathbf{b} = \begin{pmatrix} 10 \\ 16 \\ 8 \end{pmatrix}$$

$$\mathbf{c} = \begin{pmatrix} 2 & 3 & -1 & 4 & 1 & -3 & 0 & 0 & 0 \end{pmatrix}$$

Clearly, x_7, x_8, and x_9 form a convenient starting basic, which results in the following:

$$\mathbf{B} = (\mathbf{a}_7, \mathbf{a}_8, \mathbf{a}_9) = \mathbf{I}$$

$$\mathbf{c}_B = (c_7, c_8, c_9) = (0 \quad 0 \quad 0)$$

$$\boldsymbol{\pi} = \mathbf{c}_B \mathbf{B}^{-1} = (0 \quad 0 \quad 0)$$

$$\boldsymbol{\beta} = \mathbf{B}^{-1}\mathbf{b} = \mathbf{b}$$

$$z = 0$$

$$\mathbf{x}_B = \begin{pmatrix} x_{B,1} \\ x_{B,2} \\ x_{B,3} \end{pmatrix} = \begin{pmatrix} x_7 \\ x_8 \\ x_9 \end{pmatrix} = \begin{pmatrix} 10 \\ 16 \\ 8 \end{pmatrix}$$

Thus, the initial revised simplex tableau can be written as in Table 5.4.

TABLE 5.4

				RHS
z	0	0	0	0
x_7	1	0	0	10
x_8	0	1	0	16
x_9	0	0	1	8

STEP 1. Compute $z_j - c_j$ for the nonbasic variables.

$$z_1 - c_1 = \boldsymbol{\pi}\mathbf{a}_1 - c_1 = (0 \quad 0 \quad 0) \begin{pmatrix} 1 \\ 1 \\ 2 \end{pmatrix} - 2 = -2$$

$$z_2 - c_2 = \boldsymbol{\pi}\mathbf{a}_2 - c_2 = (0 \quad 0 \quad 0) \begin{pmatrix} -2 \\ 1 \\ \frac{1}{2} \end{pmatrix} - 3 = -3$$

$$z_3 - c_3 = \boldsymbol{\pi}\mathbf{a}_3 - c_3 = (0 \quad 0 \quad 0) \begin{pmatrix} 0 \\ 3 \\ -1 \end{pmatrix} - (-1) = 1$$

$$z_4 - c_4 = \boldsymbol{\pi}\mathbf{a}_4 - c_4 = (0 \quad 0 \quad 0) \begin{pmatrix} 1 \\ 2 \\ 0 \end{pmatrix} - 4 = -4$$

$$z_5 - c_5 = \boldsymbol{\pi}\mathbf{a}_5 - c_5 = (0 \quad 0 \quad 0) \begin{pmatrix} 4 \\ 1 \\ 2 \end{pmatrix} - 1 = -1$$

$$z_6 - c_6 = \boldsymbol{\pi}\mathbf{a}_6 - c_6 = (0 \quad 0 \quad 0) \begin{pmatrix} \frac{1}{2} \\ -1 \\ 5 \end{pmatrix} - (-3) = 3$$

Choose x_4 as the entering variable because x_4 has the most negative $z_j - c_j$.

STEP 2. Compute the updated x_4 column.

$$\boldsymbol{\alpha}_4 = \mathbf{B}^{-1}\mathbf{a}_4 = \begin{pmatrix} 1 & 0 & 0 \\ 0 & 1 & 0 \\ 0 & 0 & 1 \end{pmatrix} \begin{pmatrix} 1 \\ 2 \\ -1 \end{pmatrix} = \begin{pmatrix} 1 \\ 2 \\ -1 \end{pmatrix}$$

$\boldsymbol{\alpha}_4$ does not give an unboundedness indication, therefore continue with Step 3.

STEP 3. Append the updated x_4 column to the tableau and perform the minimum ratio test. From Table 5.5, we see that

$$\frac{\beta_2}{\alpha_{2,2}} = \text{minimum} \left\{ \frac{10}{1}, \frac{16}{2} \right\} = 8$$

Thus, $r = 2$ and the departing variable is in row 2. That is, $x_{B,2} = x_8$ is the departing variable.

TABLE 5.5

				RHS		x_4
z	0	0	0	0		-4
x_7	1	0	0	10		1
x_8	0	1	0	16		2
x_9	0	0	1	8		-1

STEP **4.** (a) x_4 is the new basic variable in row $r = 2$.

(b) Pivoting on $\alpha_{2,4} = 2$ results in Table 5.6.

TABLE 5.6

				RHS
z	0	2	0	32
x_7	1	$-\frac{1}{2}$	0	2
x_4	0	$\frac{1}{2}$	0	8
x_9	0	$\frac{1}{2}$	1	8

(c) Return to Step 1.

STEP **1.** Compute $z_j - c_j$ for the nonbasic variables.

$$z_1 - c_1 = \pi \mathbf{a}_1 - c_1 = (0 \quad 2 \quad 0) \begin{pmatrix} 1 \\ 1 \\ 2 \end{pmatrix} - 2 = 0$$

$$z_2 - c_2 = \pi \mathbf{a}_2 - c_2 = (0 \quad 2 \quad 0) \begin{pmatrix} -2 \\ 1 \\ \frac{1}{2} \end{pmatrix} - 3 = -1$$

$$z_3 - c_3 = \pi \mathbf{a}_3 - c_3 = (0 \quad 2 \quad 0) \begin{pmatrix} 0 \\ 3 \\ -1 \end{pmatrix} - (-1) = 7$$

$$z_5 - c_5 = \pi \mathbf{a}_5 - c_5 = (0 \quad 2 \quad 0) \begin{pmatrix} 4 \\ 1 \\ 2 \end{pmatrix} - 1 = 1$$

$$z_6 - c_6 = \pi \mathbf{a}_6 - c_6 = (0 \quad 2 \quad 0) \begin{pmatrix} \frac{1}{2} \\ -1 \\ 5 \end{pmatrix} - (-3) = 1$$

$$z_8 - c_8 = \pi a_8 - c_8 = (0 \quad 2 \quad 0) \begin{pmatrix} 0 \\ 1 \\ 0 \end{pmatrix} - 0 = 2$$

x_2 as the entering variable.

STEP 2. Compute the updated x_2 column.

$$\alpha_2 = B^{-1} a_2 = \begin{pmatrix} 1 & -\frac{1}{2} & 0 \\ 0 & \frac{1}{2} & 0 \\ 0 & \frac{1}{2} & 1 \end{pmatrix} \begin{pmatrix} -2 \\ 1 \\ \frac{1}{2} \end{pmatrix} = \begin{pmatrix} -\frac{5}{2} \\ \frac{1}{2} \\ 1 \end{pmatrix}$$

α_2 does not give an unboundedness indication; therefore go to Step 3.

STEP 3. From Table 5.7, the minimum ratio results in

$$\frac{\beta_3}{\alpha_{3,2}} = \text{minimum} \left\{ \frac{8}{\frac{1}{2}}, \frac{8}{1} \right\} = 8$$

The departing variable is $x_{B,3} = x_9$.

TABLE 5.7

				RHS		x_2
z	0	2	0	32		-1
x_7	1	$-\frac{1}{2}$	0	2		$-\frac{5}{2}$
x_4	0	$\frac{1}{2}$	0	8		$\frac{1}{2}$
x_9	0	$\frac{1}{2}$	1	8		1

STEP 4. (a) x_2 is the new basic variable in row $r = 3$.
      (b) Pivoting on $\alpha_{3,2} = 1$ results in Table 5.8.

TABLE 5.8

				RHS
z	0	$\frac{5}{2}$	1	40
x_7	1	$-\frac{1}{2}$	$\frac{5}{2}$	22
x_4	0	$\frac{1}{2}$	$-\frac{1}{2}$	12
x_2	0	$\frac{1}{2}$	1	8

(c) Return to Step 1.

STEP 1. Compute $z_j - c_j$ for the nonbasic variables.

$$z_1 - c_1 = \pi a_1 - c_1 = (0 \quad \frac{5}{2} \quad 1) \begin{pmatrix} 1 \\ 1 \\ 2 \end{pmatrix} - 2 = \frac{5}{2}$$

$$z_3 - c_3 = \boldsymbol{\pi}\mathbf{a}_3 - c_3 = (0 \quad \tfrac{5}{2} \quad 1) \begin{pmatrix} 0 \\ 3 \\ -1 \end{pmatrix} - (-1) = \tfrac{15}{2}$$

$$z_5 - c_5 = \boldsymbol{\pi}\mathbf{a}_5 - c_5 = (0 \quad \tfrac{5}{2} \quad 1) \begin{pmatrix} 4 \\ 1 \\ 2 \end{pmatrix} - 1 = \tfrac{7}{2}$$

$$z_6 - c_6 = \boldsymbol{\pi}\mathbf{a}_6 - c_6 = (0 \quad \tfrac{5}{2} \quad 1) \begin{pmatrix} \tfrac{1}{2} \\ -1 \\ 5 \end{pmatrix} - (-3) = \tfrac{11}{2}$$

$$z_8 - c_8 = \boldsymbol{\pi}\mathbf{a}_8 - c_8 = (0 \quad \tfrac{5}{2} \quad 1) \begin{pmatrix} 0 \\ 1 \\ 0 \end{pmatrix} - 0 = \tfrac{5}{2}$$

$$z_9 - c_9 = \boldsymbol{\pi}\mathbf{a}_9 - c_9 = (0 \quad \tfrac{5}{2} \quad 1) \begin{pmatrix} 0 \\ 0 \\ 1 \end{pmatrix} - 0 = 1$$

Clearly, the current basic feasible solution is optimal because $z_j - c_j \geq 0$, for all j. Thus, the optimal solution is as follows:

$$z^* = 40$$
$$x_1^* = 0$$
$$x_2^* = 8$$
$$x_3^* = 0$$
$$x_4^* = 12$$
$$x_5^* = 0$$
$$x_6^* = 0$$
$$x_7^* = 22$$
$$x_8^* = 0$$
$$x_9^* = 0$$

Note that it was never necessary to update the columns associated with the nonbasic variables, x_1, x_3, x_5, and x_6. That is, these variables played no role in the solution process. However, if the standard simplex procedure had been used, these columns would have been updated and stored at every iteration, wasting both computational time as well as computer storage.

Suboptimization

In large-scale problems, where there are potentially thousands of variables, instead of computing $z_j - c_j$ for all nonbasic variables at each iteration, a technique referred to as *suboptimization* is often used. The basic idea can be described as follows. At the first iteration, all $z_j - c_j$ are computed and a set S is formed of the (10 to 30) variables with the most negative $z_j - c_j$. On subsequent iterations, *only* the $z_j - c_j$ of the variables in S are computed, and the entering variable is chosen from this set. This process continues until $z_j - c_j \geq 0$, for all $x_j \in S$. When this occurs, the problem has been optimized over this subset of variables. *All $z_j - c_j$* are then computed, a new set S is formed as before, and the process is repeated. Of course, when all $z_j - c_j \geq 0$, then a new set S cannot be formed, and the algorithm stops as usual with the optimal solution. This technique or a variation thereof is generally used in all large-scale applications of the simplex algorithm.

THE PRODUCT FORM OF THE INVERSE

In the preceding discussion of the revised simplex method, we assumed that the basis inverse, $\mathbf{B}^{-1}$, was updated at each iteration and stored explicitly. However, this is not always the case, and we now discuss an alternative implementation of the revised simplex procedure that uses the *product form* of the inverse.

Suppose that the current basis matrix is given by

$$\mathbf{B} = (\mathbf{b}_1, \mathbf{b}_2, \ldots, \mathbf{b}_{r-1}, \mathbf{b}_r, \mathbf{b}_{r+1}, \ldots, \mathbf{b}_m) \tag{5.6}$$

In addition, suppose that the nonbasic variable x_k is the entering variable, whereas $x_{B,r}$ is the leaving variable. Then the new basis matrix following this exchange will be given by

$$\mathbf{B}_{\text{new}} = (\mathbf{b}_1, \mathbf{b}_2, \ldots, \mathbf{b}_{r-1}, \mathbf{a}_k, \mathbf{b}_{r+1}, \ldots, \mathbf{b}_m) \tag{5.7}$$

Let e_j represent the jth column of the identity, that is,

$$\mathbf{I} = (\mathbf{e}_1, \mathbf{e}_2, \ldots, \mathbf{e}_m) \tag{5.8}$$

Then, clearly, $\mathbf{B}\mathbf{e}_j = \mathbf{b}_j$, for all $j = 1, \ldots, m$, and (5.7) can be rewritten as

$$\mathbf{B}_{\text{new}} = (\mathbf{B}\mathbf{e}_1, \mathbf{B}\mathbf{e}_2, \ldots, \mathbf{B}\mathbf{e}_{r-1}, \mathbf{B}(\mathbf{B}^{-1}\mathbf{a}_k), \mathbf{B}\mathbf{e}_{r+1}, \ldots, \mathbf{B}\mathbf{e}_m)$$

$$= \mathbf{B}(\mathbf{e}_1, \mathbf{e}_2, \ldots, \mathbf{e}_{r-1}, (\mathbf{B}^{-1}\mathbf{a}_k), \mathbf{e}_{r+1}, \ldots, \mathbf{e}_m) \tag{5.9}$$

$$= \mathbf{B}(\mathbf{e}_1, \mathbf{e}_2, \ldots, \mathbf{e}_{r-1}, \boldsymbol{\alpha}_k, \mathbf{e}_{r+1}, \ldots, \mathbf{e}_m)$$

Therefore, the new basis inverse is given by

$$\mathbf{B}_{\text{new}}^{-1} = [\mathbf{B}(\mathbf{e}_1, \mathbf{e}_2, \ldots, \mathbf{e}_{r-1}, \boldsymbol{\alpha}_k, \mathbf{e}_{r+1}, \ldots, \mathbf{e}_m)]^{-1}$$

$$= (\mathbf{e}_1, \mathbf{e}_2, \ldots, \mathbf{e}_{r-1}, \boldsymbol{\alpha}_k, \mathbf{e}_{r+1}, \ldots, \mathbf{e}_m)^{-1}\mathbf{B}^{-1} \tag{5.10}$$

$$= \mathbf{E}\mathbf{B}^{-1}$$

where

$$\mathbf{E} = (\mathbf{e}_1, \mathbf{e}_2, \ldots, \mathbf{e}_{r-1}, \boldsymbol{\alpha}_k, \mathbf{e}_{r+1}, \ldots, \mathbf{e}_m)^{-1} \tag{5.11}$$

Now, writing $\mathbf{E}$ in expanding form yields

$$
\mathbf{E} =
\begin{pmatrix}
1 & \cdots & \alpha_{1,k} & \cdots & 0 \\
\vdots & \ddots & \vdots & \ddots & \vdots \\
0 & \cdots & \alpha_{r,k} & \cdots & 0 \\
\vdots & \ddots & \vdots & \ddots & \vdots \\
0 & \cdots & \alpha_{m,k} & \cdots & 1
\end{pmatrix}^{-1}
\tag{5.12}
$$

and using standard methods for matrix inversion, it is easy to verify that $\mathbf{E}$ is defined as follows:

$$
\mathbf{E} =
\begin{pmatrix}
1 & \cdots & -\alpha_{1,k}/\alpha_{r,k} & \cdots & 0 \\
\vdots & \ddots & \vdots & \ddots & \vdots \\
0 & \cdots & 1/\alpha_{r,k} & \cdots & 0 \\
\vdots & \ddots & \vdots & \ddots & \vdots \\
0 & \cdots & -\alpha_{m,k}/\alpha_{r,k} & \cdots & 1
\end{pmatrix}
\tag{5.13}
$$

Note that matrix $\mathbf{E}$ differs from the identity matrix by precisely one column, with this column being derived from the entering column $\boldsymbol{\alpha}_k = \mathbf{B}^{-1}\mathbf{a}_k$. $\mathbf{E}$ is often referred to as an *elementary matrix,* and because it differs from the identity by only one column, it can be stored as a single column vector along with an additional integer indicating its column position in the identity. Stored in this vector format, matrix $\mathbf{E}$ is generally referred to as an *eta vector.* Thus, given a pivot operation, the new basis inverse can be computed from the current basis inverse by premultiplying by an elementary matrix that can be quickly computed from the entering column.

Now consider a sequence of basis matrices, where the basis inverse resulting from iteration p is denoted by $\mathbf{B}_p^{-1} = \mathbf{E}_p\mathbf{B}_{p-1}^{-1}$ with the initial basis inverse $\mathbf{B}_0^{-1} = \mathbf{I}$. Then

$$
\mathbf{B}_0^{-1} = \mathbf{I}
\tag{5.14}
$$

$$
\mathbf{B}_1^{-1} = \mathbf{E}_1\mathbf{B}_0^{-1} = \mathbf{E}_1
\tag{5.15}
$$

$$
\mathbf{B}_2^{-1} = \mathbf{E}_2\mathbf{B}_1^{-1} = \mathbf{E}_2\mathbf{E}_1
\tag{5.16}
$$

$$
\vdots
$$

$$
\mathbf{B}_p^{-1} = \mathbf{E}_p\mathbf{B}_{p-1}^{-1} = \mathbf{E}_p\mathbf{E}_{p-1}\cdots\mathbf{E}_2\mathbf{E}_1
\tag{5.17}
$$

Thus, each basis inverse is expressed as the *product* of elementary matrices. Before presenting an example utilizing the product form of the inverse, let us examine the special nature of matrix operations with elementary matrices.

Operations with Elementary Matrices

For simplicity, let an elementary matrix $\mathbf{E}$ be given by

$$\mathbf{E} = (\mathbf{e}_1, \mathbf{e}_2, \ldots, \mathbf{e}_{r-1}, \mathbf{w}_r, \mathbf{e}_{r+1}, \ldots, \mathbf{e}_m)$$

$$= \begin{pmatrix} 1 & 0 & \cdots & w_{1,r} & \cdots & 0 \\ 0 & 1 & \cdots & w_{2,r} & \cdots & 0 \\ \vdots & \vdots & & \vdots & \ddots & \vdots \\ 0 & 0 & \cdots & w_{m,r} & \cdots & 1 \end{pmatrix} \qquad (5.18)$$

and let the vectors $\mathbf{u}$ and $\mathbf{v}$ be given by

$$\mathbf{u} = (u_1, u_2, \ldots, u_m) \qquad (5.19)$$

$$\mathbf{v} = \begin{pmatrix} v_1 \\ v_2 \\ \vdots \\ v_m \end{pmatrix} \qquad (5.20)$$

Then, the product $\mathbf{uE}$ can be computed as follows:

$$\mathbf{uE} = (u_1, \ldots, u_r, \ldots, u_m) \begin{pmatrix} 1 & \cdots & w_{1,r} & \cdots & 0 \\ 0 & \cdots & w_{2,r} & \cdots & 0 \\ \vdots & \ddots & \vdots & \ddots & \vdots \\ 0 & \cdots & w_{m,r} & \cdots & 1 \end{pmatrix}$$

$$= \left(u_1, u_2, \ldots, u_{r-1}, \sum_{i=1}^{m} u_i w_{i,r}, u_{r+1}, \ldots, u_m \right) \qquad (5.21)$$

Similarly,

$$\mathbf{Ev} = \begin{pmatrix} 1 & \cdots & w_{1,r} & \cdots & 0 \\ 0 & \cdots & w_{2,r} & \cdots & 0 \\ \vdots & \ddots & \vdots & \ddots & \vdots \\ 0 & \cdots & w_{m,r} & \cdots & 1 \end{pmatrix} \begin{pmatrix} v_1 \\ \vdots \\ v_r \\ \vdots \\ v_m \end{pmatrix}$$

$$= \begin{pmatrix} v_1 + w_{1,r}v_1 \\ \vdots \\ 0 + w_{r,r}v_r \\ \vdots \\ v_m + w_{m,r}v_m \end{pmatrix} = \begin{pmatrix} v_1 \\ \vdots \\ 0 \\ \vdots \\ v_m \end{pmatrix} + v_r \begin{pmatrix} w_{1,r} \\ \vdots \\ w_{r,r} \\ \vdots \\ w_{m,r} \end{pmatrix} \qquad (5.22)$$

Thus, matrix operations involving elementary matrices take on the simple forms provided by (5.21) and (5.22). We now illustrate the revised simplex with the product form of the inverse via a simple example.

Example 5.2: Product Form of the Inverse

$$\text{maximize } z = 6x_1 + 4x_2 + x_3 \tag{5.23}$$

subject to

$$x_1 - x_2 + x_3 + x_4 = 12 \tag{5.24}$$

$$x_1 + x_2 + x_3 + x_5 = 5 \tag{5.25}$$

$$2x_2 + x_2 + x_3 + x_6 = 8 \tag{5.26}$$

$$x_j \geq 0, \qquad \text{for all } j \tag{5.27}$$

The data for the problem can be summarized as follows:

$$\mathbf{A} = \begin{pmatrix} 1 & -1 & 1 & 1 & 0 & 0 \\ 1 & 1 & 1 & 0 & 1 & 0 \\ 2 & 1 & 1 & 0 & 0 & 1 \end{pmatrix}$$

$$\mathbf{b} = \begin{pmatrix} 12 \\ 5 \\ 8 \end{pmatrix}$$

$$\mathbf{c} = (6 \quad 4 \quad 1 \quad 0 \quad 0 \quad 0)$$

Clearly, $\mathbf{a}_4$, $\mathbf{a}_5$, and $\mathbf{a}_6$ form a convenient starting basis, which results in the following initial basic feasible solution:

Initial Solution

$$\mathbf{B} = (\mathbf{a}_4, \mathbf{a}_5, \mathbf{a}_6) = \mathbf{I}$$

$$\mathbf{c}_B = (c_4, c_5, c_6) = (0 \quad 0 \quad 0)$$

$$\mathbf{x}_B = \begin{pmatrix} x_{B,1} \\ x_{B,2} \\ x_{B,3} \end{pmatrix} = \begin{pmatrix} x_4 \\ x_5 \\ x_6 \end{pmatrix} = \beta = \begin{pmatrix} 12 \\ 5 \\ 8 \end{pmatrix}$$

$$\mathbf{x}_N = \begin{pmatrix} x_1 \\ x_2 \\ x_3 \end{pmatrix} \begin{pmatrix} 0 \\ 0 \\ 0 \end{pmatrix}$$

$$z = 0$$

$$\pi = \mathbf{c}_B \mathbf{B}^{-1} = \mathbf{c}_B \mathbf{I} = (0 \quad 0 \quad 0)$$

Computing $z_j - c_j$ for the nonbasic variables yields

$$z_1 - c_1 = \pi \mathbf{a}_1 - c_1 = -6$$

$$z_2 - c_2 = \pi \mathbf{a}_2 - c_2 = -4$$

$$z_3 - c_3 = \pi \mathbf{a}_3 - c_3 = -1$$

Choosing x_1 as the entering variable, we compute

$$\alpha_1 = \mathbf{B}^{-1} \mathbf{a}_1 = \mathbf{a}_1 = \begin{pmatrix} 1 \\ 1 \\ 2 \end{pmatrix}$$

Now apply the minimum ratio test:

$$\text{minimum} \left\{ \frac{\beta_1}{\alpha_{1,1}}, \frac{\beta_2}{\alpha_{2,1}}, \frac{\beta_3}{\alpha_{3,1}} \right\} = \text{minimum} \left\{ \frac{12}{1}, \frac{5}{1}, \frac{8}{2} \right\} = 4$$

Therefore, $x_{B,3} = x_6$ is the departing variable. From (5.13), the elementary matrix $\mathbf{E}_1$ is given by

$$\mathbf{E}_1 = \begin{pmatrix} 1 & 0 & -\frac{1}{2} \\ 0 & 1 & -\frac{1}{2} \\ 0 & 0 & \frac{1}{2} \end{pmatrix}$$

which can be stored as the eta vector $\boldsymbol{\eta}_1$, where the entry 3 represents the column location.

$$\boldsymbol{\eta}_1 = \begin{pmatrix} -\frac{1}{2} \\ -\frac{1}{2} \\ \frac{1}{2} \\ 3 \end{pmatrix}$$

Iteration 1

Now, compute the new solution. Using (5.22), we see that

$$\boldsymbol{\beta} = \mathbf{B}^{-1}\mathbf{b} = \mathbf{E}_1\mathbf{b} = \begin{pmatrix} 12 \\ 5 \\ 0 \end{pmatrix} + 8 \begin{pmatrix} -\frac{1}{2} \\ -\frac{1}{2} \\ \frac{1}{2} \end{pmatrix} = \begin{pmatrix} 8 \\ 1 \\ 4 \end{pmatrix}$$

Therefore,

$$\mathbf{x}_B = \begin{pmatrix} x_{B,1} \\ x_{B,2} \\ x_{B,3} \end{pmatrix} = \begin{pmatrix} x_4 \\ x_5 \\ x_1 \end{pmatrix} = \begin{pmatrix} 8 \\ 1 \\ 4 \end{pmatrix}$$

$$\mathbf{x}_N = \begin{pmatrix} x_6 \\ x_2 \\ x_3 \end{pmatrix} = \begin{pmatrix} 0 \\ 0 \\ 0 \end{pmatrix}$$

$$z = 0 - (z_1 - c_1)x_1 = 0 - (-6)4 = 24$$

Also, using (5.21), yields

$$\boldsymbol{\pi} = \mathbf{c}_B\mathbf{E}_1 = (0 \quad 0 \quad 6)\mathbf{E}_1 = (0 \quad 0 \quad -(\tfrac{1}{2})(0) - (\tfrac{1}{2})(0) + (\tfrac{1}{2})(6)) = (0 \quad 0 \quad 3)$$

Computing $z_j - c_j$ for the nonbasic variables yields

$$z_2 - c_2 = \boldsymbol{\pi}\mathbf{a}_2 - c_2 = -1$$

$$z_3 - c_3 = \boldsymbol{\pi}\mathbf{a}_3 - c_3 = 2$$

$$z_5 - c_5 = \boldsymbol{\pi}\mathbf{a}_5 - c_5 = 3$$

Choosing x_2 as the entering variable, we compute

$$\boldsymbol{\alpha}_2 = \mathbf{E}_1\mathbf{a}_2 = \begin{pmatrix} -1 \\ 1 \\ 0 \end{pmatrix} + 1 \begin{pmatrix} -\frac{1}{2} \\ -\frac{1}{2} \\ \frac{1}{2} \end{pmatrix} = \begin{pmatrix} -\frac{3}{2} \\ \frac{1}{2} \\ \frac{1}{2} \end{pmatrix}$$

Now apply the minimum ratio test:

$$\text{minimum} \left\{ \frac{\beta_2}{\alpha_{2,2}}, \frac{\beta_3}{\alpha_{3,2}} \right\} = \text{minimum} \left\{ \frac{1}{\frac{1}{2}}, \frac{4}{\frac{1}{2}} \right\} = 2$$

Therefore, $x_{B,2} = x_5$ is the departing variable. From (5.13), the elementary matrix $\mathbf{E}_2$ is given by

$$\mathbf{E}_2 = \begin{pmatrix} 1 & 3 & 0 \\ 0 & 2 & 0 \\ 0 & -1 & 1 \end{pmatrix}$$

which can be stored as the eta vector:

$$\boldsymbol{\eta}_2 = \begin{pmatrix} 3 \\ 2 \\ -1 \\ 2 \end{pmatrix}$$

Iteration 2

Compute the updated solution. Using (5.22), we see that

$$\boldsymbol{\beta} = \mathbf{E}_2\mathbf{E}_1\mathbf{b} = \mathbf{E}_2 \begin{pmatrix} 8 \\ 1 \\ 4 \end{pmatrix} = \begin{pmatrix} 8 \\ 0 \\ 4 \end{pmatrix} + 1 \begin{pmatrix} 3 \\ 2 \\ -1 \end{pmatrix} = \begin{pmatrix} 11 \\ 2 \\ 3 \end{pmatrix}$$

Therefore,

$$\mathbf{x}_B = \begin{pmatrix} x_{B,1} \\ x_{B,2} \\ x_{B,3} \end{pmatrix} = \begin{pmatrix} x_4 \\ x_2 \\ x_1 \end{pmatrix} = \begin{pmatrix} 11 \\ 2 \\ 3 \end{pmatrix}$$

$$\mathbf{x}_N = \begin{pmatrix} x_6 \\ x_5 \\ x_3 \end{pmatrix} = \begin{pmatrix} 0 \\ 0 \\ 0 \end{pmatrix}$$

$$z = 24 - (z_2 - c_2)x_2 = 24 - (-1)2 = 26$$

Also, using (5.21) yields

$$\boldsymbol{\pi} = \mathbf{c}_B\mathbf{E}_2\mathbf{E}_1 = (0 \quad 4 \quad 6)\mathbf{E}_2\mathbf{E}_1$$

$$= (0 \quad 3(0) + 2(4) - 1(6) \quad 6)\mathbf{E}_1$$

$$= (0 \quad 2 \quad 6)\mathbf{E}_1$$

$$= (0 \quad 2 \quad -(\tfrac{1}{2})(0) - (\tfrac{1}{2})(2) + (\tfrac{1}{2})(6))$$

$$= (0 \quad 2 \quad 2)$$

Computing $z_j - c_j$ for the nonbasic variables yields

$$z_3 - c_3 = \boldsymbol{\pi}\mathbf{a}_3 - c_3 = 3$$

$$z_5 - c_5 = \boldsymbol{\pi}\mathbf{a}_5 - c_5 = 2$$

$$z_6 - c_6 = \boldsymbol{\pi}\mathbf{a}_6 - c_6 = 2$$

Because all $z_j - c_j \geq 0$, the optimal solution is

$$z^* = 26$$

$$x_1^* = 3$$

$$x_2^* = 2$$

$$x_3^* = 0$$

$$x_4^* = 11$$

$$x_5^* = 0$$

$$x_6^* = 0$$

THE BOUNDED-VARIABLES SIMPLEX METHOD

Frequently, in developing a linear programming model, constraints of the form

$$x_j \geq l_j$$

$$x_j \leq u_j$$

are encountered. In this context, l_j and u_j represent simple lower and upper bounds, respectively, on the variable x_j. Now, letting

$$\mathbf{l} = \begin{pmatrix} l_1 \\ l_2 \\ \vdots \\ l_n \end{pmatrix}$$

$$\mathbf{u} = \begin{pmatrix} u_1 \\ u_2 \\ \vdots \\ u_n \end{pmatrix}$$

a linear program with bounded variables can be written as follows:

$$(\text{BVLP}) \qquad \text{maximize } z = \mathbf{cx}$$

$$\text{subject to}$$

$$\mathbf{Ax} = \mathbf{b} \qquad\qquad (5.28)$$

$$\mathbf{l} \leq \mathbf{x} \leq \mathbf{u}$$

As we saw in Chapter 4, it is possible to transform the lower-bound restrictions $\mathbf{x} \geq \mathbf{l}$ to simple nonnegativity restrictions by using the substitution $\mathbf{x}' = \mathbf{x} - \mathbf{l}$. Unfortunately, there is not a straightforward procedure for dealing with the upper bounds. Of course, it is possible to consider all the bounds as explicit constraints, however, this would effectively increase the size of the basis matrix from $m \times m$ to $(m + 2n) \times (m + 2n)$. Because operations involving the basis and basis inverse represent the largest part of the computational and storage overhead, this is a very inefficient approach. The basic idea of the bounded-variables simplex method is to handle the simple bounds on the variables in an implicit manner (in a manner analogous to the handling of the nonnegativity restrictions in the standard simplex method). This allows us to maintain a standard $m \times m$ basis matrix, which is generally referred to as the *working basis*.

In the standard simplex method, nonbasic variables are those variables that are fixed at their lower-bound value of zero. However, in the bounded-variables simplex method, a nonbasic variable represents a variable that is either fixed at its lower bound or upper bound. That is, the vector $\mathbf{x}$ will be partitioned into the basic variables $\mathbf{x}_B$, the nonbasic variables at their lower bound $\mathbf{x}_{N_l}$, and the nonbasic variables at their upper bound $\mathbf{x}_{N_u}$. A basic feasible solution will then be characterized mathematically by partitioning the coefficient matrix $\mathbf{A}$ into a nonsingular working basis matrix $\mathbf{B}$ and the matrices of nonbasic columns $\mathbf{N}_l$ and $\mathbf{N}_u$. That is,

$$\mathbf{A} = (\mathbf{B} : \mathbf{N}_l : \mathbf{N}_u) \tag{5.29}$$

Now, following the development in Chapter 4, the linear system $\mathbf{A}\mathbf{x} = \mathbf{b}$ can be rewritten to yield

$$\mathbf{B}\mathbf{x}_B + \mathbf{N}_l\mathbf{x}_{N_l} + \mathbf{N}_u\mathbf{x}_{N_u} = \mathbf{b} \tag{5.30}$$

This simplifies to

$$\mathbf{x}_B = \mathbf{B}^{-1}\mathbf{b} - \mathbf{B}^{-1}\mathbf{N}_l\mathbf{x}_{N_l} - \mathbf{B}^{-1}\mathbf{N}_u\mathbf{x}_{N_u} \tag{5.31}$$

Now, setting $\mathbf{x}_{N_l} = \mathbf{l}_{N_l}$ and $\mathbf{x}_{N_u} = \mathbf{u}_{N_u}$, we see that (5.31) results in

$$\mathbf{x}_B = \overline{\boldsymbol{\beta}} = \mathbf{B}^{-1}\mathbf{b} - \mathbf{B}^{-1}\mathbf{N}_l\mathbf{l}_{N_l} - \mathbf{B}^{-1}\mathbf{N}_u\mathbf{u}_{N_u} \tag{5.32}$$

The solution

$$\mathbf{x} = \begin{pmatrix} \mathbf{x}_B \\ \mathbf{x}_{N_l} \\ \mathbf{x}_{N_u} \end{pmatrix} = \begin{pmatrix} \overline{\boldsymbol{\beta}} \\ \mathbf{l}_{N_l} \\ \mathbf{u}_{N_u} \end{pmatrix} \tag{5.33}$$

is called a basic solution. If, in addition, $\mathbf{l}_B \leq \mathbf{x}_B \leq \mathbf{u}_B$, then the solution is a basic feasible solution.

Now consider the objective function $z = \mathbf{c}\mathbf{x}$. By partitioning the cost vector $\mathbf{c}$ as $\mathbf{c} = (\mathbf{c}_B, \mathbf{c}_{N_l}, \mathbf{c}_{N_u})$, the objective function can be rewritten as

$$z = \mathbf{c}_B\mathbf{x}_B + \mathbf{c}_{N_l}\mathbf{x}_{N_l} + \mathbf{c}_{N_u}\mathbf{x}_{N_u} \tag{5.34}$$

Now, substituting the expression for $\mathbf{x}_B$ defined in (5.34) into (5.31) and simplifying yield

$$z = \mathbf{c}_B\mathbf{B}^{-1}\mathbf{b} - (\mathbf{c}_B\mathbf{B}^{-1}\mathbf{N}_l - \mathbf{c}_{N_l})\mathbf{x}_{N_l} - (\mathbf{c}_B\mathbf{B}^{-1}\mathbf{N}_u - \mathbf{c}_{N_u})\mathbf{x}_{N_u} \tag{5.35}$$

And setting $\mathbf{x}_{N_l} = \mathbf{l}_{N_l}$, $\mathbf{x}_{N_u} = \mathbf{u}_{N_u}$, we see that (5.35) results in

$$z = \bar{z} = \mathbf{c}_B \mathbf{B}^{-1}\mathbf{b} - (\mathbf{c}_B \mathbf{B}^{-1}\mathbf{N}_l - \mathbf{c}_{N_l})\mathbf{l}_{N_l} - (\mathbf{c}_B \mathbf{B}^{-1}\mathbf{N}_u - \mathbf{c}_{N_u})\mathbf{u}_{N_u} \qquad (5.36)$$

which is the objective value corresponding to the current basic feasible solution.

Now letting J_l, J_u denote the index sets of the variables that are nonbasic at their lower bounds, upper bounds, respectively, (5.31) and (5.35) can be rewritten as follows:

$$z = \mathbf{c}_B \mathbf{B}^{-1}\mathbf{b} - \sum_{j \in J_l}(z_j - c_j)x_j - \sum_{j \in J_u}(z_j - c_j)x_j \qquad (5.37)$$

$$\mathbf{x}_B = \mathbf{B}^{-1}\mathbf{b} - \sum_{j \in J_l}\boldsymbol{\alpha}_j x_j - \sum_{j \in J_u}\boldsymbol{\alpha}_j x_j \qquad (5.38)$$

It is important to note from (5.32) and (5.36) that the values of $\bar{z}$ and $\bar{\boldsymbol{\beta}}$ depend on not only $\mathbf{c}_B$, $\mathbf{B}^{-1}$, and $\mathbf{b}$ (as in the standard simplex method), but also on the actual values of the nonbasic variables. As such, updating the values of $\bar{z}$ and $\bar{\boldsymbol{\beta}}$ cannot be achieved by solely updating $\mathbf{B}^{-1}$ as before. The values must be computed by directly applying (5.32) and (5.36). This will be further illustrated in the statement of the algorithm and the illustrative example that follows.

Checking for Optimality

First, consider a nonbasic variable x_j, where $j \in J_l$, that is, x_j is nonbasic and its current value is $x_j = l_j$. Because x_j is currently at its lower bound, it can only be increased. Thus, from (5.37) and by following the same logic as in Chapter 4, increasing x_j will improve (increase) the objective provided that $z_j - c_j < 0$.

Similarly, if x_j is currently nonbasic at its upper bound (i.e., $j \in J_u$), then we can only decrease x_j from its current value of u_j. Because the rate of change of z with respect to x_j is given by $-(z_j - c_j)$, decreasing x_j will increase the objective function if and only if $z_j - c_j > 0$. Therefore, the optimality conditions for problem (BVLP) can be summarized as follows.

Optimality conditions (bounded-variables maximization problem). The basic feasible solution represented by (5.33) will be optimal to (BVLP) if

$$z_j - c_j \geq 0, \qquad \text{for all } j \in J_l$$

and

$$z_j - c_j \leq 0, \qquad \text{for all } j \in J_u$$

If $z_j - c_j > 0$, for all $j \in J_l$, and $z_j - c_j < 0$, for all $j \in J_u$, then the current basic feasible solution will be the unique optimal solution. However, if some nonbasic variable x_k has $z_k - c_k = 0$, then in the absence of degeneracy, there will be alternative optimal solutions.

Determining the Entering Variable

From the preceding discussion, if the current basic feasible solution is not optimal, then we either chose to increase some x_k that is currently nonbasic at its lower bound and $z_k - c_k < 0$, or we decrease some x_k that is currently nonbasic at its upper bound and $z_k - c_k > 0$. As before, we generally choose to enter that nonbasic variable that forces the greatest rate of change of the objective. Thus, the entering variable x_k will be that variable with $z_k - c_k$ defined as follows:

$$z_k - c_k = \text{maximum}\{\underset{j \in J_l}{\text{maximum}} - (z_j - c_j), \underset{j \in J_u}{\text{maximum}}(z_j - c_j)\} \qquad (5.39)$$

There are two cases to consider when determining the departing variable. These cases correspond to whether the entering variable is being increased from its lower bound or decreased from its upper bound.

Increasing a Nonbasic Variable x_k From Its Lower Bound l_k

Suppose that $z_k - c_k < 0$ and x_k is currently nonbasic at its lower bound l_k. Then the solution can be improved by increasing x_k. Let $\Delta_k \geq 0$ be the increase in x_k, that is, the new value of x_k will be given by

$$x_k = l_k + \Delta_k \qquad (5.40)$$

Because all other nonbasic variables remain fixed at either their lower or upper bounds, substituting into (5.31) and (5.35) yields

$$\begin{aligned} \mathbf{x}_B &= \mathbf{B}^{-1}\mathbf{b} - \mathbf{B}^{-1}\mathbf{N}_l\mathbf{l}_{N_l} - \mathbf{B}^{-1}\mathbf{N}_u\mathbf{u}_{N_u} - \mathbf{B}^{-1}\mathbf{a}_k\Delta_k \\ &= \overline{\boldsymbol{\beta}} - \boldsymbol{\alpha}_k\Delta_k \end{aligned} \qquad (5.41)$$

and

$$\begin{aligned} z &= \mathbf{c}_B\mathbf{B}^{-1}\mathbf{b} - (\mathbf{c}_B\mathbf{B}^{-1}\mathbf{N}_l - \mathbf{c}_{N_l})\mathbf{l}_{N_l} - (\mathbf{c}_B\mathbf{B}^{-1}\mathbf{N}_u - \mathbf{c}_{N_u})\mathbf{u}_{N_u} - (z_k - c_k)\Delta_k \\ &= \overline{z} - (z_k - c_k)x_k \end{aligned} \qquad (5.42)$$

To maintain feasibility, the value of Δ_k must be chosen to satisfy the following conditions:

$$l_k \leq x_k \leq u_k \Rightarrow l_k \leq l_k + \Delta_k \leq u_k \qquad (5.43)$$

$$\mathbf{l}_B \leq \mathbf{x}_B \leq \mathbf{u}_B \Rightarrow \mathbf{l}_B \leq \overline{\boldsymbol{\beta}} - \boldsymbol{\alpha}_k\Delta_k \leq \mathbf{u}_B$$

$$\Rightarrow l_{B,i} \leq \overline{\beta}_i - \alpha_{k,i}\Delta_k \leq u_{B,i}, \qquad \text{for all } i = 1, \ldots, m \qquad (5.44)$$

Because $\Delta_k \geq 0$, it follows from (5.43) that

$$\Delta_k \leq u_k - l_k \qquad (5.45)$$

Now consider (5.44). If $\alpha_{k,i} > 0$, then the basic variable $x_{B,i}$ is decreasing and it follows that we must enforce

$$l_{B,i} \leq \bar{\beta}_i - \alpha_{k,i}\Delta_k, \qquad \text{for all } i \text{ such that } \alpha_{k,i} > 0 \qquad (5.46)$$

which yields

$$\Delta_k \leq \frac{\bar{\beta}_i - l_{B,i}}{\alpha_{k,i}}, \qquad \text{for all } i \text{ such that } \alpha_{k,i} > 0 \qquad (5.47)$$

On the other hand, if $\alpha_{k,i} < 0$, then the basic variable $x_{B,i}$ is increasing and we must enforce

$$\bar{\beta}_i - \alpha_{k,i}\Delta_k \leq u_{B,i}, \qquad \text{for all } i = 1, \ldots, m \qquad (5.48)$$

and, thus,

$$\Delta_k \leq \frac{u_{B,i} - \bar{\beta}_i}{-\alpha_{k,i}}, \qquad \text{for all } i \text{ such that } \alpha_{k,i} < 0 \qquad (5.49)$$

Therefore, to determine the largest value of Δ_k that will result in a feasible solution, we use the following relationship:

$$\Delta_k = \text{minimum } \{\delta_1, \delta_2, u_k - l_k\} \qquad (5.50)$$

where

$$\delta_1 = \begin{cases} \infty, & \text{if } \alpha_k \leq 0 \\ \text{minimum} \left\{ \dfrac{\bar{\beta}_i - l_{B,i}}{\alpha_{k,i}} : \alpha_{k,i} > 0 \right\}, & \text{otherwise} \end{cases} \qquad (5.51)$$

$$\delta_2 = \begin{cases} \infty, & \text{if } \alpha_k \leq 0 \\ \text{minimum} \left\{ \dfrac{u_{B,i} - \bar{\beta}_i}{-\alpha_{k,i}} : \alpha_{k,i} < 0 \right\}, & \text{otherwise} \end{cases} \qquad (5.52)$$

Note that if $\Delta_k = \delta_1 = (\bar{\beta}_r - l_{B,r})/\alpha_{k,r}$, then the departing variable is $x_{B,r}$ which becomes nonbasic at its lower bound. Similarly, if $\Delta_k = \delta_2 = (u_{B,r} - \bar{\beta}_r)/(-\alpha_{k,r})$, then $x_{B,r}$ departs at its upper bound. Finally, if $\Delta_k = u_k - l_k$, then the entering variable x_k blocks itself and x_k moves from nonbasic at its lower bound to nonbasic at its upper bound. In this last case, the basis matrix remains the same; the only changes are $\bar{z}$ and $\bar{\beta}$ according to (5.41) and (5.42). Of course, if these computations result in $\Delta_k = \infty$, then x_k can be increased without bound, and, consequently, no finite optimal solution exists.

Decreasing a Nonbasic Variable x_k From Its Upper Bound u_k

Now consider the case when $z_k - c_k > 0$ and x_k is currently nonbasic at its upper bound u_k. Then decreasing x_k will improve the objective value. Let $\Delta_k \geq 0$ be the amount by which x_k is decreased, that is, the new value of x_k will be given by

$$x_k = u_k - \Delta_k \qquad (5.53)$$

Now, as in the previous case, (5.31) and (5.35) yield

$$\begin{aligned}
\mathbf{x}_B &= \mathbf{B}^{-1}\mathbf{b} - \mathbf{B}^{-1}\mathbf{N}_l\mathbf{l}_{N_l} - \mathbf{B}^{-1}\mathbf{N}_u\mathbf{u}_{N_u} + \mathbf{B}^{-1}\mathbf{a}_k\Delta_k \\
&= \overline{\boldsymbol{\beta}} + \boldsymbol{\alpha}_k\Delta_k
\end{aligned} \qquad (5.54)$$

and

$$\begin{aligned}
\mathbf{z} &= \mathbf{c}_B\mathbf{B}^{-1}\mathbf{b} - (\mathbf{c}_B\mathbf{B}^{-1}\mathbf{N}_l - \mathbf{c}_{N_l})\mathbf{l}_{N_l} - (\mathbf{c}_B\mathbf{B}^{-1}\mathbf{N}_u - \mathbf{c}_{N_u})\mathbf{u}_{N_u} + (z_k - c_k)\Delta_k \\
&= \overline{z} + (z_k - c_k)x_k
\end{aligned} \qquad (5.55)$$

Now, to maintain feasibility, the value of Δ_k must be chosen to satisfy the following conditions:

$$l_k \le x_k \le u_k \Rightarrow l_k \le u_k - \Delta_k \le u_k \qquad (5.56)$$

$$\begin{aligned}
\mathbf{l}_B \le \mathbf{x}_B \le \mathbf{u}_B &\Rightarrow \mathbf{l}_B \le \overline{\boldsymbol{\beta}} + \boldsymbol{\alpha}_k\Delta_k \le \mathbf{u}_B \\
&\Rightarrow l_{B,i} \le \overline{\beta}_i + \alpha_{k,i}\Delta_k \le u_{B,i}, \qquad \text{for all } i = 1, \ldots, m
\end{aligned} \qquad (5.57)$$

Following the same logic as before, we see that Δ_k is defined as follows:

$$\Delta_k = \text{minimum}\{\delta_1, \delta_2, u_k - l_k\} \qquad (5.58)$$

where

$$\delta_1 = \begin{cases} \infty, & \text{if } \boldsymbol{\alpha}_k \le \mathbf{0} \\ \text{minimum}\left\{\dfrac{u_{B,i} - \overline{\beta}_i}{\alpha_{k,i}} : \alpha_{k,i} > 0\right\}, & \text{otherwise} \end{cases} \qquad (5.59)$$

$$\delta_2 = \begin{cases} \infty, & \text{if } \boldsymbol{\alpha}_k \ge \mathbf{0} \\ \text{minimum}\left\{\dfrac{\overline{\beta}_i - l_{B,i}}{-\alpha_{k,i}} : \alpha_{k,i} < 0\right\}, & \text{otherwise} \end{cases} \qquad (5.60)$$

If $\Delta_k = \delta_1 = (u_{B,r} - \overline{\beta}_r)/\alpha_{k,r}$, then the departing variable is $x_{B,r}$, which becomes nonbasic at its upper bound. Similarly, if $\Delta_k = \delta_2 = (\overline{\beta}_r - l_{B,r})/(-\alpha_{k,r})$, then $x_{B,r}$ departs at its lower bound. If $\Delta_k = u_k - l_k$, then the entering variable x_k blocks itself and x_k moves from nonbasic at its upper bound to nonbasic at its lower bound with the basis remaining the same, and $\overline{z}, \overline{\boldsymbol{\beta}}$ being updated according to (5.54) and (5.55).

As before, if these computations result in $\Delta_k = \infty$, then x_k can be decreased without bound, and, consequently, no finite optimal solution exists. We now summarize the basic steps of the bounded-variables simplex method and then conclude with an illustrative example.

The bounded-variables simplex algorithm (maximization problem)

STEP 1. *Check for possible improvement.* Examine the $z_j - c_j$ values for the nonbasic variables. If $z_j - c_j \geq 0$, for all $j \in J_l$, and $z_j - c_j \leq 0$, for all $j \in J_u$, then the current basic feasible solution is optimal; stop. Otherwise, select the nonbasic variable x_k as the entering variable with

$$z_k - c_k = \text{maximum}\{\underset{j \in J_l}{\text{maximum}} - (z_j - c_j), \underset{j \in J_u}{\text{maximum}}(z_j - c_j)\}$$

Ties in the selection of x_k may be broken arbitrarily. If x_k is currently at its lower bound (i.e., $k \in J_l$), then go to Step 2. If x_k is currently at its upper bound (i.e., $k \in J_u$), then go to Step 3.

STEP 2. *Increase x_k from its current value of l_k.* Let $x_k = l_k + \Delta_k$.
   (a) Compute Δ_k using (5.50–5.52). If $\Delta_k = \infty$, then the problem has an unbounded objective value; stop.
   (b) If $\Delta_k = u_k - l_k$, then x_k becomes nonbasic at its upper bound. Update the right-hand side of the tableau using the relationships defined by (5.41) and (5.42). The basis does not change and the remainder of the tableau remains the same. Return to Step 1.
   (c) If $\Delta_k = \delta_1$, then the departing variable $x_{B,r}$ becomes nonbasic at its lower bound. If $\Delta_k = \delta_2$, then the departing variable $x_{B,r}$ becomes nonbasic at its upper bound. The entering variable x_k is the new basic variable in row r with value $x_k = l_k + \Delta_k$. Update the remainder of the right-hand side using the relationships defined by (5.41) and (5.42). Update the remainder of the tableau by pivoting in the usual manner on $\alpha_{r,k}$. Return to Step 1.

STEP 3. *Decrease x_k from its current value of u_k.* Let $x_k = u_k - \Delta_k$.
   (a) Compute Δ_k using (5.58–5.60). If $\Delta_k = \infty$, then the problem has an unbounded objective value; stop.
   (b) If $\Delta_k = u_k - l_k$, then x_k becomes nonbasic at its lower bound. Update the right-hand side of the tableau using the relationships defined by (5.54) and (5.55). The basis does not change and the remainder of the tableau remains the same. Return to Step 1.
   (c) If $\Delta_k = \delta_1$, then the departing variable $x_{B,r}$ becomes nonbasic at its upper bound. If $\Delta_k = \delta_2$, then the departing variable $x_{B,r}$ becomes nonbasic at its lower bound. The entering variable x_k is the new basic variable in row r with value $x_k = u_k - \Delta_k$. Update the remainder of the right-hand side using the relationships defined by (5.54) and (5.55). Update the remainder of the tableau by pivoting in the usual manner on $\alpha_{r,k}$. Return to Step 1.

Example 5.3: Bounded Variables

$$\text{maximize } z = 2x_1 + 3x_2 \tag{5.61}$$

subject to

$$x_1 + 2x_2 \leq 22 \tag{5.62}$$

$$x_1 - x_2 \leq 1 \tag{5.63}$$

$$-1 \leq x_1 \leq 6 \tag{5.64}$$

$$2 \leq x_2 \leq 10 \tag{5.65}$$

Adding slack variables x_3 and x_4, the problem can be recast in the following form:

$$\text{maximize } z = 2x_1 + 3x_2 \tag{5.66}$$

subject to

$$x_1 + 2x_2 + x_3 = 22 \tag{5.67}$$

$$x_1 - x_2 + x_4 = 1 \tag{5.68}$$

$$-1 \le x_1 \le 6 \tag{5.69}$$

$$2 \le x_2 \le 10 \tag{5.70}$$

$$0 \le x_3 \le \infty \tag{5.71}$$

$$0 \le x_4 \le \infty \tag{5.72}$$

Notice that the coefficient matrix contains an imbedded identity, and thus it is *possible* that a nice starting basis is available. But, first, we must fix x_1 and x_2 at either of their bounds and compute $\overline{\boldsymbol{\beta}}$. Arbitrarily set $x_1 = -1$ (lower bound) and $x_2 = 2$ (lower bound). Then

$$\overline{\boldsymbol{\beta}} = \begin{pmatrix} \overline{\beta}_1 \\ \overline{\beta}_2 \end{pmatrix} = \begin{pmatrix} 22 - (-1) - 2(2) \\ 1 - (-1) + 2 \end{pmatrix} = \begin{pmatrix} 19 \\ 4 \end{pmatrix} \ge \mathbf{0}$$

Because $\overline{\boldsymbol{\beta}} \ge \mathbf{0}$, then x_3 and x_4 form a convenient starting basis with

$$\mathbf{x}_B = \begin{pmatrix} x_{B,1} \\ x_{B,2} \end{pmatrix} = \begin{pmatrix} x_3 \\ x_4 \end{pmatrix} = \begin{pmatrix} 19 \\ 4 \end{pmatrix}$$

If $\overline{\boldsymbol{\beta}}$ had not been nonnegative, then it would have been necessary to add artificial variables to form a starting basis. Phase 1 could then be applied in an attempt to drive the artificial variables to zero. (For an example of getting started under these conditions, see Example 5.4.)

The current value of the objective can be computed from (5.66):

$$\overline{z} = 2(-1) = 3(2) = 4$$

The initial tableau is depicted in Table 5.9. Note that the nonbasic variables have been labeled to identify that they are presently nonbasic at their lower bounds.

TABLE 5.9

	z	$\overset{l}{x_1}$	$\overset{l}{x_2}$	x_3	x_4	RHS
z	1	-2	-3	0	0	4
x_3	0	1	2	1	0	19
x_4	0	1	-1	0	1	4

STEP 1. The current solution is clearly not optimal because x_1, x_2 are nonbasic at their lower bound and $z_1 - c_1 < 0$ and $z_2 - c_2 < 0$. By using (5.39), x_2 is chosen as

the entering variable (i.e., $k = 2$). Because x_2 is currently at its lower bound, go to Step 2.

STEP 2. Let $x_2 = l_2 + \Delta_2 = 2 + \Delta_2$.

(a) Compute Δ_2 using (5.50–5.52).

$$\delta_1 = \frac{\bar{\beta}_1 - l_{B,1}}{\alpha_{1,2}} = \frac{19 - 0}{2} = \frac{19}{2}$$

$$\delta_2 = \frac{u_{B,2} - \bar{\beta}_2}{-a_{2,2}} = \frac{\infty - 4}{-(-1)} = \infty$$

$$u_2 - l_2 = 10 - 2 = 8$$

$$\Delta_2 = \text{minimum}\{\tfrac{19}{2}, 8\} = 8$$

(b) $\Delta_2 = u_2 - l_2 = 8$; therefore, x_2 goes from nonbasic at its lower bound to nonbasic at its upper bound (i.e., $x_2 = 2 + 8 = 10$). Update the right-hand side using (5.41) and (5.42).

$$\bar{z} = 4 - (z_2 - c_2)\Delta_2 = 4 - (-3)8 = 28$$

$$\bar{\beta} = \begin{pmatrix} 19 \\ 4 \end{pmatrix} - \alpha_2\Delta_2 = \begin{pmatrix} 19 \\ 4 \end{pmatrix} - 8 \begin{pmatrix} 2 \\ -1 \end{pmatrix} = \begin{pmatrix} 3 \\ 12 \end{pmatrix}$$

The updated tableau is shown in Table 5.10. Note that the basic did not change. Return to Step 1.

TABLE 5.10

	z	x_1	x_2	x_3	x_4	RHS
		l	u			
z	1	-2	-3	0	0	28
x_3	0	1	2	1	0	3
x_4	0	1	-1	0	1	12

STEP 1. Select x_1 as the entering variable because $z_1 - c_1 < 0$ and x_1 is nonbasic at its lower bound. Go to Step 2.

STEP 2. Let $x_1 = l_1 + \Delta_1 = -1 + \Delta_1$.

(a) Compute Δ_1 using (5.50–5.52).

$$\delta_1 = \text{minimum}\left(\frac{\bar{\beta}_1 - l_{B,1}}{\alpha_{1,1}}, \frac{\bar{\beta}_2 - l_{B,2}}{\alpha_{2,1}}\right) = \left(\frac{3 - 0}{1}, \frac{12 - 0}{1}\right) = 3$$

$$\delta_2 = \infty$$

$$u_1 - l_1 = 6 - (-1) = 7$$

$$\Delta_1 = \text{minimum}\ \{3, 7\} = 3$$

(c) $\Delta_1 = \delta_1 = 3$; therefore, the departing variable is $x_{B,1} = x_3$, which becomes nonbasic at its lower bound. x_1 becomes the basic variable in row 1.

$$x_1 = -1 + \Delta_1 = -1 + 3 = 2$$

$$\bar{z} = 28 - (z_1 - c_1)\Delta_1 = 28 - (-2)3 = 34$$

$$\bar{\beta}_2 = 12 - \alpha_{1,1}\Delta_1 = 12 - 1(3) = 9$$

The remainder of the tableau is updated by performing a standard pivot operation on $\alpha_{1,1} = 1$. Table 5.11 summarizes the results. Return to Step 1.

TABLE 5.11

	z	x_1	u x_2	l x_3	x_4	RHS
z	1	0	1	2	0	34
x_1	0	1	2	1	0	2
x_4	0	0	-3	-1	1	9

STEP 1. Select x_2 as the entering variable because $z_2 - c_2 > 0$ and x_2 is nonbasic at its upper bound. Go to Step 3.

STEP 3. Let $x_2 = u_2 - \Delta_2 = 10 - \Delta_2$.

(a) Compute Δ_2 using (5.58–5.60).

$$\delta_1 = \frac{u_{B,1} - \bar{\beta}_1}{\alpha_{1,2}} = \frac{6-2}{2} = 2$$

$$\delta_2 = \frac{\bar{\beta}_2 - l_{B,2}}{-\alpha_{2,2}} = \frac{9-0}{-(-3)} = 3$$

$$u_2 - l_2 = 10 - 2 = 8$$

$$\Delta_2 = \text{minimum}\{2, 3, 7\} = 2$$

(c) $\Delta_2 = \delta_1 = 2$; therefore, the departing variable is $x_{B,1} = x_1$, which becomes nonbasic at its upper bound, and x_2 becomes basic in row 1.

$$x_2 = 10 - \Delta_2 = 10 - 2 = 8$$

$$\bar{z} = 34 + (z_2 - c_2)\Delta_2 = 34 + (1)2 = 36$$

$$\bar{\beta}_2 = 9 - \alpha_{1,2}\Delta_2 = 9 - 3(2) = 3$$

The remainder of the tableau is updated by performing a standard pivot operation on $\alpha_{1,2} = 2$. Table 5.12 summarizes the results. Return to Step 1.

TABLE 5.12

	z	u x_1	x_2	l x_3	x_4	RHS
z	1	$-\frac{1}{2}$	0	$\frac{3}{2}$	0	36
x_2	0	$\frac{1}{2}$	1	$\frac{1}{2}$	0	8
x_4	0	$\frac{3}{2}$	0	$\frac{1}{2}$	1	3

STEP 1. Table 5.12 represents the optimal solution, which can be summarized as follows:

$$z^* = 36$$

$$x_1^* = 6 \quad \text{(nonbasic at upper bound)}$$

$$x_2^* = 8 \quad \text{(basic)}$$

$$x_3^* = 0 \quad \text{(nonbasic at lower bound)}$$

$$x_4^* = 3 \quad \text{(basic)}$$

Example 5.4: Finding an Initial Basic Feasible Solution

$$\text{maximize } z = 8x_1 + 10x_2 \tag{5.73}$$

subject to

$$-x_1 + x_2 \leq 4 \tag{5.74}$$

$$x_1 + x_2 \leq 9 \tag{5.75}$$

$$-x_1 + 2x_2 \geq 1 \tag{5.76}$$

$$1 \leq x_1 \leq 4 \tag{5.77}$$

$$2 \leq x_2 \leq 6 \tag{5.78}$$

Now, placing the problem in standard form by adding slack variables yields

$$\text{maximize } z = 8x_1 + 10x_2 \tag{5.79}$$

subject to

$$-x_1 + x_2 + x_3 = 4 \tag{5.80}$$

$$x_1 + x_2 + x_4 = 9 \tag{5.81}$$

$$-x_1 + 2x_2 - x_5 = 1 \tag{5.82}$$

$$1 \leq x_1 \leq 4 \tag{5.83}$$

$$2 \leq x_2 \leq 6 \tag{5.84}$$

$$0 \leq x_3 \leq \infty \tag{5.85}$$

$$0 \leq x_4 \leq \infty \tag{5.86}$$

$$0 \leq x_5 \leq \infty \tag{5.87}$$

For illustration purposes, suppose we set $x_1 = 1$ (nonbasic at lower bound) and $x_2 = 6$ (nonbasic at upper bound). Let us now investigate the possibility of using the slack/surplus variables as a starting basis. Setting $x_1 = 1$ and $x_2 = 6$ in (5.80–5.82) yields

$$x_3 = 4 + 1 - 6 = \bar{\beta}_1 = -1 \tag{5.88}$$

$$x_4 = 9 - 1 - 6 = \bar{\beta}_2 = 2 \tag{5.89}$$

$$-x_5 = 1 + 1 - 2(6) = \bar{\beta}_3 = -10 \tag{5.90}$$

Because $\bar{\beta}_1$ and $\bar{\beta}_3$ are negative, we can multiply the first and third constraints by -1 to force $\bar{\beta}_1$ and $\bar{\beta}_3$ to become positive. Note that x_3 is negative (and thus infeasible), whereas in the resulting system, x_4 and x_5 are both positive and provide part of a starting basis. Therefore, we need to add an artificial variable (x_6) to the constraint (5.80) after multiplying by -1. These operations result in the following phase I problem.

$$\text{maximize } Z = -x_6 \tag{5.91}$$

subject to

$$x_1 - x_2 - x_3 + x_6 = -4 \tag{5.92}$$

$$x_1 + x_2 + x_4 = 9 \tag{5.93}$$

$$x_1 - 2x_2 + x_5 = -1 \tag{5.94}$$

$$1 \le x_1 \le 4 \tag{5.95}$$

$$2 \le x_2 \le 6 \tag{5.96}$$

$$0 \le x_3 \le \infty \tag{5.97}$$

$$0 \le x_4 \le \infty \tag{5.98}$$

$$0 \le x_5 \le \infty \tag{5.99}$$

$$0 \le x_6 \le \infty \tag{5.100}$$

Now, letting $x_1 = 1$ (lower bound), $x_2 = 6$ (upper bound), and $x_3 = 0$ (lower bound) yield

$$\bar{\beta} = \begin{pmatrix} \bar{\beta}_1 \\ \bar{\beta}_2 \\ \bar{\beta}_2 \end{pmatrix} = \begin{pmatrix} -4 - 1 + 6 + 0 \\ 9 - 1 - 6 \\ -1 - 1 + 2(6) \end{pmatrix} = \begin{pmatrix} 1 \\ 2 \\ 10 \end{pmatrix} \ge \mathbf{0}$$

Because $\bar{\beta} \ge \mathbf{0}$, then x_6, x_4, and x_5 form a convenient starting basis with

$$\mathbf{x}_B = \begin{pmatrix} x_{B,1} \\ x_{B,2} \\ x_{B,3} \end{pmatrix} = \begin{pmatrix} x_6 \\ x_4 \\ x_5 \end{pmatrix} = \begin{pmatrix} 1 \\ 2 \\ 10 \end{pmatrix}$$

The initial objective can be computed from (5.91):

$$\bar{Z} = -1$$

The initial phase I tableau is shown in Table 5.13. The reader is invited to complete the details of this problem. Note that the original problem can be solved graphically in the x_1–x_2 plane, thus making it easy to check your solution.

TABLE 5.13

	Z	l x_1	u x_2	l x_3	x_4	x_5	x_6	RHS
Z	1	-1	1	1	0	0	0	-1
x_6	0	1	-1	-1	0	0	1	1
x_4	0	1	1	0	1	0	0	2
x_5	0	1	-2	0	0	1	0	10

In addition to the bounded-variables simplex method described before, there are additional techniques for handling constraints such as

$$x_2 + x_5 \leq 12$$

$$x_1 + x_7 + x_8 + x_{10} \leq 40$$

These techniques are known generally as GUB, for generalized upper bounding. Lasdon (1970) provides a good summary of generalized upper bounding.

DECOMPOSITION

One of the more appealing ways, at least in theory, of dealing with large linear programming problems is to decompose the model into smaller subproblems, whereby the solutions to these subproblems may somehow be used to generate the solution to the overall problem.

One quite unsophisticated and rather brute-force approach to decomposition is to

1. Select a portion (k) of the (m) constraints (where $k < m$) and, for the moment, disregard the others. The k constraints for consideration may be selected by guess, intuitively, or on the basis of a somewhat more rational exercise.

2. Form the linear subproblem with only the k constraints and solve. If the solution obtained *also* satisfies the ($m - k$) constraints that were not included in the reduced model, we are finished. Otherwise, go to Step 3.

3. Pick some of the most violated of the constraints that were not previously included in the reduced model and place them into the new, reduced model. We might also eliminate some of the constraints from the reduced model that were found to be easily satisfied (e.g., having large slack values).

4. The process continues by repeating steps 2 and 3 until a satisfactory answer is obtained. Although a bit naive, the approach has been used to solve quite large problems in practice with fairly good results.

The approach just described is not systematic, nor does it take advantage of problem structure. One of the best known of the systematic decomposition procedures is due to Dantzig and Wolfe (1960, 1961) and operates on problems having the following structure:

$$\text{maximize } z = \sum_{j=1}^{r} \mathbf{c}_j \mathbf{x}_j$$

subject to

$$
\begin{pmatrix}
\mathbf{A}_1 & \mathbf{0} & \mathbf{0} & \cdots & \mathbf{0} \\
\mathbf{0} & \mathbf{A}_2 & \mathbf{0} & \cdots & \mathbf{0} \\
\vdots & \vdots & \vdots & \ddots & \vdots \\
\mathbf{0} & \mathbf{0} & \mathbf{0} & \cdots & \mathbf{A}_r \\
\mathbf{A}_{r+1} & \mathbf{A}_{r+2} & \mathbf{A}_{r+3} & \cdots & \mathbf{A}_{2r}
\end{pmatrix}
\begin{pmatrix}
\mathbf{x}_1 \\
\mathbf{x}_2 \\
\vdots \\
\mathbf{x}_r
\end{pmatrix}
=
\begin{pmatrix}
\mathbf{b}_1 \\
\mathbf{b}_2 \\
\vdots \\
\mathbf{b}_r
\end{pmatrix}
$$

$$\mathbf{x}_j \geq \mathbf{0}, \text{ for all } r$$

Note that $\mathbf{c}_j$, $\mathbf{x}_j$, and $\mathbf{b}_j$ are *vectors*, whereas $\mathbf{A}_i$ are *matrices* and the constraint matrix is of block diagonal form. If one finds a large problem having such a structure, the use of the Dantzig–Wolfe decomposition algorithm may provide a more efficient approach to its solution and, as such, should be considered. Details of this algorithm, together with numerous other approaches to large-scale problems, appear in the text by Lasdon (1970).

SUMMARY

Two additional implementations of the simplex algorithm were introduced in this chapter. The revised simplex method provides for more efficient computer implementations through decreased storage and computational requirements. The bounded-variables simplex method generalizes the standard simplex method so that simple lower and upper bounds on the variables can be handled implicitly rather than as explicit constraints. This reduces the size of the required basis matrix and improves computational efficiency.

EXERCISES

5.1. Consider the following linear programming problem.

$$\text{maximize } z = 2x_1 + 6x_2$$

subject to

$$-x_1 + 2x_2 \le 12$$

$$x_1 + 2x_2 \le 22$$

$$x_1 - x_2 \le 8$$

$$x_1, x_2 \ge 0$$

  (a) Solve this problem graphically.
  (b) Solve this problem using the standard simplex algorithm.
  (c) Solve this problem using the revised simplex method. Compare the steps of the algorithm with part (b) and plot the path taken by the algorithm on the graph.

5.2. Solve the following problem by the revised simplex algorithm.

$$\text{maximize } z = 2x_1 + 4x_2 + 7x_3 - 3x_4 - 5x_5$$

subject to

$$3x_1 + 5x_2 + x_3 - x_4 + x_5 \le 72$$

$$2x_1 + x_2 + x_3 + 2x_4 - x_5 \le 18$$

$$x_j \ge 0, \qquad \text{for all } j$$

5.3. Solve the following problem by the revised simplex algorithm.

$$\text{maximize } z = x_1 + 2x_2 + 5x_3 + x_4 - x_5 + x_6 - x_7$$

subject to

$$x_1 + 2x_2 + x_3 - x_4 + x_5 = 20$$

$$-x_1 + x_2 + x_3 + x_4 + x_6 = 12$$

$$2x_1 + x_2 + x_3 - x_4 + x_7 = 30$$

$$x_j \ge 0, \qquad \text{for all } j$$

5.4. Solve the following problem by the revised simplex method along with the two-phase method.

$$\text{minimize } z = 7x_1 + 3x_2 + 4x_3$$

subject to

$$x_1 + x_2 \ge 10$$

$$2x_1 + 2x_2 + x_3 \le 40$$

$$-2x_1 + x_2 - x_3 = 22$$

$$x_1, x_2, x_3 \ge 0$$

5.5. Solve the following problem by the revised simplex algorithm.

$$\text{maximize } z = 3x_1 + 4x_2 + 2x_3$$

subject to

$$x_1 + 2x_2 - 2x_3 \geq 18$$
$$x_1 - 2x_2 - x_3 \leq 8$$
$$x_1 + x_2 + x_3 \geq 12$$
$$x_1, x_2, x_3 \geq 0$$

5.6. Solve the following problem by the revised simplex algorithm.

$$\text{maximize } z = 3x_1 + 2x_2 + x_3$$

subject to

$$x_1 + x_2 + 2x_3 \leq 30$$
$$x_1 - 3x_2 - x_3 \geq -6$$
$$x_1 + x_2 \leq 12$$
$$2x_1 + 2x_2 - x_3 \leq 20$$
$$x_1, x_2, x_3 \geq 0$$

5.7. Solve Exercise 5.1 by the revised simplex algorithm with the product form of the inverse.

5.8. Solve Exercise 5.2 by the revised simplex algorithm with the product form of the inverse.

5.9. Solve Exercise 5.6 by the revised simplex algorithm with the product form of the inverse.

5.10. Consider the following problem.

$$\text{maximize } z = 2x_1 + x_2$$

subject to

$$x_1 - 4x_2 \leq 8$$
$$x_1 + x_2 \leq 29$$
$$0 \leq x_1 \leq 20$$
$$0 \leq x_2 \leq 12$$

(a) Solve this problem graphically.
(b) Consider the upper-bound constraints, $x_1 \leq 20$, $x_2 \leq 12$, as explicit constraints. Solve the resulting problem with four constraints and nonnegative variables by the standard simplex method. Identify the optimal basis matrix.
(c) Starting with x_1 and x_2 nonbasic at their lower bounds, solve this problem by the bounded-variables simplex method. Identify the optimal working-basis matrix. Compare the solution procedure with that of part (b).

5.11. Consider the following problem.

$$\text{maximize } z = x_1 + 4x_2$$

subject to

$$x_1 + 2x_2 \le 13$$
$$x_1 - x_2 \le 8$$
$$-x_1 + x_2 \le 2$$
$$-3 \le x_1 \le 8$$
$$-5 \le x_2 \ge 4$$

(a) Sketch the feasible region.
(b) For each extreme point of the feasible region, identify the variables that are nonbasic at their lower bound, nonbasic at their lower bound, and basic. Also identify the working-basis matrix for each extreme point.
(c) Starting with x_1 and x_2 nonbasic at their lower bounds, solve this problem by the bounded-variables simplex method.

5.12. Starting with all variables nonbasic at their lower bounds, use the bounded-variables simplex method along with the two-phase method to solve the following problem.

$$\text{maximize } z = 2x_1 - x_2 + 3x_3 + x_4$$

subject to

$$-x_1 - 2x_2 + x_3 + 2x_4 \le 6$$
$$x_1 - x_3 + 2x_4 \ge 2$$
$$2x_1 + x_2 - x_3 - x_4 \le 8$$
$$0 \le x_j \le 5, \qquad \text{for all } j$$

5.13. Solve the following problem by the bounded-variables simplex method.

$$\text{minimize } z = x_1 - 2x_2 + 2x_3 - x_4$$

subject to

$$-x_1 - x_2 + x_3 + x_4 \le 6$$
$$x_2 + x_3 - x_4 \le 12$$
$$x_1 - x_2 + x_3 \ge 6$$
$$x_1 \le 4$$
$$0 \le x_2 \le 7$$
$$2 \le x_3 \le 12$$
$$1 \le x_4 \le 9$$

5.14. Solve the following problem by the bounded-variables simplex method.

$$\text{maximize } z = 4x_1 - 2x_2 + x_3 + 2x_4 + x_5$$

subject to

$$-x_1 - 2x_2 + x_3 + 2x_4 - x_5 \leq 3$$

$$x_1 + x_2 + x_3 + x_4 + 2x_5 \leq 4$$

$$0 \leq x_j \leq 1, \qquad \text{for all } j$$

5.15. Consider the following interval linear programming problem.

$$\text{maximize } z = \mathbf{cx}$$

subject to

$$\mathbf{d} \leq \mathbf{Ax} \leq \mathbf{b}$$

$$\mathbf{0} \leq \mathbf{x} \leq \mathbf{u}$$

(a) Show that this problem can be transformed into the following equivalent bounded-variables linear programming problem.

$$\text{maximize } z = \mathbf{cx}$$

subject to

$$\mathbf{Ax} + \mathbf{s} = \mathbf{b}$$

$$\mathbf{0} \leq \mathbf{x} \leq \mathbf{u}$$

$$\mathbf{0} \leq \mathbf{s} \leq \mathbf{b} - \mathbf{d}$$

(b) What is the major advantage of the formulation in part (a) as compared to the original formulation?

(c) Using the transformation provided in part (a) and the bounded-variables simplex method, solve the following interval linear programming problem.

$$\text{maximize } z = 2x_1 - 3x_2$$

subject to

$$-2 \leq -x_1 + 2x_2 \leq 12$$

$$6 \leq x_1 + x_2 \leq 22$$

$$0 \leq x_1 \leq 18$$

$$0 \leq x_2 \leq 6$$

5.16. Consider the following problem in which a_j, c_j, and $b > 0$. Note that besides the bounds on the variables, there is a only *one* constraint.

$$\text{maximize } z = \sum_{j=1}^{n} c_j x_j$$

subject to

$$\sum_{j=1}^{n} a_j x_j = b$$

$$0 \le x_j \ge u_j, \qquad \text{for all } j$$

(a) Develop a generalized closed-form solution for this problem.

(b) Using the method of part (a), solve the following problem.

maximize $z = 3x_1 + 10x_2 + 5x_3 + 11x_4 + 6x_5 + 14x_6$

subject to

$$x_1 + 8x_2 + 3x_3 + 4x_4 + 2x_5 + 5x_6 = 24$$

$$0 \le x_1 \ge 2$$

$$0 \le x_2 \ge 6$$

$$0 \le x_3 \ge 3$$

$$0 \le x_4 \ge 4$$

$$0 \le x_5 \ge 1$$

$$0 \le x_6 \ge 2$$

DUALITY AND SENSITIVITY ANALYSIS

CHAPTER OVERVIEW

In this chapter, we discuss one of the most important (if not the most important) aspects of linear programming. During the execution of the simplex algorithm, we not only derive the optimal solution for a given linear program, but we also solve a companion problem called the *dual* problem. In this context, the original problem is generally referred to as the *primal* problem. This primal-dual relationship is far more than a curious relationship. In linear programming, we use duality in a wide variety of both theoretical and practical ways. Included among these are the following:

1. In some cases, it may be easier (less iterations, etc.) to solve the dual than the primal.

2. The dual variables provide important economic interpretations of the results obtained when solving a linear programming problem.

3. Duality is used as an aid when investigating changes in the coefficients or formulation of a given linear programming problem (i.e., in *sensitivity analysis*).

4. Duality will be utilized to allow us to employ the simplex method to solve problems in which the initial basis is *infeasible* (the technique itself is known as the *dual simplex*).

5. Duality is used to develop a number of important theoretical results in linear programming.

This chapter begins with a discussion of the formulation of the *canonical* form of the linear programming dual. We then discuss some of the major primal-dual relationships, and take advantage of such properties to develop the dual simplex algorithm. Finally, we use the properties of duality to present economic interpretations and an approach for investigating the impact of changes in the linear programming model structure on the resulting problem solution. Such changes can be broken into two separate classes: discrete changes and continuous changes. Whereas the study of discrete changes is generally referred to as *sensitivity analysis*, the study of continuous systematic changes is called *parametric programming*.

FORMULATION OF THE LINEAR PROGRAMMING DUAL

Associated with each (*primal*) linear programming problem is a companion problem called the *dual*. The formulation of the dual problem is actually a mechanical, straightforward process. However, the mechanics differ somewhat according to the form of the primal. We begin our discussion of duality by considering the canonical form, followed by duality relationships for more general problems.

The Canonical Form of the Dual

The canonical form of a linear programming problem is one in which the objective is to be maximized, all constraints are of the ($\leq$) form, and all variables are restricted to nonnegative values. Thus, the *primal*, in canonical form is

$$\text{(P)} \quad \text{maximize } z = \mathbf{cx} \qquad (6.1)$$

$$\text{subject to}$$

$$\mathbf{Ax} \leq \mathbf{b}$$

$$\mathbf{x} \geq \mathbf{0}$$

Notice in particular that, in the canonical form, elements of the right-hand-side vector (**b**) may be *negative*. Consequently, *any* linear programming problem may be placed into the form (6.1).

If (6.1) is specified to be the primal (P), its *dual* (D) is given by (6.2):

$$\text{(D)} \quad \text{minimize } Z = \mathbf{\pi b} \qquad (6.2)$$

$$\text{subject to}$$

$$\mathbf{\pi A} \geq \mathbf{c}$$

$$\mathbf{\pi} \geq \mathbf{0}$$

where

$$\boldsymbol{\pi} = (\pi_1, \pi_2, \ldots, \pi_m)$$

is the vector of dual variables.

Note that in specifying the dual from the primal, several simple rules have been followed.

1. The objective of the primal is to be maximized; the objective of the dual is to be minimized.
2. The maximization problem must have all ($\leq$) constraints and the minimization problem has all ($\geq$) constraints.
3. All primal and dual variables must be nonnegative.
4. Each *constraint* in one problem corresponds to a *variable* (and vice versa) in the other. For example, given m primal constraints, there are m dual variables, and, given n primal variables, there are n dual constraints. Consequently, if one problem is of order $m \times n$, the other is of order $n \times m$.
5. The elements of the right-hand side of the constraints in one problem are the respective coefficients of the objective function in the other problem.
6. The matrix of constant coefficients for one problem is the transpose of the matrix of constant coefficients for the other problem.

The basic characteristics of the canonical primal and dual are summarized in Table 6.1. As we shall see when discussing general duality, these relationships can be extended to *any* linear programming problem.

TABLE 6.1 CANONICAL PRIMAL-DUAL
RELATIONSHIPS

Maximization problem		Minimization problem
m Constraints		*m Variables*
$\leq$	$\leftrightarrow$	≥ 0
n Variables		*n Constraints*
≥ 0	$\leftrightarrow$	$\geq$

Example 6.1: Formulation of the Canonical Dual

Find the dual of the following problem:

$$\text{maximize } c_1 x_1 + c_2 x_2 + c_3 x_3 \tag{6.3}$$

subject to

$$a_{1,1} x_1 + a_{1,2} x_2 + a_{1,3} x_3 \leq b_1 \tag{6.4}$$

$$a_{2,1}x_1 + a_{2,2}x_2 + a_{2,3}x_3 \leq b_2 \tag{6.5}$$

$$\mathbf{x} \geq \mathbf{0} \tag{6.6}$$

First note that the primal problem is in canonical form. That is, the primal problem is a maximization problem with all constraints of the ($\leq$) form and all variables are nonnegative. Because there are two primal constraints, there must be two dual variables, and, because there are three primal variables, there must be three dual constraints. Further, the coefficients of the primal objective, $\mathbf{c} = (c_1, c_2, c_3)$, become the right-hand-side values of the dual. The right-hand-side values of the primal

$$\mathbf{b} = \begin{pmatrix} b_1 \\ b_2 \end{pmatrix}$$

become the coefficients in the dual objective function. Finally, note that the dual must have a minimizing objective with all constraints of the ($\geq$) form. Thus, the dual can be written as

$$\text{minimize } (\pi_1, \pi_2) \begin{pmatrix} b_1 \\ b_2 \end{pmatrix} \tag{6.7}$$

subject to

$$(\pi_1, \pi_2) \begin{pmatrix} a_{1,1} & a_{1,2} & a_{1,3} \\ a_{2,1} & a_{2,2} & a_{2,3} \end{pmatrix} \geq (c_1, c_2, c_3) \tag{6.8}$$

$$\boldsymbol{\pi} \geq \mathbf{0} \tag{6.9}$$

which simplifies to

$$\text{minimize } b_1\pi_1 + b_2\pi_2 \tag{6.10}$$

subject to

$$a_{1,1}\pi_1 + a_{2,1}\pi_2 \geq c_1 \tag{6.11}$$

$$a_{1,2}\pi_1 + a_{2,2}\pi_2 \geq c_2 \tag{6.12}$$

$$a_{1,3}\pi_1 + a_{2,3}\pi_2 \geq c_3 \tag{6.13}$$

$$\boldsymbol{\pi} \geq \mathbf{0} \tag{6.14}$$

Example 6.2: Formulation of the Canonical Dual

Find the dual of the following problem:

$$\text{maximize } 4x_1 + 2x_2 \tag{6.15}$$

subject to

$$x_1 + x_2 \geq 2 \tag{6.16}$$

$$x_1 + 2x_2 \leq 15 \tag{6.17}$$

$$2x_1 - x_2 \leq 12 \qquad (6.18)$$

$$\mathbf{x} \geq \mathbf{0} \qquad (6.19)$$

Note that the problem is a maximization problem with nonnegative variables; however, the first constraint is not of the ($\leq$) form. Thus, the proper canonical form for the example is

$$\text{maximize } 4x_1 + 2x_2 \qquad (6.20)$$

subject to

$$-x_1 - x_2 \leq -2 \qquad (6.21)$$

$$x_1 + 2x_2 \leq 15 \qquad (6.22)$$

$$2x_1 - x_2 \leq 12 \qquad (6.23)$$

$$\mathbf{x} \geq \mathbf{0} \qquad (6.24)$$

Because there are three primal constraints, there must be three dual variables, and, because there are two primal variables, there must be two dual constraints. Denoting the dual variables by $\boldsymbol{\pi} = (\pi_1, \pi_2, \pi_3)$, the dual problem can be written as follows:

$$\text{minimize } -2\pi_1 + 15\pi_2 + 12\pi_3 \qquad (6.25)$$

subject to

$$-\pi_1 + \pi_2 + 2\pi_3 \geq 4 \qquad (6.26)$$

$$-\pi_1 + 2\pi_2 - \pi_3 \geq 2 \qquad (6.27)$$

$$\boldsymbol{\pi} \geq \mathbf{0} \qquad (6.28)$$

While introducing the concept of duality, we denoted the maximization problem as the primal and the minimization problem as the dual. However, as the following result illustrates, this is a completely arbitrary choice. That is, if (6.1) is the primal, then (6.2) is its dual. However, we could just as well designate (6.2) as the primal, and then (6.1) is its dual. For this reason, (6.1) and (6.2) are often referred to as a primal-dual pair. That is, if one problem is designated as the primal, then the other problem is its dual.

Theorem 6.1
The dual of the dual is the primal.

Proof
Consider again the problem given by

$$\text{minimize } \boldsymbol{\pi}\mathbf{b} \qquad (6.29)$$

subject to

$$\boldsymbol{\pi}\mathbf{A} \geq \mathbf{c}$$

$$\boldsymbol{\pi} \geq \mathbf{0}$$

Now, the first step in writing the dual of this problem is to transform it into the form of the canonical primal. This results in

$$\text{maximize } -\mathbf{b}'\boldsymbol{\pi}^t \tag{6.30}$$

subject to

$$-\mathbf{A}^t\boldsymbol{\pi}^t \leq -\mathbf{c}^t$$

$$\boldsymbol{\pi}^t \geq \mathbf{0}$$

Now, letting $\mathbf{w} = (w_1, \ldots, w_n)$ represent the dual variables of this transformed problem, writing the canonical dual yields

$$\text{minimize } \mathbf{w}(-\mathbf{c}^t) \tag{6.31}$$

subject to

$$\mathbf{w}(-\mathbf{A}^t) \geq -\mathbf{b}^t$$

$$\mathbf{w} \geq \mathbf{0}$$

But the problem specified in (6.31) can be rewritten as

$$\text{maximize } \mathbf{c}\mathbf{w}^t \tag{6.32}$$

subject to

$$\mathbf{A}\mathbf{w}^t \leq \mathbf{b}$$

$$\mathbf{w}^t \geq \mathbf{0}$$

Finally, letting $\mathbf{x} = \mathbf{w}^t$ yields precisely the primal problem defined in (6.1) and the proof is complete. □

It is not necessary to use the canonical form when formulating the dual. However, because of the symmetry that exists, some analysts do tend to prefer this approach. In fact, in succeeding sections, we will use the canonical form to establish some important duality results and relationships. But before proceeding with that analysis, let us consider the formulation of the dual from the *general* form of the primal.

General Duality

There are few requirements as to the general form of a linear programming problem. The objective may be either of a maximizing or minimizing form, variables may be restricted or unrestricted, and the constraints may be of any form ($\leq$, $\geq$, $=$) and of any mixture of the forms. To help motivate the rules for finding the dual from the general form, consider the following linear program.

$$\text{(P)} \quad \text{maximize } c_1x_1 + c_2x_2 + c_3x_3 \tag{6.33}$$

subject to

$$a_{1,1}x_1 + a_{1,2}x_2 + a_{1,3}x_3 \leq b_1 \tag{6.34}$$

$$a_{2,1}x_1 + a_{2,2}x_2 + a_{2,3}x_3 \geq b_2 \tag{6.35}$$

$$a_{3,1}x_1 + a_{3,2}x_2 + a_{3,3}x_3 = b_3 \tag{6.36}$$

$$x_1 \text{ unrestricted} \tag{6.37}$$

$$x_2 \geq 0 \tag{6.38}$$

$$x_3 \leq 0 \tag{6.39}$$

Notice that this linear program contains one constraint of each form as well as variables of different types. Let us derive the general dual of this problem in steps. First, we will transform the problem to canonical form, obtain the canonical dual, and then rewrite the dual in the corresponding general form. This will serve to illustrate the origin of the rules for general duality. After summarizing the general rules, we will illustrate their direct use via additional examples.

In order to transform the earlier linear program into canonical form, several things must be done. Constraint (6.35) must be multiplied by -1 to reverse the direction of the inequality. It is also necessary to rewrite equality constraint (6.36) as two inequality constraints. Unrestricted variable x_1 must be replaced by the difference of two nonnegative variables, that is, $x_1 = x_1^+ - x_1^-$, where $x_1^+, x_1^- \geq 0$. Finally, the nonpositive variable x_3 must be replaced by the nonnegative variable $x_3' = -x_3$. These changes can be summarized as follows:

$$\text{maximize } c_1 x_1^+ - c_1 x_1^- + c_2 x_2 - c_3 x_3' \tag{6.40}$$

subject to

$$a_{1,1}x_1^+ - a_{1,1}x_1^- + a_{1,2}x_2 - a_{1,3}x_3' \leq b_1 \tag{6.41}$$

$$-a_{2,1}x_1^+ + a_{2,1}x_1^- - a_{2,2}x_2 + a_{2,3}x_3' \leq -b_2 \tag{6.42}$$

$$a_{3,1}x_1^+ - a_{3,1}x_1^- + a_{3,2}x_2 - a_{3,3}x_3' \leq b_3 \tag{6.43}$$

$$-a_{3,1}x_1^+ + a_{3,1}x_1^- - a_{3,2}x_2 + a_{3,3}x_3' \leq -b_3 \tag{6.44}$$

$$x_1^+, x_1^-, x_2, x_3' \geq 0 \tag{6.45}$$

Now, letting ω_1, ω_2, ω_3, and ω_4 denote the dual variables corresponding to constraints (6.41–6.44), respectively, the canonical dual problem can be written as

$$\text{minimize } b_1\omega_1 - b_2\omega_2 + b_3\omega_3 - b_3\omega_4 \tag{6.46}$$

subject to

$$a_{1,1}\omega_1 - a_{2,1}\omega_2 + a_{3,1}\omega_3 - a_{3,1}\omega_4 \geq c_1 \tag{6.47}$$

$$-a_{1,1}\omega_1 + a_{2,1}\omega_2 - a_{3,1}\omega_3 + a_{3,1}\omega_4 \geq -c_1 \tag{6.48}$$

$$a_{1,2}\omega_1 - a_{2,2}\omega_2 + a_{3,2}\omega_3 - a_{3,2}\omega_4 \geq c_2 \tag{6.49}$$

$$-a_{1,3}\omega_1 + a_{2,3}\omega_2 - a_{3,3}\omega_3 + a_{3,3}\omega_4 \geq -c_3 \tag{6.50}$$

$$\omega_1, \omega_2, \omega_3, \omega_4 \geq 0 \tag{6.51}$$

Rearranging, we have

$$\text{minimize } b_1\omega_1 - b_2\omega_2 + b_3(\omega_3 - \omega_4) \tag{6.52}$$

subject to

$$a_{1,1}\omega_1 - a_{2,1}\omega_2 + a_{3,1}(\omega_3 - \omega_4) \geq c_1 \tag{6.53}$$

$$a_{1,1}\omega_1 - a_{2,1}\omega_2 + a_{3,1}(\omega_3 - \omega_4) \leq c_1 \tag{6.54}$$

$$a_{1,2}\omega_1 - a_{2,2}\omega_2 + a_{3,2}(\omega_3 - \omega_4) \geq c_2 \tag{6.55}$$

$$a_{1,3}\omega_1 - a_{2,3}\omega_2 + a_{3,3}(\omega_3 - \omega_4) \leq c_3 \tag{6.56}$$

$$\omega_1, \omega_2, \omega_3, \omega_4 \geq 0 \tag{6.57}$$

Finally, we can write the general dual of the original primal (P) by letting $\pi_1 = \omega_1$, $\pi_2 = -\omega_2$, and $\pi_3 = \omega_3 - \omega_4$, and converting constraints (6.53) and (6.54) into a single equality constraint. Note that $\pi_2 \leq 0$ because $\omega_2 \geq 0$, and π_3 is an unrestricted variable because it is the difference of two nonnegative variables. The resulting dual problem is

$$\text{(D)} \quad \text{minimize } b_1\pi_1 + b_2\pi_2 + b_3\pi_3 \tag{6.58}$$

subject to

$$a_{1,1}\pi_1 + a_{2,1}\pi_2 + a_{3,1}\pi_3 = c_1 \tag{6.59}$$

$$a_{1,2}\pi_1 + a_{2,2}\pi_2 + a_{3,2}\pi_3 \geq c_2 \tag{6.60}$$

$$a_{1,3}\pi_1 + a_{2,3}\pi_2 + a_{3,3}\pi_3 \leq c_3 \tag{6.61}$$

$$\pi_1 \geq 0 \tag{6.62}$$

$$\pi_2 \leq 0 \tag{6.63}$$

$$\pi_3 \text{ unrestricted} \tag{6.64}$$

The reader should carefully compare this dual problem with the primal linear program (P) defined in (6.33–6.39). The variables π_1, π_2, and π_3 are the dual variables for constraints (6.34–6.36), respectively. Likewise, x_1, x_2, and x_3 correspond to constraints (6.59–6.61), respectively. In particular, note that there is a definite relationship between the type of variable and the form of the corresponding constraint. For example, each unrestricted variable corresponds to an equality constraint. The relationships between the primal and dual problem are summarized in Table 6.2.

By utilizing the relationships in Table 6.2, it is possible to write the dual problem for a given linear program *without* going through the intermediate step of

TABLE 6.2   PRIMAL-DUAL RELATIONSHIPS

Maximization problem		Minimization problem
Constraints		*Variables*
$\leq$	$\leftrightarrow$	≥ 0
$\geq$	$\leftrightarrow$	≤ 0
$=$	$\leftrightarrow$	unrestricted
Variables		*Constraints*
≥ 0	$\leftrightarrow$	$\geq$
≤ 0	$\leftrightarrow$	$\leq$
unrestricted	$\leftrightarrow$	$=$

transforming the problem to canonical form. To help illustrate the use of these rules, consider the following examples.

Example 6.3: Formulation of the General Dual

Find the dual of

$$\text{maximize } 4x_1 + 2x_2 - x_3 \tag{6.65}$$

subject to

$$x_1 + x_2 + x_3 = 20 \tag{6.66}$$

$$2x_1 - x_2 \geq 6 \tag{6.67}$$

$$3x_1 + 2x_2 + x_3 \leq 40 \tag{6.68}$$

$$x_1, x_2 \geq 0 \tag{6.69}$$

$$x_3 \text{ unrestricted} \tag{6.70}$$

By letting π_1, π_2, and π_3 denote the dual variables for constraints (6.66–6.68), respectively, the results of Table 6.2 yield the following dual problem:

$$\text{minimize } 20\pi_1 + 6\pi_2 + 40\pi_3 \tag{6.71}$$

subject to

$$\pi_1 + 2\pi_2 + 3\pi_3 \geq 4 \tag{6.72}$$

$$\pi_1 - \pi_2 + 2\pi_3 \geq 2 \tag{6.73}$$

$$\pi_1 + 2\pi_2 + 3\pi_3 = -1 \tag{6.74}$$

$$\pi_1 \text{ unrestricted} \tag{6.75}$$

$$\pi_2 \leq 0 \tag{6.76}$$

$$\pi_3 \geq 0 \tag{6.77}$$

Example 6.4: Formulation of the General Dual

Find the dual of

$$\text{minimize } 5x_1 + 3x_2 \tag{6.78}$$

subject to

$$x_1 - 6x_2 \geq 2 \tag{6.79}$$

$$5x_1 + 7x_2 = -4 \tag{6.80}$$

$$x_1 + 2x_2 \leq 10 \tag{6.81}$$

$$x_1 \text{ unrestricted} \tag{6.82}$$

$$x_2 \geq 0 \tag{6.83}$$

By letting π_1, π_2, and π_3 denote the dual variables for constraints (6.79–6.81), respectively, the general form of the dual can be written as follows:

$$\text{maximize } 2\pi_1 - 4\pi_2 + 10\pi_3 \tag{6.84}$$

subject to

$$\pi_1 + 5\pi_2 + \pi_3 = 5 \tag{6.85}$$

$$-6\pi_1 + 7\pi_2 + 2\pi_3 \leq 3 \tag{6.86}$$

$$\pi_1 \geq 0 \tag{6.87}$$

$$\pi_2 \text{ unrestricted} \tag{6.88}$$

$$\pi_3 \leq 0 \tag{6.89}$$

The Standard Form

There is one other form of the primal that is commonly used. This is the form in which the constraints of the primal are *all* equalities. A problem in this form is said to be in the *standard* form. The rules for the general form may be applied to the standard form, and thus it simply represents a special case of the general form. Because all constraints in the primal are equalities, all dual variables will be unrestricted. By utilizing Table 6.2, the standard form can be written as follows:

$$\text{(P)} \quad \text{maximize } \mathbf{cx} \tag{6.90}$$

subject to

$$\mathbf{Ax} = \mathbf{b}$$

$$\mathbf{x} \geq \mathbf{0}$$

$$\text{(D)} \quad \text{minimize } \boldsymbol{\pi}\mathbf{b} \tag{6.91}$$

subject to

$$\boldsymbol{\pi}\mathbf{A} \geq \mathbf{c}$$

$$\boldsymbol{\pi} \text{ unrestricted}$$

RELATIONSHIPS IN DUALITY

There are a number of useful primal-dual relationships that one often needs in either problem solving or theoretical work. Let us begin our discussion with one of the most fundamental results concerning the objective values of the respective problems. This result leads, in turn, to other useful duality results and properties. The concept of *weak duality* essentially states that the objective value associated with any feasible solution to the maximization problem is less than or equal to the objective value of any feasible solution to the minimization problem.

Theorem 6.2 (Weak Duality)
Consider the following primal-dual pair.

$$(P) \quad \text{maximize } z = \mathbf{cx} \tag{6.92}$$

$$\text{subject to}$$

$$\mathbf{Ax} \leq \mathbf{b}$$

$$\mathbf{x} \geq \mathbf{0}$$

$$(D) \quad \text{minimize } Z = \boldsymbol{\pi}\mathbf{b} \tag{6.93}$$

$$\text{subject to}$$

$$\boldsymbol{\pi}\mathbf{A} \geq \mathbf{c}$$

$$\boldsymbol{\pi} \geq \mathbf{0}$$

Let $\bar{\mathbf{x}}$ be a feasible solution to the maximization problem (P) and let $\bar{\boldsymbol{\pi}}$ be a feasible solution to the minimization problem (D). Then, $z = \mathbf{c}\bar{\mathbf{x}} \leq \bar{\boldsymbol{\pi}}\mathbf{b} = Z$.

Proof
Because $\bar{\mathbf{x}}$ is feasible to (P), it follows from (6.92) that

$$\mathbf{A}\bar{\mathbf{x}} \leq \mathbf{b} \tag{6.94}$$

$$\bar{\mathbf{x}} \geq \mathbf{0} \tag{6.95}$$

Similarly, because $\bar{\boldsymbol{\pi}}$ is feasible to (D), we have

$$\bar{\boldsymbol{\pi}}\mathbf{A} \geq \mathbf{c} \tag{6.96}$$

$$\bar{\boldsymbol{\pi}} \geq \mathbf{0} \tag{6.97}$$

Now because $\bar{\mathbf{x}}$ is nonnegative, we can multiply both sides of (6.96) by $\bar{\mathbf{x}}$ to obtain

$$\bar{\boldsymbol{\pi}}\mathbf{A}\bar{\mathbf{x}} \geq \mathbf{c}\bar{\mathbf{x}} \tag{6.98}$$

In an analogous manner, (6.97) and (6.94) yield

$$\bar{\boldsymbol{\pi}}\mathbf{A}\bar{\mathbf{x}} \leq \bar{\boldsymbol{\pi}}\mathbf{b} \tag{6.99}$$

Finally combining the results of (6.98) and (6.99), we get

$$\mathbf{c\bar{x}} \leq \boldsymbol{\pi} \mathbf{A\bar{x}} \leq \boldsymbol{\pi}\mathbf{b} \tag{6.100}$$

and the proof is complete.   □

This is a very important result that forms the foundation for several other duality relationships. First of all, notice that each feasible solution to the maximization problem provides a lower bound for the objective of the minimization problem, and, likewise, each feasible solution to the minimization problem provides an upper bound for the objective of the maximization problem. These observations lead directly to the following results.

Corollary 6.3
If the primal objective is unbounded, then the dual problem is infeasible.

Corollary 6.4
If the dual objective is unbounded, then the primal problem is infeasible.

At first glance, the reader may also try to conclude that the converse of each corollary is true. However, this is not the case, because if one problem is infeasible, it is also possible for the other to be infeasible. This is illustrated via the following example.

Example 6.5: Infeasible Primal and Dual
Consider the following canonical primal-dual pair:

$$\text{(P)} \quad \text{maximize } x_1 + 2x_2 \tag{6.101}$$

$$\text{subject to}$$

$$-x_1 + 2x_2 \leq -2 \tag{6.102}$$

$$x_1 - 2x_2 \leq -2 \tag{6.103}$$

$$x_1, x_2 \geq 0 \tag{6.104}$$

$$\text{(D)} \quad \text{minimize } -2\pi_1 - 2\pi_2 \tag{6.105}$$

$$\text{subject to}$$

$$-\pi_1 + \pi_2 \geq 1 \tag{6.106}$$

$$2\pi_1 - 2\pi_2 \geq 2 \tag{6.107}$$

$$\pi_1, \pi_2 \geq 0 \tag{6.108}$$

Upon graphing, it is clear from Figure 6.1 that neither the primal nor the dual possesses a feasible solution. Thus, based on the observations in the previous example and Theorem 6.2, we can state the following:

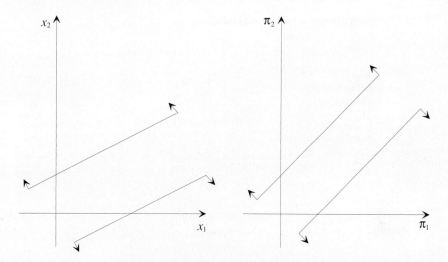

Figure 6.1   Infeasible primal and dual.

Corollary 6.5
If the primal is infeasible, then the dual is either infeasible or has an unbounded objective.

Corollary 6.6
If the dual is infeasible, then the primal is either infeasible or has an unbounded objective.

There is a very helpful diagram that often clarifies some of the foregoing relationships. Figure 6.2 depicts the primal-dual space for the canonical form.

Notice the infeasible regions for the primal and dual. A value of z in the infeasible region of the primal is "superoptimal" because it is greater than z^*. However, the constraints of the primal are violated. The same thing holds true for the dual except that we refer to minimization rather than maximization. Finally, note that the *only* time *both* problems are simultaneously feasible is when *both* are optimal. This observation is summarized in the following corollary.

Corollary 6.7
If $\bar{\mathbf{x}}$ is feasible to (P), and $\bar{\bar{\pi}}$ is feasible to (D), and $\mathbf{c}\bar{\mathbf{x}} = \bar{\bar{\pi}}\mathbf{b}$, then $\bar{\mathbf{x}}$ is an optimal solution to (P) and $\bar{\bar{\pi}}$ is an optimal solution to (D).

We now consider further relationships between the canonical primal and dual, as well as an actual method for computing the dual solution.

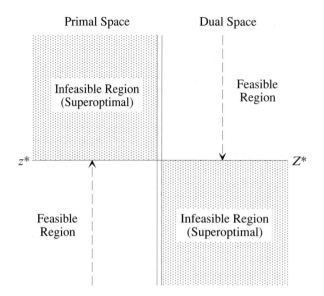

Figure 6.2   The primal-dual space.

Theorem 6.8 (Strong Duality)
Consider the following canonical primal-dual pair:

$$\text{(P)}\quad \text{maximize } \mathbf{cx} \qquad\qquad (6.109)$$

$$\text{subject to}$$

$$\mathbf{Ax} \leq \mathbf{b}$$

$$\mathbf{x} \geq \mathbf{0}$$

$$\text{(D)}\quad \text{minimize } \boldsymbol{\pi}\mathbf{b} \qquad\qquad (6.110)$$

$$\text{subject to}$$

$$\boldsymbol{\pi}\mathbf{A} \geq \mathbf{c}$$

$$\boldsymbol{\pi} \geq \mathbf{0}$$

If (P) has a feasible solution and (D) has a feasible solution, then both problems have finite optimal solutions with equal objectives.

Proof
Because (P) is feasible, it follows from Corollary 6.3 that (D) has a finite optimal solution. Likewise, because (D) is feasible, Corollary 6.4 implies that (P) has a finite optimal solution.

Now because (P) has a finite optimal solution, (P) has an extreme-point optimal solution corresponding to at least one basic feasible solution. Let **B** be an optimal basis matrix for problem (P) in standard form:

$$\text{maximize } \mathbf{cx} \qquad (6.111)$$

subject to

$$\mathbf{Ax} + \mathbf{x}_s = \mathbf{b}$$

$$\mathbf{x}, \mathbf{x}_s \geq \mathbf{0}$$

Then, $z = \mathbf{cx}^* = \mathbf{c}_B\mathbf{B}^{-1}\mathbf{b}$. By Corollary 6.7, the proof will be complete if we can produce a feasible solution to (D), which has the same objective value. Consider $\overline{\pi} = \mathbf{c}_B\mathbf{B}^{-1}$. Clearly, $\mathbf{cx}^* = \overline{\pi}\mathbf{b} = \mathbf{c}_B\mathbf{B}^{-1}\mathbf{b}$. Thus, it only remains to be shown that $\overline{\pi}$ is a feasible solution to (D).

Consider the initial simplex tableau corresponding to (6.111), as shown in Table 6.3. Now, because $\mathbf{B}$ is an optimal basis matrix, it follows from the results of Chapter 4 that the optimal tableau will be as in Table 6.4. Because Table 6.4 is the optimal tableau, it follows that the optimality conditions are satisfied, that is,

$$\mathbf{c}_B\mathbf{B}^{-1}\mathbf{A} - c \geq \mathbf{0} \qquad (6.112)$$

and

$$\mathbf{c}_B\mathbf{B}^{-1}\mathbf{I} - \mathbf{0} = \mathbf{c}_B\mathbf{B}^{-1} \geq \mathbf{0} \qquad (6.113)$$

But recall that $\overline{\pi} = \mathbf{c}_B\mathbf{B}^{-1}$. Now, substituting into (6.112) and (6.113) yields

$$\overline{\pi}\mathbf{A} - c \geq \mathbf{0} \qquad (6.114)$$

$$\overline{\pi} \geq \mathbf{0} \qquad (6.115)$$

TABLE 6.3 ORIGINAL SIMPLEX TABLEAU

	z	$\mathbf{x}$	$\mathbf{x}_s$	RHS
z	1	$-\mathbf{c}$	$\mathbf{0}$	$\mathbf{0}$
$\mathbf{x}_B$	0	$\mathbf{A}$	$\mathbf{I}$	$\mathbf{b}$

TABLE 6.4 OPTIMAL SIMPLEX TABLEAU

	z	$\mathbf{x}$	$\mathbf{x}_s$	RHS
z	1	$\mathbf{c}_B\mathbf{B}^{-1}\mathbf{A} - \mathbf{c}$	$\mathbf{c}_B\mathbf{B}^{-1}\mathbf{I} - \mathbf{0}$	$\mathbf{c}_B\mathbf{B}^{-1}\mathbf{b}$
$\mathbf{x}_B$	0	$\mathbf{B}^{-1}\mathbf{A}$	$\mathbf{B}^{-1}\mathbf{I}$	$\mathbf{B}^{-1}\mathbf{b}$

which are precisely the dual feasibility conditions. Thus, $\bar{\pi}$ is dual feasible and $\bar{\pi}\mathbf{b} = \mathbf{c}\mathbf{x}^*$. $\square$

First of all, note that dual feasibility conditions are precisely the same as primal optimality conditions. In an analogous manner, it can be shown that primal feasibility conditions are exactly the same as dual optimality conditions. Also observe that Theorem 6.8 provides a method for computing the values of the dual variables. That is, whereas the primal solution can be written as

$$\mathbf{x}_N = \mathbf{0} \tag{6.116}$$

$$\mathbf{x}_B = \mathbf{B}^{-1}\mathbf{b} = \boldsymbol{\beta} \tag{6.117}$$

the dual solution is given by

$$\boldsymbol{\pi} = \mathbf{c}_B \mathbf{B}^{-1} \tag{6.118}$$

$$\boldsymbol{\lambda} = \boldsymbol{\pi}\mathbf{A} - \mathbf{c} = \mathbf{c}_B\mathbf{B}^{-1}\mathbf{A} - \mathbf{c} \tag{6.119}$$

where $\boldsymbol{\lambda}$ is the vector of dual surplus variables. Finally, the objective value of both problems is

$$z = \mathbf{c}\mathbf{x} = \boldsymbol{\pi}\mathbf{b} = \mathbf{c}_B\mathbf{B}^{-1}\mathbf{b} \tag{6.120}$$

Thus, given a basis matrix $\mathbf{B}$, the solutions to both problems can be determined directly from $\mathbf{B}^{-1}$, $\mathbf{b}$, and $\mathbf{c}$. In the following section, we establish some additional relationships between the primal and dual variables. These relationships further lead to conditions that are referred to as *complementary slackness* conditions. Complementary slackness conditions have important theoretical implications as well as valuable economic interpretations.

Primal-Dual Tableau Relationships

Note that the tableaux depicted in Tables 6.3 and 6.4 also establish some relationships between the primal and dual variables. To see this more clearly, let us rewrite Table 6.4 utilizing the fact that

$$\boldsymbol{\pi} = \mathbf{c}_B\mathbf{B}^{-1} \tag{6.121}$$

$$\boldsymbol{\lambda} = \boldsymbol{\pi}\mathbf{A} - \mathbf{c} = \mathbf{c}_B\mathbf{B}^{-1}\mathbf{A} - \mathbf{c} \tag{6.122}$$

This results in Table 6.5.

TABLE 6.5   PRIMAL SIMPLEX TABLEAU

	z	$\mathbf{x}$	$\mathbf{x}_s$	RHS
z	1	$\boldsymbol{\lambda}$	$\boldsymbol{\pi}$	$\mathbf{c}_B\mathbf{B}^{-1}\mathbf{b}$
$\mathbf{x}_B$	0	$\mathbf{B}^{-1}\mathbf{A}$	$\mathbf{B}^{-1}$	$\mathbf{B}^{-1}\mathbf{b}$

First, note that the $z_j - c_j$ values for the primal decision variables **x** are given by the dual surplus variables $\boldsymbol{\lambda}$. Just as $\mathbf{B}^{-1}$ resides in the portion of the tableau that was occupied by the original identity, $\boldsymbol{\pi} = \mathbf{c}_B\mathbf{B}^{-1}$ is located in the top row immediately above $\mathbf{B}^{-1}$. However, as we saw in Table 6.4, this is only true if the original objective coefficients of the corresponding slack variables are zero. Thus, the $z_j - c_j$ values for the zero-cost primal slack variables $\mathbf{x}_s$ are given by the dual decision variables $\boldsymbol{\pi}$. (The reader should recall that $\boldsymbol{\pi} = \mathbf{c}_B\mathbf{B}^{-1}$ is precisely what is stored in the revised simplex tableau.)

Thus, given a simplex tableau, it is possible to read the solutions to both problems directly from the tableau. This idea is demonstrated further via the following example.

Example 6.6: Tableau Relationships

Consider the following problem, which we shall arbitrarily designate as the primal. Its final tableau is given in Table 6.6.

$$\text{maximize } z = 6x_1 + 3x_2 + 4x_3 \tag{6.123}$$

$$\text{subject to}$$

$$x_1 + 2x_2 + x_3 \leq 8 \tag{6.124}$$

$$5x_1 + 4x_2 + 3x_3 \leq 25 \tag{6.125}$$

$$x_1, x_2, x_3 \geq 0 \tag{6.126}$$

TABLE 6.6

	z	x_1	x_2	x_3	x_4	x_5	RHS
z	1	0	3	0	1	1	33
x_1	0	1	-1	0	$-\frac{3}{2}$	$\frac{1}{2}$	$\frac{1}{2}$
x_3	0	0	3	1	$\frac{5}{2}$	$-\frac{1}{2}$	$\frac{15}{2}$

Slack variable x_3 has been added to constraint (6.124) and slack variable x_4 was added to constraint (6.125).

The tableau of Table 6.6 indicates that the optimal primal solution is given by

$$z^* = 33$$

$$x_1^* = \tfrac{1}{2}$$

$$x_2^* = 0$$

$$x_3^* = \tfrac{15}{2}$$

$$x_4^* = 0$$

$$x_5^* = 0$$

Now, denote the dual decision variables by π_1 and π_2 corresponding to constraints (6.124) and (6.125), respectively. Also, let λ_1, λ_2, and λ_3 represent the respective surplus variables for the three dual constraints. Then, by using the tableau relationships established in Table 6.5, the top row of the tableau will be in the following form:

	z	x_1	x_2	x_3	x_4	x_5	RHS
z	1	λ_1	λ_2	λ_3	π_1	π_2	33

By comparing this with Table 6.6, it immediately follows that the dual solution is given by

$$Z^* = 33$$

$$\pi_1^* = z_4 - c_4 = 1$$

$$\pi_2^* = z_5 - c_5 = 1$$

$$\lambda_1^* = z_1 - c_1 = 0$$

$$\lambda_2^* = z_2 - c_2 = 3$$

$$\lambda_3^* = z_3 - c_3 = 0$$

Note that in the previous example x_1 was a basic variable in the optimal tableau and the associated dual variable $\lambda_1 = z_1 - c_1$ was equal to zero. Similarly, x_3 was basic and $\lambda_3 = z_3 - c_3$ was also zero. Finally, note that π_1, π_2, and λ_2 are all positive in the optimal tableau, and, consequently, the associated variables x_4, x_5, and x_2, respectively, are all nonbasic with value zero. That is, when a particular variable (whether a primal or a dual variable) has a positive value, then its associated dual variable has the value zero. This phenomenon is not a coincidence but, in fact, an important duality relationship called *complementary slackness*. This concept is formally defined and its significance is discussed in the following section.

Complementary Slackness

Consider again the following canonical primal-dual pair.

$$(P) \quad \text{maximize } \mathbf{cx} \qquad\qquad (6.127)$$

$$\text{subject to}$$

$$\mathbf{Ax} \leq \mathbf{b}$$

$$\mathbf{x} \geq \mathbf{0}$$

$$(D) \quad \text{minimize } \boldsymbol{\pi}\mathbf{b} \qquad\qquad (6.128)$$

$$\text{subject to}$$

$$\boldsymbol{\pi}\mathbf{A} \geq \mathbf{c}$$

$$\boldsymbol{\pi} \geq \mathbf{0}$$

Recall that $\boldsymbol{\pi} = (\pi_1, \ldots, \pi_m)$, where π_i is the dual variable corresponding to the constraint $\mathbf{a}^i\mathbf{x} \leq b_i$ (where the notation $\mathbf{a}^i$ represents the ith row of matrix $\mathbf{A}$). Similarly, x_j is the dual variable for the constraint $\boldsymbol{\pi}\mathbf{a}_j \geq c_j$. The *complementary slackness conditions* can then be stated as follows:

$$\pi_i(b_i - \mathbf{a}^i\mathbf{x}) = 0, \qquad \text{for all } i = 1, \ldots, m \qquad (6.129)$$

$$(\boldsymbol{\pi}\mathbf{a}_j - c_j)x_j = 0, \qquad \text{for all } j = 1, \ldots, n \qquad (6.130)$$

In words, these conditions state that if a constraint is nonbinding, then the associated dual variable is zero. Or, equivalently, if a dual variable is positive, then the corresponding constraint is binding. This is an important theoretical concept, and, as we will see in the following section, it also has a meaningful interpretation from an economic point of view.

Now, let $x_{s,i}$ be the slack variable in primal constraint i and let λ_j be the surplus variable in dual constraint j. That is,

$$\mathbf{a}^i\mathbf{x} + x_{s,i} = b_i \qquad (6.131)$$

$$\boldsymbol{\pi}\mathbf{a}_j - \lambda_j = c_j \qquad (6.132)$$

or, in matrix form

$$\mathbf{A}\mathbf{x} + \mathbf{x}_s = \mathbf{b} \qquad (6.133)$$

$$\boldsymbol{\pi}\mathbf{A} - \boldsymbol{\lambda} = \mathbf{c} \qquad (6.134)$$

Then, the complementary slackness conditions can be written as

$$\pi_i x_{s,i} = 0, \qquad \text{for all } i = 1, \ldots, m \qquad (6.135)$$

$$\lambda_j x_j = 0, \qquad \text{for all } j = 1, \ldots, n \qquad (6.136)$$

And because $\mathbf{x}$, $\mathbf{x}_s$, $\boldsymbol{\pi}$, and $\boldsymbol{\lambda}$ are all nonnegative, the complementary slackness conditions can also be written in the compact form:

$$\boldsymbol{\pi}\mathbf{x}_s = \sum_{i=1}^{m} \pi_i x_{s,i} = 0 \qquad (6.137)$$

$$\boldsymbol{\lambda}\mathbf{x} = \sum_{j=1}^{n} \lambda_j x_j = 0 \qquad (6.138)$$

or

$$\boldsymbol{\pi}\mathbf{x}_s + \boldsymbol{\lambda}\mathbf{x} = 0 \qquad (6.139)$$

The relationship of complementary slackness to the optimal solution of a linear programming problem is now derived via the following theorem.

Theorem 6.9 (Complementary Slackness)

Consider the following primal-dual pair that has been converted to standard form by adding the appropriate slack/surplus variables.

$$\text{(P)} \quad \text{maximize } \mathbf{cx} \tag{6.140}$$

$$\text{subject to}$$

$$\mathbf{Ax} + \mathbf{x}_s = \mathbf{b}$$

$$\mathbf{x}, \mathbf{x}_s \geq \mathbf{0}$$

$$\text{(D)} \quad \text{minimize } \boldsymbol{\pi}\mathbf{b} \tag{6.141}$$

$$\text{subject to}$$

$$\boldsymbol{\pi}\mathbf{A} - \boldsymbol{\lambda} = \mathbf{c}$$

$$\boldsymbol{\pi}, \boldsymbol{\lambda} \geq \mathbf{0}$$

Let $(\bar{\mathbf{x}}, \bar{\mathbf{x}}_s)$ be feasible to (P) and let $(\bar{\boldsymbol{\pi}}, \bar{\boldsymbol{\lambda}})$ be feasible to (D). Then $(\bar{\mathbf{x}}, \bar{\mathbf{x}}_s)$ is optimal to (P) and $(\bar{\boldsymbol{\pi}}, \bar{\boldsymbol{\lambda}})$ is optimal to (D) if and only if complementary slackness holds.

Proof

Because $(\bar{\mathbf{x}}, \bar{\mathbf{x}}_s)$ is feasible to (P), it follows from (6.140) that

$$\mathbf{A}\bar{\mathbf{x}} + \bar{\mathbf{x}}_s = \mathbf{b} \tag{6.142}$$

$$\bar{\mathbf{x}}, \bar{\mathbf{x}}_s \geq \mathbf{0} \tag{6.143}$$

Similarly, $(\bar{\boldsymbol{\pi}}, \bar{\boldsymbol{\lambda}})$ a feasible solution to (D) implies that

$$\bar{\boldsymbol{\pi}}\mathbf{A} - \bar{\boldsymbol{\lambda}} = \mathbf{c} \tag{6.144}$$

$$\bar{\boldsymbol{\pi}}, \bar{\boldsymbol{\lambda}} \geq \mathbf{0} \tag{6.145}$$

Now, multiplying (6.142) by $\bar{\boldsymbol{\pi}}$ and multiplying (6.144) by $\bar{\mathbf{x}}$ yield the following:

$$\bar{\boldsymbol{\pi}}\mathbf{A}\bar{\mathbf{x}} + \bar{\boldsymbol{\pi}}\bar{\mathbf{x}}_s = \bar{\boldsymbol{\pi}}\mathbf{b} \tag{6.146}$$

$$\bar{\boldsymbol{\pi}}\mathbf{A}\bar{\mathbf{x}} - \bar{\boldsymbol{\lambda}}\bar{\mathbf{x}} = \mathbf{c}\bar{\mathbf{x}} \tag{6.147}$$

Subtracting (6.147) from (6.146), we get

$$\bar{\boldsymbol{\pi}}\bar{\mathbf{x}}_s + \bar{\boldsymbol{\lambda}}\bar{\mathbf{x}} = \bar{\boldsymbol{\pi}}\mathbf{b} - \mathbf{c}\bar{\mathbf{x}} \tag{6.148}$$

Note that because all variables are nonnegative, it follows that the left side of Equation (6.148) is zero if and only if complementary slackness holds. Thus, from (6.148), we see that $\bar{\boldsymbol{\pi}}\mathbf{b} - \mathbf{c}\bar{\mathbf{x}} = 0$ if and only if complementary slackness holds. That is, the primal and dual solutions have the same objective value ($\bar{\boldsymbol{\pi}}\mathbf{b} = \mathbf{c}\bar{\mathbf{x}}$) if and only if complementary slackness holds. The theorem then follows directly from Corollary 6.7.   □

    The previous theorem establishes that in linear programming, primal and dual solutions satisfy the complementary slackness conditions if and only if they have the same objective value. Thus, the optimality conditions for a linear programming problem consists of three important components: *primal feasibility, dual feasibility,* and *complementary slackness.* These conditions occur not only in linear programming, but also in nonlinear programming. When these conditions are viewed in a more mathematical context in nonlinear programming, they are generally referred to as the *Karush–Kuhn–Tucker Conditions* (Karush, 1939; Kuhn and Tucker, 1950). The interested reader is referred to Fiacco and McCormick (1968), Mangasarian (1969), and Bazaraa, Sherali, and Shetty (1993).

Example 6.7: Using the Dual to Solve the Primal

Consider the following linear programming problem:

$$\text{maximize } 10x_1 + 6x_2 - 4x_3 + x_4 + 12x_5 \tag{6.149}$$

subject to

$$2x_1 + x_2 + x_3 + 3x_5 \leq 18 \tag{6.150}$$

$$x_1 + x_2 - x_3 + x_4 + 2x_5 \leq 6 \tag{6.151}$$

$$x_1, x_2, x_3, x_4, x_5 \geq 0 \tag{6.152}$$

By denoting the dual variables by $\boldsymbol{\pi} = (\pi_1, \pi_2)$, the dual problem is

$$\text{minimize } 18\pi_1 + 6\pi_2 \tag{6.153}$$

subject to

$$2\pi_1 + \pi_2 \geq 10 \tag{6.154}$$

$$\pi_1 + \pi_2 \geq 6 \tag{6.155}$$

$$\pi_1 - \pi_2 \geq -4 \tag{6.156}$$

$$\pi_2 \geq 1 \tag{6.157}$$

$$3\pi_1 + 2\pi_2 \geq 12 \tag{6.158}$$

$$\pi_1, \pi_2 \geq 0 \tag{6.159}$$

Note that the dual problem involves only two decision variables, and, thus, unlike the primal, the dual problem can be solved graphically. The feasible region is shown in Figure 6.3, and we see that the optimal solution of the dual is $\pi_1^* = 2$, $\pi_2^* = 6$, with $Z^* = 72$.

Note from the graph that constraints (6.154) and (6.156) are binding at the optimal solution, whereas all other dual constraints are nonbinding at the optimal solution. Thus, by complementary slackness, the primal variables (x_2, x_4, x_5) corresponding to the nonbinding dual constraints, (6.155), (6.157), and (6.158), are zero. Also, because $\pi_1 > 0$ and $\pi_2 > 0$, the corresponding primal constraints, (6.150) and (6.151), must be binding. Therefore, using these observations regarding complementary slackness, the primal constraint set reduces to

$$2x_1 + x_3 = 18 \tag{6.160}$$

$$x_1 - x_3 = 6 \tag{6.161}$$

Solving this simple system of linear equations yields the unique solution, $x_1 = 8$ and $x_3 = 2$, which is precisely the optimal primal solution. Also note that $z = 10(8) +$

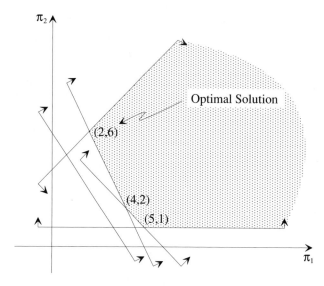

Figure 6.3 Dual feasible region for Example 6.7.

$6(0) - 4(2) + 1(0) + 12(0) = 72$, which is exactly the same objective value as generated by the dual. Thus, we have solved the dual problem graphically and used the optimal dual solution along with the complementary slackness conditions to derive the optimal solution to the primal.

*GEOMETRIC INTERPRETATION OF OPTIMALITY CONDITIONS

As discussed in the previous section, the optimality conditions for a linear programming problem consist of three distinct components: primal feasibility, dual feasibility, and complementary slackness. Thus, for the canonical primal problem:

$$(P) \quad \text{maximize } \mathbf{cx} \tag{6.162}$$

$$\text{subject to}$$

$$\mathbf{Ax} \leq \mathbf{b}$$

$$\mathbf{x} \geq \mathbf{0}$$

the (Karush–Kuhn–Tucker) optimality conditions are

$$\text{Primal Feasibility:} \quad \mathbf{Ax} \leq \mathbf{b} \tag{6.163}$$

$$\mathbf{x} \geq \mathbf{0} \tag{6.164}$$

---

* Starred sections may be omitted in an introductory course with no appreciable loss of continuity.

Dual Feasibility: $\quad\pi\mathbf{A} \geq \mathbf{c}$ $\hfill$ (6.165)

$$\pi \geq \mathbf{0} \hfill (6.166)$$

Complementary Slackness: $\quad\pi_i(b_i - \mathbf{a}^i\mathbf{x}) = 0,$

$$\text{for all } i = 1, \dots, m \hfill (6.167)$$

$$(\pi\mathbf{a}_j - c_j)x_j = 0,$$

$$\text{for all } j = 1, \dots, n \hfill (6.168)$$

If we can find vectors

$$\mathbf{x} = \begin{pmatrix} x_1 \\ x_2 \\ \vdots \\ x_n \end{pmatrix}$$

$$\pi = (\pi_1, \pi_2, \dots, \pi_m)$$

that satisfy conditions (6.163–6.168), then, by Theorem 6.9, $\mathbf{x}$ solves problem (P) (and, of course, π solves the corresponding dual problem). So that we may illustrate an important graphical interpretation of these optimality conditions, consider the following example.

Example 6.8: Geometric Interpretation of Optimality Conditions

Consider the following linear programming problem:

$$\text{maximize } 2x_1 + x_2 \hfill (6.169)$$

subject to

$$x_1 + x_2 \leq 8 \hfill (6.170)$$

$$x_1 + 3x_2 \leq 18 \hfill (6.171)$$

$$x_1 - x_2 \leq 4 \hfill (6.172)$$

$$x_1, x_2 \geq 0 \hfill (6.173)$$

Then, by following (6.163–6.168), the (Karush–Kuhn–Tucker) optimality conditions are

Primal Feasibility: $\quad x_1 + x_2 \leq 8 \hfill (6.174)$

$$x_1 + 3x_2 \leq 18 \hfill (6.175)$$

$$x_1 - x_2 \leq 4 \hfill (6.176)$$

$$x_1, x_2 \geq 0 \hfill (6.177)$$

Dual Feasibility: $\quad \pi_1 + \pi_2 + \pi_3 \geq 2 \hfill (6.178)$

$$\pi_1 + 3\pi_2 - \pi_3 \geq 1 \hfill (6.179)$$

$$\pi_1, \pi_2, \pi_3 \geq 0 \hfill (6.180)$$

Complementary Slackness: $\pi_1(8 - x_1 - x_2) = 0$ (6.181)

$$\pi_2(18 - x_1 - 3x_2) = 0 \qquad (6.182)$$

$$\pi_3(4 - x_1 + x_2) = 0 \qquad (6.183)$$

$$x_1(\pi_1 + \pi_2 + \pi_3 - 2) = 0 \qquad (6.184)$$

$$x_2(\pi_1 + 3\pi_2 - \pi_3 - 1) = 0 \qquad (6.185)$$

Now, in order to better interpret these optimality conditions geometrically, let us introduce surplus variables into the dual problem. That is, let

$$\lambda_1 = \pi_1 + \pi_2 + \pi_3 - 2 \qquad (6.186)$$

$$\lambda_2 = \pi_1 + 3\pi_2 - \pi_3 - 1 \qquad (6.187)$$

Then, after also rewriting the primal nonnegativity conditions, the optimality conditions can be written in the alternative form:

Primal Feasibility: $x_1 + x_2 \le 8$ (6.188)

$$x_1 + 3x_2 \le 18 \qquad (6.189)$$

$$x_1 - x_2 \le 4 \qquad (6.190)$$

$$-x_1 \le 0 \qquad (6.191)$$

$$-x_2 \le 0 \qquad (6.192)$$

Dual Feasibility: $\pi_1 + \pi_2 + \pi_3 - \lambda_1 = 2$ (6.193)

$$\pi_1 + 3\pi_2 - \pi_3 - \lambda_2 = 1 \qquad (6.194)$$

$$\pi_1, \pi_2, \pi_3, \lambda_1, \lambda_2 \ge 0 \qquad (6.195)$$

Complementary Slackness: $\pi_1(8 - x_1 - x_2) = 0$ (6.196)

$$\pi_2(18 - x_1 - 3x_2) = 0 \qquad (6.197)$$

$$\pi_3(4 - x_1 + x_2) = 0 \qquad (6.198)$$

$$\lambda_1(-x_1) = 0 \qquad (6.199)$$

$$\lambda_2(-x_2) = 0 \qquad (6.200)$$

First, note that the dual feasibility conditions (6.193–6.195) can be recast in vector form:

$$\pi_1 \begin{pmatrix} 1 \\ 1 \end{pmatrix} + \pi_2 \begin{pmatrix} 1 \\ 3 \end{pmatrix} + \pi_3 \begin{pmatrix} 1 \\ -1 \end{pmatrix} + \lambda_1 \begin{pmatrix} -1 \\ 0 \end{pmatrix} + \lambda_2 \begin{pmatrix} 0 \\ -1 \end{pmatrix} = \begin{pmatrix} 2 \\ 1 \end{pmatrix} \qquad (6.201)$$

$$\pi_1, \pi_2, \pi_3, \lambda_1, \lambda_2 \ge 0 \qquad (6.202)$$

Now, let $\mathbf{a}^1 = (1 \quad 1)$ and note that $\mathbf{a}^1$ is the gradient of primal constraint (6.188). Similarly, the gradients of the remaining primal constraints (6.189–6.192), respectively, are given by $\mathbf{a}^2 = (1 \quad 3)$, $\mathbf{a}^3 = (1 \quad -1)$, $\mathbf{a}^4 = (-1 \quad 0)$, and $\mathbf{a}^5 = (0 \quad -1)$. Finally, noting that $\mathbf{c} = (2 \quad 1)$, the dual feasibility conditions can be written as

$$\pi_1\mathbf{a}^1 + \pi_2\mathbf{a}^2 + \pi_3\mathbf{a}^3 + \lambda_1\mathbf{a}^4 + \lambda_2\mathbf{a}^5 = \mathbf{c} \qquad (6.203)$$

$$\pi_1, \pi_2, \pi_3, \lambda_1, \lambda_2 \ge 0 \qquad (6.204)$$

Thus, the dual feasibility conditions are specifying that we write the objective gradient $\mathbf{c}$ as a nonnegative linear combination of the gradients of the primal constraints. However, note, from the complementary slackness conditions (6.196–6.200) that each dual variable corresponds to precisely one of the primal constraints (6.188–6.192). In particular, note that if a primal constraint is nonbinding, then the corresponding dual variable equals zero.

   Therefore, in conclusion, the optimality conditions (6.188–6.200) are specifying that we find a feasible solution to the primal at which the gradient of the objective function can be written as a nonnegative linear combination of the gradients of the binding constraints. Or, equivalently, find a feasible solution to the primal where the gradient of the objective function lies in the cone generated by the gradients of the binding constraints. This concept is illustrated graphically in Figure 6.4. The feasible region corresponding to (6.188–6.192) has been graphed and the gradients of the binding constraints have been noted at each extreme point. Observe that extreme point R is the only feasible point where the gradient of the objective function lies in the cone generated by the gradients of the binding constraints. Thus, extreme point R satisfies the optimality conditions and, hence, is the optimal solution. The reader should compare this interpretation with the idea of extreme directions presented in Chapter 3.

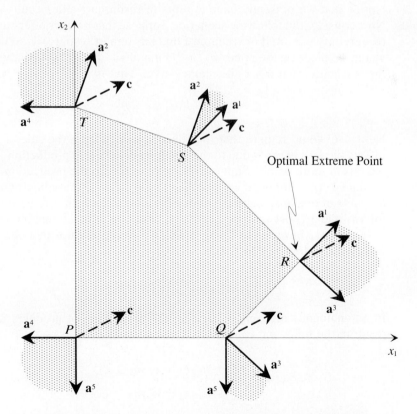

Figure 6.4   Geometric interpretation of optimality conditions.

ECONOMIC INTERPRETATION OF THE DUAL

At first, the reader may find the foregoing mechanical formulation of the dual and the accompanying theoretical development lacking motivation from a practical point of view. However, as mentioned earlier, the dual problem does, indeed, have valuable economic interpretations. To help motivate the foregoing formulation of the dual and to begin introducing some of its practical aspects, consider again Example 6.1 in the context of a simple product-mix problem. That is, suppose that we are computing the optimal mix of three products (x_1, x_2, x_3) subject to two resource constraints.

$$\text{maximize } c_1 x_1 + c_2 x_2 + c_3 x_3 \qquad \text{(Profit in \$)} \tag{6.205}$$

subject to

$$a_{1,1} x_1 + a_{1,2} x_2 + a_{1,3} x_3 \leq b_1 \qquad \text{(Units of Resource 1)} \tag{6.206}$$

$$a_{2,1} x_1 + a_{2,2} x_2 + a_{2,3} x_3 \leq b_2 \qquad \text{(Units of Resource 2)} \tag{6.207}$$

$$\mathbf{x} \geq \mathbf{0} \tag{6.208}$$

Notice that we presently have b_1 units of resource 1 and b_2 units of resource 2. Now consider the following scenario. Suppose that a buyer offers to purchase our resources for $\$\pi_1$/unit of resource 1 and $\$\pi_2$/unit of resource 2. What restrictions would we place on the prices π_1, π_2? First of all, quite obviously, we would place the restriction that π_1, π_2 be nonnegative. That is,

$$\pi_1, \pi_2 \geq 0 \tag{6.209}$$

Next, we would require that the revenue gained from the sale of the resources offsets the loss in profit due to the resulting decrease in production. Let us examine this in more detail. Suppose we were to forego the production of one unit of product 1 (x_1). Although we lose $\$c_1$ in profit, the decrease in production releases $a_{1,1}$ units of resource 1 and $a_{2,1}$ units of resource 2, which may be sold to the buyer for $\$\pi_1$ and $\$\pi_2$ per unit, respectively, resulting in additional income of $a_{1,1}\pi_1 + a_{2,1}\pi_2$. Thus, in order to break even, we would require that the prices π_1, π_2 satisfy the following constraint:

$$a_{1,1}\pi_1 + a_{2,1}\pi_2 \geq c_1 \tag{6.210}$$

In an analogous manner, decreasing the production of products 2 and 3 would result in the following constraints:

$$a_{1,2}\pi_1 + a_{2,2}\pi_2 \geq c_2 \tag{6.211}$$

$$a_{1,3}\pi_1 + a_{2,3}\pi_3 \geq c_3 \tag{6.212}$$

Observe that the restrictions generated (6.209–6.212) are precisely the constraints of the dual problem. Thus, if the prices π_1, π_2 satisfy (6.209–6.212), we will lose no money by accepting the buyer's offer.

Now consider the buyer's point of view. Based on the offer to purchase our resources, the buyer will pay us an amount given by

$$Z = b_1\pi_1 + b_2\pi_2 \tag{6.213}$$

Clearly, the buyer would want to minimize the amount paid. Thus, both parties will be satisfied if π_1 and π_2 solve the following problem:

$$\text{minimize } Z = b_1\pi_1 + b_2\pi_2 \tag{6.214}$$

subject to

$$a_{1,1}\pi_1 + a_{2,1}\pi_2 \geq c_1 \tag{6.215}$$

$$a_{1,2}\pi_1 + a_{2,2}\pi_2 \geq c_2 \tag{6.216}$$

$$a_{1,3}\pi_1 + a_{2,3}\pi_2 \geq c_3 \tag{6.217}$$

$$\boldsymbol{\pi} \geq \mathbf{0} \tag{6.218}$$

But (6.214–6.218) are precisely the dual problem. Therefore, the dual problem can be thought of as the problem of determining fair market prices of the resources. For this reason, the dual variables are often referred to as *shadow prices* or *dual prices*.

The Dual Variables as Rates of Change

Consider the canonical form of the primal and dual in summation form:

$$(P) \quad \text{maximize } z = \sum_{j=1}^{n} c_j x_j \tag{6.219}$$

subject to

$$\sum_{j=1}^{n} a_{i,j}x_j \leq b_i, \qquad \text{for } i = 1, \ldots, m \tag{6.220}$$

$$x_j \geq 0, \qquad \text{for } j = 1, \ldots, n \tag{6.221}$$

$$(D) \quad \text{minimize } Z = \sum_{i=1}^{m} b_i\pi_i \tag{6.222}$$

subject to

$$\sum_{i=1}^{m} a_{i,j}\pi_j \geq c_j, \qquad \text{for } j = 1, \ldots, n \tag{6.223}$$

$$\pi_i \geq 0, \qquad \text{for } i = 1, \ldots, n \tag{6.224}$$

Notice that in the primal, we may define the dimensions of each parameter as follows:

$$z = \text{return}$$

$$x_j = \text{units of variable } j$$

$$c_j = \text{return/(unit of variable } j)$$

$$b_i = \text{units of resource } i$$

$$a_{i,j} = \text{(units of resource } i)/(\text{unit of variable } j)$$

The only new parameters introduced in the dual formulation are Z and $\boldsymbol{\pi} = (\pi_1, \ldots, \pi_m)$. It is obvious that because $z^* = Z^*$, the dimension of Z is in terms of "return." The question remaining is: What are the dimensions associated with the dual variables, π_i?

Noting that the dual objective is in the form

$$Z = b_1\pi_1 + b_2\pi_2 + \cdots + b_m\pi_m \tag{6.225}$$

we see that

$$\frac{\partial Z}{\partial b_i} = \pi_i \tag{6.226}$$

That is, π_i is the rate of change of the objective with respect to b_i. Thus, the dual variable is expressed in terms of the *return per unit of resource i*.

There is an important implication of this result. Given an optimal solution to a linear programming problem, the dual variable then indicates the per unit contribution of the ith resource toward the increase in the (presently) optimal value of the objective. For example, if $\pi_3 = 9$, we interpret this to mean that for every unit (*up to a limit* to be discussed later) of resource 3 (the resource associated with constraint 3 in the primal), the objective value (z) will increase by 9 units.

Not only is this a useful interpretation, it is often of greater importance and interest than the optimal solution. The reason is that most companies wish to improve on the status quo. The optimal *solution* ($\mathbf{z}^*$ and $\mathbf{x}^*$) to a linear programming problem tells them only how to best allocate their resources for their *present* state. The dual variables, on the other hand, provide the company with the information needed to expand and to increase profit, if one knows how to interpret them. Those who do not take advantage of this information are overlooking one of the most important factors of linear programming.

Now, consider again the complementary slackness conditions that were introduced in (6.167–6.168) and are repeated here for convenience.

$$\pi_i(b_i - \mathbf{a}^i\mathbf{x}) = 0, \qquad \text{for all } i = 1, \ldots, m \tag{6.227}$$

$$(\boldsymbol{\pi}\mathbf{a}_j - c_j)x_j = 0, \qquad \text{for all } j = 1, \ldots, n \tag{6.228}$$

Previously, we showed that the dual variables give the rate of change of the objective function with respect to the right-hand side of the associated primal constraint. Now suppose that primal constraint i is nonbinding at the optimal solution, that is, $\mathbf{a}^i\mathbf{x} < \mathbf{b}_i$. Then we have an excess of resource i, and acquiring more of this resource will have no affect on the objective because we were not able to use all of our current supply. Therefore, the rate of change of the objective with respect to resource i is zero, that is, $\pi_i = 0$. On the other hand, suppose $\partial z/\partial b_i = \pi_i > 0$. Then acquiring more of this resource will improve the objective function. However, because the current solution is optimal, the current supply of the resource must be depleted, and thus the associated constraint is binding, that is, $\mathbf{a}^i\mathbf{x} = b_i$. Note that these economic interpretations correspond precisely to the complementary slackness conditions in (6.227). A parallel argument could be made for the conditions in (6.228).

Example 6.9: Economic Interpretation of the Dual

A company manufactures two products. Forecasting indicates that the maximum weekly demand for Product A is 900 units and for Product B is 600 units. The manufacture of each product requires raw material and two basic operations. The per unit consumption of the raw material and resources is summarized in Table 6.7. Finally, the profit per unit of Product A is $7 and the profit per unit of Product B is $12.

TABLE 6.7 DATA FOR EXAMPLE 6.9

Product	Forging (hr)	Machining (hr)	Raw material
A	0.15	0.10	2.5
B	0.20	0.20	4.0
Available per Week	200 hours	140 hours	3200 pounds

Now, let

x_1 = number of units of Product A manufactured per week

x_2 = number of units of Product B manufactured per week

Then the linear programming model for this problem is as follows, with the optimal tableau shown in Table 6.8.

maximize $z = 7x_1 + 12x_2$	(Weekly profit in $)	(6.229)
subject to		
$0.15x_1 + 0.2x_2 \leq 200$	(Forging-capacity limit)	(6.230)
$0.1x_1 + 0.2x_2 \leq 140$	(Machining-capacity limit)	(6.231)
$2.5x_1 + 4.0x_2 \leq 3200$	(Raw-material limit)	(6.232)
$x_1 \leq 900$	(Demand limit for Product A)	(6.233)
$x_2 \leq 600$	(Demand limit for Product B)	(6.234)
$x_1, x_2 \geq 0$		(6.235)

TABLE 6.8

	z	x_1	x_2	x_3	x_4	x_5	x_6	x_7	RHS
z	1	0	0	0	20	2	0	0	9200
x_3	0	0	0	1	1	−0.1	0	0	20
x_1	0	1	0	0	−40	2	0	0	800
x_2	0	0	1	0	25	−1	0	0	300
x_6	0	0	0	0	40	−2	1	0	100
x_7	0	0	0	0	−25	1	0	1	300

Now, by letting π_1, π_2, π_3, π_4, and π_5 denote the dual variables corresponding to constraints (6.230–6.234), respectively, the optimal primal and dual solutions can be read from Table 6.8 as follows:

$x_1^* = 800$ units of Product A

$x_2^* = 300$ units of Product B

$z^* = Z^* = \$92,000$ (weekly profit)

$\pi_1^* = 0$ per unit of resource 1 ($/hr of forging)

$\pi_2^* = \$20$ per unit of resource 2 ($/hr of machining)

$\pi_3^* = \$2$ per unit of resource 3 ($/lb of raw material)

$\pi_4^* = 0$ per unit of resource 4 ($/unit of demand of Product A)

$\pi_5^* = 0$ per unit of resource 5 ($/unit of demand of Product B)

Notice that because $\pi_2 = 20$, the company may increase weekly profit by $20 for each additional hour of machining time. Thus, if it is possible to purchase additional machining time for less than $20 per hour, then profit will increase as a result. That is, if an additional hour of machining time actually costs $\$\alpha$, then the company will receive $\$(20 - \alpha)$ for each additional hour purchased. Obviously, there has to be an upper limit to this result; otherwise, the amount of additional profit is unlimited. We shall see how to determine this limit when discussing sensitivity analysis and parametric programming.

In addition, we see that the value of raw materials, over the maximum of 3200 pounds now available, is $2 per pound per week. That is, the shadow price of raw materials is $2/pound. Thus, if the company can purchase additional raw material (up to a certain limit) for less than $2 per pound, it will increase profits.

As may be seen, the dual variables play an important part in problem analysis. However, until we pursue the aspect of sensitivity analysis and parametric programming, this example provides only a hint of what can and should be done. In order to further study these important concepts, we must first consider an alternative method for optimizing a linear programming problem called the dual simplex algorithm.

THE DUAL SIMPLEX ALGORITHM

Duality is a *property*. Arbitrarily designating a given linear programming problem as the "primal," we know that through certain transformations, we may arrive at a related linear programming problem that we call the "dual." Because both problems are linear programming problems, both may be solved via the simplex algorithm; but, because the solution of one problem also gives the solution of the other, we usually select the problem (i.e., the primal or the dual) that appears easiest to solve. It is further noted that every iteration of one problem is associated with a corresponding iteration of the other. Therefore, when we are solving one problem, be it the primal or the dual, we may imagine that the other is also being simultaneously solved, in its own "dimension."

In this section, we take advantage of such properties to develop the dual simplex algorithm. The dual simplex algorithm was initially developed by Lemke (1954) and, as we shall see, is a useful tool for dealing with sensitivity analysis in linear programming (and in aiding certain integer programming algorithms).

From this point on, we shall refer to the original simplex algorithm of Chapter 4 as the *primal simplex* method to differentiate between the two methods. We now present the dual simplex algorithm. For convenience in discussion, we normally refer to the problem that has been placed into the simplex tableau as the primal. The primal simplex method is then an algorithm that *always* deals with a basic feasible primal solution. We terminate the primal simplex algorithm as soon as the primal optimality conditions ($z_j - c_j \geq 0$) are satisfied or until an indication that the objective is unbounded. Recall from the previous section that primal optimality is precisely the same as dual feasibility. Thus, the primal simplex method can also be thought of as an algorithm that maintains primal feasibility and complementary slackness throughout, and tries to restore dual feasibility. These conditions for the primal simplex algorithm are illustrated on our mapping of the primal and dual space in Figure 6.5.

The dual simplex method also addresses the primal and its tableau. However, although it is the primal problem that we see in the tableau before us, it is its dual that is actually being operated on. Such an approach is possible only if the current solution is *dual feasible*. That is, primal optimality conditions are satisfied. Figure 6.6 indicates the conditions for the dual simplex method.

The basic thrust of the dual simplex algorithm is quite simple. One always attempts to retain dual feasibility (i.e., primal optimality, $z_j - c_j \geq 0$, for all j) while bringing the primal back to feasibility (i.e., $x_{B,i} \geq 0$, for all i). This may be accomplished by bringing the equivalent feasible dual solution to optimality, but all computations are performed with the *primal* tableau. Thus, the dual simplex algorithm maintains dual feasibility and complementary slackness throughout its operation, while trying to achieve primal feasibility. It is actually nothing more than the primal simplex method applied to the dual problem; however, we do so while utilizing the primal tableau. The form of the dual simplex that we shall employ is given in what follows.

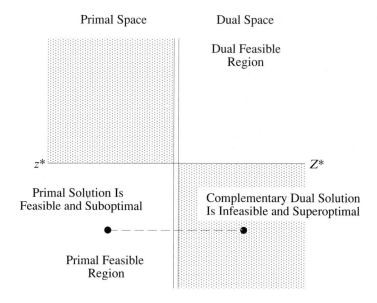

Figure 6.5   Conditions for the primal simplex algorithm.

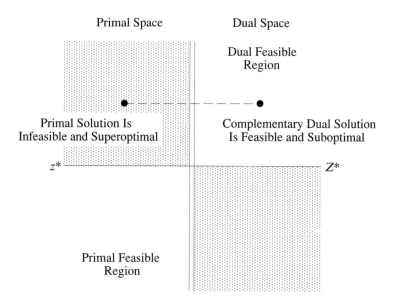

Figure 6.6   Conditions for the dual simplex algorithm.

The dual simplex algorithm

STEP 1. To employ this algorithm, the problem must be dual feasible, that is, all $z_j - c_j \geq 0$. If this condition is met, go to Step 2. (The following section will discuss one method of attaining a dual feasible solution.)

STEP 2. *Determine the departing variable.* If $\beta_i \geq 0$, for all i, then the current solution is optimal; stop. Otherwise, select the row associated with the most negative β_i. Denote this row as row r. The basic variable $x_{B,r}$ associated with this row is the departing variable.

STEP 3. *Check for primal feasibility.* If $\alpha_{r,j} \geq 0$, for all j, then the primal problem is infeasible and the dual problem has an unbounded objective; stop. Otherwise, go to Step 4.

STEP 4. *Determine the entering variable.* Use the following minimum ratio test to determine the entering basic variable. That is, let

$$\frac{z_k - c_k}{-\alpha_{r,k}} = \text{minimum} \left\{ \frac{z_j - c_j}{-\alpha_{r,j}} : \alpha_{r,j} < 0 \right\}$$

Column k is the pivot column, $\alpha_{r,k}$ is the pivot element, and the basic variable x_k associated with column k is the entering variable. Go to Step 5.

STEP 5. *Pivot and establish a new tableau*

(a) The entering variable x_k is the new basic variable in row r.

(b) Use elementary row operations on the old tableau so that the column associated with x_k in the new tableau consists of all zero elements except for a 1 at the pivot position $\alpha_{r,k}$.

(c) Return to Step 2.

The dual simplex algorithm is generally considered unequal to the task of performing as a general-purpose linear programming algorithm because of the difficulty in finding an initial basic solution that is dual feasible. Consequently, the following example obviously has been contrived so as to exploit the dual simplex properties. The reader should recognize the improbability of finding such problems in practice.

Example 6.10: The Dual Simplex Method

$$\text{minimize } z = 8x_1 + 5x_2 \tag{6.236}$$

subject to

$$x_1 + x_2 \geq 3 \tag{6.237}$$

$$2x_1 + x_2 \geq 4 \tag{6.238}$$

$$x_1, x_2 \geq 0 \tag{6.239}$$

Although we do not need to write down the dual problem to execute the dual simple algorithm, let us do so in this case so that we may track the solutions to both problems graphically. By letting π_1, π_2 designate the dual variables, the dual problem can be written as follows:

$$\text{maximize } 3\pi_1 + 4\pi_2 \tag{6.240}$$

subject to

$$\pi_1 + 2\pi_2 \le 8 \tag{6.241}$$

$$\pi_1 + \pi_2 \le 5 \tag{6.242}$$

$$\pi_1, \pi_2 \ge 0 \tag{6.243}$$

Rather than preprocessing the primal problem as usual, instead we shall change the objective to maximization form and multiply both constraints through by -1. Adding slack variables x_3 to constraint (6.241) and x_4 to constraint (6.242) yields

$$\text{maximize } z' = -8x_1 - 5x_2 \tag{6.244}$$

subject to

$$-x_1 - x_2 + x_3 = -3 \tag{6.245}$$

$$-2x_1 - x_2 + x_4 = -4 \tag{6.246}$$

$$\mathbf{x} \ge \mathbf{0} \tag{6.247}$$

The initial tableau for the resulting problem is then given in Table 6.9. Notice that the initial basis is primal infeasible ($x_3 = -3$ and $x_4 = -4$) and dual feasible (all $z_j - c_j \ge 0$). Thus, the dual simplex algorithm can be employed.

TABLE 6.9

	z'	x_1	x_2	x_3	x_4	RHS
z'	1	8	5	0	0	0
x_3	0	-1	-1	1	0	-3
x_4	0	-2	-1	0	1	-4

STEP 2. The most negative β_i is $\beta_2 = -4$. Thus, $r = 2$ and $x_{B,2} = x_4$ is the departing variable. Go to Step 3.

STEP 3. Because $\alpha_{2,1}$ and $\alpha_{2,2} < 0$, the primal infeasibility condition is not satisfied. Go to Step 4.

STEP 4. We now examine the ratios $(z_j - c_j)/(-\alpha_{2,j})$ where $\alpha_{2,j} < 0$:

$$\frac{z_1 - c_1}{-\alpha_{2,1}} = \frac{8}{-(-2)} = 4$$

$$\frac{z_2 - c_2}{-\alpha_{2,2}} = \frac{5}{-(-1)} = 5$$

Thus, $k = 1$ and the entering variable is x_1.

STEP 5. (a) Because x_1 is the entering variable and x_4 is the departing variable, x_1 replaces x_4 in $\mathbf{x}_B$ as the basic variable in row 2.

(b) Pivot as usual on $\alpha_{2,1} = -2$. This results in Table 6.10.

Note that in Table 6.10, the value of the objective (z') has *decreased*, the top row of the tableau still indicates primal optimality (dual feasibility), and we are not yet primal feasible.

TABLE 6.10

	z'	x_1	x_2	x_3	x_4	RHS
z'	1	0	1	0	4	-16
x_3	0	0	$-\frac{1}{2}$	1	$-\frac{1}{2}$	-1
x_1	0	1	$\frac{1}{2}$	0	$-\frac{1}{2}$	2

(c) Return to Step 2.

STEP 2. The only negative β_i is $\beta_1 = -1$. Thus, $r = 1$ and $x_{B,1} = x_3$ is the departing variable. Go to Step 3.

STEP 3. Because $\alpha_{1,2}$ and $\alpha_{1,4} < 0$, the primal infeasibility condition is not satisfied. Go to Step 4.

STEP 4. The ratios $(z_j - c_j)/(-\alpha_{1,j})$, where $\alpha_{1,j} < 0$ are

$$\frac{z_2 - c_2}{-\alpha_{1,2}} = \frac{1}{-(-\frac{1}{2})} = 2$$

$$\frac{z_4 - c_4}{-\alpha_{1,4}} = \frac{4}{-(-\frac{1}{2})} = 8$$

Thus, $k = 2$ and the entering variable is x_2.

STEP 5. (a) x_2 replaces x_3 in $\mathbf{x}_B$ as the basic variable in row 1.

(b) Pivot as usual on $\alpha_{1,2} = -\frac{1}{2}$ to obtain Table 6.11.

Table 6.11 represents the optimal solution because both primal and dual feasibility are satisfied. Note that both the complementary primal and dual solutions can be read from the optimal tableau. These solutions are listed in Table 6.12 (λ_1 and λ_2 are the respective dual slack variables). Observe that complementary slackness is satisfied and that both solutions correspond to an objective value $z^* = -z' = 18$.

TABLE 6.11

	z'	x_1	x_2	x_3	x_4	RHS
z'	1	0	0	2	3	-18
x_2	0	0	1	-2	1	2
x_1	0	1	0	1	-1	1

TABLE 6.12 COMPLEMENTARY PRIMAL AND DUAL SOLUTIONS FOR TABLE 6.11

Primal solution	Complementary dual solution
$x_1 = 1$	$\lambda_1 = 0$
$x_2 = 2$	$\lambda_2 = 0$
$x_3 = 0$	$\pi_1 = 2$
$x_4 = 0$	$\pi_2 = 3$

Figure 6.7 depicts the sequence of iterations followed by the primal and dual problems in their respective feasible regions. Because dual feasibility was maintained throughout the algorithm, note that the optimal solution was obtained as soon as the primal feasible region was reached.

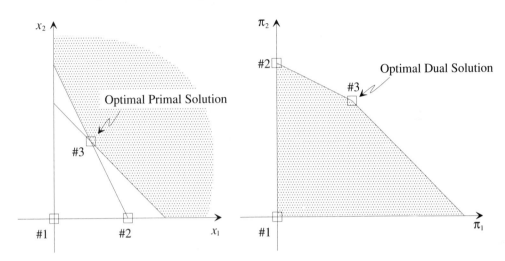

Figure 6.7 Sequence of primal and dual solutions.

An Extended Dual Simplex Algorithm

The type of problem that we encountered in Example 6.10 is, as we warned, simply not typical of what we would expect in practice. As a result, effort has been made to extend the dual simplex algorithm into a form that is more robust. One such extension has been called the "artificial constraint method," and we briefly illustrate its approach in this section.

The artificial constraint method is intended for a linear programming problem in which one or more $x_{B,i} = \beta_i < 0$ and one or more $z_j - c_j < 0$. (Recall from Chapter 4 that this problem also could be addressed using the two-phase method.) Given these conditions, we add the artificial constraint

$$\sum_{j \in P} x_j \leq M \tag{6.248}$$

where

$$P = \{j : z_j - c_j < 0\} \tag{6.249}$$

M is a positive number that should be substantially larger than any other value that is expected to be encountered in the computations.

The artificial constraint (6.248) is changed into an equality by adding a slack variable. The resulting preprocessed constraint is then included in the simplex tableau. We then select, as the *entering variable*, that variable associated with the most negative $z_j - c_j$ value, just as we did in the primal simplex. The *departing variable*, however, is *always* chosen as the slack variable associated with the artificial constraint. This initialization process will always result in *all* $z_j - c_j$ becoming nonnegative. Once the top row is nonnegative, the dual simplex algorithm may be applied in the usual manner. The entire process is demonstrated in the following example.

Example 6.11: The Artificial Constraint Method

$$\text{maximize } z = 3x_1 + 5x_2 \tag{6.250}$$

subject to

$$2x_1 + 5x_2 \geq 10 \tag{6.251}$$

$$x_1 \geq 1 \tag{6.252}$$

$$2x_1 + x_2 \leq 8 \tag{6.253}$$

$$x_1, x_2 \geq 0 \tag{6.254}$$

The feasible region of this linear programming problem is depicted graphically in Figure 6.8.

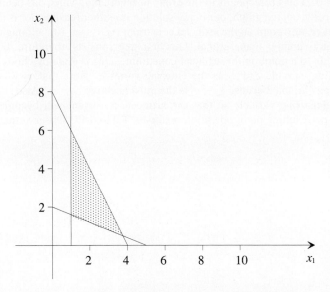

Figure 6.8 Feasible region for Example 6.11.

To preprocess, we change all constraints to ($\leq$) form and add the respective slack variables x_3, x_4, and x_5 to constraints (6.251–6.253). The problem can then be recast in the following form with the initial tableau shown in Table 6.13.

$$\text{maximize } z = 3x_1 + 5x_2 \tag{6.255}$$

subject to

$$-2x_1 - 5x_2 + x_3 = -10 \tag{6.256}$$

$$-x_1 + x_4 = -1 \tag{6.257}$$

$$2x_1 + x_2 + x_5 = 8 \qquad (6.258)$$

$$x_1, x_2, x_3, x_4, x_5 \geq 0 \qquad (6.259)$$

Because the problem is both infeasible and nonoptimal, we add the artificial constraint $x_1 + x_2 \leq 10$ (note that we have let $M = 10$ in Equation (6.248)). The resultant graph and tableau are given in Figure 6.9 and Table 6.14, respectively (x_6 is the slack variable in the artificial constraint).

TABLE 6.13

	z	x_1	x_2	x_3	x_4	x_5	RHS
z	1	-3	-5	0	0	0	0
x_3	0	-2	-5	1	0	0	-10
x_4	0	-1	0	0	1	0	-1
x_5	0	2	1	0	0	1	8

Note from Table 6.14 that the initial problem solution is specified by the nonbasic variables $x_1, x_2 = 0$. This corresponds to the origin in Figure 6.9, which has been boxed. As may be seen from the graph, our initial solution is neither feasible nor is it near optimal. However, any point on the artificial constraint ($x_1 + x_2 \leq 10$), although not feasible, is certainly greater than optimal. Our first step consists, therefore, of moving from the origin to a point on the artificial constraint. This is done by choosing x_6 as the departing variable and x_2 as the entering variable. Note that x_2 was chosen as the entering variable because $z_2 - c_2$ is the most negative $z_j - c_j$. Also, x_6 was chosen as the departing variable so that the artificial constraint will become binding or active. The resulting pivot operation results in Table 6.15. Observe that

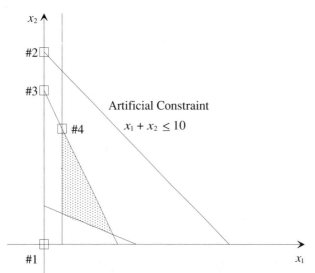

Figure 6.9 Feasible region with artificial constraint.

TABLE 6.14

	z	x_1	x_2	x_3	x_4	x_5	x_6	RHS
z	1	-3	-5	0	0	0	0	0
x_3	0	-2	-5	1	0	0	0	-10
x_4	0	-1	0	0	1	0	0	-1
x_5	0	2	1	0	0	1	0	8
x_6	0	1	1	0	0	0	1	10

TABLE 6.15

	z	x_1	x_2	x_3	x_4	x_5	x_6	RHS
z	1	2	0	0	0	0	5	50
x_3	0	3	0	1	0	0	5	40
x_4	0	-1	0	0	1	0	0	-1
x_5	0	1	0	0	0	1	-1	-2
x_2	0	1	1	0	0	0	1	10

graphically, we have moved to the point where $x_1 = 0$, $x_2 = 10$, which is *on* the artificial constraint. The dual simplex method is then applied with the results shown in Tables 6.16 and 6.17; the corresponding solutions are also traced on the graph. Note that the final solution (point 4) corresponds to the optimal extreme point of the feasible region.

TABLE 6.16

	z	x_1	x_2	x_3	x_4	x_5	x_6	RHS
z	1	7	0	0	0	5	0	40
x_3	0	8	0	1	0	5	0	30
x_4	0	-1	0	0	1	0	0	-1
x_6	0	1	0	0	0	-1	1	2
x_2	0	2	1	0	0	1	0	8

TABLE 6.17

	z	x_1	x_2	x_3	x_4	x_5	x_6	RHS
z	1	0	0	0	7	5	0	33
x_3	0	0	0	1	8	5	0	22
x_1	0	1	0	0	-1	0	0	1
x_6	0	0	0	0	-1	-1	1	3
x_2	0	0	1	0	2	1	0	6

When employing the artificial constraint method, there are three possible outcomes:

1. The associated artificial constraint slack variable is in the final basis at a *positive* value. This indicates that the artificial constraint is *nonbinding* at the final solution and thus we have reached the optimal solution.

2. The associated artificial constraint slack variable is either nonbasic or in the final basis at a *zero* value, indicating that the artificial constraint is *binding* at the final solution. In this case, either the value of M was too small or the original problem may have an unbounded objective.

3. One (or more) β_i is negative, but there is no possible pivot (for *any* of the negative β_i). That is, the corresponding $\alpha_{r,j}$ values are all nonnegative. In this case, the dual is unbounded and the primal is *infeasible*.

Outcome 1 has already been illustrated in Example 6.11. Notice that $x_6^* = 3$ and x_6 is the artificial constraint slack variable.

The second outcome may indicate either a poor choice (too small) for the value of M or a problem with an unbounded objective, or both. To illustrate, consider Example 6.11 except that the artificial constraint is given by $x_1 + x_2 \leq 5$. If the reader plots out this constraint, on Figure 6.8 for example, he or she will see that the artificial constraint serves to exclude a region (and, in this case, the optimal solution) of the feasible solution space. The same result (i.e., the artificial constraint slack variable at a zero value in the final solution) can be made to occur by dropping the constraint $2x_1 + x_2 \leq 8$ from this example. We can see from Figure 6.9 that if this is done, the problem actually has an unbounded objective but because an artificial constraint has been added, we will obtain an indication of optimality and feasibility with a "solution" at $x_1 = 0$ and $x_2 = 10$. As long as M is a finite value, the artificial constraint itself will bound the unbounded objective.

The third and final outcome occurs, as discussed, whenever the primal is infeasible. The reader can easily demonstrate this outcome by solving a simple, infeasible example such as

$$\text{maximize } z = 3x_1 + 2x_2 \tag{6.260}$$

$$\text{subject to}$$

$$x_1 + x_2 \leq 5 \tag{6.261}$$

$$x_1 + x_2 \geq 10 \tag{6.262}$$

$$x_1, x_2 \geq 0 \tag{6.263}$$

SENSITIVITY ANALYSIS IN LINEAR PROGRAMMING

Throughout the development up to this point, we have cited numerous instances in which simplifying assumptions have been made so as to ease both model development and analysis. It should be obvious that real problems will seldom, if ever, strictly satisfy all of these assumptions. Data used may be subject to error, cost and resource availabilities can change with time, and the system itself may be modified. Management must deal with the future as well as the present, and deal with various price and commodity fluctuations, so it should be apparent that an approach is needed to include such considerations within the linear programming technique. This approach is generally referred to as *sensitivity analysis* (or post-optimality analysis).

In this section we present an approach for determining the impact of discrete changes in the linear programming model structure on the resulting problem solution. In the following section, we consider continuous systematic changes via a technique called *parametric programming*.

To clarify the difference between discrete and continuous changes, consider analyzing the impact of one factor, say, c_3 (i.e., the "cost" coefficient associated with decision variable x_3). Now, if we wish to determine the impact of a change in the value of c_3 from its present value (c_3) to a specific new value (c_3'), we are dealing with a discrete change. However, if what we want to examine is the impact over the entire range of values from, say, $\bar{c}_3$ to $\hat{c}_3$, then the problem deals with a parametric change.

Before beginning our study of sensitivity analysis, let us review some basic tableau relationships and interpretations via the following example.

Example 6.12

We shall refer to the following example frequently throughout this section.

$$\text{maximize } z = 10x_1 + 7x_2 + 6x_3 \qquad \text{(Return in \$)} \qquad (6.264)$$

subject to

$$3x_1 + 2x_2 + x_3 \leq 36 \qquad \text{(Units of Resource 1)} \qquad (6.265)$$

$$x_1 + x_2 + 2x_3 \leq 32 \qquad \text{(Units of Resource 2)} \qquad (6.266)$$

$$2x_1 + x_2 + x_3 \leq 22 \qquad \text{(Units of Resource 3)} \qquad (6.267)$$

$$x_1, x_2, x_3 \geq 0 \qquad (6.268)$$

By letting x_4, x_5, and x_6 denote the slack variables for constraints (6.265–6.267), respectively, the data for the problem in standard form can be summarized as follows:

$$\mathbf{A} = \begin{pmatrix} 3 & 2 & 1 & 1 & 0 & 0 \\ 1 & 1 & 2 & 0 & 1 & 0 \\ 2 & 1 & 1 & 0 & 0 & 1 \end{pmatrix}$$

$$\mathbf{b} = \begin{pmatrix} 36 \\ 32 \\ 22 \end{pmatrix}$$

$$\mathbf{c} = (10 \quad 7 \quad 6 \quad 0 \quad 0 \quad 0)$$

The optimal tableau is shown in Table 6.18.

TABLE 6.18

	z	x_1	x_2	x_3	x_4	x_5	x_6	RHS
z	1	3	0	0	1	0	5	146
x_2	0	1	1	0	1	0	−1	14
x_5	0	−2	0	0	1	1	−3	2
x_3	0	1	0	1	−1	0	2	8

First, note that the optimal primal solution is given by

$$z^* = 146$$

$$\mathbf{x}_B^* = \begin{pmatrix} x_{B,1} \\ x_{B,2} \\ x_{B,3} \end{pmatrix} = \begin{pmatrix} x_2 \\ x_5 \\ x_3 \end{pmatrix} = \mathbf{B}^{-1}\mathbf{b} = \boldsymbol{\beta} = \begin{pmatrix} 14 \\ 2 \\ 8 \end{pmatrix}$$

$$\mathbf{x}_N^* = \begin{pmatrix} x_1 \\ x_4 \\ x_6 \end{pmatrix} = \begin{pmatrix} 0 \\ 0 \\ 0 \end{pmatrix}$$

Next, observe that the original identity was associated with slack variables x_4, x_5, and x_6, that is, $\mathbf{I} = (\mathbf{a}_4, \mathbf{a}_5, \mathbf{a}_6)$. Therefore, the basis inverse $\mathbf{B}^{-1} = \mathbf{B}^{-1}\mathbf{I} = (\boldsymbol{\alpha}_4, \boldsymbol{\alpha}_5, \boldsymbol{\alpha}_6)$ and may be read directly from the optimal tableau as

$$\mathbf{B}^{-1} = \begin{pmatrix} 1 & 0 & -1 \\ 1 & 1 & -3 \\ -1 & 0 & 2 \end{pmatrix}$$

Now, let $\boldsymbol{\pi} = (\pi_1, \pi_2, \pi_3)$ represent the dual variables corresponding to constraints (6.265–6.267), respectively. Then, again using the information in Table 6.18, we see that

$$\boldsymbol{\pi} = (\pi_1, \pi_2, \pi_3) = \mathbf{c}_B\mathbf{B}^{-1} = (1 \quad 0 \quad 5)$$

Note that, as explained earlier, these values are simply the $z_j - c_j$ values for the zero-cost slack variables x_4, x_5, and x_6.

Which resources in this problem are fully utilized (or scarce)? This question is answered quite readily by noting the values of the slack variables. That is, because $x_4 = x_6 = 0$, we see that resources 1 and 3 are scarce resources. However, $x_5 = 2$ and thus we have 2 units of resource 2 remaining. Note also that π_1, π_2, and π_3 are the respective shadow prices of resources 1, 2, and 3. As expected $\pi_2 = 0$ because resource 2 is not fully utilized. That is, acquiring more of resource 2 will not affect the value of the objective function. However, the rate of change of the objective with respect to additional units of resources 1 and 3 are given by the shadow prices, $\pi_1 = 1$ and $\pi_3 = 5$. That is, for each additional unit of resource 1, the objective value (return) will increase by $1. It is important to realize that this rate of change is only valid for relatively small increases in the quantity of resource 1. More specifically, it is only valid as long as the current basis remains optimal; a method for determining these limits will be discussed in detail later. Likewise, for each additional unit of resource 3 up to a certain limit, the objective value will increase by $5.

Changes in the Objective Coefficients

The first discrete change to be considered will be a change in the value of an objective function coefficient (i.e., a c_j). We handle such changes differently according to whether the respective c_j of interest is associated with a basic or nonbasic variable.

A change in the c_j of a nonbasic variable

Let us first consider a change in the cost coefficient c_k of a nonbasic variable x_k. Note that $\mathbf{c}_B$ is not affected; thus, the only impact of such a change is on the single tableau element, $z_k - c_k$. By letting c'_k be the *new* value of c_k, then $z_k - c'_k$ will replace $z_k - c_k$ in the optimal tableau. Of course, if $z_k - c'_k$ remains nonnegative, then the current basis remains optimal. However, if $z_k - c'_k < 0$, then dual feasibility (primal optimality) has been lost and must be restored by using the primal simplex method. The value of $z_k - c'_k$ can be computed quite easily using the following relationship.

$$z_k - c'_k = \mathbf{c}_B \mathbf{B}^{-1}\mathbf{a}_j - c'_j$$

$$= \mathbf{c}_B \mathbf{B}^{-1}\mathbf{a}_j - c_j + c_j - c'_j = (z_k - c_k) + (c_k - c'_k) \qquad (6.269)$$

A change in the c_j of a basic variable

Now consider a change in the cost coefficient c_k associated with a basic variable $x_{B,i} = x_k$. Because x_k is a basic variable, a change in c_k results in a change in the $\mathbf{c}_B$ vector. Thus, such a change can affect any or all of the $z_j - c_j$ elements and the value of z. Let c'_k be the *new* value of c_k and let $\mathbf{c}'_B$ denote the revised $\mathbf{c}_B$. The $z_j - c_j$ elements associated with the basic variables will remain zero, so we only need to update the $z_j - c_j$ for the nonbasic variables as follows:

$$z'_j - c_j = \mathbf{c}'_B \mathbf{B}^{-1}\mathbf{a}_j - c_j = \mathbf{c}'_B \boldsymbol{\alpha}_j - c_j, \text{ for all nonbasic variables } x_j \qquad (6.270)$$

In addition, the updated value of the objective function is given by

$$z' = \mathbf{c}'_B \mathbf{B}^{-1}\mathbf{b} = \mathbf{c}'_B \boldsymbol{\beta} \qquad (6.271)$$

If some $z'_j - c_j$ is negative, then dual feasibility (primal optimality) must be restored by using the primal simplex method.

Example 6.13: A Change in a Nonbasic c_j

Consider again the problem of Example 6.12, which, for convenience, is reproduced here, along with the optimal tableau (Table 6.19).

$$\text{maximize } z = 10x_1 + 7x_2 + 6x_3 \qquad \text{(Return in \$)} \qquad (6.272)$$

subject to

$$3x_1 + 2x_2 = x_3 \le 36 \qquad \text{(Units of Resource 1)} \qquad (6.273)$$

$$x_1 + x_2 + 2x_3 \le 32 \qquad \text{(Units of Resource 2)} \qquad (6.274)$$

$$2x_1 + x_2 + x_3 \leq 22 \qquad \text{(Units of Resource 3)} \qquad (6.275)$$

$$x_1, x_2, x_3 \geq 0 \qquad\qquad\qquad (6.276)$$

TABLE 6.19

	z	x_1	x_2	x_3	x_4	x_5	x_6	RHS
z	1	3	0	0	1	0	5	146
x_2	0	1	1	0	1	0	-1	14
x_5	0	-2	0	0	1	1	-3	2
x_3	0	1	0	1	-1	0	2	8

Notice that x_1 is nonbasic in this final tableau. Let us assume that the unit return of x_1 changes from its present value of $c_1 = 10$ to $c_1' = 14$. Then, from (6.269),

$$z_1 - c_1' = (z_1 - c_1) + (c_1 - c_1') = 3 + (10 - 14) = -1$$

or, equivalently, one could compute

$$z_1 - c_1' = \mathbf{c}_B \mathbf{B}^{-1} \mathbf{a}_1 - c_1' = \mathbf{c}_B \boldsymbol{\alpha}_1 - c_1' = (7 \quad 0 \quad 6) \begin{pmatrix} 1 \\ -2 \\ 1 \end{pmatrix} - 14 = -1$$

Thus, primal optimality (dual feasibility) has been lost. Therefore, we must perform at least one primal simplex pivot to restore optimality. The first pivot would have x_1 as the entering variable and x_3 as the leaving variable; this results in the optimal tableau shown in Table 6.20.

TABLE 6.20

	z	x_1	x_2	x_3	x_4	x_5	x_6	RHS
z	1	0	0	1	0	0	7	154
x_2	0	0	1	-1	2	0	-3	6
x_5	0	0	0	2	-1	1	1	18
x_1	0	1	0	1	-1	0	2	8

Example 6.14: A Change in a Basic c_j

Given the original linear programming model of Example 6.12, let us assume that the value of c_3 should be 5 rather than 6. Thus, a change in the coefficient associated with a *basic* variable has been made. That is, $c_3' = 5 = c_{B,3}'$. From (6.270) and (6.271), we then obtain

$$z_1' - c_1 = \mathbf{c}_B' \mathbf{B}^{-1} \mathbf{a}_1 - c_1 = \mathbf{c}_B' \boldsymbol{\alpha}_1 - c_1 = (7 \quad 0 \quad 5) \begin{pmatrix} 1 \\ -2 \\ 1 \end{pmatrix} - 10 = 2$$

$$z_4' - c_4 = \mathbf{c}_B' \mathbf{B}^{-1} \mathbf{a}_4 - c_4 = \mathbf{c}_B' \boldsymbol{\alpha}_4 - c_4 = (7 \quad 0 \quad 5) \begin{pmatrix} 1 \\ 1 \\ -1 \end{pmatrix} - 0 = 2$$

$$z_6' - c_6 = \mathbf{c}_B' \mathbf{B}^{-1} \mathbf{a}_6 - c_6 = \mathbf{c}_B' \boldsymbol{\alpha}_6 - c_6 = (7 \quad 0 \quad 5) \begin{pmatrix} -1 \\ -3 \\ 2 \end{pmatrix} - 0 = 3$$

$$z' = \mathbf{c}_B' \mathbf{B}^{-1} \mathbf{b} = \mathbf{c}_B' \boldsymbol{\beta} = (7 \quad 0 \quad 5) \begin{pmatrix} 14 \\ 2 \\ 8 \end{pmatrix} = 138$$

In this example, although the values of the entire top row changed, the optimal basis remains the same, and thus no further processing is required. Note, however, that the optimal value of the objective has changed.

Changes in the Right-Hand Side

If a change in a particular b_i is made, there is an impact on both the $\mathbf{x}_B$ vector and the value of z. Recalling that $\mathbf{x}_B$ is given simply by $\mathbf{B}^{-1}\mathbf{b}$ and recalling that $\mathbf{B}^{-1}$ can be found from the tableau by a proper arrangement of the $\boldsymbol{\alpha}_j$ column vectors, we have

$$\mathbf{x}_B' = \mathbf{B}^{-1}\mathbf{b}' \tag{6.277}$$

where

$\mathbf{x}_B'$ = new values of the basic variables in the tableau of interest

$\mathbf{B}^{-1}$ = inverse of the present basis matrix

$\mathbf{b}'$ = new set of right-hand side constants

Also

$$z' = \mathbf{c}_B \mathbf{x}_B' = \mathbf{c}_B \mathbf{B}^{-1}\mathbf{b}' \tag{6.278}$$

Equation (6.278) should be studied carefully. The basis inverse $\mathbf{B}^{-1}$ may contain negative elements, and thus there is always a possibility that $\mathbf{x}_B'$ may include some negative elements. That is, there is always a possibility that we may lose primal feasibility. However, because dual feasibility is not affected, this presents no real problem because the dual simplex algorithm may be used to regain primal feasibility. This is illustrated in Example 6.15.

Example 6.15: A Change in a b_i

Consider again the original problem given in Example 6.12 when b_3 is changed from 22 to 24. That is, $b_3' = 24$.

Table 6.19 provided the final tableau to the original problem. As we observed in Example 6.12, the inverse basis matrix is given by

$$\mathbf{B}^{-1} = \begin{pmatrix} 1 & 0 & -1 \\ 1 & 1 & -3 \\ -1 & 0 & 2 \end{pmatrix}$$

Thus, from (6.277) and (6.278), we have

$$\mathbf{x}'_B = \begin{pmatrix} 1 & 0 & -1 \\ 1 & 1 & -3 \\ -1 & 0 & 2 \end{pmatrix} \begin{pmatrix} 36 \\ 32 \\ 24 \end{pmatrix} = \begin{pmatrix} 12 \\ -4 \\ 12 \end{pmatrix}$$

$$z' = (7 \quad 0 \quad 6) \begin{pmatrix} 12 \\ -4 \\ 12 \end{pmatrix} = 156$$

The new updated tableau is given in Table 6.21. Note that we have lost primal feasibility, and thus the dual simplex is used to obtain the new final tableau of Table 6.22.

TABLE 6.21

	z	x_1	x_2	x_3	x_4	x_5	x_6	RHS
z	1	3	0	0	1	0	5	156
x_2	0	1	1	0	1	0	-1	12
x_5	0	-2	0	0	1	1	-3	-4
x_3	0	1	0	1	-1	0	2	12

TABLE 6.22

	z	x_1	x_2	x_3	x_4	x_5	x_6	RHS
z	1	0	0	0	$\frac{5}{2}$	$\frac{3}{2}$	$\frac{1}{2}$	140
x_2	0	0	1	0	$\frac{3}{2}$	$\frac{1}{2}$	$-\frac{1}{2}$	10
x_1	0	1	0	0	$-\frac{1}{2}$	$-\frac{1}{2}$	$\frac{3}{2}$	2
x_3	0	0	0	1	$-\frac{1}{2}$	$\frac{1}{2}$	$\frac{1}{2}$	10

Changes in the Technological Coefficients $a_{i,j}$

Discrete changes in the technological coefficients are relatively easy to handle if the $a_{i,j}$ to be changed are associated with a *nonbasic* variable. However, a change in an $a_{i,j}$ associated with a *basic* variable is considerably more involved, and thus, for such a case, we shall resort to simply resolving the problem from the beginning.

Restricting our attention then to changes in the technological coefficients of nonbasic variables, we note that any change in the $\mathbf{a}_k$ column for a nonbasic variable x_k will directly affect the associated $\boldsymbol{\alpha}_k$ vector (and, indirectly, the value of $z_k - c_k$). At any iteration, the $\boldsymbol{\alpha}_k$ column vector is given by $\mathbf{B}^{-1}\mathbf{a}_k$, so we have

$$\boldsymbol{\alpha}_k' = \mathbf{B}^{-1}\mathbf{a}_k' \tag{6.279}$$

where

$\mathbf{B}^{-1} = $ inverse of the present basis matrix

$\mathbf{a}_k' = $ *new* vector of technological coefficients associated with nonbasic variable x_k

$\boldsymbol{\alpha}_k' = $ *new* updated vector corresponding to x_k in the final simplex tableau

The updated $z_k - c_k$ is given by

$$z_k' - c_k = \mathbf{c}_B\boldsymbol{\alpha}_k' - c_k = \mathbf{c}_B\mathbf{B}^{-1}\mathbf{a}_k' - c_k \tag{6.280}$$

and if $z_k' - c_k$ should go negative, primal optimality (dual feasibility) has been lost and the primal simplex must be applied.

Example 6.16: A Change in a Nonbasic Coefficient $a_{i,j}$

Let us assume, for the original problem given in Example 6.12, that we are informed that an error was made in data collection and that $a_{1,3}$ is really 1 rather than 2. That is, $a_{1,3}' = 1$ and thus

$$\mathbf{a}_1' = \begin{pmatrix} 3 \\ 1 \\ 1 \end{pmatrix}$$

From (6.279), we have

$$\boldsymbol{\alpha}_1' = \begin{pmatrix} 1 & 0 & -1 \\ 1 & 1 & -3 \\ -1 & 0 & 2 \end{pmatrix}\begin{pmatrix} 3 \\ 1 \\ 1 \end{pmatrix} = \begin{pmatrix} 2 \\ 1 \\ -1 \end{pmatrix}$$

and from (6.280), we obtain

$$z_1' - c_1 = (7 \quad 0 \quad 6)\begin{pmatrix} 2 \\ 1 \\ -1 \end{pmatrix} - 10 = -2$$

The resultant tableau, after the change in $a_{1,3}$, is shown in Table 6.23. The solution is no longer optimal and we apply the primal simplex algorithm. The result-

TABLE 6.23

	z	x_1	x_2	x_3	x_4	x_5	x_6	RHS
z	1	-2	0	0	1	0	5	146
x_2	0	2	1	0	1	0	-1	14
x_5	0	1	0	0	1	1	-3	2
x_3	0	-1	0	1	-1	0	2	8

ing iterations are shown in Tables 6.24 and 6.25. Note that in this case two iterations were required to restore optimality.

TABLE 6.24

	z	x_1	x_2	x_3	x_4	x_5	x_6	RHS
z	1	0	0	0	3	2	-1	150
x_2	0	0	1	0	-1	-2	5	10
x_1	0	1	0	0	1	1	-3	2
x_3	0	0	0	1	0	1	-1	10

TABLE 6.25

	z	x_1	x_2	x_3	x_4	x_5	x_6	RHS
z	1	0	$\frac{1}{5}$	0	$\frac{14}{5}$	$\frac{8}{5}$	0	152
x_6	0	0	$\frac{1}{5}$	0	$-\frac{1}{5}$	$-\frac{2}{5}$	1	2
x_1	0	1	$\frac{3}{5}$	0	$\frac{2}{5}$	$-\frac{1}{5}$	0	8
x_3	0	0	$\frac{1}{5}$	1	$-\frac{1}{5}$	$\frac{3}{5}$	0	12

Addition of a New Variable

Let us assume that once we have solved a linear programming problem for a company, they ask us to consider the impact of the introduction of a new product. This new product must be represented by a new decision variable (i.e., a new x_j).

The introduction of a new decision variable will either affect the optimality (dual feasibility) of the present solution or else have no effect at all. In the first case, we must introduce the new x_j into the basis. In the second, the new x_j stays nonbasic (i.e., we should not introduce the new product into the market).

Let us assume that x_k is the new decision variable and the data associated with x_k is given by c_k and $\mathbf{a}_k$. That is, c_k is the objective coefficient (or unit return) of x_k and the vector $\mathbf{a}_k$ specifies the coefficients of x_k in the existing constraints. Because $\mathbf{a}_k$ specifies the rates at which x_k will be consuming the various resources, $\mathbf{a}_k$ is often referred to as a consumption vector. The process of checking the optimality condition of x_k is actually quite straightforward. We simply compute $z_k - c_k$ in the usual manner, that is,

$$z_k - c_k = \mathbf{c}_B \mathbf{B}^{-1} \mathbf{a}_k - c_k = \boldsymbol{\pi} \mathbf{a}_k - c_k \qquad (6.281)$$

If $z_k - c_k \geq 0$, then the present solution remains optimal. However, if $z_k - c_k < 0$, then x_k should be introduced into the basis by performing a primal simplex pivot in the updated x_k column given by

$$\boldsymbol{\alpha}_k = \mathbf{B}^{-1} \mathbf{a}_k \qquad (6.282)$$

The reader should realize that this addition of a variable to the primal corresponds to the addition of a constraint to the dual. Hence, if this constraint that is added to the dual results in the current dual solution becoming infeasible, we lose primal optimality and must update with a primal pivot operation.

Example 6.17: Addition of a New Variable

Recall the original model of Example 6.12:

$$\text{maximize } z = 10x_1 + 7x_2 + 6x_3 \tag{6.283}$$

subject to

$$3x_1 + 2x_2 + x_3 + x_4 = 36 \tag{6.284}$$

$$x_1 + x_2 + 2x_3 + x_5 = 32 \tag{6.285}$$

$$2x_1 + x_2 + x_3 + x_6 = 22 \tag{6.286}$$

$$x_1, x_2, x_3, x_4, x_5, x_6 \geq 0 \tag{6.287}$$

The final tableau was given previously in Table 6.19. Recall, from Example 6.12, that the optimal dual solution from this table is

$$\boldsymbol{\pi} = (\pi_1, \pi_2, \pi_3) = \mathbf{c}_B \mathbf{B}^{-1} = (1 \quad 0 \quad 5)$$

Let us now assume that a new decision variable, say, x_7, is to be evaluated. Also, assume that the data for x_7 is given by

$$c_7 = 8$$

$$\mathbf{a}_7 = \begin{pmatrix} 2 \\ 3 \\ 1 \end{pmatrix}$$

Thus, the unit return for x_7 is \$8 and each unit of x_7 consumes 2 units of resource 1, 3 units of resource 2, and 1 unit of resource 3. Now checking primal optimality (dual feasibility), we see that, using (6.281),

$$z_7 - c_7 = \boldsymbol{\pi}\mathbf{a}_7 - c_7 = (1 \quad 0 \quad 5)\begin{pmatrix} 2 \\ 3 \\ 1 \end{pmatrix} - 8 = -1$$

Because $z_7 - c_7 < 0$, primal optimality (dual feasibility) has been lost. Therefore, we need to compute the updated x_7 column, append it to the current tableau, and restore optimality using the primal simplex method. (Note that if $z_7 - c_7$ had been nonnegative, the current solution would still be optimal and the analysis would be complete.) Now using (6.282), we compute the updated x_7 column as

$$\boldsymbol{\alpha}_7 = \mathbf{B}^{-1}\mathbf{a}_7 = \begin{pmatrix} 1 & 0 & -1 \\ 1 & 1 & -3 \\ -1 & 0 & 2 \end{pmatrix}\begin{pmatrix} 2 \\ 3 \\ 1 \end{pmatrix} = \begin{pmatrix} 1 \\ 2 \\ 0 \end{pmatrix}$$

This results in the tableau shown in Table 6.26. The reader is invited to complete this example by using the primal simplex to restore optimality.

TABLE 6.26

	z	x_1	x_2	x_3	x_4	x_5	x_6	x_7	RHS
z	1	3	0	0	1	0	5	-1	146
x_2	0	1	1	0	1	0	-1	1	14
x_5	0	-2	0	0	1	1	-3	2	2
x_3	0	1	0	1	-1	0	2	0	8

Addition of a New Constraint

A rather common error in model development is to forget about one or more constraints. It is also not atypical to find that once the mathematical model has been solved, a new constraint (such as the passage of a new clear air act) comes into being. The consideration of the impact of such new constraints is relatively straightforward.

It is easy to determine whether a new constraint has an impact (i.e., without yet evaluating the actual measure of the impact). We simply evaluate the constraint at the present basic feasible solution ($\mathbf{x}^*$). If the constraint is satisfied, there is *no* impact and we go no further. (Note that we are simply checking primal feasibility as we checked dual feasibility in the previous section.) However, if the constraint is violated at $\mathbf{x}^*$, the constraint *will* have an impact on the solution and we proceed to the next phase of our analysis.

Having determined that a new constraint will affect the present solution, we proceed to incorporate this new constraint into the previous final tableau. The slack variable for the new constraint will enter the basis and a new row must be added to the tableau. However, when adding this new row to the tableau, we must return the tableau to canonical form by "eliminating" the coefficients of any variables in the new constraint that are basic in the previous final tableau. We accomplish this through simple matrix row operations that are illustrated via the following example.

Example 6.18: Addition of a New Constraint

Returning again to our problem of Example 6.12, let us evaluate the impact of adding the following constraint:

$$x_1 + x_2 + x_3 \leq 20$$

Examining Table 6.18, we see that $x_1^* = 0$, $x_2^* = 14$, and $x_3^* = 8$. Thus, $x_1 + x_2 + x_3 = 22$ at the current optimal solution, the new constraint is violated and will have an impact on the solution. We first rewrite the constraint as

$$x_1 + x_2 + x_3 + x_7 = 20$$

where x_7 is a zero-cost slack variable. Now adding this constraint to Table 6.18, we have the tableau of Table 6.27. Note, however, that adding the constraint in this manner destroys the canonical form of the tableau. That is, x_2 and x_3 are basic variables, but the columns of the tableau associated with x_2 and x_3 no longer corre-

TABLE 6.27   TABLEAU WITH NEW CONSTRAINT ADDED

	z	x_1	x_2	x_3	x_4	x_5	x_6	x_7	RHS
z	1	3	0	0	1	0	5	0	146
x_2	0	1	1	0	1	0	−1	0	14
x_5	0	−2	0	0	1	1	−3	0	2
x_3	0	1	0	1	−1	0	2	0	8
x_7	0	1	1	1	0	0	0	1	20

spond to columns of the identity. However, it is a straightforward task to restore the tableau to canonical form using simple matrix operations. This is done by multiplying constraint rows 1 and 3 by −1 and adding them to the new constraint row. This results in Table 6.28.

TABLE 6.28   TABLEAU RESTORED TO CANONICAL FORM

	z	x_1	x_2	x_3	x_4	x_5	x_6	x_7	RHS
z	1	3	0	0	1	0	5	0	146
x_2	0	1	1	0	1	0	−1	0	14
x_5	0	−2	0	0	1	1	−3	0	2
x_3	0	1	0	1	−1	0	2	0	8
x_7	0	−1	0	0	0	0	−1	1	−2

Note that, as expected, Table 6.28 indicates that we have lost primal feasibility. Therefore, we continue with the dual simplex procedure, which results in the new optimal solution in Table 6.29.

TABLE 6.29

	z	x_1	x_2	x_3	x_4	x_5	x_6	x_7	RHS
z	1	0	0	0	1	0	2	3	140
x_2	0	0	1	0	1	0	−2	1	12
x_5	0	1	0	0	1	1	5	−2	6
x_3	0	0	0	1	−1	0	1	1	6
x_1	0	0	0	0	0	0	1	−1	2

PARAMETRIC PROGRAMMING

Up to this point, we have considered only the impact of *discrete* changes in the structure of a linear programming model. In this section, we now consider the impact *over a range* of variation in model structure, but we only examine these systematic changes in the cost vector **c** and the right-hand-side vector **b**. Evaluation of other parameters, over a range, is also possible but tends to be much more involved. Fortunately, for the majority of practical cases, the c_j's and b_i's are the data of major interest.

Consider, for example, a mathematical model of a utility company. Now, if such a model were linear (or could be so approximated) and were solved by the simplex method, the results obtained are valid only for conditions reflected by the original mathematical model. The utility company's resources include basic energy sources such as oil, coal, and natural gas, and the availability of such resources varies with time (and other factors), so it would be wise to determine the impact of resource availability on the problem solution (i.e., on $\mathbf{x}^*$ and z^*). For such a problem, our interest is centered on the variation of the b_i's.

As another example, our interest might be directed toward the selection of an investment portfolio. Obviously, the return from investment opportunities, in general, can be only estimated. Recognizing the inherent risks in such estimates, it is a wise move to evaluate the portfolio choice (and total estimated return) over a variation in individual returns (i.e., a variation in each c_j).

Systematic Variation of the Cost Vector c

We consider first an evaluation of the impact on the original problem solution of a systematic variation of the objective function coefficients. This is modeled by perturbing the original objective coefficients via a scalar parameter t and a perturbation vector $\mathbf{c}'$. The resulting model can be written as follows:

$$\text{maximize } z = \mathbf{cx} + t\mathbf{c}'\mathbf{x} = (\mathbf{c} + t\mathbf{c}')\mathbf{x} \tag{6.288}$$

subject to

$$\mathbf{Ax} = \mathbf{b}$$

$$\mathbf{x} \geq \mathbf{0}$$

To help fix the basic idea, consider that a company's profit is a linear function of its sales of three products. Letting x_1, x_2, and x_3 be the number of units of products 1, 2, and 3 sold, respectively, we have

$$z = \mathbf{cx} = c_1x_1 + c_2x_2 + c_3x_3$$

where c_1, c_2, and c_3 are associated per unit profits. Now let us consider a single parameter that we identify simply as t. Such a parameter might represent calendar time, inflation rate, or it might simply be used to investigate a relationship between the profits of x_1, x_2, and x_3. For example, an increase in unit profit of one product may result in a decrease in unit profit of another. If the profit per product can be reasonably expected to vary linearly with this parameter, our method of analysis is appropriate. As an illustration, examine the following revised objective function:

$$z = (c_1 + t)x_1 + (c_2 - 2t)x_2 + c_3x_3$$

$$= (c_1, c_2, c_3)\mathbf{x} + t(1 \quad -2 \quad 0)\mathbf{x} \tag{6.289}$$

$$= \mathbf{cx} + t\mathbf{c}'\mathbf{x}$$

where

$$\mathbf{c} = (c_1, c_2, c_3)$$

$$\mathbf{c}' = (1 \quad -2 \quad 0)$$

$$\mathbf{x} = \begin{pmatrix} x_1 \\ x_2 \\ x_3 \end{pmatrix}$$

Reading the function defined in (6.289), we see that the profit of x_1 increases linearly with t. The profit of product x_2, on the other hand, decreases, and it does so as a function of two times t. Finally, the profit for x_3 is unaffected by t.

Note that because we are examining systematic changes in the $\mathbf{c}$ vector, only primal optimality (dual feasibility) will be affected. That is, the changes in $\mathbf{c}$ have no affect on primal feasibility. Let us begin our analysis of this problem by deriving the optimality conditions for the perturbed problem in (6.288).

First, recall that the optimality conditions for a linear program in the standard form

$$\text{maximize } z = \mathbf{cx} \tag{6.290}$$

$$\text{subject to}$$

$$\mathbf{Ax} = \mathbf{b}$$

$$\mathbf{x} \geq \mathbf{0}$$

are given simply by

$$z_j - c_j = \mathbf{c}_B \mathbf{B}^{-1} \mathbf{a}_j - c_j \geq 0, \qquad \text{for all } j \tag{6.291}$$

Now, note that in the perturbed problem, we have simply replaced $\mathbf{c}$ by $(\mathbf{c} + t\mathbf{c}')$. That is, each c_j has been replaced by $c_j + tc_j'$. Thus, by substituting into (6.291), the optimality conditions for problem (6.288) are given by

$$(\mathbf{c}_B + t\mathbf{c}_B')\mathbf{B}^{-1}\mathbf{a}_j - (c_j + tc_j') \geq 0, \qquad \text{for all } j \tag{6.292}$$

But, by rearranging terms, (6.292) simplifies to

$$(\mathbf{c}_B\mathbf{B}^{-1}\mathbf{a}_j - c_j) + t(\mathbf{c}_B'\mathbf{B}^{-1}\mathbf{a}_j - c_j')$$

$$= (z_j - c_j) + t(z_j' - c_j') \geq 0, \qquad \text{for all } j \tag{6.293}$$

Thus, the optimality conditions of the perturbed problem combines the $z_j - c_j$ values computed with the original costs with those computed using the perturbation costs (i.e., $z_j' - c_j'$). Also not that the objective value of the perturbed problem is given by

$$z = (\mathbf{c}_B + t\mathbf{c}_B')\mathbf{B}^{-1}\mathbf{b} = \mathbf{c}_B\mathbf{B}^{-1}\mathbf{b} + t\mathbf{c}_B'\mathbf{B}^{-1}\mathbf{b} \tag{6.294}$$

We will handle this information in the tableau by adding an additional row to the top of the tableau. Finally, note that, as expected, the optimality conditions and objective value for the perturbed problem reduce to the original unperturbed problem when $t = 0$. A systematic procedure for performing an analysis of the impact of variations in the c_j's can be summarized as follows.

Parametric programming procedure: c vector (maximization problem)

STEP 1. Set the parameter $t = 0$ and find an optimal solution to the original problem.

STEP 2. Add an additional top row to the optimal tableau containing the $z_j' - c_j'$, which are computed using $z_j' - c_j' = \mathbf{c}_B' \mathbf{B}^{-1} \mathbf{a}_j - c_j'$. The contribution to the objective function value is given by $\mathbf{c}_B' \mathbf{B}^{-1} \mathbf{b}$.

STEP 3. Determine the parameter range over which the tableau is optimal by examining the optimality conditions

$$(z_j - c_j) + t(z_j' - c_j') \geq 0, \qquad \text{for all } j$$

Let this range be given by $l \leq t \leq u$, where l is the lower bound and u is the upper bound on parameter t. (Note that the values of l and u need not be finite.)

STEP 4. If l is finite, determine which nonbasic variable has $(z_j - c_j) + t(z_j' - c_j') = 0$ when $t = l$. Enter this variable into the basis by performing a primal simplex pivot. This will possibly result in a new tableau that is optimal for additional values of t.

Similarly, if u is finite, determine which nonbasic variable has $(z_j - c_j) + t(z_j' - c_j') = 0$ when $t = u$. Enter this variable into the basis by performing a primal simplex pivot. This will possibly result in a new tableau that is optimal for additional values of t.

STEP 5. Repeat Steps 3 and 4 until all the appropriate ranges of the parameter have been investigated.

We now illustrate the solution procedure via the following simple example.

Example 6.19: Parametric Programming: c Vector

$$\text{maximize } z = x_1 + 4x_2 + t(x_1 + x_2) = (1 + t)x_1 + (4 + t)x_2 \qquad (6.295)$$

subject to

$$2x_1 + x_2 \leq 10 \qquad (6.296)$$

$$x_1 + x_2 \leq 6 \qquad (6.297)$$

$$x_2 \leq 4 \qquad (6.298)$$

$$x_1, x_2 \geq 0 \qquad (6.299)$$

Considering the original problem with $t = 0$ and denoting the respective slack variables by x_3, x_4, and x_5 yield the optimal tableau shown in Table 6.30.

Note, from (6.295), that $\mathbf{c}$ and $\mathbf{c}'$ are given by

$$\mathbf{c} = (1 \quad 4 \quad 0 \quad 0 \quad 0)$$

$$\mathbf{c}' = (1 \quad -1 \quad 0 \quad 0 \quad 0)$$

Now, add an additional top row to the optimal tableau containing $z_j' - c_j' = \mathbf{c}_B' \mathbf{B}^{-1} \mathbf{a}_j - c_j'$ with objective value $\mathbf{c}_B' \mathbf{B}^{-1} \mathbf{b}$. The updated tableau is shown in Table 6.31.

TABLE 6.30

	z	x_1	x_2	x_3	x_4	x_5	RHS
z	1	0	0	0	1	3	18
x_3	0	0	0	1	-2	1	2
x_1	0	1	0	0	1	-1	2
x_2	0	0	1	0	0	1	4

TABLE 6.31

	z	x_1	x_2	x_3	x_4	x_5	RHS
		0	0	0	1	-2	-2
z	1	0	0	0	1	3	18
x_3	0	0	0	1	-2	1	2
x_1	0	1	0	0	1	-1	2
x_2	0	0	1	0	0	1	4

Examining Table 6.31, we see that the present solution is $x = (2, 4, 2, 0, 0)$ and $z = 18 - 2t$. Now we use the optimality conditions given in (6.293) to determine for what range of the parameter t the current solution is optimal. The optimality conditions are

$$(z_4 - c_4) + t(z_4' - c_4') = 1 + t \geq 0$$
$$(z_5 - c_5) + t(z_5' - c_5') = 3 - 2t \geq 0$$

which result in

$$-1 \leq t \leq \tfrac{3}{2}$$

Thus, the current tableau is optimal for $-1 \leq t \leq \tfrac{3}{2}$. Note that $(z_4 - c_4) + t(z_4' - c_4') = 0$ when $t = -1$. Therefore, an alternative optimal solution exists for this tableau, which may be found by entering x_4 via a primal simplex pivot. The departing variable is x_1 and the new tableau is shown in Table 6.32.

TABLE 6.32

	z	x_1	x_2	x_3	x_4	x_5	RHS
		-1	0	0	0	-1	-4
z	1	-1	0	0	0	4	16
x_3	0	2	0	1	0	-1	6
x_4	0	1	0	0	1	-1	2
x_2	0	0	1	0	0	1	4

From Table 6.32, we see that it is optimal if

$$-1 - t \geq 0$$

$$4 - t \geq 0$$

That is, for $-\infty \leq t \leq -1$, $\mathbf{x} = (0, 4, 6, 2, 0)$ and $z = 16 - 4t$. Note that because we have already examined the bound $t = -1$, no finite bounds remain to be examined for the tableau of Table 6.32.

Returning to Table 6.31, we now examine the case of $t = \frac{3}{2}$. In this case, we perform a primal pivot entering x_5 into the basis. Table 6.33 shows the resulting tableau.

TABLE 6.33

	z	x_1	x_2	x_3	x_4	x_5	RHS
		0	0	2	-3	0	2
z	1	0	0	-3	7	0	12
x_5	0	0	0	1	-2	1	2
x_1	0	1	0	1	-1	0	4
x_2	0	0	1	-1	2	0	2

Examining Table 6.33, we find that $\mathbf{x} = (4, 2, 0, 0, 2)$ and $z = 12 + 2t$ for $\frac{3}{2} \leq t \leq \frac{7}{3}$. This results from the optimality conditions

$$-3 + 2t \geq 0$$

$$7 - 3t \geq 0$$

There are two finite bounds on the parameter t at this point, but we have already examined $t = \frac{3}{2}$. Thus, we look at $t = \frac{7}{3}$. For this case, we use a primal pivot to enter x_4 in Table 6.33. The resulting tableau is shown in Table 6.34.

TABLE 6.34

	z	x_1	x_2	x_3	x_4	x_5	RHS
		0	$\frac{3}{2}$	$\frac{1}{2}$	0	0	5
z	1	0	$-\frac{7}{2}$	$\frac{1}{2}$	0	0	5
x_5	0	0	1	0	0	1	4
x_1	0	1	$\frac{1}{2}$	$\frac{1}{2}$	0	0	5
x_4	0	0	$\frac{1}{2}$	$-\frac{1}{2}$	1	0	1

Table 6.34 indicates optimality if

$$-\tfrac{7}{2} + (\tfrac{3}{2})t \geq 0$$

$$\tfrac{1}{2} + (\tfrac{1}{2})t \geq 0$$

This results in the solution $\mathbf{x} = (5, 0, 0, 1, 4)$, $z = 5 + 5t$ for $\tfrac{7}{3} \leq t \leq \infty$.

At this point, no *finite* bounds on t remain to be examined. Thus, we may summarize the results, for the entire range of t (from $-\infty$ to ∞), as shown in Table 6.35. The results of Table 6.35 pertaining to the objective value z can also be viewed graphically, as shown in Figure 6.10.

TABLE 6.35 RESULTS OF EXAMPLE 6.19

Range of t	Optimal solution	Optimal objective
$-\infty \leq t \leq -1$	$\mathbf{x} = (0, 4, 6, 2, 0)$	$z = 16 - 4t$
$-1 \leq t \leq \tfrac{3}{2}$	$\mathbf{x} = (2, 4, 2, 0, 0)$	$z = 18 - 2t$
$\tfrac{3}{2} \leq t \leq \tfrac{7}{3}$	$\mathbf{x} = (4, 2, 0, 0, 2)$	$z = 12 + 2t$
$\tfrac{7}{3} \leq t \leq \infty$	$\mathbf{x} = (5, 0, 0, 1, 4)$	$z = 5 + 5t$

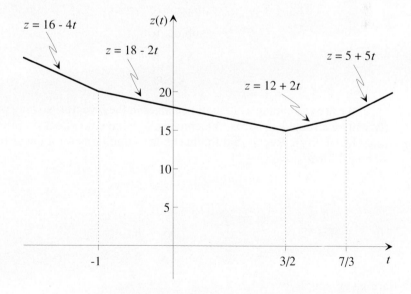

Figure 6.10 Graph for Example 6.19.

Despite the rather trivial nature of this problem, it does serve to illustrate the general procedure that is to be carried out for any problem in which variations in the cost vector **c** are to be analyzed. The use of such an analysis can yield significant information to the organization. Consider, for example, the problem investigated in Example 6.19. Let us assume that at present, the value of t is $-\frac{1}{2}$. At such a value, the optimal solution is to produce 4 units of x_2 and none of x_1. Thus, the "production line" for x_1 need not be built. However, if our forecasters predict a likelihood in the near future of t exceeding -1, we should plan for the future production of x_1, because it enters the optimal solution at that time. Furthermore, if it is anticipated that t will exceed $\frac{7}{3}$, note that it is optimal to produce 5 units of x_1 and to stop production of x_2.

Systematic Variation of the Right-Hand-Side Vector b

The only other systematic variation to be considered is that associated with the right-hand-side values of the constraints (e.g., the units of resource availabilities or limits on legal restrictions). Because a change in **b** is equivalent, *in terms of the dual*, to a change in **c**, the process used to analyze a systematic variation in **b** is the dual of that process used to analyze a systematic variation in **c**. Consider the following model in which the vector **b** is perturbed using a vector **b**′ and the scalar parameter t.

$$\text{maximize } z = \mathbf{cx} \tag{6.300}$$

$$\text{subject to}$$

$$\mathbf{Ax} = \mathbf{b} + t\mathbf{b}'$$

$$\mathbf{x} \geq \mathbf{0}$$

We are examining systematic changes in the **b** vector, so only primal feasibility will be affected. That is, the changes in **b** have no affect on primal optimality (dual feasibility). Recall that the feasibility conditions for a linear program in the standard form

$$\text{maximize } z = \mathbf{cx} \tag{6.301}$$

$$\text{subject to}$$

$$\mathbf{Ax} = \mathbf{b}$$

$$\mathbf{x} \geq \mathbf{0}$$

are given simply by

$$\mathbf{x}_B = \boldsymbol{\beta} = \mathbf{B}^{-1}\mathbf{b} \geq 0 \tag{6.302}$$

Noting that we have simply replaced $\mathbf{b}$ by $(\mathbf{b} + t\mathbf{b}')$, the primal feasibility conditions for the perturbed problem are given by

$$\mathbf{x}_B = \mathbf{B}^{-1}(\mathbf{b} + t\mathbf{b}') = \mathbf{B}^{-1}\mathbf{b} + t\mathbf{B}^{-1}\mathbf{b}' = \boldsymbol{\beta} + t\boldsymbol{\beta}' \geq 0 \qquad (6.303)$$

Also note that the objective value of the perturbed problem is given by

$$z = \mathbf{c}_B\mathbf{B}^{-1}(\mathbf{b} + t\mathbf{b}') = \mathbf{c}_B\mathbf{B}^{-1}\mathbf{b} + t\mathbf{c}_B\mathbf{B}^{-1}\mathbf{b}' = \mathbf{c}_B\boldsymbol{\beta} + t\mathbf{c}_B\boldsymbol{\beta}' \qquad (6.304)$$

This additional information will be displayed in the tableau by adding an additional right-hand-side column (RHS′) to the tableau. Also observe that the feasibility conditions and objective value for the perturbed problem reduce to the original unperturbed problem when $t = 0$. A systematic procedure for performing an analysis of the impact of systematic variations in the b_i's can be summarized as follows.

Parametric programming procedure: b vector (maximization problem)

STEP 1. Set the parameter $t = 0$ and find an optimal solution to the original problem.

STEP 2. Add an additional right-hand-side column to the optimal tableau containing the objective value $\mathbf{c}_B\mathbf{B}^{-1}\mathbf{b}'$ and the vector $\boldsymbol{\beta}' = \mathbf{B}^{-1}\mathbf{b}'$.

STEP 3. Determine the parameter range over which the tableau is primal feasible by examining the conditions

$$\mathbf{x}_B = \mathbf{B}^{-1}\mathbf{b} + t\mathbf{B}^{-1}\mathbf{b}' = \boldsymbol{\beta} + t\boldsymbol{\beta}' \geq 0$$

Let this range be given by $l \leq t \leq u$, where l is the lower bound and u is the upper bound on the parameter t. (Note that the values of l and u need not be finite.)

STEP 4. If l is finite, determine which basic variable $x_{B,i}$ has $\beta_i + t\beta_i' = 0$ when $t = l$. Perform a dual simplex pivot, choosing this variable as the departing variable. This will possibly result in a new tableau that is feasible (and optimal) for additional values of t. Similarly, if u is finite, determine which basic variable $x_{B,i}$ has $\beta_i + t\beta_i' = 0$ when $t = u$. Choose this as the departing variable and perform a dual simplex pivot. This will possibly result in a new tableau that is feasible (and optimal) for additional values of t.

STEP 5. Repeat Steps 3 and 4 until all the appropriate ranges of the parameter have been investigated.

The solution procedure is now illustrated with an example.

Example 6.20: Parametric Programming: b Vector

$$\text{maximize } z = x_1 + 4x_2 \qquad (6.305)$$

subject to

$$2x_1 + x_2 \leq 10 \qquad (6.306)$$

$$x_1 + x_2 \leq 6 + t \qquad (6.307)$$

$$x_2 \leq 4 - t \qquad (6.308)$$

$$x_1, x_2 \geq 0 \qquad (6.309)$$

For this simple example, note that

$$\mathbf{b} = \begin{pmatrix} 10 \\ 6 \\ 4 \end{pmatrix}$$

$$\mathbf{b}' = \begin{pmatrix} 0 \\ 1 \\ -1 \end{pmatrix}$$

Resource 2 (associated with constraint 2) is being increased at a rate that is directly proportional to t whereas resource 3 is being decreased at a rate proportional to t. Thus, the combined total amount of resources 2 and 3 remains the same, and we may be investigating a trade-off between these two resources. The question to be answered is: What are the optimal solutions and associated objective value as these resources change?

This is the same basic example as that of Example 6.19, and the optimal solution with $t = 0$ is shown in Table 6.30. Computing the additional right-hand-side column as specified in (6.303) and (6.304) results in Table 6.36.

TABLE 6.36

	z	x_1	x_2	x_3	x_4	x_5	RHS	RHS'
z	1	0	0	0	1	3	18	-2
x_3	0	0	0	1	-2	1	2	-3
x_1	0	1	0	0	1	-1	2	2
x_2	0	0	1	0	0	1	4	-1

Examining Table 6.36, we see that the present solution is given by

$$\mathbf{x}_B = \begin{pmatrix} x_{B,1} \\ x_{B,2} \\ x_{B,3} \end{pmatrix} = \begin{pmatrix} x_3 \\ x_1 \\ x_2 \end{pmatrix} = \boldsymbol{\beta} + t\boldsymbol{\beta}' = \begin{pmatrix} 2 - 3t \\ 2 + 2t \\ 4 - t \end{pmatrix}$$

$$z = 18 - 2t$$

Thus, feasibility (and hence optimality) will occur when

$$x_3 = 2 - 3t \geq 0$$

$$x_1 = 2 + 2t \geq 0$$

$$x_2 = 4 - t \geq 0$$

or, equivalently, when $-1 \leq t \leq \frac{2}{3}$.

We are now ready for Step 4 of the process. Arbitrarily, we shall first examine the lower bound on $t(t = -1)$. When $t = -1$, x_1 is zero in the tableau of Table 6.36 and thus, using the dual simplex, we choose x_1 as the departing variable to obtain the tableau shown in Table 6.37.

TABLE 6.37

	z	x_1	x_2	x_3	x_4	x_5	RHS	RHS'
z	1	3	0	0	4	0	24	4
x_3	0	1	0	1	−1	0	4	−1
x_5	0	−1	0	0	−1	1	−2	−2
x_2	0	1	1	0	1	0	6	1

The basis in this tableau is feasible only when

$$x_3 = 4 - t \geq 0$$

$$x_5 = -2 - 2t \geq 0$$

$$x_2 = 6 + t \geq 0$$

which reduces to the condition $-6 \leq t \leq -1$.

At this point, observe that x_2 would be chosen as the departing variable when considering the lower limit of $t = -6$ in Table 6.37. However, note from Table 6.37, that we *cannot* use the dual simplex to obtain a new feasible basis as we did earlier because there is no possible dual pivot for the row associated with x_2 in Table 6.37. This result should not be surprising. Look at the original problem formulation in (6.306–6.308). Notice in constraint (6.307) that the resources associated with this constraint go negative at $t < -6$ (i.e., we run out of resources). This fact is what causes us to be unable to pivot on the x_2 row in Table 6.36, and it is a logical result when interpreted from a physical point of view. As a result, the problem is infeasible for $-\infty \leq t < -6$.

Thus, we have now examined the parameter t from $-\infty$ to $\frac{2}{3}$. If we return to Table 6.36, we may next examine t at its upper limit of $\frac{2}{3}$. In Table 6.36, we see that x_3 becomes zero at $t = \frac{2}{3}$, and a dual pivot in the x_3 row results in Table 6.38.

TABLE 6.38

	z	x_1	x_2	x_3	x_4	x_5	RHS	RHS'
z	1	0	0	$\frac{1}{2}$	0	$\frac{7}{2}$	19	$-\frac{7}{2}$
x_4	0	0	0	$-\frac{1}{2}$	1	$-\frac{1}{2}$	−1	$\frac{3}{2}$
x_1	0	1	0	$\frac{1}{2}$	0	$-\frac{1}{2}$	3	$\frac{1}{2}$
x_2	0	0	1	0	0	1	4	−1

The basis in Table 6.38 is feasible when

$$x_4 = -1 + (\tfrac{3}{2})t \geq 0$$

$$x_1 = 3 + (\tfrac{1}{2})t \geq 0$$

$$x_2 = 4 - t \geq 0$$

which results in the range $\frac{2}{3} \le t \le 4$. Again, the reader should note that the problem is infeasible for $t > 4$, because there is no possible dual pivot in the x_2 row of Table 6.38.

As with variations in $\mathbf{c}$, we can now summarize our results in tabular form, as shown in Table 6.39.

TABLE 6.39 RESULTS OF EXAMPLE 6.20

Range of t	Optimal solution	Optimal objective
$-\infty \le t < -6$	Problem is infeasible.	
$-6 \le t \le -1$	$x_1 = 0$ $x_2 = 6 + t$ $x_3 = 4 - t$ $x_4 = 0$ $x_5 = -2 - 2t$	$z = 24 + 4t$
$-1 \le t \le \frac{2}{3}$	$x_1 = 2 + 2t$ $x_2 = 4 - t$ $x_3 = 2 - 3t$ $x_4 = 0$ $x_5 = 0$	$z = 18 - 2t$
$\frac{2}{3} \le t \le 4$	$x_1 = 3 + (\frac{1}{2})t$ $x_2 = 4 - t$ $x_3 = 0$ $x_4 = -1 + (\frac{3}{2})t$ $x_5 = 0$	$z = 19 - (\frac{7}{2})t$
$4 < t \le \infty$	Problem is infeasible.	

The results of Table 6.20 are also summarized graphically in Figure 6.11. Note that the maximum of this concave piecewise linear function occurs at $t = -1$. Thus, given a choice, we should choose $t = -1$ because this results in the maximum value of the objective. Note that this choice of $t = -1$ corresponds to $(6 + t) = 6 - 1 = 5$ units of resource 2 and $(4 - t) = 4 + 1 = 5$ units of resource 3.

Resource Values and Ranges

Recall that for every primal constraint there is a dual decision variable. That is, given a primal linear programming problem with m constraints, there will be m dual variables. Thus, π_i, the ith dual variable, is associated with the ith primal constraint. However, not only is there an "association," but there is also an economic interpretation. That is, as discussed earlier, the value of π_i represents the change in the objective that may be obtained for every extra unit of resource i, *within a specified range of values for resource i*. Mathematically, π_i represents

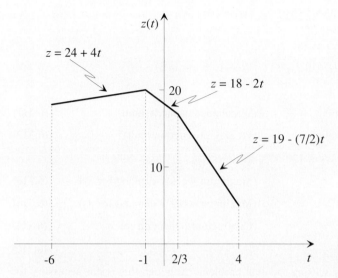

Figure 6.11   Graph for Example 6.20.

the rate of change of the objective as long as the current basis remains optimal. The same ideas that were used in parametric programming may be used to determine this range.

To illustrate the manner in which this is done, consider the following example.

Example 6.21: Ranges for b_i

A chemical company produces two products. Product 1, known as "StripEasy," is a paint and finish remover for tough refinishing jobs (and is quite thick). Product 2, sold under the brand name "RenewIt," is less viscous and intended for easier refinishing (or to be used after one application of StripEasy). Every can of StripEasy returns a profit of $0.50, and the return for RenewIt is $0.30 per can.

Because these products can only be shipped out at the end of each week, the total amount produced must fit within the company warehouse, which has a capacity of 400,000 cubic feet. Each container (for either product) consumes 2 cubic feet of warehouse space.

Both products are produced through the same system of pipes, vessels, and processors. Production rates are 2000 and 3000 cans per hour for StripEasy and RenewIt, respectively, with a total of 120 hours per week available.

Marketing has determined that under the present market state and advertising level, the maximum weekly amounts that can be sold are 250,000 cans of StripEasy and 300,000 cans of RenewIt. Finally, because of a previous contract, the company must furnish at least 60,000 can per week of RenewIt to a particular customer.

Letting x_1 be the number of cans per week of StripEasy and x_2 be the number of cans of RenewIt, we may formulate this problem as the following linear program.

Find x_1 and x_2 so as to

maximize $z = 0.5x_1 + 0.3x_2$ (Weekly profit in \$)   (6.310)

subject to

$2x_1 + 2x_2 \leq 400{,}000$ (Warehouse capacity limit)   (6.311)

$3x_1 + 2x_2 \leq 720{,}000$ (Weekly production limit)   (6.312)

(or $(\frac{1}{2000})x_1 + (\frac{1}{3000})x_2 \leq 120$)

$x_1 \leq 250{,}000$ (Maximum weekly demand for x_1)   (6.313)

$x_2 \leq 300{,}000$ (Maximum weekly demand for x_2)   (6.314)

$x_1 \geq 60{,}000$ (Contractual obligation for x_2)   (6.315)

$x_1, x_2 \geq 0$ (6.316)

Let us begin by writing the dual problem. Although this is not necessary for the succeeding analysis, it will serve to reinforce some of the earlier concepts. Denoting the dual variables corresponding to constraints (6.311–6.315) by $\pi_1, \pi_2, \pi_3, \pi_4$, and π_5, respectively, the dual problem can be written as follows. In particular, note that whereas π_1, π_2, π_3, and π_4 are nonnegative variables, the dual variable π_5 is nonpositive.

minimize Z
$$= 400{,}000\pi_1 + 720{,}000\pi_2 + 250{,}000\pi_3 + 300{,}000\pi_4 + 60{,}000\pi_5 \qquad (6.317)$$

subject to

$$2\pi_1 + 3\pi_2 + \pi_3 \geq 0.5 \qquad (6.318)$$

$$2\pi_1 + 2\pi_2 + \pi_4 + \pi_5 \geq 0.3 \qquad (6.319)$$

$$\pi_1, \pi_2, \pi_3, \pi_4 \geq 0 \qquad (6.320)$$

$$\pi_5 \leq 0 \qquad (6.321)$$

Let us now solve the primal problem for the optimal, weekly production program. Note that because constraint (6.315) is of the ($\geq$) form, the initial tableau will be neither primal nor dual feasible. Therefore, we will need to use either the two-phase method described in Chapter 4 or the artificial constraint method and the dual simplex procedure addressed earlier in this chapter. We will assume that the problem is solved by the two-phase method. Let us denote the respective slack variables for constraints (6.311–6.314) by x_3, x_4, x_5, and x_6. Also, let x_7 be the surplus variable associated with constraint (6.315), and let x_8 be the required artificial variable in this constraint. Thus, the original identity matrix is given by $\mathbf{I} = (\mathbf{a}_3, \mathbf{a}_4, \mathbf{a}_5, \mathbf{a}_6, \mathbf{a}_8)$ and, for any subsequent tableau, the basis inverse will be given by $\mathbf{B}^{-1} = (\boldsymbol{\alpha}_3, \boldsymbol{\alpha}_4, \boldsymbol{\alpha}_5, \boldsymbol{\alpha}_6, \boldsymbol{\alpha}_8)$. Therefore, so that we may easily access both $\mathbf{B}^{-1}$ and information regarding the dual variables, we will carry the artificial variable column into the phase-II tableau.

Of course, this is not necessary for the solution of the problem, but it retains valuable information. The reader should realize that we would *never*, under any circumstances, choose an artificial variable as the entering variable in phase II, even though doing so would improve the objective. In fact, it is very likely that entering an artificial variable into the basis would lead to an improved objective value, because the resulting solution would be *infeasible*, and, hence, *superoptimal*.

The final phase-two tableau containing the optimal, weekly production program is given in Table 6.40. Note, again, that the tableau contains the column associated with the artificial variable x_8.

TABLE 6.40

	z	x_1	x_2	x_3	x_4	x_5	x_6	x_7	x_8	RHS
z	1	0	0	0.25	0	0	0	0.2	-0.2	85,000
x_1	0	1	0	$\frac{1}{2}$	0	0	0	1	-1	140,000
x_4	0	0	0	$-\frac{3}{2}$	1	0	0	-1	1	180,000
x_5	0	0	0	$-\frac{1}{2}$	0	1	0	-1	1	110,000
x_6	0	0	0	0	0	0	1	1	-1	240,000
x_2	0	0	1	0	0	0	0	-1	1	60,000

From Table 6.40, we see that the optimal primal solution is

$$z^* = \$85,000$$

$$x_1^* = 140,000$$

$$x_2^* = 60,000$$

$$x_3^* = 0$$

$$x_4^* = 180,000$$

$$x_5^* = 110,000$$

$$x_6^* = 240,000$$

$$x_7^* = 0$$

Also, the optimal basis inverse can be read from the tableau as

$$\mathbf{B}^{-1} = (\alpha_3, \alpha_4, \alpha_5, \alpha_6, \alpha_8) = \begin{pmatrix} \frac{1}{2} & 0 & 0 & 0 & -1 \\ -\frac{3}{2} & 1 & 0 & 0 & 1 \\ -\frac{1}{2} & 0 & 1 & 0 & 1 \\ 0 & 0 & 0 & 1 & -1 \\ 0 & 0 & 0 & 0 & 1 \end{pmatrix}$$

Recall that the values of the dual decision variables will be located in the top row of the tableau above $\mathbf{B}^{-1}$, provided that the objective coefficients of the associated primal variables are zero. Because this is the case, the optimal dual solution is

$$Z^* = \$85,000$$

$$\pi_1^* = z_3 - c_3 = 0.25$$

$$\pi_2^* = z_4 - c_4 = 0$$

$$\pi_3^* = z_5 - c_5 = 0$$

$$\pi_4^* = z_6 - c_6 = 0$$

$$\pi_5^* = z_8 - c_8 = -0.2$$

Note that as expected from the dual formulation, $\pi_1, \pi_2, \pi_3, \pi_4 \geq 0$, and $\pi_5 \leq 0$. The reader may also verify that $\boldsymbol{\pi} = (\pi_1, \pi_2, \pi_3, \pi_4, \pi_5)$ can be computed as follows:

$$\boldsymbol{\pi} = \mathbf{c}_B \mathbf{B}^{-1} = (0.5 \quad 0 \quad 0 \quad 0 \quad 0.3) \begin{pmatrix} \frac{1}{2} & 0 & 0 & 0 & -1 \\ -\frac{3}{2} & 1 & 0 & 0 & 1 \\ -\frac{1}{2} & 0 & 1 & 0 & 1 \\ 0 & 0 & 0 & 1 & -1 \\ 0 & 0 & 0 & 0 & 1 \end{pmatrix} = (0.25 \quad 0 \quad 0 \quad 0 \quad -0.2)$$

Now, if one simply stops at this point and tells the company that its best plan is to produce 150,000 cans of StripEasy and 50,000 cans of RenewIt per week (for a weekly profit of $90,000), he or she has failed to exploit the full potential of the simplex method. Organizations are interested not just in the optimal plan for the present organizational status, they also (and often of more significance) wish to know what they should do in order to grow, increase profits, and so forth. Such information is contained within the simplex tableau.

Notice first that the only nonzero dual decision variables are $\pi_1 = 0.25$ and $\pi_5 = -0.2$, associated with the first constraint (6.311) and fifth constraint (6.315). This means that under the present program, we may increase profit by $0.25 for every extra cubic foot of warehouse space. Similarly, we may increase profit by $0.20 for every unit *less* of product x_2 (RenewIt) that we ship, per our contract, to our customer. Or, equivalently, shipping an additional unit of x_2 to our customer decreases profit by $0.20.

All other dual decision variables are zero, implying that for the present basis, the worth of any additional "resources" 2, 3, or 4 is zero. This is obvious as the slack variables (x_4, x_5, and x_6) associated with these resources are positive-valued in the optimal solution (i.e., there are idle units of production capacity and we have not yet reached the weekly demand limits on x_1 and x_2).

With such information, we have a *portion* of what we need to answer such questions as the following:

1. What is a reasonable price to pay for additional warehouse space?
2. What is a reasonable penalty to pay for not completely satisfying the contract for 60,000 cans of x_2 (RenewIt)?

The partial answer to question 1 is a maximum of $0.25 per cubic foot of warehouse space rented, per week (i.e., the value of π_1). The partial answer to question 2 is a maximum of $0.20 for every can of x_2 below the 60,000 units that we have contracted.

To find *complete* answers to both questions, we must determine the ranges of either b_1 or b_5 over which the values of π_1 and π_5 remain constant. This, of course, calls for parametric analysis of the right-hand-side vector. Let us begin by discussing how such an analysis may be performed for b_1.

Because we only want to consider changes in the right-hand-side of constraint 1 (warehouse capacity), the analysis can be done by perturbing the **b** vector as follows:

$$\mathbf{b} + t\mathbf{b}' = \begin{pmatrix} 400,000 \\ 720,000 \\ 250,000 \\ 300,000 \\ 60,000 \end{pmatrix} + t \begin{pmatrix} 1 \\ 0 \\ 0 \\ 0 \\ 0 \end{pmatrix}$$

Thus, as was discussed previously, the current basis will remain optimal provided that

$$\mathbf{x}_B = \mathbf{B}^{-1}\mathbf{b} + t\mathbf{B}^{-1}\mathbf{b}' = \boldsymbol{\beta} + t\boldsymbol{\beta}' \geq \mathbf{0}$$

Computing $\boldsymbol{\beta}'$, we find that $\boldsymbol{\beta}'$ is simply the first column of $\mathbf{B}^{-1}$, that is,

$$\boldsymbol{\beta}' = \begin{pmatrix} \frac{1}{2} & 0 & 0 & 0 & -1 \\ -\frac{3}{2} & 1 & 0 & 0 & 1 \\ -\frac{1}{2} & 0 & 1 & 0 & 1 \\ 0 & 0 & 0 & 1 & -1 \\ 0 & 0 & 0 & 0 & 1 \end{pmatrix} \begin{pmatrix} 1 \\ 0 \\ 0 \\ 0 \\ 0 \end{pmatrix} = \begin{pmatrix} \frac{1}{2} \\ -\frac{3}{2} \\ -\frac{1}{2} \\ 0 \\ 0 \end{pmatrix}$$

Then we can find the range of t for which the current basis remains optimal by examining

$$\mathbf{x}_B = \boldsymbol{\beta} + t\boldsymbol{\beta}' = \begin{pmatrix} 140,000 \\ 180,000 \\ 110,000 \\ 240,000 \\ 60,000 \end{pmatrix} + t \begin{pmatrix} \frac{1}{2} \\ -\frac{3}{2} \\ -\frac{1}{2} \\ 0 \\ 0 \end{pmatrix} \geq \mathbf{0}$$

This results in the following feasibility conditions:

$$x_1 = 140,000 + (\tfrac{1}{2})t \geq 0$$

$$x_4 = 180,000 - (\tfrac{3}{2})t \geq 0$$

$$x_5 = 110,000 - (\tfrac{1}{2})t \geq 0$$

$$x_6 = 240,000 + 0t \geq 0$$

$$x_2 = 60,000 + 0t \geq 0$$

and this is true if

$$-280,000 \leq t \leq 120,000$$

Returning to constraint (6.311), we see that this means that as long as b_1 lies between $(400,000 - 280,000) = 120,000$ and $(400,000 + 120,000) = 520,000$ cubic feet, and no other change is made in the original problem data, the present basis is optimal. Thus, we may consider renting up to 120,000 extra cubic feet of warehouse space before its associated shadow price (π_1) changes from $0.25 cubic foot per week.

For example, if we are given the opportunity of renting, say, 100,000 cubic feet of space at a rate of $0.15 per cubic foot per week, we should take it and expect an increase in weekly profits of

$$(0.25 - 0.15)100,000 = \$10,000$$

In a similar manner, we can determine ranges for the remaining b_i's. For example, to determine the range for b_5 for which the current basis remains optimal, we would perturb the right-hand-side vector $\mathbf{b}$ as follows:

$$\mathbf{b} + t\mathbf{b}' = \begin{pmatrix} 400,000 \\ 720,000 \\ 250,000 \\ 300,000 \\ 60,000 \end{pmatrix} + t \begin{pmatrix} 0 \\ 0 \\ 0 \\ 0 \\ 1 \end{pmatrix}$$

Upon examining the feasibility conditions, $\mathbf{x}_B = \mathbf{B}^{-1}\mathbf{b} + t\mathbf{B}^{-1}\mathbf{b}'$, we find that the current basis remains feasible (and optimal) if $-60,000 \le t \le 140,000$. Examining the fifth constraint (6.315), we see that the current basis is still optimal if b_5 lies between 0 and 200,000. This means that from a physical point of view, our maximum, per unit weekly penalty of $0.20 per can holds true all the way down to a total elimination (a reduction of 60,000 units of x_2) of this contract. Thus, if the customer is willing to modify the contract, for a penalty of $5,000 per week, so as only to hold the company to 25,000 cans per week of RenewIt, we should accept this, as our total weekly profit will increase by

$$0.20(60,000 - 25,000) - 5000 = \$2000$$

The range for each b_i is summarized in Table 6.41.

TABLE 6.41   RANGES FOR THE b_i OF EXAMPLE 6.21

Resource	Current value	Allowable decrease before basis change	Allowable increase before basis change
b_1	400,000	280,000	120,000
b_2	720,000	180,000	∞
b_3	250,000	110,000	∞
b_4	300,000	240,000	∞
b_5	60,000	60,000	140,000

Objective Coefficients and Ranges

In the preceding section, we examined the right-hand-side vector $\mathbf{b}$ and determined the range for each "resource" that leaves the current basis optimal. In an analogous manner, it is possible to examine each objective coefficient c_j to determine a range of values over which the current basis remains optimal. This process will be illustrated by considering again the problem of Example 6.21.

Example 6.22: Ranges for c_j

Consider again the linear programming model defined by

$$\text{maximize } z = 0.5x_1 + 0.3x_2 \tag{6.322}$$

subject to

$$2x_1 + 2x_2 \leq 400{,}000 \tag{6.323}$$

$$3x_1 + 2x_2 \leq 720{,}000 \tag{6.324}$$

$$x_1 \leq 250{,}000 \tag{6.325}$$

$$x_2 \leq 300{,}000 \tag{6.326}$$

$$x_1 \geq 60{,}000 \tag{6.327}$$

$$x_1, x_2 \geq 0 \tag{6.328}$$

Recall that the optimal solution for this problem is given in Table 6.40. Now suppose that we wish to determine the range of c_1, the objective coefficient of x_1, over which the current basis remains optimal. This can be done by doing a simple parametric analysis of the cost vector.

We only want to consider changes in the objective coefficient of x_1 (StripEasy), so we can proceed with the analysis by perturbing the $\mathbf{c}$ vector as follows:

$$\mathbf{c} + t\mathbf{c}' = (0.5 \quad 0.3 \quad 0 \quad 0 \quad 0 \quad 0 \quad 0 \quad 0) + t(1 \quad 0 \quad 0 \quad 0 \quad 0 \quad 0 \quad 0 \quad 0)$$

We are only changing the $\mathbf{c}$ vector, so only dual feasibility (primal optimality) will be affected. Thus, the current basis will remain optimal provided that

$$(z_j - c_j) + t(z_j' - c_j') = (\mathbf{c}_B\mathbf{B}^{-1}\mathbf{a}_j - c_j) + t(\mathbf{c}_B'\mathbf{B}^{-1}\mathbf{a}_j - c_j') \geq 0, \qquad \text{for all } j$$

Note that $\mathbf{c}_B' = (1 \quad 0 \quad 0 \quad 0 \quad 0)$. Of course, $(z_j - c_j) + t(z_j' - c_j') = 0$ for all basic variables, therefore we only need to examine the optimality conditions for the nonbasic variables x_3 and x_7. Utilizing the information in Table 6.40, we readily compute the optimality conditions

$$(z_3 - c_3) + t(z_3' - c_3') = 0.25 + t\left[(1 \quad 0 \quad 0 \quad 0 \quad 0)\begin{pmatrix}\frac{1}{2}\\-\frac{3}{2}\\-\frac{1}{2}\\0\\0\end{pmatrix} - 0\right] = 0.25 - (\tfrac{1}{2})t \geq 0$$

$$(z_7 - c_7) + t(z_7' - c_7') = 0.2 + t\left[(1 \quad 0 \quad 0 \quad 0 \quad 0)\begin{pmatrix}1\\-1\\-1\\1\\-1\end{pmatrix} - 0\right] = 0.2 + t \geq 0$$

which simplify to $t \geq -0.2$. Observing that the current value of c_1 is 0.50, we see that as long as $c_1 \geq (0.50 - 0.20) = 0.30$ and the remaining problem data do not change, the current basis remains optimal. Thus, the unit profit of x_1 would need to drop below \$0.30 before we would stop producing StripEasy.

In a similar manner, it can be determined that the current basis will remain optimal as long as $c_2 \leq 0.5$. These results are summarized in Table 6.42.

TABLE 6.42 RANGES FOR THE c_j OF EXAMPLE 6.22

Cost coefficient	Current value	Allowable decrease before basis change	Allowable increase before basis change
c_1	0.5	0.2	∞
c_2	0.3	∞	0.2

SUMMARY

In this chapter, we have attempted to give an essentially complete picture of duality. We have seen that each linear programming problem, regardless of form, has a companion problem, called the dual, which provides important theoretical information as well as essential economic interpretations. The fundamental concepts of duality were used to develop the dual simplex algorithm, which proved to be an essential tool in sensitivity analysis and parametric programming. One of the lessons to be learned is that the "solution" of a linear programming problem by the simplex algorithm is only one phase in problem analysis. The successful analyst will always consider the sensitivity of the results to possible changes in the model and the potential for extra profit (or reduced cost) that may exist via constraint relaxation. This last analysis is often the most important of all to both the company and the analyst. It shows what can be done to improve the company's position, and it highlights the potential role of the analyst in the actual determination of high-level company policies and plans.

EXERCISES

6.1. Consider the following problem.

$$\text{maximize } z = 3x_1 + 2x_2$$

subject to

$$x_1 + 2x_2 \leq 11$$

$$x_1 - 3x_2 \leq 1$$

$$x_1, x_2 \geq 0$$

  (a) Solve this problem graphically.
  (b) Write the canonical dual and solve the dual graphically. Compare the optimal objective values of the two problems.

6.2. Write the canonical dual of the following linear programming problem.

$$\text{maximize } z = 4x_1 - 3x_2 + 5x_3$$

subject to

$$-x_1 + x_2 \leq 8$$
$$x_1 + 2x_2 + x_3 \leq 30$$
$$2x_1 - x_2 - 2x_3 \leq -6$$
$$x_1 + x_2 + 2x_3 \leq 20$$
$$x_1, x_2, x_3 \geq 0$$

6.3. Consider the following linear programming problem.

$$\text{maximize } z = x_1 + 2x_2 - 3x_3$$

subject to

$$-3x_1 + x_2 + 2x_3 = 16$$
$$2x_1 + 4x_2 + 3x_3 \geq 20$$
$$x_1 \geq 0$$
$$x_2 \leq 0$$

$$x_3 \text{ unrestricted}$$

(a) Without transforming the given problem, write the general dual.
(b) Transform the given problem into the canonical form given in (6.1). Write the canonical dual of this transformed problem and verify that it is equivalent to the general dual from part (a).

6.4. Write the general dual of the following problem.

$$\text{minimize } z = 3x_1 + 2x_2 - 4x_3$$

subject to

$$5x_1 - 7x_2 + x_3 \geq 12$$
$$x_1 - x_2 + 2x_3 = 18$$
$$2x_1 - x_3 \leq 6$$
$$x_1, x_2, x_3 \geq 0$$

6.5. Write the dual of the following problems.

(a) $$\text{minimize } z = \mathbf{cx}$$

subject to

$$\mathbf{Ax} = \mathbf{b}$$
$$\mathbf{Dx} \geq \mathbf{d}$$
$$\mathbf{Ex} \leq \mathbf{g}$$
$$\mathbf{x} \geq \mathbf{0}$$

(b)   maximize $z = \mathbf{cx} + \mathbf{dy}$

subject to

$$\mathbf{Ax} + \mathbf{Dy} = \mathbf{b}$$

$$\mathbf{Ex} + \mathbf{Fy} \le \mathbf{g}$$

$$\mathbf{x} \ge \mathbf{0}$$

$\mathbf{y}$ unrestricted

6.6. Consider the following linear programming problem.

$$\text{maximize } z = 3x_1 + 10x_2 + 5x_3 + 11x_4 + 6x_5 + 14x_6$$

subject to

$$x_1 + 7x_2 + 3x_3 + 4x_4 + 2x_5 + 5x_6 = 42$$

$$x_j \ge 0, \qquad \text{for all } j$$

(a) Write the dual problem.
(b) Solve the dual problem by inspection.

6.7. Consider the following problem.

$$\text{maximize } z = x_1 + 2x_2 - 9x_3 + 8x_4 - 36x_5$$

subject to

$$2x_2 - x_3 + x_4 - 3x_5 \le 40$$

$$x_1 - x_2 + 2x_4 - 2x_5 \le 10$$

$$x_j \ge 0, \qquad \text{for all } j$$

(a) Write the dual problem and solve it graphically.
(b) Using complementary slackness and the optimal dual solution found in part (a), find an optimal solution to the primal problem.

6.8. Use the concepts of duality and complementary slackness to show that $(x_1, x_2, x_3, x_4) = (10, 0, 16, 6)$ is an optimal solution to the following problem.

$$\text{maximize } z = x_1 + 2x_2 + 5x_3 + x_4$$

subject to

$$x_1 + 2x_2 + x_3 - x_4 \le 20$$

$$-x_1 + x_2 + x_3 + x_4 \le 12$$

$$2x_1 + x_2 + x_3 - x_4 \le 30$$

$$x_1, x_2, x_3, x_4 \ge 0$$

6.9. Consider the following resource allocation problem and the accompanying optimal tableau, where x_3, x_4, x_5, and x_6 are the slack variables for constraints 1 through 4, respectively.

$$\text{maximize } z = 2x_1 + 3x_2$$

subject to

$$x_1 + 2x_2 \leq 16 \qquad \text{(Resource 1)}$$
$$x_1 \leq 10 \qquad \text{(Resource 2)}$$
$$x_2 \leq 6 \qquad \text{(Resource 3)}$$
$$5x_1 + 6x_2 \leq 60 \qquad \text{(Resource 4)}$$
$$x_1, x_2 \geq 0$$

	z	x_1	x_2	x_3	x_4	x_5	x_6	RHS
z	1	0	0	$\frac{3}{4}$	0	0	$\frac{1}{4}$	27
x_4	0	0	0	$\frac{3}{2}$	1	0	$-\frac{1}{2}$	4
x_2	0	0	1	$\frac{5}{4}$	0	0	$-\frac{1}{4}$	5
x_5	0	0	0	$-\frac{5}{4}$	0	1	$\frac{1}{4}$	1
x_1	0	1	0	$-\frac{3}{2}$	0	0	$\frac{1}{2}$	6

(a) Write the dual problem in standard equality form.
(b) From the foregoing tableau, specify the optimal primal and optimal dual solutions.
(c) What is the shadow price of each resource? If you could acquire an additional unit of one of the resources, which resource would you choose? Why?

6.10. Consider the following problem.

$$\text{maximize } z = x_1 + 4x_2$$

subject to

$$2x_1 + x_2 \leq 24$$
$$-x_1 + x_2 \leq 4$$
$$3x_1 + 5x_2 \leq 60$$
$$x_1, x_2 \geq 0$$

(a) Solve this problem graphically.
(b) Write the dual problem in standard equality form. Using complementary slackness and the optimal primal solution found in part (a), find an optimal solution to the dual problem.

6.11. Consider the following problem.

$$\text{maximize } z = 3x_1 - x_2 + 6x_3$$

subject to

$$5x_1 + x_2 + 4x_3 \leq 42$$
$$2x_1 - x_2 + 2x_3 \leq 18$$
$$x_1, x_2, x_3 \geq 0$$

(a) Write the dual problem.

(b) Solve the primal problem by the primal simplex algorithm. Identify both the optimal primal and optimal dual solutions from the final tableau.

(c) At each iteration in part (b), identify the dual solution and indicate which dual constraints are violated. Also, at each iteration, identify the 2×2 primal basis matrix and the 3×3 dual basis matrix.

(d) Write the complementary slackness conditions and verify that these conditions are satisfied by the optimal solutions found in part (b).

6.12. Consider the following problem.

$$\text{minimize } z = 6x_1 + 24x_2 + 16x_3$$

subject to

$$x_1 + x_2 + 2x_3 \geq 2$$
$$-x_1 + 3x_2 + x_3 \geq 9$$
$$x_1, x_2 \geq 0$$

(a) Solve this problem by the dual simplex method.

(b) Solve this problem by the two-phase method and the primal simplex method.

6.13. Consider the following problem.

$$\text{maximize } z = -8x_1 - x_2 - x_3$$

subject to

$$2x_1 + x_2 \geq 20$$
$$-2x_1 + x_2 - x_3 \geq 12$$
$$x_1 + x_3 \leq 7$$
$$x_1 - x_2 + x_3 \geq 5$$
$$x_1, x_2, x_3 \geq 0$$

(a) Use the dual simplex method to show that this problem is infeasible.

(b) What does this indicate about the dual problem?

6.14. Suppose that the problem to maximize $\mathbf{cx}$ subject to $\mathbf{Ax} = \mathbf{b}$, $\mathbf{x} \geq \mathbf{0}$, has a finite optimal solution. Let $\mathbf{d}$ be an arbitrary vector in E^m. Show that if the problem to maximize $\mathbf{cx}$ subject to $\mathbf{Ax} = \mathbf{d}$, $\mathbf{x} \geq \mathbf{0}$, has a feasible solution, then it has a finite optimal solution.

6.15. Solve the following problem by the dual simplex method.

$$\text{maximize } z = -8x_1 - 2x_2 - 12x_3$$

subject to

$$2x_1 + x_2 + 4x_3 \geq -1$$
$$x_1 - 2x_2 + 2x_3 \geq 6$$
$$-x_2 + x_3 \leq 2$$
$$x_1, x_2, x_3 \geq 0$$

6.16. Consider the following linear programming problem.

$$\text{maximize } z = 2x_1 + 5x_2$$

subject to

$$x_1 - x_2 \le 4$$
$$x_1 + x_2 \le 14$$
$$-x_1 + 3x_2 \le 24$$
$$x_1, x_2 \ge 0$$

(a) Illustrate graphically that at the optimal extreme point, the gradient of the objective function lies in the cone generated by the gradients of the binding constraints.

(b) Write the objective gradient as a nonnegative linear combination of the gradients of the binding constraints. Interpret the coefficients in this nonnegative linear combination.

6.17. Solve the following problem by the dual simplex method.

$$\text{minimize } z = 3x_1 + x_2 + 2x_3$$

subject to

$$3x_1 + 2x_2 + x_3 \ge 8$$
$$x_1 - 4x_2 + 2x_3 \ge 20$$
$$4x_1 + 6x_2 + 2x_3 \le 30$$
$$x_1, x_2, x_3 \ge 0$$

6.18. Consider the following problem.

$$\text{maximize } z = 2x_1 + 3x_2$$

subject to

$$x_1 + x_2 \ge 2$$
$$x_1 + 2x_2 \le 10$$
$$x_1, x_2 \ge 0$$

(a) Solve this problem by the artificial constraint method.
(b) Plot the results graphically.

6.19. Consider the following problem.

$$\text{maximize } z = 2x_1 + 6x_2$$

subject to

$$2x_1 + x_2 \ge 6$$
$$-x_1 + 2x_2 \ge 2$$
$$x_1, x_2 \ge 0$$

(a) Solve this problem by the artificial constraint method.

(b) Plot the results graphically.

6.20. Consider the following linear programming problem where $\mathbf{A}$ is $m \times n$, $\mathbf{b}$ is $m \times 1$, and $\mathbf{c}$ is $n \times 1$.

$$\text{minimize } z = \mathbf{b'w} - \mathbf{cx}$$

subject to

$$\mathbf{Ax} \leq \mathbf{b}$$

$$\mathbf{A'w} \geq \mathbf{c'}$$

$$\mathbf{x} \geq \mathbf{0}$$

$$\mathbf{w} \geq \mathbf{0}$$

Show that this problem has an optimal objective value of zero or is infeasible.

6.21. Consider the following optimal tableau for a maximization problem with ($\leq$) constraints. Let x_4 and x_5 be the slack variables in the first and second constraints, respectively.

	z	x_1	x_2	x_3	x_4	x_5	RHS
z	1	0	$\frac{5}{3}$	0	$\frac{1}{3}$	$\frac{7}{3}$	d
x_3	0	0	$\frac{2}{3}$	1	$\frac{1}{3}$	$\frac{1}{3}$	14
x_1	0	1	$\frac{1}{3}$	0	$-\frac{1}{3}$	$\frac{2}{3}$	16

(a) Specify $\mathbf{B}^{-1}$ corresponding to this tableau.

(b) What is the rate of change of the objective with respect to the x_5 (i.e., $\partial z / \partial x_5$)?

(c) What is the rate of change of x_3 with respect to the x_2 (i.e., $\partial x_3 / \partial x_2$)?

(d) What is the rate of change of the objective with respect to the right-hand side of the first constraint (i.e., $\partial z / \partial b_1$)?

(e) Find the optimal objective value d.

6.22. Consider the following resource-allocation problem and the accompanying optimal tableau (x_5, x_6, and x_7 are the respective slack variables).

$$\text{maximize } z = 15x_1 + 8x_2 + 10x_3 + 12x_4 \qquad \text{(Profit \$)}$$

subject to

$$x_1 + 2x_2 + x_4 \leq 20 \qquad \text{(Resource 1)}$$

$$x_1 + x_2 + x_3 + x_4 \leq 54 \qquad \text{(Resource 2)}$$

$$2x_1 + x_3 + x_4 \leq 36 \qquad \text{(Resource 3)}$$

$$x_1, x_2, x_3, x_4 \geq 0$$

	z	x_1	x_2	x_3	x_4	x_5	x_6	x_7	RHS
z	1	9	0	0	2	4	0	10	440
x_2	0	$\frac{1}{2}$	1	0	$\frac{1}{2}$	$\frac{1}{2}$	0	0	10
x_6	0	$-\frac{3}{2}$	0	0	$-\frac{1}{2}$	$-\frac{1}{2}$	1	-1	8
x_3	0	2	0	1	1	0	0	1	36

(a) Write the dual problem and specify the optimal dual solution from the foregoing tableau.

(b) What are the shadow prices of the resources? If you were to choose between increasing the amount of resource 1, 2, or 3, which would you choose to increase and why?

(c) Suppose that the coefficient of x_4 in the objective function changes from 12 to 16. Use sensitivity analysis to find the new optimal solution.

(d) Suppose that the available amount of resource 1 changes from 20 to 40. Using sensitivity analysis, find the new optimal solution.

(e) If the constraint $x_1 \geq 10$ is added to the problem, use sensitivity analysis to find the new optimal solution.

(f) Suppose that the constraint $3x_1 + 2x_2 + 2x_3 + x_4 \leq 80$ is added to the problem. Using sensitivity analysis, find the new optimal solution.

(g) Suppose that a new product is proposed with objective coefficient 16 and consumption vector $(1 \quad 2 \quad 1)^t$. Using sensitivity analysis, find the new optimal solution.

6.23. Three products, A, B, and C, are made using two manufacturing processes. The unit production times in hours are given in the accompanying table.

Product	Unit production times (hours)	
	Process 1	Process 2
1	2	2
2	2	1
3	1	3

The time available for Process 1 is 36 hours, and for Process 2 is 48 hours. Products A, B, C sell for $9, $8, and $6, respectively. Let x_1, x_2, and x_3 represent the amounts of Products 1, 2, 3, respectively, and let x_4 and x_5 be the slack variables for the respective process constraints. Then the following tableau gives the optimal product mix.

	z	x_1	x_2	x_3	x_4	x_5	RHS
z	1	0	$\frac{1}{4}$	0	$\frac{15}{4}$	$\frac{3}{4}$	171
x_1	0	1	$\frac{5}{4}$	0	$\frac{3}{4}$	$-\frac{1}{4}$	15
x_3	0	0	$-\frac{1}{2}$	1	$-\frac{1}{2}$	$\frac{1}{2}$	6

(a) Suppose the unit profit of Product 1 changes from $9 to $8. Use sensitivity analysis to find the new optimal solution.

(b) Suppose there is the additional requirement that at least 10 units of Product 3 must be produced. Using sensitivity analysis, find the new optimal solution.

(c) For what range of values on the total time available for Process 1 will the current basis remain optimal?

(d) If an additional hour of Process 1 could be purchased for $3, would it be profitable to do so? Explain.

(e) Find the range of values on the unit profit of Product 2 for which the current basis remains optimal.

(f) If the time available for Process 2 changes from 48 to 32 hours, use sensitivity analysis to find the new optimal solution.

6.24. Consider the following problem.

$$\text{maximize } z = 20x_1 + 12x_2$$

subject to

$$2x_1 + x_2 \leq 24$$

$$x_1 + x_2 \leq 15$$

$$x_1, x_2 \geq 0$$

(a) Solve this problem by the simplex method.

(b) Consider the parameterized objective to maximize $z = (20x_1 + 12x_2) + t(2x_1 + 3x_2)$. Find optimal solutions for all values of t.

6.25. Consider the problem of Exercise 6.24. Suppose that the vector

$$\mathbf{b} = \begin{pmatrix} 24 \\ 15 \end{pmatrix}$$

is replaced by

$$\begin{pmatrix} 24 \\ 15 \end{pmatrix} + t \begin{pmatrix} 1 \\ -1 \end{pmatrix}$$

Find optimal solutions for all values of t.

6.26. Consider the following resource-allocation problem and the accompanying optimal tableau.

$$\text{maximize } z = 4x_1 + 6x_2 + 5x_3 \qquad \text{(Profit \$)}$$

subject to

$$3x_1 + x_2 + 2x_3 \leq 64 \qquad \text{(Resource 1)}$$

$$x_1 + x_2 + x_3 \leq 20 \qquad \text{(Resource 2)}$$

$$x_1 + 2x_2 + 3x_3 \leq 30 \qquad \text{(Resource 3)}$$

$$x_1, x_2, x_3 \geq 0$$

	z	x_1	x_2	x_3	x_4	x_5	x_6	RHS
z	1	0	0	3	0	2	2	100
x_4	0	0	0	3	1	−5	2	24
x_2	0	0	1	2	0	−1	1	10
x_1	0	1	0	−1	0	2	−1	10

(a) Determine the range on the unit profit of Product 1 (c_1) for which the current basis remains optimal.

(b) A new product is proposed. It is estimated that each unit of this new product consumes 4 units of Resource 1, 2 units of Resource 2, and 1 unit of Resource 3. What should be the unit profit on this new product for it to be profitable to manufacture?

(c) What is the range on the available units of Resource 3 (b_3) for which the current basis remains optimal?

(d) Suppose that an additional 5 units of Resource 3 could be acquired for a total cost of $6.00. Would it be profitable to do so?

(e) Suppose that the production requirements for Product 3 are modified so that each unit of Product 3 now requires 2 units of Resource 1, 1 unit of Resource 2, and 1 unit of Resource 3. Using sensitivity analysis, find the new optimal solution.

6.27. Consider Exercise 6.26. Find optimal solutions for all values of t based on the objective function to maximize $z = (4x_1 + 6x_2 + 5x_3) + t(x_1 - x_3)$.

6.28. Consider Exercise 6.26. Find optimal solutions for all values of t based on the parameterized resource vector

$$\begin{pmatrix} 64 \\ 20 \\ 30 \end{pmatrix} + t \begin{pmatrix} 1 \\ 0 \\ -1 \end{pmatrix}$$

6.29. Consider the following result, which is known as Farkas' Lemma. It is one of a class of several theorems that are referred to as theorems of the alternative.

Exactly one of the following two systems has a solution:

System 1: $\mathbf{Ax} = \mathbf{b}, \mathbf{x} \geq \mathbf{0}$, for some $\mathbf{x} \in E^n$

System 2: $\mathbf{yA} \leq \mathbf{0}, \mathbf{yb} > \mathbf{0}$, for some $\mathbf{y} \in E^m$

(a) Suppose that $\mathbf{A} = (\mathbf{a}_1, \mathbf{a}_2, \ldots, \mathbf{a}_n)$ and each $\mathbf{a}_j \in E^2$. Give a geometric interpretation of Farkas' Lemma.

(b) Prove Farkas' Lemma using the concepts of duality. (Hint: Consider the problem to minimize $\mathbf{0x}$ subject to $\mathbf{Ax} = \mathbf{b}, \mathbf{x} \geq \mathbf{0}$.)

6.30. Consider the canonical primal-dual pair given in (6.1) and (6.2). After adding a slack vector $\mathbf{s}$ to the primal and a surplus vector $\boldsymbol{\lambda}$ to the dual, the problems can be written as follows.

(P)	maximize $\mathbf{cx}$	(D)	minimize $\boldsymbol{\pi}\mathbf{b}$
	subject to		subject to
	$\mathbf{Ax} + \mathbf{s} = \mathbf{b}$		$\boldsymbol{\pi}\mathbf{A} - \boldsymbol{\lambda} = \mathbf{c}$
	$\mathbf{x} \geq \mathbf{0}$		$\boldsymbol{\pi} \geq \mathbf{0}$
	$\mathbf{s} \geq \mathbf{0}$		$\boldsymbol{\lambda} \geq \mathbf{0}$

Prove the following:

(a) x_j is unbounded in the primal feasible region if and only if its complementary dual variable λ_j is bounded above in the dual feasible region.

(b) x_j is zero in all optimal feasible solutions to the primal if and only if its complementary dual variable λ_j is strictly positive in some optimal feasible solution to the dual.

(c) x_j is strictly positive for some primal feasible solution if and only if its complementary dual variable λ_j is bounded above in the set of all optimal dual feasible solutions.

(d) x_j is unbounded on the set of all optimal primal feasible solutions if and only if its complementary dual variable λ_j is zero for all dual feasible solutions.

(e) Replace x_j by s_i, replace λ_j by π_i, and repeat parts (a) through (d).

*ALTERNATIVES TO THE SIMPLEX ALGORITHM

CHAPTER OVERVIEW

This chapter discusses alternative algorithms for solving linear programming problems and introduces the basic notion of computational complexity. Although the simplex algorithm works well in practice, the algorithm requires the enumeration of a potentially very large subset of extreme points. In fact, it has been shown that in the worst case, the computational effort required by the simplex algorithm increases exponentially with problem size. On problems in which this exponential growth occurs, the simplex method can fail because computational time increases so rapidly as problem size increases. The alternative algorithms, most notably *Khachian's ellipsoid algorithm* and *Karmarkar's projective algorithm*, use techniques that are radically different from the simplex algorithm, and have been proven to exhibit polynomial worst-case running time. Thus, theoretically, both algorithms are superior to the simplex method. In reality, however, Khachian's algorithm has failed to be of significant computational value. On the other hand, Karmarkar's algorithm and its variants have shown much more promise and, in fact, there are now commercial codes available that utilize these algorithmic techniques.

* Starred sections may be omitted in an introductory course with no appreciable loss of continuity.

COMPUTATIONAL COMPLEXITY

The theory of computational complexity was developed to classify algorithms in terms of their computational efficiency. Algorithms are generally classified as polynomial algorithms or exponential algorithms. An algorithm is classified as a polynomial time algorithm if the number of computational steps required by the algorithm to solve a problem is bounded above by a polynomial in terms of the size of the problem. In this context, the problem size generally refers to the number of bits, L, required to represent the problem in the computer. Thus, a polynomial time algorithm's running time is never greater than some fixed power of L, regardless of the problem that is solved. For example, a polynomial algorithm may be one whose worst-case performance is proportional to L^3. In contrast, the worst-case performance of an exponential algorithm grows at least as fast as an exponential function of the problem size. An example of an exponential function would be 2^L. Polynomial algorithms are generally considered to be "good" algorithms, whereas algorithms that experience exponential growth are not.

Given a positive scalar quantity α, the number of binary bits required to store this value is given by $\lceil \log(1 + \alpha) \rceil$, where the log is base 2, and the notation $\lceil \beta \rceil$ represents the rounded up integer value of β. For example, if $\alpha = 25$, then $\lceil \log(1 + 25) \rceil = 5$, and the number 25 can be represented by 5 binary bits. In fact, the binary representation of 25 is 11001. In general, we would need an additional bit to record the sign of an arbitrary real number. Thus, the number of bits required to store the linear programming problem

$$\text{maximize } \mathbf{cx} \tag{7.1}$$

subject to

$$\mathbf{Ax} \le \mathbf{b}$$

$$\mathbf{x} \ge \mathbf{0}$$

is given by

$$L = \{1 + \lceil \log(1 + m) \rceil\} + \{1 + \lceil \log(1 + n) \rceil\}$$
$$+ \sum_{i=1}^{m} \sum_{j=1}^{n} \{1 + \lceil \log(1 + |a_{i,j}|) \rceil\} + \sum_{i=1}^{m} \{1 + \lceil \log(1 + |b_i|) \rceil\} \tag{7.2}$$
$$+ \sum_{j=1}^{n} \{1 + \lceil \log(1 + |c_j|) \rceil\}$$

Klee and Minty (1972) have constructed pathological examples that clearly demonstrate that the simplex algorithm is not a polynomial time algorithm. That is, in the worst case, the computational time required can grow exponentially with the problem size L. To see this more clearly, consider a polynomial algorithm whose computational performance is proportional to L^3 and an exponential algorithm whose worst-case performance is 2^L. A comparison of the growth of these two functions is given in Table 7.1. Note that as the problem size increases, the difference in the worst-case performance of the two algorithms widens dramatically.

TABLE 7.1 COMPARISON OF POLYNOMIAL AND EXPONENTIAL FUNCTIONS

L	L^3	2^L
10	1.00×10^3	1.02×10^3
20	8.00×10^3	1.05×10^6
50	1.25×10^5	1.13×10^{15}
100	1.00×10^6	1.27×10^{30}
200	1.25×10^8	1.61×10^{60}

However, despite this worst-case performance of the simplex algorithm, it performs quite well in practice, and the number of pivots required to solve real problems is generally a linear function of the number of constraints. Consequently, the lack of a polynomial bound on the simplex method is more of theoretical interest than of practical importance.

In the following section, we discuss the polynomial time algorithm of Khachian (1979). Khachian's algorithm is generally referred to as the ellipsoid algorithm, and before its development, it was unclear whether linear programming problems were in the class of "easy" problems for which a polynomial time algorithm could be constructed. Therefore, it was of significant theoretical importance in that it showed that linear programs do indeed fall in the class of easy problems. However, in practice, its computational performance proved to be much worse than that of the simplex method.

In subsequent sections, we discuss affine scaling algorithms and Karmarkar's projective algorithm. Karmarkar's algorithm (1984) was another significant theoretical breakthrough. Not only was it a polynomial time algorithm with a substantially better worst-case performance than Khachian's algorithm, but it also showed much more computational promise. Whereas the simplex algorithm moves from extreme point to adjacent extreme point in search of an optimal solution, Karmarkar's algorithm and its variants, affine scaling algorithms, are "interior-point" algorithms. That is, they move through the interior of the feasible region until an optimal point on the boundary is reached.

KHACHIAN'S ELLIPSOID ALGORITHM

In 1979, the Russian mathematician, L. G. Khachian published a paper proving that linear programming problems could be solved in polynomial time. The new algorithm proposed by Khachian, called the *ellipsoid algorithm,* actually solved the problem of finding a feasible solution to a system of linear inequalities of the form $\mathbf{Gy} \leq \mathbf{h}$, where $\mathbf{G}$ is a $p \times q$ matrix and $\mathbf{y}$ and $\mathbf{h}$ are column vectors. Note that there are no nonnegativity restrictions on $\mathbf{y}$ and there is no explicit objective function to be minimized or maximized. Thus, it is first necessary to convert a linear programming problem into an equivalent system of inequalities having the same solution set as the linear program. To illustrate this process, consider the following linear programming problem (P) and its dual problem (D).

$$\text{(P)} \quad \text{maximize } \mathbf{cx} \tag{7.3}$$

$$\text{subject to}$$

$$\mathbf{Ax} \leq \mathbf{b}$$

$$\mathbf{x} \geq \mathbf{0}$$

$$\text{(D)} \quad \text{minimize } \boldsymbol{\pi}\mathbf{b} \tag{7.4}$$

$$\text{subject to}$$

$$\boldsymbol{\pi}\mathbf{A} \geq \mathbf{c}$$

$$\boldsymbol{\pi} \geq \mathbf{0}$$

Now, consider the system of inequalities:

$$\mathbf{Ax} \leq \mathbf{b} \tag{7.5}$$

$$-\mathbf{x} \leq \mathbf{0} \tag{7.6}$$

$$-\mathbf{A}^t\boldsymbol{\pi}^t \leq -\mathbf{c}^t \tag{7.7}$$

$$-\boldsymbol{\pi}^t \leq \mathbf{0} \tag{7.8}$$

$$-\mathbf{cx} + \mathbf{b}^t\boldsymbol{\pi}^t \leq 0 \tag{7.9}$$

First, observe that the inequalities (7.5–7.9) can be placed in the form $\mathbf{Gy} \leq \mathbf{h}$ by letting

$$\mathbf{y} = \begin{pmatrix} \mathbf{x} \\ \boldsymbol{\pi}^t \end{pmatrix} \tag{7.10}$$

$$\mathbf{G} = \begin{pmatrix} \mathbf{A} & \mathbf{0} \\ -\mathbf{I} & \mathbf{0} \\ \mathbf{0} & -\mathbf{A}^t \\ \mathbf{0} & -\mathbf{I} \\ -\mathbf{c} & \mathbf{b}^t \end{pmatrix} \tag{7.11}$$

$$\mathbf{h} = \begin{pmatrix} \mathbf{b} \\ \mathbf{0} \\ -\mathbf{c}^t \\ \mathbf{0} \\ \mathbf{0} \end{pmatrix} \tag{7.12}$$

Next, note that (7.5–7.6) are the primal feasibility conditions, (7.7–7.8) are the dual feasibility conditions, and (7.9) is equivalent to $\boldsymbol{\pi}\mathbf{b} \le \mathbf{cx}$. Thus, by the weak duality theorem (Theorem 6.2), any solution to the system (7.5–7.9) will yield optimal solutions to both (P) and (D). Observe that if $\mathbf{A}$ is $m \times n$, that is, Problem (P) has m constraints and n variables, then the system of inequalities (7.5–7.9) consists of $p = 2m + 2n + 1$ inequalities and $q = m + n$ variables. Finally, note that for linear programs having a unique solution, the feasible space for the corresponding inequality system given in (7.5–7.9) consists of a single point.

Let $S = \{\mathbf{y} : \mathbf{Gy} \le \mathbf{h}\}$. The ellipsoid algorithm starts with an ellipsoid (Euclidean ball) centered at the origin that is sufficiently large to contain the set S. If the center of the ellipsoid is in S (i.e., is feasible to $\mathbf{Gy} \le \mathbf{h}$), then the feasibility problem is solved. Otherwise, the center of the ellipsoid must violate at least one of the inequality constraints. Choose that constraint that is most violated. The chosen inequality constraint would act as a separating hyperplane that separates the center from the set S, as shown in Figure 7.1. Because S must lie in one halfspace of this separating hyperplane, another ellipsoid with smaller volume containing S can be constructed, as indicated in Figure 7.1. Repeated iterations of the algorithm yield smaller and smaller ellipsoids until the centers of the ellipsoids converge to the feasible region.

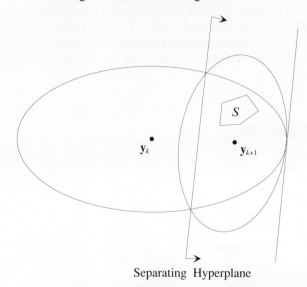

Separating Hyperplane

Figure 7.1 Khachian's ellipsoid algorithm.

Khachian proved that in the worst case, the number of iterations required is proportional to q^2L. In fact, Khachian showed that in at most $6(q + 1)^2L$ iterations, the algorithm will find a feasible solution. Thus, if a feasible solution is not found in $6(q + 1)^2L$ iterations, then we can conclude that set S is empty. Also, the number of elementary arithmetical operations required by each iteration of the algorithm is proportional to q^2. Therefore, the overall complexity of the algorithm is proportional to q^4L.

Although Khachian's algorithm was a significant theoretical breakthrough in complexity theory, it proved to be far less efficient in solving practical linear programming problems than the simplex method. Thus, Khachian's algorithm is an example of an algorithm that exhibits good theoretical worst-case performance but poor performance from an empirical viewpoint.

AFFINE SCALING ALGORITHMS

Affine scaling algorithms, variants of Karmarkar's algorithm, were proposed independently by Barnes (1986), Cavalier and Soyster (1985), and Vanderbei, Meketon, and Freedman (1986) following the development of Karmarkar's algorithm (1984). However, affine scaling algorithms actually find their origin in the earlier work of Dikin (1967, 1974). As with Karmarkar's original approach, affine scaling algorithms are interior-point algorithms. That is, one attempts to find successive improving directions while remaining in the strict interior of the feasible region. In Figure 7.2, this is contrasted with the simplex method, which moves along the boundary of the feasible region from one extreme point to an adjacent extreme point until an optimal extreme point is found.

The fundamental idea is to transform the solution space so that the current solution is centrally located in the transformed feasible region. In this way, it is possible to make a substantial move in an improving direction before encountering the boundary of the feasible region. To better understand this fundamental idea,

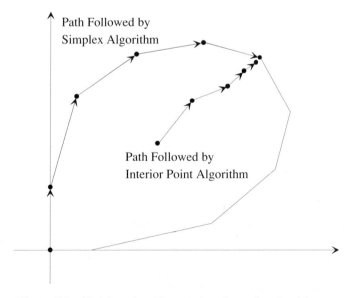

Path Followed by
Simplex Algorithm

Path Followed by
Interior Point Algorithm

Figure 7.2   Simplex algorithm vs. interior-point algorithm.

consider the feasible region depicted in Figure 7.3. Note that by moving in the improving direction **d** at $\bar{x}$, we can make a substantial improvement in the value of the objective. However, when moving in the same direction, starting at $\tilde{x}$, we encounter the boundary of the feasible region before much improvement in the objective is possible.

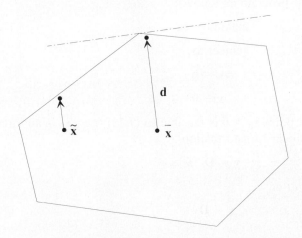

Figure 7.3 Differences in possible improvement.

Thus, the basic strategy is to transform the feasible region so that the current interior solution is near the center of the transformed region without altering the actual structure of the problem. By starting at this new point, a move is then made in the direction of steepest ascent, stopping just before encountering the boundary of the feasible region so that the point remains an interior point of the feasible region. The space is transformed again so that this new point is centrally located and the process is repeated. Affine scaling algorithms accomplish this with a simple scaling of the problem data, whereas Karmarkar's algorithm uses scaling followed by a projection.

Affine scaling algorithms have received a vast amount of attention in the literature and several affine variants have been derived as a result. These include primal affine scaling algorithms, dual affine scaling algorithms, and primal-dual path-following variants. In general, computer implementations of affine scaling variants have proven to be more fruitful than implementations of Karmarkar's original approach. And, in fact, there are now commercially available codes based on affine scaling algorithms. Most notably, AT&T has released a software package under the trademark KORBX that implements primal affine, dual affine, and primal-dual affine scaling variants of Karmarkar's algorithm.

Due to this more widespread usage and their slightly less complicated nature, we have chosen to discuss affine scaling algorithms first followed by a summary of Karmarkar's original approach.

Convergence of affine scaling algorithms has been investigated by several researchers, including Barnes (1986), Vanderbei et al. (1986), Kortanek and Shi

(1987), and Sherali (1987), and the algorithms have been shown to converge under various assumptions.

A Primal Affine Scaling Algorithm

Consider a standard linear program as follows:

$$\text{maximize } \mathbf{cx} \tag{7.13}$$

subject to

$$\mathbf{Ax} = \mathbf{b}$$

$$\mathbf{x} \geq \mathbf{0}$$

Suppose that some initial point $\mathbf{x}_0 > \mathbf{0}$ is feasible to (7.13) (i.e., $\mathbf{x}_0$ is an interior point), and consider the linear transformation

$$\mathbf{y} = \mathbf{D}^{-1}\mathbf{x} \tag{7.14}$$

and its inverse transformation

$$\mathbf{x} = \mathbf{Dy} \tag{7.15}$$

where

$$\mathbf{D} = \text{diag}(x_{0,1}, x_{0,2}, \ldots, x_{0,n}) = \begin{pmatrix} x_{0,1} & 0 & \cdots & 0 \\ 0 & x_{0,2} & \cdots & 0 \\ \vdots & \vdots & \ddots & \vdots \\ 0 & 0 & \cdots & x_{0,n} \end{pmatrix} \tag{7.16}$$

Note that $\mathbf{D}$ is a diagonal matrix with strictly positive diagonal elements. Then, by using (7.15), the linear programming problem (7.13) can be transformed into

$$\text{maximize } (\mathbf{cD})\mathbf{y} \tag{7.17}$$

subject to

$$(\mathbf{AD})\mathbf{y} = \mathbf{b}$$

$$\mathbf{y} \geq \mathbf{0}$$

Notice that the initial point $\mathbf{x}_0$ is transformed into

$$\mathbf{y}_0 = \mathbf{D}^{-1}\mathbf{x}_0 = \begin{pmatrix} 1/x_{0,1} & 0 & \cdots & 0 \\ 0 & 1/x_{0,2} & \cdots & 0 \\ \vdots & \vdots & \ddots & \vdots \\ 0 & 0 & \cdots & 1/x_{0,n} \end{pmatrix} \begin{pmatrix} x_{0,1} \\ x_{0,2} \\ \vdots \\ x_{0,n} \end{pmatrix} = \begin{pmatrix} 1 \\ 1 \\ \vdots \\ 1 \end{pmatrix} = \mathbf{1} \tag{7.18}$$

and $\mathbf{y}_0$ is a feasible solution to (7.17) because $\mathbf{x}_0$ is a feasible solution to (7.13). Observe also that $\mathbf{y}_0$ has the interesting property that it is a unit distance from all the hyperplanes defining the nonnegative orthant, that is, y_0 is centrally located in the feasible space. Thus, if **d** is an improving feasible direction at y_0, we can move a substantial distance in the direction **d** before reaching the boundary.

Let us now find an improving feasible direction **d** at y_0. That is, find a direction **d** such that $y_0 + \theta d$, $\theta \geq 0$, is feasible for sufficiently small values of θ, and the objective improves (increases) moving in direction **d**.

First, we will address the issue of feasibility. In order for $y_0 + \theta d$, $\theta \geq 0$, to be feasible, the constraint set associated with (7.17) must be satisfied; that is,

$$(\mathbf{AD})(y_0 + \theta d) = b \tag{7.19}$$

$$y_0 + \theta d \geq 0 \tag{7.20}$$

However, from (7.19), we see that

$$(\mathbf{AD})(y_0 + \theta d) = (\mathbf{AD})y_0 + \theta(\mathbf{AD})d = b + \theta(\mathbf{AD})d = b \tag{7.21}$$

and, thus, it must be true that

$$(\mathbf{AD})d = 0 \tag{7.22}$$

That is, **d** must lie in the null space of **AD**. Now consider (7.20) in component form:

$$y_{0,j} + \theta d_j \geq 0, \qquad \text{for all } j = 1, \ldots, n \tag{7.23}$$

Note that if $d_j \geq 0$, then $y_{0,j} + \theta d_j = 1 + \theta d_j > 0$, for all $\theta \geq 0$. However, if $d_j < 0$, then $y_{0,j} + \theta d_j \geq 0$ provided that $0 \leq \theta \leq -y_{0,j}/d_j = -1/d_j$. Thus, letting

$$\theta = \begin{cases} \infty, & \text{if } \mathbf{d} \geq \mathbf{0} \\ \text{minimum} \left\{ \dfrac{-1}{d_j} : d_j < 0 \right\}, & \text{otherwise} \end{cases} \tag{7.24}$$

the new vector $y_0 + \theta d$ would be nonnegative and, thus, feasible to (7.17) provided that $(\mathbf{AD})d = 0$. Furthermore, if $\alpha \in (0, 1)$, then $y_0 + \alpha\theta d$ would be strictly positive in addition to being feasible to (7.17).

For **d** to be an improving direction, we would need that

$$(\mathbf{cD})(y_0 + \theta d) > (\mathbf{cD})y_0 \tag{7.25}$$

But expanding the left side of (7.25) yields

$$(\mathbf{cD})y_0 + \theta(\mathbf{cD})d > (\mathbf{cD})y_0 \tag{7.26}$$

which implies that

$$(\mathbf{cD})d > 0 \tag{7.27}$$

That is, vectors **cD** and **d** form a strict acute angle. Thus, observe that we can find an improving feasible direction by simply projecting the vector **cD** onto the null space of the matrix $\mathbf{B} = \mathbf{AD}$. Let us denote this projection vector by c_p. Then, representing the null space of $\mathbf{B} = \mathbf{AD}$ by $S = \{d : \mathbf{B}d = 0\}$, we want to find that element of S that is a minimum distance from $\mathbf{cD}$, as illustrated graphically in Figure 7.4. This results in the following *quadratic* programming problem:

$$\text{minimum } \|\mathbf{cD} - \mathbf{d}^t\|^2 \tag{7.28}$$

subject to

$$\mathbf{Bd} = \mathbf{0}$$

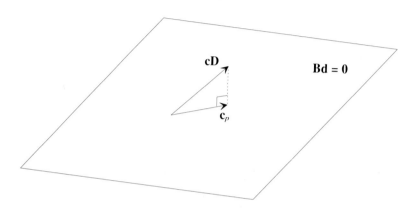

Figure 7.4   Finding the projection $\mathbf{c}_p$.

It can be shown that applying the Karush–Kuhn–Tucker optimality conditions to (7.28) results in the closed-form solution

$$\mathbf{c}_p = \mathbf{d}^* = [\mathbf{I} - \mathbf{B}^t(\mathbf{BB}^t)^{-1}\mathbf{B}]\mathbf{Dc}^t \tag{7.29}$$

provided that the $m \times m$ matrix $(\mathbf{BB}^t)$ is invertible. And, in fact, it is easily shown that $(\mathbf{BB}^t)$ is invertible if $\mathbf{B}$ has full row rank or, equivalently, if $\mathbf{A}$ has full row rank.

Note that the computation of $\mathbf{c}_p$ requires inverting an $m \times m$ matrix or, equivalently, finding the solution of a system of m equations in m unknowns. This is by far the most computationally intensive step of the algorithm, and much research has been done to develop efficient methods for computing this projection. These include using approximate projection methods and exploiting the symmetric structure of matrix $\mathbf{BB}^t$.

Let us now rewrite the expression for $\mathbf{c}_p$ in (7.29) so that we can compute $\mathbf{c}_p$ in a stepwise fashion by first finding the solution of a system of linear equations. This method is generally preferable to matrix inversion. Expanding the right side of (7.29) yields

$$\begin{aligned} \mathbf{c}_p &= \mathbf{Dc}^t - \mathbf{B}^t(\mathbf{BB}^t)^{-1}\mathbf{BDc}^t \\ &= \mathbf{DC}^t - \mathbf{B}^t\mathbf{w} \end{aligned} \tag{7.30}$$

where

$$\mathbf{w} = (\mathbf{BB}^t)^{-1}\mathbf{BDc}^t \tag{7.31}$$

Thus, vector $\mathbf{w}$ can be found by exploiting the special structure of matrix $(\mathbf{BB})^t$ and solving the linear system

$$(\mathbf{BB}^t)\mathbf{w} = \mathbf{BDc}^t$$

The new solution of the transformed problem (7.17) would then be given by

$$\mathbf{y}_1 = \mathbf{y}_0 + \alpha\theta\mathbf{c}_p = \mathbf{1} + \alpha\theta\mathbf{c}_p \tag{7.33}$$

and, consequently, the new solution of the original problem (7.13) is

$$\mathbf{x}_1 = \mathbf{D}\mathbf{y}_1 = \mathbf{D}(\mathbf{y}_0 + \alpha\theta\mathbf{c}_p) = \mathbf{D}\mathbf{y}_0 + \alpha\theta\mathbf{D}\mathbf{c}_p = \mathbf{x}_0 + \alpha\theta\mathbf{D}\mathbf{c}_p \tag{7.34}$$

Comparing the objective values of the iterations, notice that

$$\mathbf{c}\mathbf{x}_1 = \mathbf{c}(\mathbf{x}_0 + \alpha\theta\mathbf{D}\mathbf{c}_p) = \mathbf{c}\mathbf{x}_0 + \alpha\theta(\mathbf{c}\mathbf{D})\mathbf{c}_p \tag{7.35}$$

and, thus,

$$\mathbf{c}\mathbf{x}_1 > \mathbf{c}\mathbf{x}_0 \tag{7.36}$$

because $\alpha > 0$, $\theta > 0$, and $(\mathbf{c}\mathbf{D})\mathbf{c}_p > 0$. If $(\mathbf{c}\mathbf{D})\mathbf{c}_p = 0$, then $\mathbf{c}\mathbf{D}$ is orthogonal to the null space of $\mathbf{B} = \mathbf{AD}$ and the optimal solution would be at hand.

One can thus define an iterative process. That is, given a feasible solution $\mathbf{x}_k > \mathbf{0}$, a new problem

$$\text{maximize } (\mathbf{c}\mathbf{D})\mathbf{y} \tag{7.37}$$

$$\text{subject to}$$

$$(\mathbf{AD})\mathbf{y} = \mathbf{b}$$

$$\mathbf{y} \geq \mathbf{0}$$

if formulated, where $\mathbf{D} = \text{diag}(x_{k,1}, \ldots, x_{k,n})$, and $\mathbf{y}_k = \mathbf{D}^{-1}\mathbf{x}_k = \mathbf{1} = (1, 1, \ldots, 1)^t$ is a centrally located feasible solution. An improved solution for (7.17) is then

$$\mathbf{y}_{k+1} = \mathbf{y}_k + \alpha\theta\mathbf{c}_p = \mathbf{1} + \alpha\theta\mathbf{c}_p \tag{7.38}$$

and the corresponding solution of (7.13) is

$$\mathbf{x}_{k+1} = \mathbf{D}\mathbf{y}_{k+1} = \mathbf{x}_k + \alpha\theta\mathbf{D}\mathbf{c}_p \tag{7.39}$$

where $\mathbf{c}_p$ is the projection of $\mathbf{c}\mathbf{D}$ onto the null space of $\mathbf{AD}$, θ is as defined in (7.24), and $\alpha \in (0, 1)$, say, for example, $\alpha = 0.9$.

This is a very simple ascent procedure that is repeated until some suitable convergence criterion is satisfied. What distinguishes this procedure from other algorithms for linear programming is that it avoids the problem of what to do when one hits the boundaries of the nonnegative orthant. That is, by including the parameter $\alpha \in (0, 1)$, one never reaches the boundary and always remains interior to the feasible region.

Primal affine scaling algorithm (maximization problem)

STEP 1. Set the iteration counter $k = 0$ and choose an initial solution $x_k > 0$ such that $\mathbf{A}x_k = \mathbf{b}$. (A method for finding an initial solution will be discussed later.)

STEP 2. Define $\mathbf{D} = \mathrm{diag}(x_{k,1}, \ldots, x_{k,n})$.

STEP 3. Compute $\mathbf{BB}^t = (\mathbf{AD})(\mathbf{AD})^t = \mathbf{AD}^2\mathbf{A}^t$ and compute $\mathbf{BDc}^t = \mathbf{AD}^2\mathbf{c}^t$.

STEP 4. Solve the linear system $(\mathbf{BB}^t)\mathbf{w} = \mathbf{BDc}^t$ for $\mathbf{w}$. (Note that $\mathbf{w} = (\mathbf{BB}^t)^{-1}\mathbf{BDc}^t$.)

STEP 5. Compute $\mathbf{c}_p = [\mathbf{I} - \mathbf{B}^t(\mathbf{BB}^t)^{-1}\mathbf{B}]\mathbf{Dc}^t = \mathbf{Dc}^t - \mathbf{B}^t\mathbf{w}$.

STEP 6. Compute θ using (7.24). If $\theta = \infty$, then the objective is unbounded; stop. Otherwise, if $\theta < \infty$, continue with Step 7.

STEP 7. Compute $\mathbf{x}_{k+1} = \mathbf{x}_k + \alpha\theta\mathbf{Dc}_p$, where $\alpha \in (0, 1)$. (For example, $\alpha = 0.9$.)

STEP 8. Check for convergence. If satisfied, stop. Otherwise, replace k by $k + 1$ and go to Step 2.

There are several stopping criteria that may be used. For example, assuming that the objective value is nonzero, we may check to see if the relative change in the objective value is less than some prescribed tolerance, say, $\varepsilon = 10^{-4}$. That is, if $|(\mathbf{cx}_{k+1} - \mathbf{cx}_k)/\mathbf{cx}_k| < \varepsilon$, then we would stop with $\mathbf{x}^* = \mathbf{x}_{k+1}$. Another termination method relies on obtaining an approximation of the dual solution. Todd and Burrell (1986) showed that under suitable assumptions (namely, nondegeneracy), the vector $\mathbf{w}$ computed in Step 4 is an approximate solution to the dual problem, and, in fact, the sequence of $\mathbf{w}$'s generated by the algorithm converge to the optimal dual solution. For this reason, another stopping criterion is $|(\mathbf{w}^t\mathbf{b} - \mathbf{cx})/\mathbf{cx}| < \varepsilon$.

Let us now illustrate the algorithmic process by way of an example.

Example 7.1: Primal Affine Scaling Algorithm

$$\text{maximize } 4x_1 + 5x_2 \tag{7.40}$$

subject to

$$-x_1 + 2x_2 + x_3 = 8 \tag{7.41}$$

$$2x_1 + x_2 + x_4 = 14 \tag{7.42}$$

$$x_j \geq 0, \qquad \text{for all } j \tag{7.43}$$

Before determining a solution using the affine scaling algorithm, note that the optimal simplex tableau is as shown in Table 7.2. That is, the optimal primal solution is

$$\mathbf{z}^* = 46$$

$$\mathbf{x}^* = \begin{pmatrix} x_1^* \\ x_2^* \\ x_3^* \\ x_4^* \end{pmatrix} = \begin{pmatrix} 4 \\ 6 \\ 0 \\ 0 \end{pmatrix}$$

Also, from Table 7.2, the optimal values of the dual decision variables corresponding to constraints (7.41) and (7.42), respectively, are

$$\pi_1^* = \tfrac{6}{5} = 1.2$$

$$\pi_2^* = \tfrac{13}{5} = 2.6$$

TABLE 7.2

	z	x_1	x_2	x_3	x_4	RHS
z	1	0	0	$\frac{6}{5}$	$\frac{13}{5}$	46
x_2	0	0	1	$\frac{2}{5}$	$\frac{1}{5}$	6
x_1	0	1	0	$-\frac{1}{5}$	$\frac{2}{5}$	4

So that we may compare the trajectories generated by the simplex algorithm and the affine scaling algorithm, let us choose as our initial solution a point near the origin in x_1–x_2 space. That is, let

$$\mathbf{x}_0 = \begin{pmatrix} x_{0,1} \\ x_{0,2} \\ x_{0,3} \\ x_{0,4} \end{pmatrix} = \begin{pmatrix} 0.1 \\ 0.1 \\ 7.9 \\ 13.7 \end{pmatrix} > \mathbf{0}$$

Then

$$\mathbf{D} = \text{diag}(0.1, 0.1, 7.9, 13.7) = \begin{pmatrix} 0.1 & 0 & 0 & 0 \\ 0 & 0.1 & 0 & 0 \\ 0 & 0 & 7.9 & 0 \\ 0 & 0 & 0 & 13.7 \end{pmatrix}$$

and

$$\mathbf{B} = \mathbf{AD} = \begin{pmatrix} -1 & 2 & 1 & 0 \\ 2 & 1 & 0 & 1 \end{pmatrix} \begin{pmatrix} 0.1 & 0 & 0 & 0 \\ 0 & 0.1 & 0 & 0 \\ 0 & 0 & 7.9 & 0 \\ 0 & 0 & 0 & 13.7 \end{pmatrix} = \begin{pmatrix} -0.1 & 0.2 & 7.9 & 0 \\ 0.2 & 0.1 & 0 & 13.7 \end{pmatrix}$$

Now, summarizing the sequence of computations for the first iteration, we have

$$\mathbf{BB}^t = \begin{pmatrix} -0.1 & 0.2 & 7.9 & 0 \\ 0.2 & 0.1 & 0 & 13.7 \end{pmatrix} \begin{pmatrix} -0.1 & 0.2 \\ 0.2 & 0.1 \\ 7.9 & 0 \\ 0 & 13.7 \end{pmatrix} = \begin{pmatrix} 62.4600 & 0 \\ 0 & 187.7400 \end{pmatrix}$$

$$\mathbf{Dc}^t = \begin{pmatrix} 0.1 & 0 & 0 & 0 \\ 0 & 0.1 & 0 & 0 \\ 0 & 0 & 7.9 & 0 \\ 0 & 0 & 0 & 13.7 \end{pmatrix} \begin{pmatrix} 4 \\ 5 \\ 0 \\ 0 \end{pmatrix} = \begin{pmatrix} 0.4 \\ 0.5 \\ 0 \\ 0 \end{pmatrix}$$

$$\mathbf{BDc}^t = \begin{pmatrix} -0.1 & 0.2 & 7.9 & 0 \\ 0.2 & 0.1 & 0 & 13.7 \end{pmatrix} \begin{pmatrix} 0.4 \\ 0.5 \\ 0 \\ 0 \end{pmatrix} = \begin{pmatrix} 0.06 \\ 0.13 \end{pmatrix}$$

Now, solving the linear system $(\mathbf{BB}^t)\mathbf{w} = \mathbf{BDc}^t$ for $\mathbf{w}$, we get

$$\mathbf{w} = (\mathbf{BB}^t)^{-1}\mathbf{BDc}^t = \begin{pmatrix} 0.000961 \\ 0.000692 \end{pmatrix}$$

which yields the improving feasible direction

$$
\mathbf{c}_p = \mathbf{Dc}^t - \mathbf{B}^t\mathbf{w} = \begin{pmatrix} 0.4 \\ 0.5 \\ 0 \\ 0 \end{pmatrix} - \begin{pmatrix} -0.1 & 0.2 \\ 0.2 & 0.1 \\ 7.9 & 0 \\ 0 & 13.7 \end{pmatrix} \begin{pmatrix} 0.000961 \\ 0.000692 \end{pmatrix} = \begin{pmatrix} 0.399958 \\ 0.499739 \\ -0.007589 \\ -0.009487 \end{pmatrix}
$$

Computing θ from (7.24), we get

$$
\theta = \min\{1/0.007589,\ 1/0.009487\} = 105.412690
$$

Finally, using a step length parameter $\alpha = 0.9$, (7.39) yields

$$
\mathbf{x}_1 = \mathbf{x}_0 + \alpha\theta\mathbf{Dc}_p
$$

$$
= \begin{pmatrix} 0.1 \\ 0.1 \\ 7.9 \\ 13.7 \end{pmatrix} + (0.9)(105.412690) \begin{pmatrix} 0.1 & 0 & 0 & 0 \\ 0 & 0.1 & 0 & 0 \\ 0 & 0 & 7.9 & 0 \\ 0 & 0 & 0 & 13.7 \end{pmatrix} \begin{pmatrix} 0.399958 \\ 0.499739 \\ -0.007589 \\ -0.009487 \end{pmatrix}
$$

$$
= \begin{pmatrix} 3.894454 \\ 4.841091 \\ 2.212272 \\ 1.370000 \end{pmatrix}
$$

Redefining diagonal matrix $\mathbf{D}$ as

$$
\mathbf{D} = \mathrm{diag}(3.894454,\ 4.841091,\ 2.212272,\ 1.370000)
$$

the process is now repeated and results in

$$
\mathbf{x}_2 = \mathbf{x}_1 + \alpha\theta\mathbf{Dc}_p
$$

$$
= \begin{pmatrix} 3.894454 \\ 4.841091 \\ 2.212272 \\ 1.370000 \end{pmatrix}
$$

$$
+ (0.9)(0.286041) \begin{pmatrix} 3.894454 & 0 & 0 & 0 \\ 0 & 4.841091 & 0 & 0 \\ 0 & 0 & 2.212272 & 0 \\ 0 & 0 & 0 & 1.37 \end{pmatrix} \begin{pmatrix} 0.201536 \\ 0.665093 \\ -2.556050 \\ -3.495999 \end{pmatrix}
$$

$$
= \begin{pmatrix} 4.096510 \\ 5.669981 \\ 0.756548 \\ 0.137000 \end{pmatrix}
$$

The first seven iterations are summarized in Table 7.3. Note that at iteration 7, the solution has essentially converged to the optimal solution $\mathbf{x}^*$ given in Table 7.2. Also observe that the sequence of $\mathbf{w}$'s generated has converged to the optimal dual solution provided in Table 7.2.

TABLE 7.3 RESULTS OF EXAMPLE 7.1

k	$\mathbf{x}_k$	$\mathbf{cx}_k$	$\mathbf{w}$	$\|\mathbf{c}_p\|$
0	0.1 0.1 7.9 13.7	0.9	0.000961 0.000692	0.640197
1	3.894454 4.841091 2.212272 1.370000	39.783274	1.155396 2.551823	4.386155
2	4.096595 5.669981 0.756548 0.137000	44.735943	1.195076 2.601032	0.973525
3	3.979769 5.952057 0.075655 0.0884404	45.679363	1.200008 2.599791	0.247122
4	4.006216 5.978728 0.048579 0.008840	45.918504	1.199981 2.600006	0.062863
5	3.998689 5.996907 0.004876 0.005715	45.979290	1.200000 2.599999	0.015969
6	4.000401 5.998627 0.003148 0.000571	45.994737	1.200000 2.600000	0.004059
7	3.999915 5.999800 0.000315 0.000369	45.998662	1.200000 2.600000	0.001031

If we regard x_3 and x_4 as slack variables in the original formulation of the problem, we can view the progress of the algorithm in the x_1–x_2 plane. The feasible region of the problem is graphed in Figure 7.5, and the paths followed by both the affine scaling algorithm and the simplex method are noted. Observe that while the simplex method approaches the optimal solution by moving from extreme point to extreme point, the iterations of the affine scaling algorithm move through the interior of the feasible region. In this simple example, the affine scaling algorithm required more iterations than the simplex method; however, in large problems, it requires only a fraction of as many and its computational advantages increase with problem size.

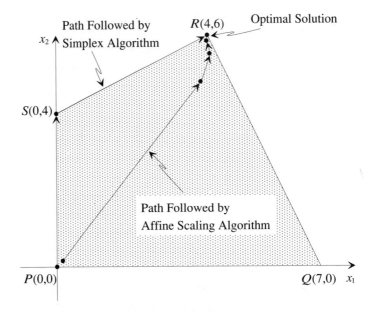

Figure 7.5 Graph for Example 7.1.

In the previous example, we were able to view the progress of the algorithm through the interior of the feasible region. From Table 7.3, observe also that the Euclidean norm of $\mathbf{c}_p$, $\|\mathbf{c}_p\|$, is approaching zero. This is because as we approach the optimal solution, vector $\mathbf{c}\mathbf{D}$ is becoming normal to the null space of $\mathbf{B} = \mathbf{A}\mathbf{D}$. Let us now look at another example in which it is possible to examine how the feasible region is actually transformed by matrix $\mathbf{D}$.

Example 7.2: Primal Affine Scaling Algorithm

Consider the following single-constraint linear programming problem in three variables.

$$\text{maximize } 2x_1 + x_2 + 3x_3 \tag{7.44}$$

subject to

$$x_1 + x_2 + x_3 = 3 \tag{7.45}$$

$$x_j \geq 0, \qquad \text{for all } j \tag{7.46}$$

Obviously, by inspection, the optimal solution to this linear programming problem is

$$\mathbf{x}^* = \begin{pmatrix} x_1^* \\ x_2^* \\ x_3^* \end{pmatrix} = \begin{pmatrix} 0 \\ 0 \\ 3 \end{pmatrix}$$

Let us start the algorithmic process with the initial solution

$$\mathbf{x}_0 = \begin{pmatrix} x_{0,1} \\ x_{0,2} \\ x_{0,3} \end{pmatrix} = \begin{pmatrix} 1 \\ 1 \\ 1 \end{pmatrix}$$

Then the initial transformation matrix is

$$\mathbf{D} = \begin{pmatrix} 1 & 0 & 0 \\ 0 & 1 & 0 \\ 0 & 0 & 1 \end{pmatrix}$$

and the resulting y-space problem is

$$\text{maximize } 2y_1 + y_2 + 3y_3$$

$$\text{subject to}$$

$$y_1 + y_2 + y_3 = 3$$

$$y_j \geq 0, \qquad \text{for all } j$$

Now, computing the $\mathbf{c}_p$ vector as in the previous example by projecting $\mathbf{cD} = (2, 1, 3)$ onto the null space of $\mathbf{AD} = \mathbf{0}$ (i.e., $y_1 + y_2 + y_3 = 0$) results in

$$\mathbf{c}_p = \mathbf{Dc}^t - \mathbf{B}^t(\mathbf{BB}^t)^{-1}\mathbf{BDc}^t = \begin{pmatrix} 0 \\ -1 \\ 1 \end{pmatrix}$$

Therefore, by using (7.38) and (7.39), the next iteration is given by

$$\mathbf{y}_1 = \begin{pmatrix} 1 \\ 1 \\ 1 \end{pmatrix} + \alpha\theta\mathbf{c}_p = \begin{pmatrix} 1 \\ 1 \\ 1 \end{pmatrix} + (0.9)(1.0)\begin{pmatrix} 0 \\ -1 \\ 1 \end{pmatrix} = \begin{pmatrix} 1 \\ 0.1 \\ 1.9 \end{pmatrix}$$

$$\mathbf{x}_1 = \mathbf{Dy}_1 = \begin{pmatrix} 1 & 0 & 0 \\ 0 & 1 & 0 \\ 0 & 0 & 1 \end{pmatrix}\begin{pmatrix} 1 \\ 0.1 \\ 1.9 \end{pmatrix} = \begin{pmatrix} 1 \\ 0.1 \\ 1.9 \end{pmatrix}$$

Because $\mathbf{D}$ was the identity matrix for this iteration, the movements in x-space and y-space were identical and are illustrated in Figure 7.6.

We now redefine the matrix $\mathbf{D}$ using the $\mathbf{x}_1$, that is,

$$\mathbf{D} = \begin{pmatrix} 1 & 0 & 0 \\ 0 & 0.1 & 0 \\ 0 & 0 & 1.9 \end{pmatrix}$$

This results in the transformed problem

$$\text{maximize } 2y_1 + 0.1y_2 + 5.7y_3$$

$$\text{subject to}$$

$$y_1 + 0.1y_2 + 1.9y_3 = 3$$

$$y_j \geq 0, \qquad \text{for all } j$$

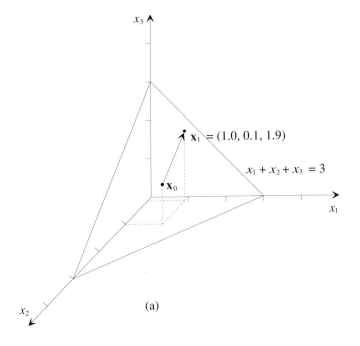

(a)

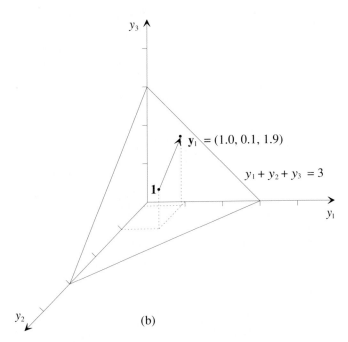

(b)

Figure 7.6   Iteration 1 of Example 7.2.

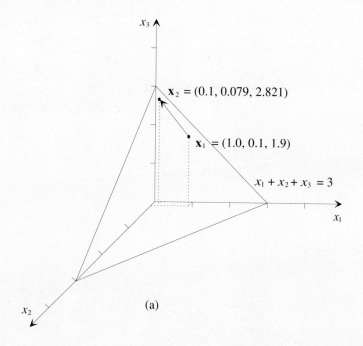

(a)

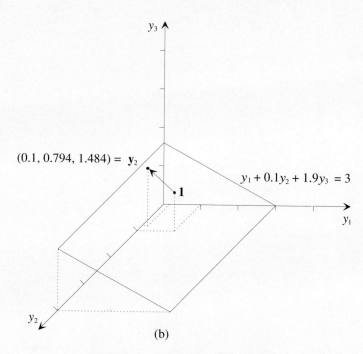

(b)

Figure 7.7   Iteration 2 of Example 7.2.

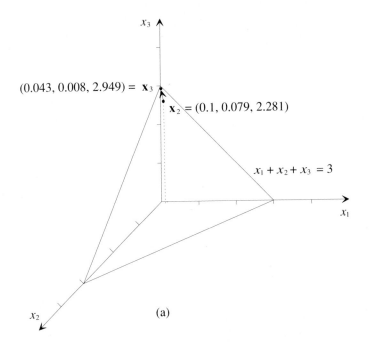

(a)

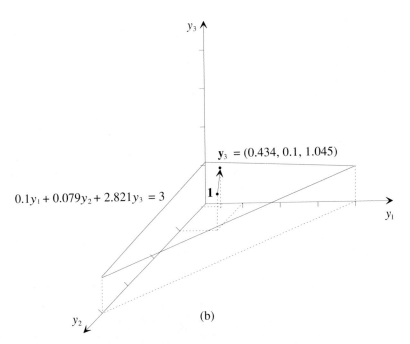

(b)

Figure 7.8 Iteration 3 of Example 7.2.

Again, using (7.38) and (7.39) results in

$$
\mathbf{y}_2 = \begin{pmatrix} 1 \\ 1 \\ 1 \end{pmatrix} + \alpha\theta\mathbf{c}_p = \begin{pmatrix} 1 \\ 1 \\ 1 \end{pmatrix} + (0.9)(1.283333)\begin{pmatrix} -0.779220 \\ -0.177922 \\ 0.419485 \end{pmatrix} = \begin{pmatrix} 0.1000 \\ 0.7945 \\ 1.4845 \end{pmatrix}
$$

$$
\mathbf{x}_2 = \mathbf{D}\mathbf{y}_2 = \begin{pmatrix} 1 & 0 & 0 \\ 0 & 0.1 & 0 \\ 0 & 0 & 1.9 \end{pmatrix}\begin{pmatrix} 0.1 \\ 0.7945 \\ 1.4845 \end{pmatrix} = \begin{pmatrix} 0.10000 \\ 0.07945 \\ 2.82055 \end{pmatrix}
$$

The iteration is again illustrated graphically in Figure 7.7. Note that the feasible region was transformed dramatically by matrix $\mathbf{D}$ so that the iteration in y-space begins at the centrally located point $\mathbf{1}$. This transformation becomes even more pronounced in the next iteration, in which $\mathbf{D}$ is defined by

$$
\mathbf{D} = \begin{pmatrix} 0.10000 & 0 & 0 \\ 0 & 0.07945 & 0 \\ 0 & 0 & 2.82055 \end{pmatrix}
$$

Note, in Figure 7.8, that a substantial move in y-space actually results in a very small move in x-space. This would become even more apparent in subsequent iterations, as the vector $\mathbf{cD}$ becomes nearly normal to the null space of $\mathbf{AD}$.

Although the primal affine scaling algorithm is conceptually elegant, simple, and works effectively, it is not without its problems. First, it is inherently difficult to computationally maintain feasibility when making a step using a constraint set of the form $\mathbf{Ax} = \mathbf{b}$. This difficulty arises because one must find a direction vector $\mathbf{d}$ such that $\mathbf{Ad} = \mathbf{0}$. Computationally, this actually means that $\mathbf{Ad} = \boldsymbol{\varepsilon}$, where $\boldsymbol{\varepsilon}$ is a vector of residual errors. Thus, $\mathbf{A}(\mathbf{x} + \alpha\theta\mathbf{d}) = \mathbf{b} + \alpha\theta\boldsymbol{\varepsilon}$, and errors tend to propagate from iteration to iteration resulting in a solution which is actually infeasible. The dual affine scaling algorithm was designed to overcome this problem. The dual affine scaling algorithm operates on the dual problem in much the same way that the primal affine scaling algorithm operates on the primal. The distinct advantage is that the dual problem of (7.13) has unrestricted variables and inequality constraints. Thus, it is much easier to correct for feasibility problems as well as allow approximate projections. The interested reader is referred to Alder et al. (1986) and Marsten et al. (1988).

Another problem that may arise is linear programming problems that have alternative optimal solutions. While the simplex algorithm always generates a basic feasible solution corresponding to one of the alternative optimal extreme points, affine scaling algorithms may converge to an optimal solution that is not an extreme point and thus not a basic feasible solution. If a basic feasible solution is desired (e.g., for sensitivity analysis), then a corrective measure, such as a purification algorithm, must be taken. See, for example, Kortanek and Strojwas (1984).

Finding an Initial Strictly Positive Solution

Let us now address the problem of getting started. That is, how does one find an initial strictly positive solution to the following linear programming problem, assuming that such a solution exists?

$$\text{maximize } \mathbf{cx} \tag{7.47}$$

$$\text{subject to}$$

$$\mathbf{Ax} = \mathbf{b}$$

$$\mathbf{x} \geq \mathbf{0}$$

As suggested by Karmarkar (1984), we may introduce a single artificial variable, say, λ, and set up the following phase I linear program to find a strictly positive solution:

$$\text{minimize } \lambda \text{ (or maximize } -\lambda) \tag{7.48}$$

$$\text{subject to}$$

$$\mathbf{Ax} + (\mathbf{b} - \mathbf{A1})\lambda = \mathbf{b}$$

$$\mathbf{x} \geq \mathbf{0}$$

$$\lambda \geq 0$$

Note that $\mathbf{x} = \mathbf{1}$, $\lambda = 1$ is a strictly positive solution that is feasible to (7.48), and, thus, we can apply the affine scaling algorithm directly to this problem to find a feasible point. As indicated by Lustig (1985), λ will generally become the blocking variable in a relatively few iterations and by setting the step length parameter $\alpha = 1$ (instead of, e.g., 0.9), a strictly positive solution is generated, assuming that the feasible region has a nonempty interior. By dropping the artificial variable λ, the resulting solution can be used along with the original objective to restart the primal affine scaling algorithm.

KARMARKAR'S PROJECTIVE ALGORITHM

As mentioned earlier, Karmarkar's algorithm (1984) is an interior-point algorithm in which, fundamentally, one attempts to find successive improving directions in the strict interior of the nonnegative orthant. Karmarkar's algorithm is designed to solve a linear programming problem of the following form.

$$\text{minimize } \mathbf{cx} \tag{7.49}$$

$$\text{subject to}$$

$$\mathbf{Ax} = \mathbf{0}$$

$$\mathbf{1x} = 1$$

$$\mathbf{x} \geq \mathbf{0}$$

where $\mathbf{A}$ is an $m \times n$ matrix with rank($\mathbf{A}$) = m, $\mathbf{A}$ and $\mathbf{c}$ are all integer, and $\mathbf{1}$ = $(1, 1, \ldots, 1) \in E^n$. In addition, it is assumed that the optimal value of the objective function is zero and that $\mathbf{x}_0 > \mathbf{0}$ is a feasible solution to (7.49).

Define the diagonal matrix $\mathbf{D}$ as follows

$$\mathbf{D} = \text{diag}(x_{0,1}, x_{0,2}, \ldots, x_{0,n}) = \begin{pmatrix} x_{0,1} & 0 & \cdots & 0 \\ 0 & x_{0,2} & \cdots & 0 \\ \vdots & \vdots & \ddots & \vdots \\ 0 & 0 & \cdots & x_{0,n} \end{pmatrix} \tag{7.50}$$

and consider the nonlinear transformation defined by

$$\mathbf{y} = \frac{\mathbf{D}^{-1}\mathbf{x}}{\mathbf{1}\mathbf{D}^{-1}\mathbf{x}} \tag{7.51}$$

In particular, note that

$$\mathbf{y}_0 = \frac{\mathbf{D}^{-1}\mathbf{x}_0}{\mathbf{1}\mathbf{D}^{-1}\mathbf{x}_0} = \begin{pmatrix} 1/n \\ 1/n \\ \vdots \\ 1/n \end{pmatrix} \tag{7.52}$$

That is, the transformation (7.51) maps $\mathbf{x}_0$ into the "center" of the simplex $\mathbf{1}\mathbf{x} = \sum_{j=1}^{n} x_j = 1$. The inverse mapping (with domain $\mathbf{1}\mathbf{y} = 1$) results in

$$\mathbf{x} = \frac{\mathbf{D}\mathbf{y}}{\mathbf{1}\mathbf{D}\mathbf{y}} \tag{7.53}$$

Substituting (7.53) into (7.49) and recalling that the optimal objective value is zero result in

$$\text{minimize } (\mathbf{c}\mathbf{D})\mathbf{y} \tag{7.54}$$

$$\text{subject to}$$

$$(\mathbf{A}\mathbf{D})\mathbf{y} = \mathbf{0}$$

$$\mathbf{1}\mathbf{y} = 1$$

$$\mathbf{y} \geq \mathbf{0}$$

The central step of Karmarkar's algorithm is then to find an improving feasible direction by projecting the objective vector $\mathbf{c}\mathbf{D}$ onto the null space of

$$\mathbf{B} = \begin{pmatrix} \mathbf{A}\mathbf{D} \\ \mathbf{1} \end{pmatrix} \tag{7.55}$$

That is, as described in the previous section, we calculate

$$\mathbf{c}_p = [\mathbf{I} - \mathbf{B}^t(\mathbf{B}\mathbf{B}^t)^{-1}\mathbf{B}]\mathbf{D}\mathbf{c}^t \tag{7.56}$$

Next, a step is taken in the normalized direction $-\mathbf{c}_p/\|\mathbf{c}_p\|$ according to

$$\mathbf{y}_1 = \mathbf{y}_0 - \alpha r \mathbf{c}_p/\|\mathbf{c}_p\| \tag{7.57}$$

where r is the radius of the largest sphere one can inscribe in the n-dimensional simplex, $\mathbf{1x} = 1$, and $\alpha \in (0, 1)$ is a step-length parameter. (It can be easily shown that $r = 1/[n(n - 1)]^{1/2}$.) Finally, compute $\mathbf{x}_1$ according to (7.53), that is,

$$\mathbf{x}_1 = \frac{\mathbf{Dy}_1}{\mathbf{1Dy}_1} \tag{7.58}$$

The process is then repeated with the new diagonal matrix $\mathbf{D}$ determined by $\mathbf{x}_1$.

An iteration of Karmarkar's algorithm is illustrated graphically in Figure 7.9 and the algorithm is now summarized.

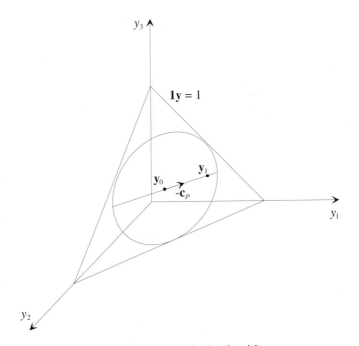

Figure 7.9   Karmarkar's algorithm.

Karmarkar's Algorithm (minimization problem)

STEP 1. Set the iteration counter $k = 0$ and choose an initial solution $\mathbf{x}_k > \mathbf{0}$ such that $\mathbf{Ax}_k = \mathbf{0}$, $\mathbf{1x}_k = 1$.

STEP 2. Define $\mathbf{D} = \mathrm{diag}(x_{k,1}, \ldots, x_{k,n})$ and let

$$\mathbf{B} = \begin{pmatrix} \mathbf{AD} \\ \mathbf{1} \end{pmatrix}$$

STEP 3. Compute $\mathbf{c}_p = [\mathbf{I} - \mathbf{B}^t(\mathbf{BB}^t)^{-1}\mathbf{B}]\mathbf{Dc}^t$.

STEP **4.** Compute

$$\mathbf{y}_{k+1} = \mathbf{y}_k - \alpha r\mathbf{c}_p/\|\mathbf{c}_p\| = \begin{pmatrix} 1/n \\ 1/n \\ \vdots \\ 1/n \end{pmatrix} - \alpha r\mathbf{c}_p/\|\mathbf{c}_p\|$$

where $\alpha \in (0, 1)$, and $r = 1/[n(n-1)]^{1/2}$.

STEP **5.** Compute

$$\mathbf{x}_{k+1} = \frac{\mathbf{D}\mathbf{y}_{k+1}}{\mathbf{1}\mathbf{D}\mathbf{y}_{k+1}}$$

STEP **6.** Check for convergence. If satisifed, stop. Otherwise, replace k by $k + 1$ and go to Step 2.

Karmarkar showed that the theoretical computational time required for each iteration of the projective algorithm is proportional to $n^{2.5}L$. In addition, he showed that the number of iterations required by the algorithm is proportional to nL, so that the overall computational effort is on the order of $n^{3.5}L$. He did this by ingeniously introducing the "potential function,"

$$f(\mathbf{x}) = \sum_{j=1}^{n} \ln\left(\frac{\mathbf{c}\mathbf{x}}{x_j}\right)$$

and proving that for a suitable step size α, each iteration of the algorithm reduced the potential function by a fixed amount. The reader interested in the theoretical details is referred to Karmarkar (1984), Hooker (1986), and Padberg (1986).

SUMMARY

In this chapter, we have presented a brief introduction to computational complexity and introduced some alternatives to the simplex algorithm. These alternative algorithms perform in a manner quite different from the simplex method, and under suitable assumptions, it can be shown that their worst-case performance is polynomial. In particular, the affine scaling variants of Karmarkar's algorithm are now commercially available alternatives to the simplex algorithm.

EXERCISES

7.1. Use the solution defined by (7.29) to find the point that is closest to (6, 1) and lies on the line $2x_1 + x_2 = 0$. Illustrate your results graphically.

7.2. Consider the following problem.

$$\text{maximize } x_1 + x_2$$

subject to

$$2x_1 + x_2 \le 4$$

$$x_1, x_2 \ge 0$$

(a) Solve this problem graphically.
(b) Starting with the initial solution $(x_1, x_2) = (0.1, 0.1)$, solve this problem using the affine scaling algorithm.
(c) Illustrate the progress of the algorithm on the graph in x_1–x_2 space.

7.3. Consider the following problem.

$$\text{maximize } x_1 + 4x_2$$

subject to

$$x_1 + x_2 \le 10$$

$$-x_1 + x_2 \le 2$$

$$x_1, x_2 \ge 0$$

(a) Solve this problem using the simplex algorithm.
(b) Starting with the initial solution $(x_1, x_2) = (0.1, 0.1)$, solve this problem using the affine scaling algorithm.
(c) Illustrate the progress of the algorithm on the graph in x_1–x_2 space.
(d) Write the dual problem and sketch the feasible region. Illustrate graphically that the sequence of $\mathbf{w}$'s generated in Step 4 of the algorithm are converging to the optimal dual solution.

7.4. Consider the following problem.

$$\text{maximize } x_1 + x_2$$

subject to

$$2x_1 + x_2 + x_3 = 1$$

$$x_1, x_2, x_3 \ge 0$$

(a) Set up the phase I problem described in (7.48).
(b) Apply the affine scaling algorithm to find an initial strictly positive feasible solution.

7.5. Consider the following problem.

$$\text{minimize } x_1$$

subject to

$$x_1 + 5x_2 + 2x_3 = 3$$

$$x_1 + x_2 + x_3 = 1$$

$$x_1, x_2, x_3 \ge 0$$

(a) Solve this problem using Karmarkar's algorithm.

(b) Illustrate the progress of the algorithm on the simplex in E^3.

7.6. Write a computer program that implements the affine scaling algorithm.

7.7. Write a computer program that implements Karmarkar's algorithm.

7.8. In the main step of both Karmarkar's algorithm and the affine scaling algorithm, we are required to use $(\mathbf{BB}^t)^{-1}$ to compute the next iteration. Show that if an $m \times n$ matrix $\mathbf{B}$ has full row rank (i.e., rank($\mathbf{B}$) = m), then $\mathbf{BB}^t$ is invertible.

7.9. Consider a problem of the form to minimize $f(\mathbf{x})$ subject to $\mathbf{g}(\mathbf{x}) = (g_1(\mathbf{x}), \dots, g_m(\mathbf{x}))^t = \mathbf{0}$, where $f: E^n \to E^1$ and $\mathbf{g}: E^n \to E^m$. The Karush–Kuhn–Tucker optimality conditions for this problem are

$$\nabla f(\mathbf{x}) + \sum_{i=1}^{m} u_i \nabla g_i(\mathbf{x}) = \nabla f(\mathbf{x}) + \nabla \mathbf{g}(\mathbf{x})\mathbf{u} = \mathbf{0}$$

$$\mathbf{g}(\mathbf{x}) = \mathbf{0}$$

Use the Karush–Kuhn–Tucker conditions to show that the optimal solution of the problem to minimize $f(\mathbf{x}) = \|\mathbf{c} - \mathbf{x}\|^2$ subject to $\mathbf{g}(\mathbf{x}) = \mathbf{Bx} = \mathbf{0}$ is given by $\mathbf{x} = [\mathbf{I} - \mathbf{B}^t(\mathbf{BB}^t)^{-1}\mathbf{B}]\mathbf{c}$.

8

APPLICATIONS OF LP IN INFORMATION TECHNOLOGY

CHAPTER OVERVIEW

In the preceding chapters, as well in those to follow, we use—for the most part—the linear programming algorithm as simply a *tool for the derivation* (i.e., *construction*) *of a solution* (i.e., *program, or the optimal values of the structural variables*). And, once this solution (program) has been derived (constructed), we typically regard the numbers obtained as the end result of the exercise.

However, there is another, and extremely useful way in which LP may be utilized. Specifically, rather than focusing on just the derivation of a program, we seek to *construct a model using the results derived by the LP algorithm*. Thus, in this manner, the program derived by the LP algorithm is but an intermediate step toward our final goal.

And what, the reader may ask, is the purpose of using LP to construct models? The answer to this is that there is a significant need for the analysis of data—particularly so in the discipline known as *information technology* (or information systems, decision sciences, or decision support systems). Here, the analyst seeks to take raw data and transform them into useful information. Some of the more typical tools used in this transformation and analysis of data are those of *prediction, classification,* and *clustering.* In essence, the decision analyst relies on the development of various models into which the data are fed, and out of which information (for the support of decision making) is generated. Typically, these models for prediction, classification, and clustering rely on techniques from such areas as applied statistics (e.g., linear regression and discriminant analysis) and computer science (e.g., clustering and sorting heuristics). However, as will be illustrated, one may often use LP to develop such models—and to oftentimes achieve superior results.

PROBLEM TYPES

It has been noted by several investigators that there would appear to be but a few basic problem types—and a number of individuals have attempted to categorize these. Ten Dyke (1990) identifies four different problem categories, whereas Ignizio (1991) recognizes three (with one of these being a composite of two of the types listed by Ten Dyke). The three fundamental problem types identified by Ignizio are as follows:

- *Problems of **association**.* Such problems are characterized by the need to associate the attributes of a given object, or objects, to a specific outcome (or class, or response). Included among this type of problem are those of (1) *prediction*—wherein we seek to associate a set of input attributes to a specific outcome or response, and (2) *pattern classification*, or pattern recognition—wherein we seek to associate a given object, on the basis of its features, to a given class. It would appear that problems of association are invariably solved by substitution (i.e., the facts, or data, are substituted into the model and the outcome, or outcomes, then observed).

- *Problems of **construction**.* This type of problem involves the need to *create*, from among a set of alternatives, *a plan or configuration*. In essence, problems of construction are those that involve the development of combinations, or blends. It would appear that a common thread among problems of construction is that they are solved via a mechanism of feedback and comparison (e.g., as observed in the simplex algorithm).

- *Problems thus far beyond our abilities to solve in a scientific manner.* Among this type of problem are those tasks that involve creativity, leaps of faith, reasoning by (obscure) analogy, learning how to talk, and those problems in which the very nature of the problem itself is not defined. Humans find it difficult to characterize, if even understand, how such problems are either modeled or solved.

Examples of problems of association include those of (1) curve fitting, (2) forecasting, (3) medical diagnosis, (4) automotive troubleshooting, and (5) preventive maintenance. A few specific examples of problems of construction include (1) scheduling, (2) systems configuration, such as that of where to place the components of a computer, or aircraft, (3) facility layout and location, (4) printed circuit board configuration, (5) weapons mix, (6) product mix, and (7) cluster analysis.

Examining the applications described in the previous chapters, we may note that virtually all fall within the problem type that we have designated as problems of construction. That is, we sought to *construct* a product mix; to *construct* a production schedule; to *construct* a blend of materials; to *construct* an assignment of workers to jobs; to *construct* an investment plan; and so on. And we have not really touched on problems of the association type. This omission will be rectified in this chapter.

METHODS IN INFORMATION TECHNOLOGY

We shall make no attempt herein to try to define the discipline of information technology. Instead, we simply note that much of the effort within this sector is concerned with the transformation of data into information—specifically, information that may be used by management in support of its decision making—and that is a major, if not *the* major goal of information technology.

Typically, organizations are long on data (raw, unprocessed numbers and facts) and short on information (i.e., processed data *in a useful format*). Thus, much of what is addressed in information technology is concerned with those processes that serve to support the transformation of large volumes of data (i.e., databases) into useful information in support of decision making. As mentioned earlier, three methods play a particularly important role in this transformation process: *prediction, classification,* and *clustering.* The first two should now be recognized as problems of association, and the last is one of construction. All three shall be addressed in the material to follow. However, rather than employing the more conventional methods associated with the development of models for these techniques, we demonstrate the use of LP—and discover that, for many cases, the LP approach has significant advantages.

More specifically, we shall employ LP (which may be characterized as a rather typical method for solving construction-type problems) to derive the models to then be used in the solution of problems of association, specifically, those of prediction and classification. And we also indicate just how one might form a mathematical model for the problem of clustering.

PREDICTION VIA LINEAR PROGRAMMING

Much of what one does in life involves prediction or forecasting. For example, consider the following problem. We have been asked by a certain firm to develop a CER (i.e., a *cost-estimating relationship*, a term commonly used in both industry and government) for the production of plastic pipe. The firm involved has kept data as to the cost of producing a number of types of plastic pipes of different diameters—and these costs have been translated into equivalent 1990 dollars. Besides these data (on a total of just five different pipe diameters), all that we know is that if our cost predictions are either too high *or* too low, a penalty will be assessed by the customer—and this penalty will be *a linear function of the difference between the predicted and actual costs*. The firm, in turn, states that it would like for us to develop a CER that will accurately predict pipe production costs while minimizing the sum of any cost penalties. The reader should note carefully that our goal here is to develop a function (i.e., a model) rather than to simply solve an existing model.

When presented with such a problem as described earlier, most individuals (and precisely 100% of the students to whom we have previously provided this illustration) seem to select *conventional* regression (i.e., statistically based regression—or "least squares"—as may be found in most texts on applied statistics). Further, usually, this choice is limited to that of *linear* regression. And these choices are made almost immediately—and certainly before much thought is given to the actual problem under consideration.

However, there are numerous reasons to consider alternatives to conventional regression. First, all too often, *linear* regression is employed simply because it is easy (and/or supporting software is available)—even when the associated response may be *nonlinear*. Second, conventional regression of any form is a *parametric* tool. That is, it is assumed that we have a normal population with equal variances (or multivariate population with equal correlation matrices)—an assumption that may not be true and/or defensible for the situation under consideration. Further, if such an assumption is incorrect, then any sensitivity analysis performed on the resulting predictive function is likely to be in error. Third, conventional regression focuses solely on the minimization of the *sum of the squares* of all residuals (i.e., the sum of the squares of all differences between actual and predicted values). And it must be realized that the minimization of the sum of the squares of the residual is not necessarily the optimal measure of predictive performance—and it certainly is not so in the pipe production problem we have outlined earlier (i.e., a more appropriate measure is clearly the minimization of the sum—or weighted sum—of the absolute values of the residuals). Further, as has been well documented, "outliers" (i.e., unusual data points) can have a dramatic impact on the predictive function formed when one minimizes the sum of the squares of the residuals (Campbell and Ignizio, 1972). Fourth, conventional regression can—and does—lead to the development of irrational coefficients. One particularly interesting example of this was a regression equation developed by the U.S. Army Corps of Engineers for the prediction of water flow for the rivers passing through several western towns (as a function of winter snowfall on the mountains, sunshine, temperatures, etc.). Using linear regression, the Corps developed a predictive function that worked quite well for all *previous* data. However, when this same function was used to predict water flow for the very next year, it indicated that a *negative* flow of water was to be expected (i.e., the water would evidently flow *up* the mountains). This rather amusing result leads to a fifth drawback of conventional regression—its inability to (easily) handle side conditions.

Thus, unless the problem under consideration is actually appropriate for conventional regression, and clearly satisfies the associated assumptions, one should consider (nonparametric) alternatives for the development of the predictive function. One very powerful alternative is that of linear programming (or linear goal programming—a tool to be discussed in Part 3) and it is that approach that we advocate for the problem previously cited. If it is reasonable to assume that cost is a linear function of pipe diameter, then the CER (predictive equation) sought is of the following form:

$$y_i = ax_i + b$$

where

$$y_i = \text{predicted cost of pipe } i$$

$$x_i = \text{diameter of pipe } i$$

$$a \text{ and } b = \text{unknown constants, and unrestricted in sign}$$

Because the penalty imposed on any incorrect estimates is a linear function of either the positive or negative error, an appropriate measure of fit is that of the minimization of the absolute values of the residuals, or

$$\text{minimize} \sum_{i=1}^{m} |r_i|$$

where

$$y_i = \text{predicted cost of pipe } i$$

$$x_i = \text{diameter of pipe } i$$

We may also write each r_i as

$$r_i = C_{A(i)} - (ax_i + b)$$

where

$$C_{A(i)} = \textit{actual} \text{ cost of pipe } i$$

Using the last expression, we can establish a mathematical model that may be solved for the coefficients of the predictive equation. This model may be written as

$$\text{minimize} \sum_{i=1}^{m} |r_i| \qquad (8.1)$$

subject to

$$C_{A(i)} - (ax_i + b) = 0, \qquad \text{for all } i \qquad (8.2)$$

Unfortunately, the model is not in the form of an LP model. However, this situation may be easily rectified. First, we may transform the unrestricted variables. Let

$$a = a_1 - a_2$$
$$b = b_1 - b_2$$

where

$$a_1, a_2, b_1, b_2 \geq 0$$

Next, we modify (8.2) by replacing the unrestricted variable r_i with the differences between two nonnegative variables (η_i and ρ_i), which lead to

$$C_{A(i)} - (a_1 x_i - a_2 x_i + b_1 - b_2) + \eta_i - \rho_i = 0, \qquad \text{for all } i \qquad (8.3)$$

where

$$\eta_i = \text{negative residual (or deviation) at data point } i$$

$$\rho_i = \text{positive residual (or deviation) at data point } i$$

and both η_i and ρ_i are nonnegative.

To satisfy (8.3) we must minimize both η_i and ρ_i, which leads to an alternative form of the original objective function, that is,

$$\text{minimize} \sum_{i=1}^{m} (\eta_i + \rho_i) \qquad (8.4)$$

As a final result, we may form the complete, equivalent LP model as

$$\text{minimize} \sum_{i=1}^{m} (\eta_i + \rho_i) \qquad (8.5)$$

subject to

$$C_{A(i)} - (a_1 x_i - a_2 x_i + b_1 - b_2) + \eta_i - \rho_i = 0, \qquad \text{for all } i \qquad (8.6)$$

$$a_1, a_2, b_1, b_2 \geq 0, \qquad \text{and } \eta_i, \rho_i \geq 0, \qquad \text{for all } i \qquad (8.7)$$

Not only may we now solve the preceding model for the predictive equation (CER) coefficients, but we may also perform—independent of any assumptions whatsoever regarding the data population—a valid sensitivity analysis of the model (i.e., using the methods already covered in Chapter 6). Thus, the use of LP results in the ability to perform *nonparametric, constrained regression*—and permits the selection of a wide variety (or, by means of linear goal programming, even a combination) of measures of fit. As just one example, suppose that we wish to develop a predictive function in which the most appropriate measure of performance is that of the minimization of the *single worst residual* (i.e., the minmax, or Chebyshev measure). For example, in our pipe production problem, suppose that the firm wishes to minimize the maximum penalty that it might expect to incur. This situation is easily formulated via the modification of the model previously established in (8.5)–(8.7). Specifically, our new model will be

$$\text{minimize } \delta \qquad (8.8)$$

subject to:

$$[C_{A(i)} - (a_1 x_i - a_2 x_i + b_1 - b_2)] + \delta \geq 0, \qquad \text{for all } i \qquad (8.9)$$

$$[C_{A(i)} - (a_1 x_i - a_2 x_i + b_1 - b_2)] - \delta \leq 0, \qquad \text{for all } i \qquad (8.10)$$

$$a_1, a_2, b_1, b_2, \delta \geq 0 \qquad (8.11)$$

where δ is the maximum residual (maximum amount of deviation).

The two models just presented represent just two possibilities for the development of models for nonparametric, constrained regression via LP. In Part 3, we shall elaborate further on such models—wherein we will address multiobjective optimization and, in particular, linear goal programming. As will be apparent then, goal programming and constrained regression are (at least at the level discussed here) virtually one and the same thing (Charnes and Cooper, 1975).

We might also note that if one is really bound and determined to minimize the sum of the squares of the residuals, then this may also be accomplished via a mathematical programming formulation—resulting in a quadratic programming model that may be easily solved via the simplex algorithm. Further, sensitivity analysis on this model is valid even if the assumptions necessary for conventional regression do not hold. We leave it to the reader to develop the general form of such a model.

Returning to our original model of (8.5)–(8.7), let us demonstrate its use on a numerical example—where this example is a simplification (primarily in that we consider fewer data points) of a problem encountered in actual practice.

Example 8.1: A Predictive Function for Power Line Repairs

In this example, we assume that our purpose is to develop a model for the estimate of construction costs of various tasks associated with the repair of existing power lines—where such repair is limited to that of "wood pole" lines. For such jobs, the main differences are that of the number of poles, feet of wire, number of insulators, and such that are replaced. Table 8.1 provides a set of data on eight different jobs that we shall employ to illustrate the use of constrained regression. Note carefully that the costs listed in the table are those due solely to labor.

Note also that with just eight observations, one would most definitely be on thin ice to assume that these data come from a normal distribution with equal variances (in the actual problem, a massive number of jobs had been recorded and, in fact, it was determined that the data did *not* come from a normal distribution). Further, examine closely job 2 and compare it with either job 5 or 6. All three jobs would appear to involve essentially the same amount of repair activity, yet the cost of job 2 (i.e., a direct function of the time required to perform job 2) is considerably higher than that of job 5 or 6. We shall return to this apparent anomaly in the database later in our discussion.

TABLE 8.1 DATA FOR POWER LINE REPAIRS

Job number	x_1: Number of poles replaced	x_2: Amount of wire, in 100 ft, replaced	x_3: Number of crossarms replaced	x_4: Number of insulators replaced	x_5: Number of guy wires and guards replaced	$TC_{(A)}$: Total cost, in dollars
1	1	4	1	2	1	560
2	3	10	3	6	1	3080
3	4	24	8	12	2	2520
4	1	5	2	3	0	630
5	3	12	3	12	1	1750
6	3	12	3	8	1	1890
7	2	10	4	6	0	1120
8	4	12	8	12	0	1960

The power company wanted to develop a means to predict the cost (and time) of such tasks and they called on an outside consultant to perform this analysis. The consultant simply put all of the data into a conventional linear regression package and cranked out the resultant predictive function. Specifically, he assumed that the cost of a repair job was a linear function (which, as it turned out for this case, was not an unreasonable assumption) of the various tasks listed in Table 8.1. Thus, the general form of the regression equation used was

$$TC_{(p)} = a_1 x_1 + a_2 x_2 + a_3 x_3 + a_4 x_4 + a_5 x_5 \qquad (8.12)$$

where

$TC_{(p)}$ = predicted cost, in dollars, to complete a job

$a_1, \ldots, a_5$ = coefficients of the predictive function (values to be found)

x_j = number of units of component j in the job under consideration

The result of the employment of conventional regression was the development of the following predictive equation:

$$TC_{(P/CR)} = 1211x_1 + 259x_2 - 588x_3 - 259x_4 - 378x_5$$

where

$TC_{(P/CR)}$ = predicted cost using conventional regression

Let us consider this result. First, although there is no justification for the assumption of a normal distribution or equal variances, this does not in itself preclude the use of linear regression—but it does weaken any interpretation given to sensitivity analysis. Second, notice that if we have a job that involves just the replacement of, say, 10 insulators, then the company is predicted to make a *profit* of $2590 on the repairs, according to this function. If this were indeed true, then the company should probably purchase a rifle for everyone in the region and beg them to shoot out their insulators (actually, many repairs are due precisely to individuals who use insulators for target practice). Of course, what we are really witnessing is yet another example of the development of irrational coefficients via regression.

If, instead of using conventional regression, we employ constrained regression (i.e., linear programming), then the associated model may be developed via the extension of the model illustrated in (8.5)–(8.7). Specifically, there will be eight constraints (one for each data point) and an objective function in which the sum of all residuals (deviation variables) is to be minimized. Thus, the form of the associated model is simply

$$\text{minimize} \sum_{i=1}^{8} (\eta_i + \rho_i)$$

subject to:

$$C_{A(i)} - TC_{(P/LP)} + \eta_i - \rho_i = 0, \qquad \text{for } i = 1, 8$$

$$\text{all coefficients} \geq 0, \qquad \text{and } \eta_i, \rho_i \geq 0, \qquad \text{for all } i$$

where

$TC_{(P/LP)}$ is the predicted cost using *constrained* regression.

And notice that we have replaced all coefficients in the original regression equation with the differences between two nonnegative variables. Further, this is an LP model (more precisely, it is a linear goal programming model that may be solved via LP) and we may thus find its solution via any LP package. Using the program derived from this model, we may then form a new predictive function:

$$TC_{(P/LP)} = 533x_1 + 189x_2 + 357x_3 - 84x_4 - 245x_5$$

Although the two predictive equations (i.e., the one developed by conventional regression and the one developed by LP) are obviously different, the reader is probably saying, at this point, that constrained regression has also produced some irrational coefficients. For example, using this last function, we still predict a profit if we replace nothing but insulators. However, note that we may easily avoid negative coefficients by simply not replacing the coefficients with the differences between two new, nonnegative variables. If this is done, then all coefficients must be non-negative valued.

However, instead of proceeding with a model that does not permit negative coefficients, let us further attempt to compare objectively the two predictive equations that have been developed. Specifically, which one is the "best"? The answer to that question is simply that, should we wish to minimize the sum of the squares of the residuals, the first function is best. However, should we instead wish to minimize the sum of the absolute values of the residuals, the second function is best. Given the problem as stated earlier, this conclusion is probably not terribly satisfying. Thus, let us examine the results even further. One very effective way to do this is to set up a table of residuals. That is, for each predictive function, we list the associated residuals for all eight jobs. This results in Table 8.2. Note that the second column presents the residuals developed via conventional regression and the third column summarizes the residuals developed by means of LP.

TABLE 8.2 RESIDUALS

	Residual by	
Job	Least squares	Absolute values
1	2.9	0
2	8.0	18.5
3	0.5	0
4	1.1	1
5	2.8	0
6	9.0	1.6
7	0.3	0
8	1.3	0

If we examine the second column (i.e., that associated with conventional regression—and denoted as the "Least Squares" column) of Table 8.2, we might be tempted to conclude that job 6 is an unusual data point, or "outlier." Conversely, by examining the column under linear programming (i.e., absolute values), it would appear that it is job 2 that is (quite definitely) unusual, whereas job 6 is relatively typical. Such observations can be even better illustrated by means of a residual "spike chart," a graph of the residuals developed by each method. Such a graph is provided in Figure 8.1. Here, the cross-hatched spikes are the residuals associated with constrained regression (or AVs, for absolute values) and the solid spikes are the residuals from conventional regression (or LSs, or least squares).

Even with the benefit of the table and spike chart, there still remains the question as to which approach for model development was best for this specific example. In order to make any intelligent decision as to that question, one must investigate the data of Table 8.1 in even more detail. In the actual problem, what was

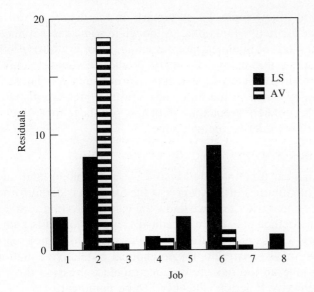

Figure 8.1 Spike chart.

done was to attempt to determine those factors that might possibly cause similar jobs to require considerably dissimilar times. As a result, several possibilities were identified, including

- different work crews (e.g., with different levels of *experience*)
- urban, suburban, or rural regions (i.e., was job *location* a factor?)
- *weather* (i.e., do local weather conditions significantly influence job times?)

By matching the data with these new factors (i.e., through an examination of weather data from the local airport), it was quickly concluded (without resorting to any formal statistical analysis) that virtually every spike for the LP-derived constrained regression function could be explained by weather alone. For example, job 2 is one that was conducted during an exceptionally cold, windy day during which it actually started to sleet while the repairs were being made. As a result, these spikes were ultimately denoted as "bad weather spikes" and it was concluded that either weather should be considered as a variable in the predictive equation or that *two* predictive equations were needed: one for good weather and one for bad. The latter alternative was ultimately selected.

Notice, however, had we used the conventional regression predictive equation, we might have been led to believe that job 6 was the unusual job when, in fact, it was not. This distortion is a result of the squaring of the residuals, an effect frequently cited for least squares. On the other hand, using absolute values, quite often (very important) information concerning the data may be readily identified. In fact, in our work, we have employed constrained regression (LP) not only as a tool for prediction, but, and even more importantly, as a tool for performing "audits" on databases. Using the method in such a manner, we have been able to trace production-line defects to various causes (e.g., impaired workers), to determine when a chemical processor needs to be taken off line and repaired, and so on.

Before moving on to a new topic, let us remark on the use of LP for the development of *nonlinear* predictive functions. Although in some cases it may be necessary to resort to methods of nonlinear programming, there is a host of other instances in which we may still employ LP—if the predictive equation may be represented as a polynomial. Consider, for example, a problem in which just two factors are assumed to play a role in the outcome—and in which we also have reason to believe that the predictive equation is of the nonlinear form that follows. If so, we may use LP to derive the solution.

$$TC_{(P)} = a_1 x_1 + a_2 x_2 + a_3 x_1^2 + a_4 x_2^2 + a_5 x_1 x_2 + a_6$$

Notice that the nonlinear function listed before is a polynomial (and, more specifically, a quadratic function). Further, whereas the function is certainly nonlinear in form, it becomes linear whenever we know the values of x_1 and x_2 and are seeking the values of the coefficients. More specifically, a polynomial is a linear function of its coefficients. Consequently, the constraints of a constrained regression (LP) model become linear constraints for polynomial predictive equations. Thus, one should never give up too quickly on constrained regression (i.e., via LP) just because the predictive function is believed to be nonlinear.

PATTERN CLASSIFICATION VIA LP

Pattern classification, like prediction and forecasting, is a vital and frequently encountered problem—across a diverse spectrum of applications. The specific name given to this area depends to a great degree on the background of the user. Thus, whereas some call the method pattern classification, others may call it *pattern recognition, discriminant analysis,* or *grouping.* Just a few examples of pattern classification include the following:

- *Granting loans or credit.* Here, as based upon the data supplied by the applicant, we seek to determine whether or not to classify the applicant as a good or bad credit risk.
- *Friend or foe identification.* Using signals from radars, lasers, sonar, and/or other detection devices, we wish to classify the object under surveillance as either friendly or unfriendly.
- *Group technology.* The goal of group technology is to expedite the manufacturing process by means of grouping "similar" parts families.
- *Medical diagnostics.* Here, we seek to associate a set of symptoms with a particular disease (i.e., class).
- *Stocks and commodities analysis.* As based upon various data, we hope to assign investment alternatives to various classes (e.g., attractive, neutral, unattractive).

Pattern classification is concerned with the "best" assignment of objects, as based upon their features (i.e., attributes), to one of a set of *predetermined* classes. And note carefully that if, instead, the classes have *not* been predetermined, then the assignment of "similar" objects to "similar" groups is a problem of *cluster analysis* rather than pattern classification—and all too often these two methods are confused. In this section, we focus exclusively on the pattern classification problem—and, in particular, on a nonparametric approach to this problem via linear programming. We shall, however, address cluster analysis in a section to follow.

The "typical" approaches to the development of models for pattern classification have been through statistically based—and usually parametric—methods such as Fisher's linear discriminant function (Young and Calvert, 1974). However, the limitations of such methods (i.e., their reliance upon very restrictive assumptions) led to the investigation and development of various nonparametric approaches. This ultimately led, in the 1960s, to interest in the development of models for the pattern classification problem via linear programming (Mangasarian, 1965). Such interest was rekindled in the 1980s through the works of, in particular, Freed and Glover (1981). And, at this time, much of the use of LP in the development of models for pattern classification has been due to the efforts of Freed and Glover. As such, we shall briefly describe the basics of their approach—but we shall use a somewhat revised model as developed by Ignizio (1986), and Ignizio and Cavalier (1986).

The approach is based upon the development of the so-called *linear* discriminant function, where this function is expressed as

$$f(x) = w_1 x_{i,1,k} + w_2 x_{i,2,k} + \cdots + w_m x_{i,n,k} + b \qquad (8.13)$$

where

$$x_{i,j,k} = \text{score achieved by object } i, \text{ of class } k, \text{ on attribute } j$$

$$w_j = \text{weight given to attribute } j$$

$$b = \text{constant (and unrestricted in sign)}$$

We now discuss just how one may employ linear programming to develop this function by means of solving for the unknown weights.

The form of the LP model employed to represent the pattern classification problem depends upon the measure of performance selected—and this choice is usually a function of the characteristics of the particular problem encountered. However, two of the most typical measures of performance are those of (1) the minimization of the sum (or weighted sum) of the misclassifications and (2) the minimization of the single worst misclassification. In order to keep the discussion simple, let us restrict our focus to training samples from just two classes (the extension of this approach to more than two classes is accomplished by means of the development of pairwise separating surfaces). Thus, the general form of the first LP model (i.e., as used to generate a function that will serve to minimize the sum of all misclassifications) is as follows.

Model I

Find **w** so as to

$$\text{minimize } z = \sum_{i=1}^{p} (\rho_i) + \sum_{i=p+1}^{m} (\eta_i) \qquad (8.14)$$

subject to:

$$\sum_{j=1}^{n} (w_j x_{i,j,k}) + b - \rho_i \leq -r, \qquad \text{for } i = 1, \ldots, p \qquad (8.15)$$

$$\sum_{j=1}^{n} (w_j x_{i,j,k}) + b + \eta_i \geq r, \qquad \text{for } i = p + 1, \ldots, m \qquad (8.16)$$

where

$$w_j = \text{weight assigned to score (attribute) } j \text{ (and unrestricted in sign)}$$

$$x_{i,j,k} = \text{score achieved by object } i, \text{ of class } k, \text{ on attribute } j$$

$$b = \text{constant (and unrestricted in sign)}$$

r = small positive constant (a value of 0.1 is employed herein)

$-1 \leq w_j \leq 1$

$i = 1, \ldots, p$ represents the indices of the objects in the *first* class

$i = p + 1, \ldots, m$ represents the indices of the objects in the *second* class

The second model (i.e., to develop a function that will minimize the single worst misclassification) is then given as follows.

Model II
Find **w** so as to

$$\text{minimize } z = \delta \tag{8.17}$$

subject to:

$$\sum_{j=1}^{n} (w_j x_{i,j,k}) + b - \delta \leq -r, \qquad \text{for } i = 1, \ldots, p \tag{8.18}$$

$$\sum_{j=1}^{n} (w_j x_{i,j,k}) + b + \delta \geq r, \qquad \text{for } i = p + 1, \ldots, m \tag{8.19}$$

where all notation, as well as the restriction on the upper and lower limits on the weights, is the same as previously defined except that here we use δ to denote the amount of misclassification—and we should note that $\delta \geq 0$.

Before we proceed to a numerical illustration of these two approaches, let us reflect for a moment on the discriminant function that is ultimately developed. Figure 8.2 depicts such a function in two dimensions. Here, we have drawn two separating lines—one to the left of the actual separating line by an amount r and the other to the right of that line by the same amount. Thus, the use of r in the right-hand side of these models should be somewhat clearer. Notice also, in this

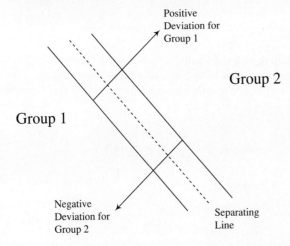

Positive
Deviation for
Group 1

Group 2

Group 1

Negative
Deviation for
Group 2

Separating
Line

Figure 8.2   Object misclassification.

figure, that group 1 is to the left of the separating line and group 2 is to the right. Thus, any misclassifications of objects from group 1 will appear as positive deviations (i.e., ρ_i), whereas any misclassifications of objects from group 2 will appear as negative deviations (i.e., η_i). As such, it should be clear that the objective function in (8.14) serves to minimize the sum of all misclassifications (i.e., the amounts of misclassification, not the number of misclassifications). Using the same figure, we may interpret δ, in the second model, to be the largest of the negative or positive deviations associated with the object that is worst misclassified (i.e., farthest away from the boundary of its proper class).

Example 8.2: Linear Discriminant Function via LP

To demonstrate the use of LP for the determination of a linear discriminant function for pattern classification, let us consider the data set listed in Table 8.3. Here, we have two classes (A and B) and five examples of objects of each class. This data set is also known as the *training set* for the pattern classification boundaries determination. By means of the construction of a linear programming model, we may show that a linear boundary may be developed that will serve to completely classify, with 100% accuracy, any training set data *that is linearly separable*.

TABLE 8.3 TWO-CLASS DATA SET

Object	Score $(x_{i,1,k})$	Score $(x_{i,2,k})$	Class
1	1	1	A
2	2	2	A
3	3	1	A
4	3	3	A
5	6	3	A
6	4	1	B
7	5	2	B
8	7	2	B
9	8	4	B
10	9	1	B

Let us first develop the LP model that may be used to derive a linear discriminant function for the minimization of the sum of all misclassifications. This model may be written, for our particular data set—and using 0.1 as the value for r, as follows.

Find **w** so as to

$$\text{minimize } z = \sum_{i=1}^{5} (\rho_i) + \sum_{i=6}^{10} (\eta_i)$$

subject to:

$$w_1 + w_2 + b - \rho_1 \leq -0.1$$
$$2w_1 + 2w_2 + b - \rho_2 \leq -0.1$$

$$3w_1 + w_2 + b - \rho_3 \leq -0.1$$
$$3w_1 + 3w_2 + b - \rho_4 \leq -0.1$$
$$6w_1 + 3w_2 + b - \rho_5 \leq -0.1$$
$$4w_1 + w_2 + b + \eta_6 \geq 0.1$$
$$5w_1 + 2w_2 + b + \eta_7 \geq 0.1$$
$$7w_1 + 2w_2 + b + \eta_8 \geq 0.1$$
$$8w_1 + 4w_2 + b + \eta_9 \geq 0.1$$
$$9w_1 + w_2 + b + \eta_{10} \geq 0.1$$

where

$$w_j \text{ and } b \text{ are unrestricted in sign}$$

$$-1 \leq w_j \leq 1, \qquad \text{for all } j$$

Solving this model, we obtain

$$w_1 = 0.4, \qquad w_2 = -0.6, \qquad b = -0.7$$

and thus the equation for the linear discriminant function that minimizes the sum of the amount of all misclassifications is

$$0.4x_{i,1,k} - 0.6x_{i,1,k} = 0.7$$

And this line has been plotted in Figure 8.3 as the dashed line that separates the As from the Bs. The fact that we have been able to completely separate the training data with the discriminant line developed is a result restricted—for the most part—to contrived, textbook examples.

Should we instead wish to minimize the single worst misclassification, we simply form the following model.

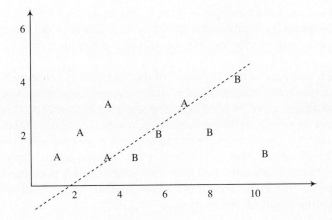

Figure 8.3 Discriminant line for Example 8.2.

Find **w** so as to

$$\text{minimize } z = \delta$$

subject to:

$$w_1 + w_2 + b - \delta \leq -0.1$$

$$2w_1 + 2w_2 + b - \delta \leq -0.1$$

$$3w_1 + w_2 + b - \delta \leq -0.1$$

$$3w_1 + 3w_2 + b - \delta \leq -0.1$$

$$6w_1 + 3w_2 + b - \delta \leq -0.1$$

$$4w_1 + w_2 + b + \delta \geq 0.1$$

$$5w_1 + 2w_2 + b + \delta \geq 0.1$$

$$7w_1 + 2w_2 + b + \delta \geq 0.1$$

$$8w_1 + 4w_2 + b + \delta \geq 0.1$$

$$9w_1 + w_2 + b + \delta \geq 0.1$$

where

w_j and b are unrestricted in sign, $\delta \geq 0$, and $-1 \leq w_j \leq 1$,   for all j.

Solving this model, we (by chance only) find precisely the same solution as that found for the previous model. Thus, the same discriminant line is appropriate, in this particular example, for *either* measure of classification performance. Such a coincidence might also occur, but with less likelihood, if the training data were *not* linearly separable.

These methods might appear, from the example data set selected, to be very efficient and effective approaches to pattern classification. Unfortunately, the data set used (as well as those employed in most descriptions of this type of approach) is of a special—if not very contrived—nature. When dealing with real-world problems, linear separability is rare—and thus a certain amount (and quite often a very significant amount) of misclassification of the training data will be encountered.

Yet another problem that may occur, for either of the two models presented, is the development of a *degenerate* solution; that is, the values of all weights and the constant (i.e., *b*) could be zero! While there are ways to circumvent this outcome, as well as others associated with these models, we shall not pursue this matter further as, quite simply, *we do not recommend the employment of such an approach to pattern classification.* And this is true even for the most recent revisions of the method. Instead, the approach that is recommended is presented in the section that follows. However, before we move to that section, let us observe one more interesting result of our analysis.

When we use LP, in the fashion illustrated, for pattern-classification problems, the final result is the determination of the coefficients of the discriminant function. We can present this result in the form of a network model—and more specifically in the form of what is sometimes termed (although very loosely) a *neural network*. To illustrate, let us use the discriminant function developed for the previous example. We may construct a network to accomplish the classification using the weights found for this function—and this network is illustrated in Figure 8.4.

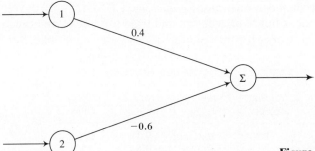

Figure 8.4 Discriminant network.

Now, using this network, we may examine what happens if we attempt to classify an object. For example, should we select an object from the original training set, such as point (8, 4), we should classify it as being in class B. Using the original discriminant function, and simply plugging in these values, we note that the left-hand side (i.e., $0.4x_{i,1,k} - 0.6x_{i,1,k}$) is greater than the right (0.7) and thus the object is class B. However, using the network of Figure 8.4, we can achieve precisely the same results. That is, a value of 8 is input to the top and leftmost node and a value of 4 is simultaneously input to the bottom, leftmost node. These signals are transmitted across the weighted branches (where these weights are simply the weights of the linear discriminant function) and then summed in the single output node. If this sum exceeds the threshold value (i.e., the value of the constant from the LP solution), then the object is in class B. Otherwise, it is assigned to class A. Next, let us assume that we wish to assign a totally new object, with scores of $4\frac{1}{4}$ and $1\frac{2}{3}$. Inputting these scores would result in a sum, at the output node, of precisely 0.7. Thus, we would not exceed the threshold and would consequently assign this object to class A. In this instance, we have an object that lies precisely on the discriminant line and thus it is really a toss-up as to which class it is assigned.

We might also note that it is just as easy to develop a network representation for predictive functions. In such cases, we simply sum the inputs to the final node (i.e., without comparing it with a threshold value) and present that sum as the final output (i.e., the prediction). The one main advantage of such network representation, for either prediction or classification, is in the speed of the network if replicated in actual hardware.

AN ENHANCED APPROACH TO PATTERN CLASSIFICATION*

Consider the data set listed in Table 8.4 and depicted graphically in Figure 8.5. Obviously, this is *not* a set of training data to which we would want to apply the limited and rather naive approach described in the previous section. Simply put, the two classes are clearly, *as in the case of most real problems,* not linearly separable and any attempt to develop a *linear* separating surface will result in a less than satisfactory result. Fortunately, however, the notions summarized in the previous section may be easily extended to classes that are not linearly separable. In fact, as we shall demonstrate, we may use LP to not only develop *nonlinear* separating surfaces but to also design and train (i.e., develop branch weights) equivalent neural networks for the solution of *any* pattern-classification problem.

TABLE 8.4   ILLUSTRATIVE TRAINING SET

Training object	Score 1 $[x_{i,1,k}]$	Score 2 $[x_{i,2,k}]$	Object class
1	0.2	0.5	A
2	0.2	0.8	A
3	0.3	0.4	A
4	0.3	0.7	A
5	0.4	0.3	A
6	0.4	0.9	A
7	0.5	0.3	A
8	0.5	0.8	A
9	0.6	0.4	A
10	0.6	0.6	A
11	0.6	0.7	A
12	0.7	0.5	A
13	0.4	0.5	B
14	0.4	0.6	B
15	0.5	0.5	B
16	0.6	0.9	B
17	0.7	0.8	B
18	0.7	0.9	B
19	0.8	0.7	B
20	0.8	0.8	B

The method to be described is that due to Baek and Ignizio (Ignizio and Baek, 1992). It is, in turn, based upon a number of earlier efforts, and in particular the work of Bennett and Mangasarian (1990) and Roy and Mukhopadhyay (1991). The essence of the Baek and Ignizio method is that of the development of quad-

---

* The approach described in this section is actually a simplication of the actual methodology. Specifically, the approach discussed herein is restricted to relatively small models with appropriately scaled data. Extension to larger problems and less structured data should be obvious.

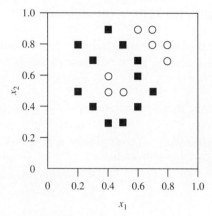

Figure 8.5 Graph of training set.

ratic functions, by means of the solution of sets of LP models, for the determination of

- separating surfaces (designated as *class supermasks*) for all of the objects within each class, and
- separating surfaces (designated as *foreign-object masks*) for all of the "foreign" objects within the supermask of a given class (i.e., where a foreign object is any object within the supermask of a given class that does not belong to that class)

We shall let $m(\mathbf{x})$ represent the general form of the quadratic function for either a supermask or foreign-object mask. To illustrate, in two dimensions, $m(\mathbf{x})$ would take on the following form.

$$m(\mathbf{x}) = m(x_{i,1,k}, x_{i,2,k})$$
$$= v_1 x_{i,1,k}^2 + v_2 x_{i,2,k}^2 + v_3 x_{i,1,k} x_{i,2,k} + v_4 x_{i,1,k} + v_5 x_{i,2,k} + v_6$$

where

v_t = weight to be assigned to term t of the quadratic function

$x_{i,j,k}$ = score achieved by object i, of class k, on attribute j

These supermasks and masks may be developed, for any pattern-classification problem, so as to assure 100% classification of the training data.* Further, we may use the results developed by the LP models to construct a completely trained neural network for pattern classification. We shall clarify this discussion via a numerical example. However, before proceeding to that example, let us state the steps of the Baek and Ignizio algorithm for pattern classification.

---

* But not, of course, of any new data—and this latter deficiency is true for *any* method for pattern classification.

Baek and Ignizio Algorithm for Pattern Classification

It is assumed that we are provided with a training set consisting of K classes of objects. Further, we shall represent the set of indices associated with the objects of class r by $\mathbf{P}_r$.

1. Set $r = 1$, where r is the index associated with the class of objects presently under consideration.

2. *Establishment of a Supermask:* To obtain the supermask of class r, denoted as $s_r(\mathbf{x})$, solve the following LP model

$$\text{minimize} \sum_{i \in \mathbf{P}_r} \rho_i$$

   subject to:

$$s_r(\mathbf{x}) - \rho_i = 0, \qquad i \in \mathbf{P}_r$$

$$v_1 + v_2 + \cdots + v_q = -1$$

   (where q is the dimensionality of the *original* pattern space, that is, 2 in the case of Figure 8.5)

$$\rho_i \geq 0, \qquad i \in \mathbf{P}_r \tag{8.20}$$

   Note that $s_r(\mathbf{x})$ is the *unnormalized* quadratic *supermasking* function for which the coefficients (i.e., the v_t's) are to be determined. The normalized supermasking function, $S_r(\mathbf{x})$, is found by simply dividing each value of v, as determined by the LP solution, by the norm of these weights.

3. *Identification of Foreign Objects:* Let $\mathbf{A}$ be the set of indices for those objects within the supermask of class r and that do not belong to class r, as defined by

$$\mathbf{A} = \left\{ i \,\middle|\, s_r(\mathbf{x}) \geq 0 \text{ and } i \in \left(\bigcup_{\substack{k=1 \\ k \neq r}}^{K} \mathbf{P}_k \right) \right\} \tag{8.21}$$

   (a) Set $u = 1$, $\mathbf{A}^{(1)} = \mathbf{A}$.

   (b) *Construction of Foreign-Object Mask:* To obtain the uth foreign object mask of class r, denoted as $m_{r,u}(\mathbf{x})$, solve the following LP model:

$$\text{minimize} \sum_{i \in \mathbf{A}^{\{u\}}} \eta_i$$

   subject to:

$$m_{r,u}(\mathbf{x}) \leq 0, \qquad i \in \mathbf{P}_r$$

$$m_{r,u}(\mathbf{x}) + \eta_i - \rho_i = 0, \qquad i \in \mathbf{A}^{\{u\}}$$

$$\rho_1 + \rho_2 + \cdots + \rho_m = 1 \tag{8.22}$$

$$\eta_i, \rho_i \geq 0, \qquad i \in \mathbf{A}^{\{u\}}$$

where $m_{r,u}(\mathbf{x})$ is the (unnormalized) quadratic masking function for which the coefficients are to be determined. Again, we normalize these weights to determine the final form of the foreign object mask, that is, $M_{r,u}(\mathbf{x})$.

(c) If the LP model of (8.22) has the solution of $\Sigma\, \eta_i = 0$, go to step 4. Otherwise, proceed to step 3(d), which follows.

(d) Identify the unseparated objects of the set $\mathbf{A}$ by

$$\mathbf{A}^{(u+1)} = \{i\,|\,\eta_i > 0\} \tag{8.23}$$

Set $u = u + 1$ and go to step 3(b).

4. If $r = K$, stop. Otherwise, set $r = r + 1$ and return to step 2.

Once we have employed this algorithm to solve for the quadratic functions (i.e., all of the supermasks and foreign-object masks), we may easily convert these results into an equivalent neural network for pattern classification. And it should be emphasized that the resultant network is fully trained. Such a network equivalent may be represented, in general form, as shown in Figure 8.6.

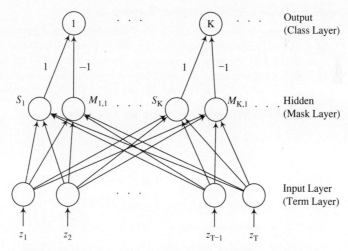

Figure 8.6 Neural network representation: general form.

Notice carefully that the equivalent neural network consists of three layers of nodes. The number of nodes in the output layer will be equal to the number of classes. The number of nodes in the input layer will be equal to the number of terms in the quadratic expression used to define the supermask or foreign-object mask, and is denoted as T. The number of hidden layer nodes will be equal to the number of supermasks and masks that had to be developed so as to define the boundaries of each class. Note further that

S_k represents the supermask node for class k ($k = 1, 2, \ldots, K$)

$M_{k,u}$ represents the uth foreign-object mask node associated with the supermask for class k

z_t represents the input signal presented to the first layer node associated with the tth term of the quadratic function employed in the analysis

Finally, the weights on the various branches are determined as follows:

- The branch weight from input node t to a supermask (or mask) node of the hidden layer is given by the coefficient of the tth term of the quadratic function for the *normalized* supermask (or mask).
- The branch weight from the hidden layer node associated with a supermask to an output layer node is given a value of $+1$.
- The branch weight from the hidden layer node associated with a foreign-object mask to an output layer node is given a value of -1.

Example 8.3: Neural Network Design/Training via LP

In this example, we shall employ the Baek and Ignizio algorithm for the development of a neural network model for the solution of the pattern-classification problem associated with the objects previously listed in Table 8.4 and depicted in Figure 8.5. By employing this algorithm, and beginning with class A (i.e., let $r = 1$, where class A is class 1 and class B is class 2), the first LP model developed is

minimize $\rho_1 + \rho_2 + \rho_3 + \rho_4 + \rho_5 + \rho_6 + \rho_7 + \rho_8 + \rho_9 + \rho_{10} + \rho_{11} + \rho_{12}$

subject to:

$$0.04v_1 + 0.25v_2 + 0.10v_3 + 0.20v_4 + 0.50v_5 + v_6 - \rho_1 = 0$$

$$0.04v_1 + 0.64v_2 + 0.16v_3 + 0.20v_4 + 0.80v_5 + v_6 - \rho_2 = 0$$

$$0.09v_1 + 0.16v_2 + 0.12v_3 + 0.30v_4 + 0.40v_5 + v_6 - \rho_3 = 0$$

$$0.09v_1 + 0.49v_2 + 0.21v_3 + 0.30v_4 + 0.70v_5 + v_6 - \rho_4 = 0$$

$$0.16v_1 + 0.09v_2 + 0.12v_3 + 0.40v_4 + 0.30v_5 + v_6 - \rho_5 = 0$$

$$0.16v_1 + 0.81v_2 + 0.36v_3 + 0.40v_4 + 0.90v_5 + v_6 - \rho_6 = 0$$

$$0.25v_1 + 0.09v_2 + 0.15v_3 + 0.50v_4 + 0.30v_5 + v_6 - \rho_7 = 0$$

$$0.25v_1 + 0.64v_2 + 0.40v_3 + 0.50v_4 + 0.80v_5 + v_6 - \rho_8 = 0$$

$$0.36v_1 + 0.16v_2 + 0.24v_3 + 0.60v_4 + 0.40v_5 + v_6 - \rho_9 = 0$$

$$0.36v_1 + 0.36v_2 + 0.36v_3 + 0.60v_4 + 0.60v_5 + v_6 - \rho_{10} = 0$$

$$0.36v_1 + 0.49v_2 + 0.42v_3 + 0.60v_4 + 0.70v_5 + v_6 - \rho_{11} = 0$$

$$0.49v_1 + 0.25v_2 + 0.35v_3 + 0.70v_4 + 0.50v_5 + v_6 - \rho_{12} = 0$$

$$v_1 + v_2 = -1 \quad \text{and} \quad v_1, v_2, v_3, \ldots, v_6 \text{ unrestricted}$$

$$\rho_i \geq 0, \qquad i = 1, 2, \ldots, 12$$

Notice carefully that, because there are 12 objects in class A, there are 12 associated linear constraints in the model. These constraints, in turn, were formed by the substitution of the scores (i.e., x values from the table) into the following quadratic function:

$$m(\mathbf{x}) = m(x_{i,1,k}, x_{i,2,k})$$
$$= v_1 x_{i,1,k}^2 + v_2 x_{i,2,k}^2 + v_3 x_{i,1,k} x_{i,2,k} + v_4 x_{i,2,k} + v_5 x_{i,2,k} + v_6$$

where v_t is the weight to be assigned to term t of the quadratic function.

Solving this LP, we obtain

$$v_1 = -0.5714, v_2 = -0.4286, v_3 = -0.2143, v_4 = 0.6214, v_5 = 0.6, v_6 = -0.2729$$

After normalizing these results, we obtain the following supermask for the objects of class A:

$$-0.4870x_1^2 - 0.3653x_2^2 - 0.1826x_1x_2 + 0.5296x_1 + 0.5114x_2 - 0.2326 \geq 0$$

Our next step is to determine if there are, within this supermask of class A, any foreign objects. If we substitute the scores of the objects in Table 8.4 into this supermask function, we will note that three of the objects in class B (i.e., objects 13, 14, and 15) are indeed within the supermask of class A (i.e., they satisfy the previous inequality). Thus, these objects are classified as *foreign objects for the class A supermask,* and a new LP model is developed—in an attempt to "filter out" these foreign objects by additional separating surfaces. The LP model for the foreign objects mask is given by

$$\text{minimize } \eta_1 + \eta_2 + \eta_3$$

subject to:

$$0.04v_1 + 0.25v_2 + 0.10v_3 + 0.20v_4 + 0.50v_5 + v_6 \leq 0$$

$$0.04v_1 + 0.64v_2 + 0.16v_3 + 0.20v_4 + 0.80v_5 + v_6 \leq 0$$

$$0.09v_1 + 0.16v_2 + 0.12v_3 + 0.30v_4 + 0.40v_5 + v_6 \leq 0$$

$$0.09v_1 + 0.49v_2 + 0.21v_3 + 0.30v_4 + 0.70v_5 + v_6 \leq 0$$

$$0.16v_1 + 0.09v_2 + 0.12v_3 + 0.40v_4 + 0.30v_5 + v_6 \leq 0$$

$$0.16v_1 + 0.81v_2 + 0.36v_3 + 0.40v_4 + 0.90v_5 + v_6 \leq 0$$

$$0.25v_1 + 0.09v_2 + 0.15v_3 + 0.50v_4 + 0.30v_5 + v_6 \leq 0$$

$$0.25v_1 + 0.64v_2 + 0.40v_3 + 0.50v_4 + 0.80v_5 + v_6 \leq 0$$

$$0.36v_1 + 0.16v_2 + 0.24v_3 + 0.60v_4 + 0.40v_5 + v_6 \leq 0$$

$$0.36v_1 + 0.36v_2 + 0.36v_3 + 0.60v_4 + 0.60v_5 + v_6 \leq 0$$

$$0.36v_1 + 0.49v_2 + 0.42v_3 + 0.60v_4 + 0.70v_5 + v_6 \leq 0$$

$$0.49v_1 + 0.25v_2 + 0.35v_3 + 0.70v_4 + 0.50v_5 + v_6 \leq 0$$

$$0.16v_1 + 0.25v_2 + 0.20v_3 + 0.40v_4 + 0.50v_5 + v_6 + \eta_1 - \rho_1 = 0$$

$$0.16v_1 + 0.36v_2 + 0.24v_3 + 0.40v_4 + 0.60v_5 + v_6 + \eta_2 - \rho_2 = 0$$

$$0.25v_1 + 0.25v_2 + 0.25v_3 + 0.50v_4 + 0.50v_5 + v_6 + \eta_3 - \rho_3 = 0$$

$$\rho_1 + \rho_2 + \rho_3 = 1$$

$$\eta_i, \rho_i \geq 0, \qquad i = 1, 2, 3$$

$$v_t \ (t = 1, 2, \ldots, 6) \text{ unrestricted}$$

The solution of this most recent LP model is given as

$$\eta_1 = \eta_2 = \eta_3 = 0$$

$$v_1 = -7.1429, \ v_2 = -28.5714, \ v_3 = 0, \ v_4 = 6.4286, \ v_5 = 28.5714, \ v_6 = -8.1429$$

As before, we must normalize the weights to produce the actual foreign-object mask. The mask of the three foreign objects within the supermask of class A is then given by

$$-0.1688x_1^2 - 0.6751x_2^2 + 0.1519x_1 + 0.6751x_2 - 0.1924 \geq 0$$

Because $\eta_1 = \eta_2 = \eta_3 = 0$, all foreign objects are masked. We may now plot the supermask and the single associated foreign mask for class A—and this is depicted in Figure 8.7. Notice carefully that a new object is assigned to class A if its test scores lie within the supermask of class A and *not* within any foreign mask for class A. Thus, in Figure 8.7, those objects that fall within the shaded region will be assigned to class A.

This procedure is then repeated for all of the remaining classes under consideration. The results of the application of the approach to class B is depicted in Figure 8.8. Here, the supermask is the ellipse encompassing all objects of class B, and the foreign object mask (encompassing the six foreign objects within the supermask) is

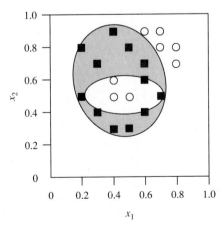

Figure 8.7 Supermask and foreign objects mask for class A.

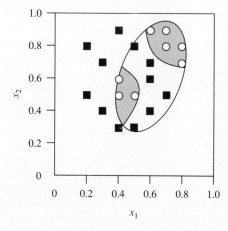

Figure 8.8 Supermask and foreign objects mask for class B.

the unshaded region within the supermask of class B. Thus, any new object within the supermask of B and *not* within the foreign mask (i.e., any object in the shaded area) will be assigned to class B.

Although we have not listed the LP models associated with the supermask and foreign-object mask for class B (because their development follows that already discussed for class A), the results obtained in the solution of those models follow.

The supermask for class B is the ellipsoid defined by

$$-0.4883x_1^2 - 0.4883x_2^2 + 0.3052x_1x_2 + 0.3601x_1 + 0.4883x_2 - 0.2490 \geq 0$$

and the mask for the foreign objects within the supermask of class B is given by

$$0.1939x_1^2 - 0.7758x_1x_2 + 0.5431x_1 + 0.1552x_2 - 0.2037 \geq 0$$

Once we have solved our series of LP models so as to develop the supermasks and associated foreign-object masks, we may then establish the associated neural network representation. The equivalent neural network for this problem is shown in Figure 8.9. Note that in this figure only the weights associated with a portion of the arcs have been specified in order to reduce complexity.

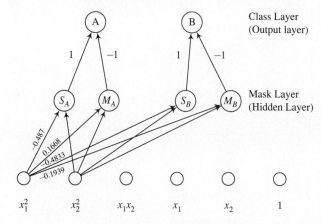

Figure 8.9 The associated neural network for Example 8.3.

Using the neural network of Figure 8.9, we would *hope* to be able to classify all future objects encountered. And it is likely that in most instances this expectation will be satisfied. However, the network (or algorithm) *as illustrated* may not be able to reach a conclusion for any objects that either (1) lie within more than one mask/supermask of different classes (e.g., an object with scores of 0.6, 0.5), or (2) lie within a foreign-object mask of one class and outside of the supermask of any other class (e.g., an object with scores of 0.3, 0.5), or (3) does not lie within any of the masks/supermasks (i.e., an object with scores of 0.4, 0.2). Such objects may be called "ambiguous" objects as it is not clear as to just where they should be assigned—*and such instances will occur regardless of the approach used to classify*. However, in such cases, the problem may be easily rectified by means of certain modifications to the network weights and the inclusion of additional network layers (designated as *maximum selector layers*) that serve to compare the output signals from various sets of nodes and then select only the maximum signal that is output from that node set (and, as a consequence, assign any "ambiguous" object to its "nearest" class). Thus, we may add such maximum selector layers to the outputs of each set of supermask and foreign-object mask nodes (i.e., one layer to the outputs of nodes S_A and M_A and another layer to the outputs of nodes S_B and M_B, in the previous figure). The outputs of these two maximum selector layers are then sent to a final maximum selector layer. However, first, we must, as mentioned, modify the network branch weights. Specifically, the weights between the input layer nodes and the supermask nodes are made the negative of their original values, and the weight on the output link from each internal maximum selector layer to the final maximum selector is set to -1. Figure 8.10 provides an illustration of such a revised network, in general form.

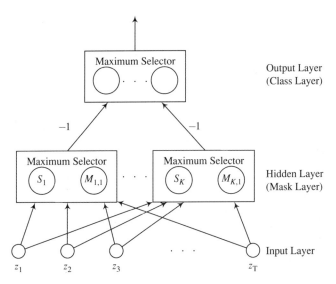

Figure 8.10 Neural network with maximum selectors.

Other Modifications

Before leaving this method, let us remark on yet some further revisions that may be easily accomplished and that may result in some improvement to the Baek and Ignizio procedure (Ignizio, 1992b). First, carefully examine Figure 8.7. Note that some of the objects within the supermask (e.g., the objects for class A) lie relatively deep within the interior of the supermask region. Actually, in most real-world situations, the number of objects of a given class that lie deep within the interior of the class boundaries will be a large number—if not the majority—and thus the situation will be much more pronounced than in our example. As such, these objects are unlikely to play any role whatsoever in the determination of the supermask coefficients.

However, also recall from the Baek and Ignizio algorithm that there is a constraint associated for every object of class r when determining the supermask of class r. Further, any objects that play no part in the determination of the supermask will be represented by nonbinding constraints. Thus, an intuitively appealing concept for the reduction of the supermask LP models is to remove all objects (and thus constraints) that are associated with objects of the class and that are also "near" the centroid of the class. In fact, albeit in but limited experimentation thus far, this concept has served to substantially reduce the size of the LP models necessary to determine a good supermask.

An analogous approach may be used to reduce the size of the LP models associated with the development of foreign-object masks. Here, there are typically but a few foreign objects (and associated foreign-object constraints in the LP model), but there are many supermask class objects (i.e., objects within the supermask that belong to the class associated with the supermask). Typically, however, only the supermask class objects "near" the foreign objects serve to help define the foreign-object mask. For example, in Figure 8.7, only those objects of class A nearest to the three foreign objects of class B serve to define the foreign-object mask illustrated. Thus, it is intuitively appealing to eliminate some of the supermask class objects from the LP model—and this may be done by eliminating those constraints associated with supermask class objects that are "far" away from the centroid of the foreign objects under consideration. Thus, in Figure 8.7, we might decide to eliminate objects 2 and 6 when developing the LP for the determination of the foreign-object masks for foreign objects 13, 14, and 15.

When applying either of the two revisions listed just described, we simply check the results achieved by the solution of our LP models so as to determine whether or not it may be necessary to reinsert any object previously deleted. For example, should we remove certain objects of class r in the determination of the supermask for class r, then once the reduced LP model is solved, we simply check (via substitution) to see if any of the eliminated constraints are violated by the solution to the reduced LP. If so, we may return the objects associated with these violated constraints into the model and solve the new model (there are more systematic and effective means than this for returning constraints, but we will not deal with them here).

There is yet another revision that may be considered. Note that the LP models to be solved have about as many variables as terms in the quadratic expression and about as many rows as objects within the class under consideration. Typically, the number of rows (objects) will greatly exceed the number of variables (terms) and thus it may be useful to solve the dual rather than the primal.

Finally, we should note that the software programs available for the solution of LP problems are now so powerful that it is unlikely that the LP models for a typical pattern classification problem will be of excessive size. Thus, although the revisions described may serve to reduce problem size and computation time, they are unlikely to be absolutely necessary in most situations.

CLUSTER ANALYSIS VIA MATHEMATICAL PROGRAMMING

As mentioned earlier, cluster analysis is often confused with pattern recognition. However, whereas pattern classification deals with the assignment of objects to *predetermined* classes, cluster analysis focuses on the determination of the number of groups into which a set of objects should be partitioned, typically on the basis of some measure of similarity between objects. Further, pattern recognition is a problem of association, whereas that of clustering is one of construction (i.e., the construction of the various clusters). For example, if we have 10 objects and 4 are red and the rest are yellow, it is clear that—on the basis of color alone—the red objects belong to one group and the yellow belong to another. Unfortunately, in most real-world applications of cluster analysis, the determination of the number of groups—and of the similarity measure or measures to employ—is nowhere near as easy as implied in the red and yellow object example. However, there is a variety of ways in which analysts do attempt to determine clusters, or groupings, of objects (or, more precisely, of the data that represent the attributes of the objects)—and the most widely employed of these are strictly heuristic (Therrien, 1989). Because our interest is in linear programming, we will not address such methods. Instead, we shall describe an approach for cluster analysis that employs mathematical models and that, for some types of cluster analysis, is far more systematic.

Most conventional methods of cluster analysis (also known as unsupervised learning) ignore side conditions. However, when employing mathematical programming, the inclusion of such side conditions is quite natural. In order to illustrate this notion, and to illustrate the approach of mathematical programming to cluster analysis, let us focus our attention on a specific example.

Example 8.4: Cluster Analysis with Side Conditions

Let us assume that there are a total of M objects that we wish to assign to clusters (where it is not yet known just how many clusters to use) according to some seemingly natural measure of similarity (e.g., age, physical condition, physical location, and shape). To this point, the problem would appear to be one of conventional cluster analysis. However, let us add some side conditions. First, each of these objects will consume some portion of the resource, or resources (e.g., space, food, and energy) of that cluster to which it is assigned—and each potential cluster will be limited in the amount of such resources. Second, we wish to limit the number of clusters used. Third, and last, we want to minimize the initial costs associated with the assignment of each object to a cluster. Based upon this description, and the data that are associated with this problem, we should be able to construct a mathematical programming model that serves to represent our desire to find the clustering of the objects under consideration. In order to demonstrate such a procedure, let us examine a specific problem of clustering: the *site-location* problem.

Assume that we are faced with the problem of determining just where to locate a number of warehouses so as to best serve a set of M customers. We have narrowed our list of candidate sites to N, and from these we wish to construct no more than K sites. Further, we assume that—*once the specific warehouse sites have been selected*—we will assign each customer to that warehouse to which it is physically nearest. At first blush, this may not sound like a clustering analysis problem to the reader and, in fact, the name originally applied by Ignizio (1968, 1971) to such problems was that of *partial cover problems* (a subset of a class of problems that he denoted as *generalized covering problems*). However, upon closer examination, it should be evident that the site-location problem is indeed a type of clustering problem. Specifically, we wish to assign (i.e., cluster) a number of objects (i.e., customers) to an unspecified number of clusters (i.e., sites).

In our site-location problem, we shall initially assume that we wish to select the locations of up to K warehouses from among N possible sites so as to

- minimize the total initial (e.g., construction) costs of all warehouses deployed
- minimize the sum of the distances from each customer to its assigned warehouse
- assign every customer to the warehouse to which it is physically nearest

We may now proceed to the mathematical formulation of this problem. This model follows.

Select *up to* K warehouse sites, from among M candidate sites, so as to

$$\text{minimize} \sum_{j=1}^{N} c_j y_j \tag{8.24}$$

$$\text{minimize} \sum_{i=1}^{M} \sum_{j=1}^{N} d_{ij} x_{ij} \tag{8.25}$$

subject to:

$$\sum_{j=1}^{N} y_j \le K \tag{8.26}$$

$$\sum_{j=1}^{N} x_{ij} = 1, \qquad \text{for all } i \tag{8.27}$$

$$x_{i,j} - y_j \le 0, \qquad \text{for all } i, j \tag{8.28}$$

where

$x_{i,j}$ = 1 if customer i is assigned to site j, and 0 otherwise

y_j = 1 if site j is selected as a warehouse site, and 0 otherwise

c_j = initial cost of a warehouse at site j

$d_{i,j}$ = distance from customer i to site j

Notice carefully that the last two constraints in this model serve to assign each customer to one site, *but only to a site that has actually been selected* (i.e., if the last constraint were omitted, the preceding constraint could be satisfied by assigning some customers to nonexisting warehouses). However, the reader may still find the resultant model somewhat unsettling as it has two objective functions rather than a single one. In Part 3, we shall learn how to deal with such multiple objective models (models that are found in many, if not most real-world situations). Here, however, we may note that either one of the objectives may be transformed into a "constraint" so as to develop a more conventional model. For example, if we set a maximum budget (B) for the initial costs of the warehouses, then the first objective may be converted into the following "constraint":

$$\sum_{j=1}^{N} c_j y_j \le B \tag{8.29}$$

Even with conversion to a single objective model, the problem is still not a conventional LP as the variables are restricted to values of just 0 or 1. In Part 2, we describe methods for solving such 0–1 models—both exact and heuristic. In fact, we shall return to this same site-selection/clustering problem in Chapter 12 and discuss just how it may be approached via the methodology of heuristic programming.

Returning to our example, we may easily include a number of other features within the model. For example, we may wish to not assign any more customers to a given warehouse than that warehouse is able to handle. Or we may wish to minimize the maximum distance from any customer to the warehouse to which it is assigned. The first situation may be handled by including a constraint of the following form:

$$\sum_{j=1}^{N} a_{ij} x_{ij} \le A_j \tag{8.30}$$

where

A_j = total amount of resource j at any warehouse to be located at site j

$a_{i,j}$ = level of resource j required by customer i

The second consideration may be modeled as

$$\text{minimize } \delta \tag{8.31}$$

subject to:

$$d_{i,j} x_{i,j} - \delta \le 0, \qquad \text{for all } i \text{ and } j \tag{8.32}$$

where δ is the maximum distance between a customer and the warehouse site to which it will be assigned.

Typically, in cluster analysis, one attempts to determine some appropriate measure, or measures, of similarity between the objects to be clustered. In the previous example, the measure used was the physical distance between the objects (customers) and the site (cluster location) that "served" these objects. A more general way in which to express the desire to cluster according to similarity is by means of a *similarity matrix*. Here, each matrix row or column is headed by an object. The elements of the matrix are then the similarity measures (which may be developed according to a host of philosophies—of which fuzzy calculus seems to predominate recent literature). For sake of illustration, let us assume that just one measure of similarity is being employed, where a value of 1 indicates the highest level (i.e., the two objects under consideration are, with respect to that measure, completely similar) and a value of 0 denotes a pair of objects that are completely dissimilar. Any values between 0 and 1 indicate varying levels of similarity between object pairs. If

$s_{i,j}$ = measure of similarity between objects i and j

$x_{i,k}$ = 1 if object i is assigned to cluster k, and 0 otherwise

Then we may let

$$S_k = \min \{s_{i,j}x_{i,k}x_{j,k}|x_{i,k}x_{j,k} = 1\} \tag{8.33}$$

$$S = \min_k \{S_k\} \tag{8.34}$$

where

S = minimum level of similarity found in any of the clusters developed

S_k = minimum level of similarity found in cluster k

It should be obvious then that we seek to maximize the value of S, that is, to maximize the minimum level of similarity encountered across all clusters that are developed (i.e., active clusters). To accomplish this, we may use the following submodel:

$$\text{minimize } \lambda = 1 - S \tag{8.35}$$

subject to:

$$s_{i,j}x_{i,k}x_{j,k} + 1 \geq x_{i,k}x_{j,k}, \qquad \text{for all } i, j, k, \text{ where } j > i \tag{8.36}$$

The solution to this submodel is that assignment of objects to clusters that serve to minimize the maximum dissimilarity (or maximize the minimum similarity) of the objects in the clusters. Thus, this submodel, along with whatever constraints may be necessary (such as that each object must be assigned to precisely one active cluster, as illustrated earlier), may be used to represent a fairly general type of clustering problem. Of course, we now have an additional element of nonlinearity (note that the formulations in the previous example were actually nonlinear as they require 0–1 variables), that is, the product terms found in the constraints of the most recent formulation.

INPUT–OUTPUT ANALYSIS AND LP

Input–output (or interindustry) analysis is a topic that predates linear programming (Leontief, 1951). However, it is an analysis that may be enhanced via LP. Further, it fits well into the area of decision science as input–output analysis is a tool for the *prediction* of the response of industries—or entire nations—to changes in one or more sectors of production. Although input–output analysis is a relatively old tool, we believe that the adaptation of LP to this methodology for forecasting fits well within the purpose of this chapter.

The fundamental idea underlying input–output (I/O) analysis is that there is a high degree of interdependence among the goods and services produced in an industrial economy. Consequently, an increase or decrease in one sector of production may have an effect on other sectors. There are two basic types of I/O models: the static model and the dynamic model. The static model, which is the one we discuss, deals with but a single time period, whereas the dynamic model investigates changes in the economy over several periods of time.

The heart of the I/O model is the *transactions table*. Each element in this table represents a total sales activity for each industry during the period of interest. Consider, for example, the transactions table of Table 8.5. For simplicity, only three industries are considered to make up the economy: industries I, II, and III. The first row of the table denotes the activities of industry I. Industry I produced a grand total of 100 (monetary) units of goods of which

- 40 units were sold to the consumer
- 10 units were sold to industry II
- 30 units were sold to industry III
- 20 units were used by industry I itself for its own production processes

Rows II and III are interpreted in a similar manner. The "households" row is, however, a "catchall" used to represent wages, capital services, and so forth. As such, it is something akin to a fourth "industry." Industries I, II, and III require 50, 70, and 20 units, respectively, of the households output for their production.

TABLE 8.5 TRANSACTIONS TABLE

From	To			Consumers	Total
	I	II	III		
I	20	10	30	40	100
II	10	20	20	60	110
III	20	10	10	40	80
Households	50	70	20	0	140
Total	100	110	80	140	

The elements of the columns represent the total purchase of an industry from the other industries so as to support its operations during the base period.

We shall next address the interior of the transactions table (the portion of the table concerned only with industries I, II, and III) and the total output row on the bottom. Dividing each column of this interior matrix by the associated total amount results in a matrix known as the technical coefficients matrix, as shown in Table 8.6. To illustrate, the coefficient in the first row and first column of Table 8.6 is found by dividing 20 by 100 (i.e., from the first column in Table 8.5).

TABLE 8.6 TECHNICAL
COEFFICIENTS MATRIX

	I	II	III
I	0.200	0.091	0.375
II	0.100	0.182	0.250
III	0.200	0.091	0.125

The elements of the technical coefficients matrix then represent the per unit contribution of each industry. For example, consider the first column and second row element in this table, which is 0.100. This indicates that, for an output of 1 unit from industry I, an input of 0.100 unit is required from industry II. For convenience, we designate the elements of the technical coefficients matrix as $a_{i,j}$, where

$\mathbf{A}$ = matrix of technical coefficients

$a_{i,j}$ = amount of industry i that is necessary to produce 1 unit of commodity j

We may also define

$$x_1 = \text{total output of industry I}$$

$$x_2 = \text{total output of industry II}$$

$$x_3 = \text{total output of industry III}$$

Thus, for the period shown in Table 8.5, the values of x_1, x_2, and x_3 are 100, 110, and 80, respectively.

The variables x_1, x_2, and x_3 make up the output vector, which we designate as $\mathbf{x}_k$, where k is the time period under consideration. Thus, for the initial period $k = 0$, then

$$\mathbf{x}_0 = \begin{bmatrix} 100 \\ 110 \\ 80 \end{bmatrix}$$

Next, consider the column under "consumers" in Table 8.5. This is the "demand" for each commodity. We let

y_1 = demand for commodity one (industry I)
y_2 = demand for commodity two (industry II)
y_3 = demand for commodity three (industry III)

Thus, a demand vector, $\mathbf{y}_k$, may be introduced, where, for our example:

$$\mathbf{y}_0 = \begin{bmatrix} 40 \\ 60 \\ 40 \end{bmatrix} = \begin{bmatrix} y_1 \\ y_2 \\ y_3 \end{bmatrix}, \qquad \text{at } k = 0$$

For the base period, we may write the representative equation of each industry (where I = 1, II = 2, and III = 3 in the notation to follow):

Industry I: $\quad x_1 - a_{1,1}x_1 - a_{1,2}x_2 - a_{1,3}x_3 = y_1$

Industry II: $\quad x_2 - a_{2,1}x_1 - a_{2,2}x_2 - a_{2,3}x_3 = y_2$

Industry III: $\quad x_3 - a_{3,1}x_1 - a_{3,2}x_2 - a_{3,3}x_3 = y_3$

Or, for the specific values in our example:

$$x_1 - 0.2x_1 - 0.091x_2 - 0.375x_3 = 40$$

$$x_2 - 0.1x_1 - 0.182x_2 - 0.250x_3 = 60$$

$$x_3 - 0.2x_1 - 0.091x_2 - 0.125x_3 = 40$$

This may be rewritten in general matrix form as

$$(\mathbf{I} - \mathbf{A})\mathbf{x} = \mathbf{y} \tag{8.37}$$

The matrix $(\mathbf{I} - \mathbf{A})$ is known as the Leontief matrix. Under the assumption that the underlying structure of the economy is linear and that the technical coefficients do not change with time, we can determine a future production vector $\mathbf{x}_k$ given a demand vector $\mathbf{y}_k$. That is, we solve

$$(\mathbf{I} - \mathbf{A})\mathbf{x} = \mathbf{y} \tag{8.38}$$

$$\mathbf{x} \geq \mathbf{0} \tag{8.39}$$

If the Leontief matrix is nonsingular, then

$$\mathbf{x} = (\mathbf{I} - \mathbf{A})^{-1}\mathbf{y} \tag{8.40}$$

Up to this point, the I/O analysis has been performed using only (very basic) linear algebra. However, by means of linear programming, we may extend the scope of the methodology. Let us assume, for example, that there is a profit c_i associated with the output of each unit of industry i. We might then wish to maximize the total profit by finding the optimal production output of all industries. That is, we seek to find $\mathbf{x}$ so as to

$$\text{maximize } \mathbf{cx}$$

$$\text{subject to:}$$

$$(\mathbf{I} - \mathbf{A})\mathbf{x} \leq \mathbf{y}$$

More realistically, the amount produced by each industry has an upper limit, u_i, and thus we want to find $\mathbf{x}$ to

$$\text{maximize } \mathbf{cx} \tag{8.41}$$

$$\text{subject to:}$$

$$(\mathbf{I} - \mathbf{A})\mathbf{x} \leq \mathbf{y} \tag{8.42}$$

$$\mathbf{x} \leq \mathbf{u} \tag{8.43}$$

where

$$\mathbf{u} = \begin{bmatrix} u_1 \\ \vdots \\ u_m \end{bmatrix}$$

Numerous other variations are possible, and this is why the LP approach to I/O analysis is so attractive.

Example 8.5: LP applied to I/O analysis

Given the economy model of Tables 8.5 and 8.6, let us assume that 3 years from now the demand and production limit estimates are given by

$$\mathbf{y}_3 = \begin{bmatrix} 70 \\ 50 \\ 50 \end{bmatrix}$$

$$\mathbf{u} = \begin{bmatrix} 120 \\ 120 \\ 95 \end{bmatrix}$$

Further, it is desired to maximize the amount of units produced by industries I and III. Thus, the associated LP model is

$$\text{maximize } x_1 + x_3$$

$$\text{subject to:}$$

$$x_1 - 0.2x_1 - 0.091x_2 - 0.375x_3 \leq 70$$

$$x_2 - 0.1x_1 - 0.182x_2 - 0.250x_3 \leq 50$$

$$x_3 - 0.2x_1 - 0.091x_2 - 0.125x_3 \leq 50$$

$$x_1 \leq 120$$

$$x_2 \leq 120$$

$$x_3 \le 95$$

$$x_1, x_2, x_3 \ge 0$$

Solving this model (an exercise left for the reader), we may then maximize the output of industries I and II while staying within the predicted demand and production limits.

SIMULATION OF CONTINUOUS PROCESSING SYSTEMS

Simulation is an important and extremely useful tool for the support of decision making. However, most methods of simulation discussed in the literature focus primarily on the simulation of *discrete* processing systems (e.g., the manufacture of cars, refrigerators, and other discrete objects) and are much less effective in dealing with *continuous* processing systems (e.g., the production of chemicals, beverages, and gasoline). Further, most of the approaches described are Monte Carlo–based procedures. However, one may develop simulations of continuous processing systems through the employment of those tools (including LP) used in the development of predictive functions. Specifically, we develop those relationships that serve to predict the response (i.e., output or outputs) of each component of the system to be simulated as a function of the input, or inputs, to that component. Thus, in place of these actual components, we now have a set of predictive functions, tied together in a manner that replicates both the flow and logic of the system to be simulated. Consider, for example, the simulation of a simple chemical processing system consisting of but two processing units and three storage tanks.

The base stock (liquid raw material) is contained within storage tank 1 and it flows, through a single pipe whose flow rate may be adjusted over a specified range, to reactor A. The flow out of reactor A may go directly to reactor B or be diverted to an intermediate storage tank, tank 2 (and then, whenever desired and possible, on to reactor B). Finally, the flow out of reactor B is sent directly to the final storage tank, tank 3. We also know that the output of each reactor is a function of a number of variables (e.g., input flow rate, reactor pressure, and reactor temperature). Thus, one may use an appropriate method (e.g., LP) to develop a function that serves to predict the reactor output attributes (e.g., output flow rate and product viscosity) as a function of the input variables. It should then be apparent that one may construct a simulation of the complete system using the predictive functions as major components of the overall concept.

SUMMARY AND CONCLUSIONS

Typically, one thinks of LP as simply a tool for the solution of problems that may be represented by a linear objective and set of linear constraints. That is, we often view LP as just a means to develop a *program*. Such a perspective is, however, self-limiting as it is possible—and often very beneficial—to use LP to develop *models,* in particular, models for prediction and classification. We have addressed several examples of this use of LP in this chapter; as well as indicating an approach to the mathematical modeling of yet another type of problem: cluster analysis. We then concluded our discussion with a brief note on the use of LP in I/O analysis, and on the extension of the use of predictive functions to the simulation of continuous processing systems. We hope that exposure to these applications will engender further interest, on the part of the reader, in the use of LP in support of the processing and analysis of data.

As one final note, we would like to point out that there is yet another departure of this chapter from those that have preceded it. Specifically, in our previous chapters, we focused on *optimization,* that is, the development of a solution better than any other for a given linear programming model. Here, however, our attention has been directed toward the development of models (i.e., for prediction, classification, and clustering), and it should be clear that the predictions, classifications, and clusters developed by such models cannot be guaranteed to be optimal. This is because, and although we certainly used an optimizing tool (i.e., linear programming), the data used to develop these models only represent a sample (and usually a very small sample) of the total population. Further, the forms of the models used (e.g., linear predictive functions, polynomial predictive functions, and quadratic separating surfaces) are but guesses as to the actual nature of such forms.

EXERCISES

8.1. A firm wishes to determine its production rate over the next 6 months. Expected demand for its product is 30, 80, 60, 50, 80, and 100 units for months 1 through 6, respectively. The product is perishable and thus cannot be stored from month to month. The loss per unit not sold is $8 and the loss per unit short is estimated to be $5. Because of the nature of the production process, production output must be maintained at some *linear* function over the time period. That is, if we begin with a production rate of, say, 50 units and end with a rate of 100 units, the production rate for any intermediate month must lie on the straight line between 50 and 100. Formulate this problem and determine its model to be developed if we wish to minimize total costs. You may assume that fractional production rates are permissible.

8.2. Formulate and solve Exercise 8.1 if we desire to minimize the maximum loss in any one month.

8.3. Compare the results obtained in Exercise 8.2 with that which would be achieved using the least squares method to fit the production rate response.

8.4. For Example 8.1, develop the CER when all coefficients of the function are required to be *nonnegative*. Compare the resulting residuals with those developed previously and comment on the results.

8.5. Use the revised version (i.e., as listed in this chapter) of the Freed and Glover linear discriminant analysis approach to develop a discriminant line for the data of Table 8.4. Comment on the results achieved and compare them with that achieved by the method of Baek and Ignizio.

8.6. Using the Baek and Ignizio algorithm for pattern classification, develop the neural network classifier that may be established for the training set shown in Table 8.7. Comment on the regions formed by the supermasks and foreign-object masks.

TABLE 8.7

Training object	Score 1 $[x_{i,1,k}]$	Score 2 $[x_{i,2,k}]$	Object class
1	0.2	1.0	A
2	0.2	0.8	A
3	0.3	0.9	A
4	0.4	0.8	A
5	0.4	1.0	A
6	0.8	0.4	A
7	0.8	0.2	A
8	0.9	0.3	A
9	1.0	0.2	A
10	1.0	0.4	A
11	0.2	0.2	B
12	0.2	0.4	B
13	0.3	0.3	B
14	0.4	0.2	B
15	0.4	0.4	B
16	0.8	0.8	B
17	0.8	1.0	B
18	0.9	0.9	B
19	1.0	0.8	B
20	1.0	1.0	B

8.7. Graph the data set for Exercise 8.6 and comment on the appropriateness of the development of a linear discriminant function for the separation of the two classes.

8.8. Using the graph developed for Exercise 8.7, assume that the class assignment—and even the number of classes—was unknown for this data set. How many clusters would you suspect existed, and for which objects, just by a visual inspection of the graph? What was the basis of your conclusion?

8.9. List a possible application for cluster analysis in the medical profession. Describe just what might be accomplished via the use of clustering.

8.10. Solve Example 8.5, the I/O model, and discuss your results.

8.11. Comment on just how the applications of LP in prediction and pattern classification differ from those in cluster analysis and I/O analysis.

PART **II** NETWORK AND INTEGER MODELS

CHAPTER ═══ **9**

THE NETWORK SIMPLEX METHOD

CHAPTER OVERVIEW

In this chapter, we introduce a class of linear programming problems with a special structure. This special structure, often referred to as a network structure, not only allows this class of problems to be viewed using a graphical representation, but also enables the problems to be solved in a very efficient manner. That is, even though these network structured problems can be solved using the standard simplex algorithm, it is far more efficient to develop an algorithm that exploits the special network structure. In fact, this special algorithm, which is called the *network simplex method,* is typically 200 to 300 times faster than the standard simplex method in solving large network structured problems. In addition, these problems exhibit a property known as *unimodularity.* A problem involving only integer data and having the unimodularity property will always yield a strictly integer solution when solved by the simplex method. That is, such a problem has extreme points (basic solutions) that are all strictly integer-valued.

Following the introduction of some basic network terminology, this chapter discusses the *minimum-cost network flow problem.* The special properties of this network model are then characterized, which leads to the development of the network simplex method. The chapter concludes with a discussion of a method for handling network flows with arc capacities and with a presentation of some typical applications.

NETWORK TERMINOLOGY

Before describing the network flow model that will be considered throughout this chapter, consider the following definitions of terms commonly employed in graph and network theory.

> **GRAPH** A *graph* $G(N, A)$ consists of a set of *nodes* (or vertices) $N = \{1, 2, \ldots, m\}$ and a set of *arcs* (or links) $A = \{(i, j), (k, l), \ldots, (p, q)\}$ connecting some of the node pairs in N. The arcs can be either *directed* or *undirected*. If all of the arcs are directed, then G is called a *directed graph* or *digraph*. If all of the arcs are undirected, then G is called an *undirected graph*. Finally, if G contains both undirected and directed arcs, then G is referred to as a *mixed graph*.

In this chapter, we will be devoting our attention to the study of directed graphs. Figure 9.1 gives an example of a directed graph. Note that the arc notation (i, j) refers to an arc that is directed *from* node i *to* node j. We say that arc (i, j) is *incident* to nodes i and j. Node i is called the *tail node* and node j is the *head node*. For example, in Figure 9.1, arc $(1, 2)$ is directed from node 1 to node 2. Finally, the *degree* of a node is the number of arcs incident to it. For example, in the graph of Figure 9.1, nodes 1 and 6 have degree 2, and nodes 2 and 5 have degree 4.

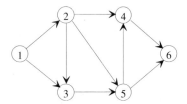

Figure 9.1 A directed graph.

> **PATH** A *path* is a sequence of nodes $i_0, i_1, i_2, \ldots, i_p$ connected by the directed arcs $(i_0, i_1), (i_1, i_2), \ldots, (i_{p-1}, i_p)$. A path is *simple* if each node appears only once in the sequence.
> **CIRCUIT** A *circuit* is a closed path. That is, a circuit consists of a path from node i_0 to node i_p along with the arc (i_p, i_0).

Figures 9.2(a) and 9.2(b) illustrate a path and a circuit, respectively; note that all arcs are oriented in the same direction.

> **CHAIN** A *chain* has the same basic structure as a path except that not all arcs are necessarily directed toward node i_p. A chain is *simple* if each node appears only once in the sequence.
> **CYCLE** A *cycle* is a closed chain.

A simple chain and a cycle are illustrated in Figures 9.2(c) and 9.2(d), respectively.

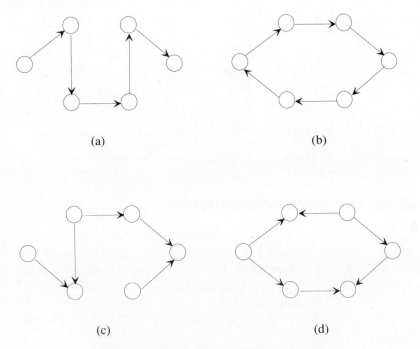

(a)   (b)

(c)   (d)

Figure 9.2   Path, circuit, chain, and cycle.

CONNECTED GRAPH   A graph $G(N, A)$ is *connected* if there exists a chain between every pair of nodes in N.

SUBGRAPH   A graph $G'(N', A')$ is a *subgraph* of $G(N, A)$ if $N' \subset N$ and $A' \subset A$. It is also understood that if arc $(i, j) \in A'$, then $i, j \in N'$.

SPANNING SUBGRAPH   A graph $G'(N', A')$ is a *spanning subgraph* of $G(N, A)$ if G' is a subgraph of G and $N' = N$, that is, G' includes all of the nodes of G.

TREE   There are numerous ways of defining a tree graph, and it is quite easily shown that the following definitions are equivalent:

 (i)   A *tree* is a connected graph with no cycles.
 (ii)   A *tree* is a graph with no cycles, m nodes, and $m - 1$ arcs.
(iii)   A *tree* is a connected graph with m nodes and $m - 1$ arcs.

SPANNING TREE   A connected spanning subgraph with no cycles is a *spanning tree*.

As we will see in subsequent sections, spanning trees play a fundamental role in the development of the network simplex method. An example of a spanning tree of a graph is illustrated in Figure 9.3. Note that tree T has $m = 6$ nodes and $m - 1 = 5$ arcs. Obviously, for a general graph, there can be a multitude of spanning trees. In fact, for a complete graph (i.e., a graph with an arc between every node pair) with m nodes, Cayley (1874) showed that there are m^{m-2} distinct spanning trees.

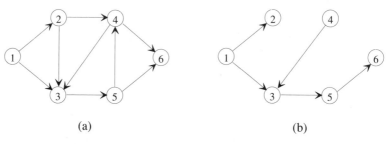

(a)   (b)

Figure 9.3 (a) Graph G and (b) spanning tree T.

MINIMUM-COST NETWORK FLOW PROBLEMS

Let $G(N, A)$ be a directed graph and assume that associated with each node $i \in N = \{1, \ldots, m\}$, there is a quantity b_i. If $b_i > 0$ for a particular node i, then we will assume that b_i represents the supply of a certain commodity at node i. Similarly, if $b_i < 0$, then b_i is the demand at node i, and, finally, if $b_i = 0$, then node i represents a transshipment node. Without loss of generality, we will assume throughout this chapter that total supply equals total demand, or, mathematically, that

$$\sum_{i=1}^{m} b_i = 0 \tag{9.1}$$

If this is not the case, we can balance supply and demand by adding a dummy demand node or a dummy supply node. For example, if $\sum_{i=1}^{m} b > 0$, then supply exceeds demand, and we would add an additional demand node $m + 1$ with $b_{m+1} = -\sum_{i=1}^{m} b_i$. This is illustrated graphically in Figure 9.4, where node 5 represents the dummy demand node. If the solution to this modified problem specifies that supply node k ships some positive quantity to the dummy demand node, then we would interpret this as excess demand that is simply stored or retained at node k.

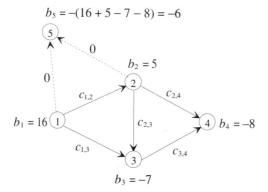

Figure 9.4 Adding a dummy demand node.

An analogous argument can be made if demand exceeds supply, except that in this case, we will encounter shortages instead of excesses.

In addition to the quantity b_i associated with each node, there is a unit cost per flow, c_{ij}, associated with each arc. The problem is to determine the least-cost flows in the network that satisfy the total demand, where the decision variables are given by

$$x_{ij} = \text{quantity of flow from node } i \text{ to node } j \text{ along arc } (i, j) \in A$$

If a dummy node has been added to balance supply/demand, then the unit costs on the additional arcs should be assigned according to the situation. For example, if only transportation costs are considered in the model, then the cost associated with each arc connected to the dummy node will be zero (as noted in Figure 9.4), because no actual shipping occurs along these arcs. However, in some instances when a dummy demand node has been added, it may be appropriate to assign a unit cost to each dummy arc that is equal to the unit inventory cost because the units "shipped" along the dummy arcs are actually held in inventory at the respective supply nodes. Similarly, the arc costs may reflect lost opportunity costs in the case of a dummy supply node.

The flow out of node i is given by $\sum_{j_{(i,j)\in A}} x_{i,j}$, whereas the flow into node i is given by $\sum_{j_{(j,i)\in A}} x_{j,i}$. By flow conservation, the flow out of node i minus the flow into node i must equal the supply (demand) at node i. Thus, for each node i, we must have

$$\sum_{\substack{j \\ (i,j)\in A}} x_{i,j} - \sum_{\substack{j \\ (j,i)\in A}} x_{j,i} = b_i \tag{9.2}$$

These flow conservation equations together with the objective of finding the least-cost solution leads directly to the linear programming formulation of the minimum-cost flow problem:

$$\text{minimize } z = \sum_{(i,j)\in A} c_{i,j} x_{i,j} \tag{9.3}$$

subject to:

$$\sum_{\substack{j \\ (i,j)\in A}} x_{i,j} - \sum_{\substack{j \\ (j,i)\in A}} x_{j,i} = b_i, \qquad \text{for all } i \in N \tag{9.4}$$

$$x_{i,j} \geq 0, \qquad \text{for all } (i, j) \in A \tag{9.5}$$

To illustrate this formulation and to begin to study the structure of the underlying problem, consider the following simple example.

Example 9.1: A Minimum-Cost Network Flow Problem

Consider the network flow problem given in Figure 9.5.

First, note that, for the graph $G(N, A)$ of Figure 9.5, $N = \{1, 2, 3, 4\}$ and $A = \{(1, 2), (1, 3), (2, 3), (2, 4), (3, 4)\}$. Next, observe that $\sum_{i=1}^{m} b_i = 10 + 0 - 3 - 7 = 0$ (i.e., supply = demand). The minimum-cost network flow problem can then be

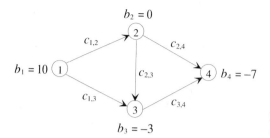

Figure 9.5 Network flow problem of Example 9.1.

written as

$$\text{minimize } z = c_{1,2}x_{1,2} + c_{1,3}x_{1,3} + c_{2,3}x_{2,3} + c_{2,4}x_{2,4} + c_{3,4}x_{3,4} \tag{9.6}$$

subject to:

$$x_{1,2} + x_{1,3} = 10 \tag{9.7}$$

$$x_{2,3} + x_{2,4} - x_{1,2} = 0 \tag{9.8}$$

$$x_{3,4} - x_{1,3} - x_{2,3} = -3 \tag{9.9}$$

$$-x_{2,4} - x_{3,4} = -7 \tag{9.10}$$

$$x_{i,j} \geq 0, \qquad \text{for all } (i, j) \in A \tag{9.11}$$

The objective function in (9.6) gives the total cost of satisfying the demands and constraints (9.7–9.10) correspond to flow conservation at nodes 1 through 4, respectively.

Now consider the constraint matrix corresponding to (9.7–9.10):

$$
\mathbf{A}^0 = \begin{array}{c} \\ node \\ 1 \\ 2 \\ 3 \\ 4 \end{array}
\begin{array}{c}
\\ (1, 2) \\ \left(\begin{array}{c} 1 \\ -1 \\ 0 \\ 0 \end{array} \right.
\end{array}
\begin{array}{c}
arc \\ (1, 3) \\ 1 \\ 0 \\ -1 \\ 0
\end{array}
\begin{array}{c}
\\ (2, 3) \\ 0 \\ 1 \\ -1 \\ 0
\end{array}
\begin{array}{c}
\\ (2, 4) \\ 0 \\ 1 \\ 0 \\ -1
\end{array}
\begin{array}{c}
\\ (3, 4) \\ 0 \\ 0 \\ 1 \\ -1
\end{array}
\left. \begin{array}{c} \\ \\ \\ \\ \end{array} \right)
$$

Note that each row of $\mathbf{A}^0$ corresponds to a node and each column corresponds to an arc. As a consequence, each column of $\mathbf{A}^0$ contains precisely two nonzero entries: $+1$ appears in the row corresponding to the tail node of the arc and -1 appears in the row corresponding to the head node. Because of this unique structure, $\mathbf{A}^0$ is referred to as a *node–arc incidence matrix*. By letting $\mathbf{a}_{i,j}$ represent the column associated with arc (i, j), note that $\mathbf{a}_{i,j}$ can be represented as

$$\mathbf{a}_{i,j} = \mathbf{e}_i - \mathbf{e}_j \tag{9.12}$$

where the notation $\mathbf{e}_k$ refers to a vector of all zeros except for a 1 in the kth position. For example, consider $\mathbf{a}_{2,3}$, the column corresponding to arc $(2, 3)$ in

Example 9.1. Then

$$\mathbf{a}_{2,3} = \begin{pmatrix} 0 \\ 1 \\ -1 \\ 0 \end{pmatrix} = \begin{pmatrix} 0 \\ 1 \\ 0 \\ 0 \end{pmatrix} - \begin{pmatrix} 0 \\ 0 \\ 1 \\ 0 \end{pmatrix} = \mathbf{e}_2 - \mathbf{e}_3$$

Observe that summing the rows of $\mathbf{A}^0$ results in the zero vector, and thus the rows of $\mathbf{A}^0$ are linearly dependent. Noting that $\mathbf{A}^0$ has m rows, then clearly rank $(\mathbf{A}^0) < m$. In fact, as we will prove in the following section, rank $(\mathbf{A}^0) = m - 1$.

However, to apply the simplex algorithm, we need to ensure that the coefficient matrix has full row rank. This can be accomplished by either deleting a row of $\mathbf{A}^0$ or adding a linearly independent column to $\mathbf{A}^0$. For our development, we will choose the later. Because each column corresponds to a variable (arc), this additional column can actually be thought of as an artificial variable. Without loss of generality, we will assume that this variable corresponds to node m with the corresponding column

$$\mathbf{e}_m = \begin{pmatrix} 0 \\ 0 \\ \vdots \\ 0 \\ 1 \end{pmatrix}$$

where 1 appears in the mth row corresponding to node m. (The choice of row m was actually an arbitrary one, and any node will work equally as well.) Because this new column has only a single nonzero entry, we can visualize it as an arc with node m as the tail node with the arc being directed into space. For the network of Example 9.1, this would be depicted as in Figure 9.6. Because of this graphical interpretation of the artificial variable, it is called the *root arc* and the associated node is called the *root node*. By assuming that the root arc is incident to node m, the associated variable will be designated by x_m.

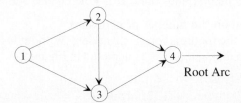

Root Arc

Figure 9.6 Network flow problem with root arc.

Also, because the network flow problem is assumed to be balanced (i.e., supply = demand), the flow on the root arc will always be zero (i.e., $x_m = 0$). Thus, we need not make a special effort to force the artificial variable x_m to zero because this will happen automatically. The sole purpose of the root arc is to enable us to extract a nonsingular $m \times m$ basis submatrix from the matrix $\mathbf{A} =$

$(\mathbf{A}^0, \mathbf{e}_m)$. Because rank $(\mathbf{A}^0) = m - 1$, the column associated with the root arc will be included in every basis submatrix of $\mathbf{A}$.

Matrix $\mathbf{A}$ for the problem of Example 9.1 can be represented as follows:

$$\mathbf{A} = \begin{array}{c} \\ 1 \\ 2 \\ 3 \\ 4 \end{array} \begin{pmatrix} (1,2) & (1,3) & (2,3) & (2,4) & (3,4) & \text{root arc} \\ 1 & 1 & 0 & 0 & 0 & 0 \\ -1 & 0 & 1 & 1 & 0 & 0 \\ 0 & -1 & -1 & 0 & 1 & 0 \\ 0 & 0 & 0 & -1 & -1 & 1 \end{pmatrix}$$

By using this notation, the minimum-cost flow problem of (9.3–9.5) can also be written in compact form:

$$\text{minimize } z = \mathbf{cx} \tag{9.13}$$

subject to:

$$\mathbf{Ax} = \mathbf{b}$$

$$\mathbf{x} \geq \mathbf{0}$$

Bases and Rooted Spanning Trees

One of the keys to developing an efficient algorithm for this class of linear programming problems is establishing a relationship between the algebraic and graphical representations of basic solutions. In particular, one of the most important relationships is the one that exists between basis matrices and rooted spanning trees. We begin our investigation of this relationship via the following theorem.

Theorem 9.1 (Every rooted spanning tree is a basis)
Let T be a rooted spanning tree for the graph $G(N, A)$. Define an $m \times m$ submatrix $\mathbf{B}$ of $\mathbf{A}$ that has the $m - 1$ columns of $\mathbf{A}$ associated with the $m - 1$ arcs of T and the artificial column associated with the root arc. Then $\mathbf{B}$ is lower triangular with nonzero diagonal elements, and, thus, $\mathbf{B}$ is nonsingular and a basis matrix.

Proof (By induction on the number of nodes m)
Let us begin by verifying the theorem for $m = 2$ nodes. Without loss of generality, assume that the rooted spanning tree T is as depicted in Figure 9.7. Then matrix $\mathbf{B}$ will be defined as

$$\mathbf{B} = \begin{array}{c} \\ 1 \\ 2 \end{array} \begin{array}{c} (1,2) \quad \text{root arc} \\ \begin{pmatrix} 1 & 0 \\ -1 & 1 \end{pmatrix} \end{array}$$

and, clearly, $\mathbf{B}$ is lower triangular nonsingular.

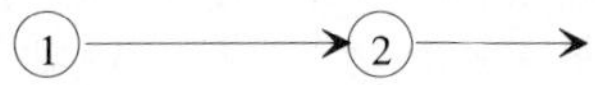

Figure 9.7 Rooted spanning tree for $m = 2$ nodes.

Now, define the induction hypothesis. That is, assume that the theorem is true for graphs with less than or equal to $m - 1$ nodes. To complete the proof, we only need to show that the theorem holds for a graph with m nodes, where $m \geq 3$.

Let G be a graph with m nodes, $m \geq 3$, and let T be a rooted spanning tree for G. Let the *end nodes* of a tree be those nodes that have degree ≤ 1. Then, T has at least one end node, say, node i_e, because an unrooted tree with $m \geq 3$ nodes has at least two end nodes.

Now, let T' be the subgraph formed when end node i_e and its connecting arc are removed from T. Clearly, T' is a rooted tree with $m - 1$ nodes.

Let G' be the subgraph of G formed when i_e is removed from G along with all incident arcs. Then T' is a rooted spanning tree for graph G'. Denote the $(m - 1) \times (m - 1)$ matrix associated with T' by $\mathbf{B'}$. That is, $\mathbf{B'}$ consists of the columns corresponding to the arcs of T'. Then by the induction hypothesis, $\mathbf{B'}$ is lower triangular nonsingular.

Now reconnect i_e to T' to form T. Denoting the matrix representation of T by $\mathbf{B}$, we may form $\mathbf{B}$ by simply adding to $\mathbf{B'}$ the column and row corresponding to node i_e and its connecting arc. This can be formed as follows.

$$
\mathbf{B} = \begin{pmatrix}
\pm 1 & 0 & 0 & \cdots & 0 \\
0 & & & & \\
\vdots & & & & \\
\pm 1 & & & \mathbf{B'} & \\
\vdots & & & & \\
\vdots & & & & \\
0 & & & &
\end{pmatrix}
$$

Recalling that $\mathbf{B'}$ is lower triangular nonsingular, it follows immediately that $\mathbf{B}$ is also lower triangular nonsingular and the proof is complete. $\square$

It is actually quite straightforward to construct the lower triangular matrix representation for a rooted spanning tree. To illustrate how this is done consider the following example.

Example 9.2: Lower Triangular Matrix Representation of a Rooted Spanning Tree

Consider the rooted spanning tree depicted in Figure 9.8.

The fundamental idea in writing down the lower triangular matrix representation is simply to work your way from the ends of the tree toward the root. For example, in Figure 9.8, there are initially three end nodes (nodes 1, 4, and 5). Let us arbitrarily choose node 4 and imagine that we are disconnecting it from the graph by removing arc (4, 3). Then noting the construction in Theorem 9.1, we would simply list node 4 as the first row and arc (4, 3) as the first column. The remaining graph now has two end nodes (nodes 1 and 5). We would now repeat the process by choosing, say, node 5 and the connecting arc (3, 5). The second row in our matrix would then be formed by node 5 with the second column corresponding to arc (3, 5). Continuing

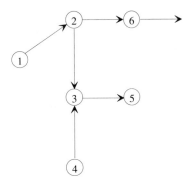

Figure 9.8   Rooted spanning tree.

this process until we reach the root would result in the following matrix representation of the rooted spanning tree T.

$$
\mathbf{B} = \begin{array}{c} \\ 4 \\ 5 \\ 1 \\ 3 \\ 2 \\ 6 \end{array}
\begin{pmatrix}
(4,3) & (3,5) & (1,2) & (2,3) & (2,6) & \text{root arc} \\
1 & 0 & 0 & 0 & 0 & 0 \\
0 & -1 & 0 & 0 & 0 & 0 \\
0 & 0 & 1 & 0 & 0 & 0 \\
-1 & 1 & 0 & -1 & 0 & 0 \\
0 & 0 & -1 & 1 & 1 & 0 \\
0 & 0 & 0 & 0 & -1 & 1
\end{pmatrix}
$$

Obviously, there are many ways to represent T as a lower triangular matrix depending on the sequence in which the end nodes are selected. The only rooted tree that would have a unique representation would be a rooted tree that consists of a single chain.

Noting that Theorem 9.1 constructs an $m \times m$ nonsingular submatrix of the matrix $\mathbf{A}$, it is quite obvious that the artificial variable corresponding to the root arc must be included in every basis. This follows directly from the fact that rank $(\mathbf{A}^0) \le m - 1$. In fact, the following corollary also follows directly from Theorem 9.1 and the definitions of $\mathbf{A}$ and $\mathbf{A}^0$.

Corollary 9.2
Rank$(\mathbf{A}) = m$ and rank$(\mathbf{A}^0) = m - 1$.

Theorem 9.1 established that every rooted spanning tree is a basis. However, before we can take advantage of this result, we must also verify that the converse is true. That is, it still remains to be shown that every basis corresponds to a rooted spanning tree. This result is proved in the following theorem thus establishing the equivalence of rooted spanning trees and basis matrices.

Theorem 9.3 (Every basis is a rooted spanning tree)
Let G be a connected graph with node–arc incidence matrix $\mathbf{A}^0$ and define $\mathbf{A} = (\mathbf{A}^0, \mathbf{e}_m)$ where $\mathbf{e}_m$ is the artificial column corresponding to the root arc with root node m. Let $\mathbf{B}$ be an $m \times m$ basis submatrix of $\mathbf{A}$ and let T be the subgraph of G

generated by **B**. That is, T consists of the m nodes of G along with the m arcs corresponding to the columns of **B**. Then T is a rooted spanning tree.

Proof

Since **B** is a basis matrix, it follows that rank(**B**) = m. Therefore, **B** must contain the artificial column corresponding to the root arc because, by Corollary 9.2, rank($\mathbf{A}^0$) = $m - 1$. Also T contains all the nodes of G, and thus T is a rooted spanning subgraph of G. Then, it remains to be shown that T is a rooted tree, that is, T has $m - 1$ arcs plus the root arc and has no cycles.

Clearly, T has $m - 1$ arcs plus the root arc because **B** consists of $m - 1$ columns of **A** along with the column associated with the root arc. Thus, the proof will be complete if we can show that T contains no cycles. By contradiction, suppose that T contains a cycle consisting of the nodes $i, j, k, l, \ldots, s, t, i$, as illustrated in Figure 9.9.

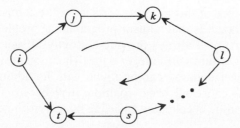

Figure 9.9   Cycle.

Now consider the columns corresponding to the arcs in this cycle. Assign an arbitrary orientation to the cycle and assign a coefficient of $+1$ to a column if the corresponding arc is in the same direction as the assigned orientation. Similarly, assign a coefficient of -1 to a column if the direction of the associated arc is in the direction opposite the orientation. Then, summing the columns by using these assigned coefficient yields

$$\mathbf{a}_{i,j} + \mathbf{a}_{j,k} - \mathbf{a}_{l,k} + \cdots + \mathbf{a}_{s,t} - \mathbf{a}_{i,t}$$
$$= (\mathbf{e}_i - \mathbf{e}_j) + (\mathbf{e}_j - \mathbf{e}_k) - (\mathbf{e}_l - \mathbf{e}_k) + \cdots + (\mathbf{e}_s - \mathbf{e}_t) - (\mathbf{e}_i - \mathbf{e}_t)$$
$$= \mathbf{0}$$

Thus, the columns associated with the arcs in the cycle are linearly dependent and could not be part of a basis. This contradicts the fact that **B** is a basis matrix, and thus T cannot contain a cycle. Hence, T is a rooted spanning tree.   □

Theorems 9.1 and 9.3 establish the equivalence between basis matrices and rooted spanning trees. Although minimum-cost network flow problems are linear programming problems and, hence, can be solved by the standard simplex algorithm of Chapter 4, the unique graphical representation of the basis matrices can be exploited to yield a much more efficient implementation of the simplex method. The details of this specialization will be developed shortly, but first let us consider one additional property of this class of problems.

Unimodularity

Let **A** be an $m \times n$ matrix with all integer entries. **A** is said to be *unimodular* if every $m \times m$ submatrix of **A** has a determinant equal to ± 1 or 0.

Now consider the coefficient matrix, $\mathbf{A} = (\mathbf{A}^0, \mathbf{e}_m)$, associated with the minimum-cost network flow problem. First, recall from Theorem 9.1 that every basis submatrix of **A** corresponds to a rooted spanning tree and can be written as a lower triangular matrix with nonzero diagonal entries. Next note that all of these diagonal entries are ± 1. Then, because the determinant of a lower triangular matrix is simply the product of the diagonal elements, we see that the determinant of every basis matrix is ± 1. Thus, every $m \times m$ nonsingular submatrix of **A** has determinant ± 1, and it follows immediately that every $m \times m$ submatrix of **A** has determinant ± 1 or 0. Therefore, we have established that $\mathbf{A} = (\mathbf{A}^0, \mathbf{e}_m)$ is a unimodular matrix.

What is the significance of this property? As we will establish in the following theorem, the major impact is that a linear programming problem that exhibits the unimodularity property will always provide a strictly integer solution when solved by the simplex algorithm with integer data. Thus, for problems with this special property, an integer solution can be found with no additional computational burden. In fact, because all basic solutions will be strictly integer, the computational process can be accelerated by using integer arithmetic rather than real arithmetic.

Theorem 9.4

Let **A** be an $m \times n$ matrix with rank(**A**) $= m \leq n$, and suppose that the entries in **A** are all integers. If **A** is unimodular, then every basic solution of $\mathbf{Ax} = \mathbf{b}$ is strictly integer for all integer **b**.

Proof

Let **b** be an integer vector, and consider the linear system $\mathbf{Ax} = \mathbf{b}$. Consider a basic solution of this system with the associated basis submatrix **B** of **A**. Then recall from Chapter 4 that this basic solution is given by $\mathbf{x}_N = \mathbf{0}$ and $\mathbf{x}_B = \mathbf{B}^{-1}\mathbf{b}$. It remains to be shown that the entries in $\mathbf{x}_B$ are all integer.

Let $\mathbf{B} = (\mathbf{b}_1, \mathbf{b}_2, \ldots, \mathbf{b}_m)$ and let $\mathbf{B}_j$ represent the matrix derived from **B** by replacing column $\mathbf{b}_j$ by vector **b**. That is,

$$\mathbf{B}_j = (\mathbf{b}_1, \mathbf{b}_2, \ldots, \mathbf{b}_{j-1}, \mathbf{b}, \mathbf{b}_{j+1}, \ldots, \mathbf{b}_m)$$

Then, by Cramer's rule, the solution of the system $\mathbf{Bx} = \mathbf{b}$ is given by

$$x_j = \frac{|\mathbf{B}_j|}{|\mathbf{B}|}, \qquad \text{for all } j = 1, \ldots, m$$

where the notation $|\mathbf{B}|$ represents the determinant of matrix **B**. But because **A** is unimodular, for the $m \times m$ basis submatrix **B**, we have $|\mathbf{B}| = \pm 1$. Also, because all the entries in **A** and **b** are integer, then $|\mathbf{B}_j|$ is integer. Therefore, x_j is integer for all j and it follows that every basic solution of the system $\mathbf{Ax} = \mathbf{b}$ is integer and thus every extreme point solution is all integer. ☐

We have established that if we begin with integer data, the simplex method will always yield an integer solution to a network flow problem. This property of unimodularity is also shared by other classes of problems. Among these are the transportation problem and the linear assignment problem, which are discussed in Chapter 10.

In the following section, we tailor the steps of the simplex method to dramatically increase computational efficiency when solving network flow problems.

THE NETWORK SIMPLEX METHOD

Before presenting the details of the network simplex method, let us begin by briefly reviewing the basic steps of the standard primal simplex algorithm as applied to a *minimization* problem in standard form:

$$\text{minimize } z = \mathbf{cx}$$

subject to:

$$\mathbf{Ax} = \mathbf{b} \tag{9.14}$$

$$\mathbf{x} \geq \mathbf{0}$$

Brief Review of the Standard Primal Simplex Method

Suppose a basic feasible solution to the linear programming problem given in (9.14) is defined by the basis matrix $\mathbf{B}$. Recall, from Chapter 4, that the canonical representation for this basic feasible solution is established by first partitioning the problem data into basic and nonbasic components, that is,

$$\mathbf{x} = \begin{pmatrix} \mathbf{x}_B \\ \mathbf{x}_N \end{pmatrix}, \qquad \mathbf{A} = (\mathbf{B} : \mathbf{N}), \qquad \text{and } \mathbf{c} = (\mathbf{c}_B, \mathbf{c}_N)$$

Solving for z and $\mathbf{x}_B$ in terms of $\mathbf{x}_N$ then yields the canonical representation

$$z = \mathbf{c}_B \mathbf{B}^{-1} \mathbf{b} - (\mathbf{c}_B \mathbf{B}^{-1} \mathbf{N} - \mathbf{c}_N) \mathbf{x}_N = \mathbf{c}_B \mathbf{B}^{-1} \mathbf{b} - \sum_{j \in J} (\mathbf{c}_B \mathbf{B}^{-1} \mathbf{a}_j - c_j) x_j \tag{9.15}$$

$$\mathbf{x}_B = \mathbf{B}^{-1} \mathbf{b} - \mathbf{B}^{-1} \mathbf{N} \mathbf{x}_N = \mathbf{B}^{-1} \mathbf{b} - \sum_{j \in J} (\mathbf{B}^{-1} \mathbf{a}_j) x_j \tag{9.16}$$

By setting the nonbasic variables equal to zero, the basic feasible solution is given by

$$\mathbf{x} = \begin{pmatrix} \mathbf{x}_B \\ \mathbf{x}_N \end{pmatrix} = \begin{pmatrix} \mathbf{B}^{-1} \mathbf{b} \\ \mathbf{0} \end{pmatrix} \geq \mathbf{0} \tag{9.17}$$

with the corresponding objective value

$$z = \mathbf{c}_B \mathbf{B}^{-1} \mathbf{b} \tag{9.18}$$

From (9.15), we see from the coefficients of the nonbasic variables that the optimality conditions for this *minimization* problem are

$$z_j - c_j = \mathbf{c}_B \mathbf{B}^{-1} \mathbf{a}_j - c_j \le 0, \qquad \text{for all } j \tag{9.19}$$

Recalling that the dual solution is given by $\boldsymbol{\pi} = \mathbf{c}_B \mathbf{B}^{-1}$, the optimality conditions can also be written in the form

$$z_j - c_j = \boldsymbol{\pi} \mathbf{a}_j - c_j \le 0, \qquad \text{for all } j \tag{9.20}$$

The main steps of the simplex algorithm can then be summarized as follows. First, optimality is checked using the conditions in (9.20). If the current basic feasible solution is not optimal, an entering nonbasic variable is chosen and increased from its current value of zero. The departing variable is then determined by using the relationship in (9.16) and the feasibility condition $\mathbf{x}_B \ge \mathbf{0}$. Assuming that the entering variable is x_k, this results in

$$\mathbf{x}_B = \mathbf{B}^{-1}\mathbf{b} + x_k(-\mathbf{B}^{-1}\mathbf{a}_k) = \boldsymbol{\beta} - x_k\boldsymbol{\alpha}_k = \begin{pmatrix} \beta_1 \\ \beta_2 \\ \vdots \\ \beta_m \end{pmatrix} - x_k \begin{pmatrix} \alpha_{1,k} \\ \alpha_{2,k} \\ \vdots \\ \alpha_{m,k} \end{pmatrix} \ge \mathbf{0} \tag{9.21}$$

which leads to the minimum ratio

$$x_k = \frac{\beta_r}{\alpha_{r,k}} = \text{minimum} \left\{ \frac{\beta_i}{\alpha_{i,k}} : \alpha_{i,k} > 0 \right\} \tag{9.22}$$

The basic variable $x_{B,r}$ associated with row r is the departing variable and the solution is updated using a pivot operation.

Now consider the minimum-cost network flow problem, which can be written in precisely the form of (9.14). Recall, however, that the elements of $\mathbf{x}$ are of the form $x_{i,j}$ and variable $x_{i,j}$ corresponds to the flow on arc (i, j). Therefore, with a slight modification in notation, we will now proceed to customize the steps of the simplex algorithm by exploiting the special network structure of this problem.

Again, assume that a basic feasible solution is currently available with basis matrix $\mathbf{B}$. (Subsequently, we will discuss a procedure for finding a starting basic feasible solution.) Recall that $\mathbf{B}$ can be written as a lower triangular matrix and is equivalent to a rooted spanning tree of the minimum-cost flow network. Thus, in this case, the basic variables are, in fact, the basic arcs comprising the rooted spanning tree.

Computing the Flows

Let us begin by computing the basic solution corresponding to basis matrix $\mathbf{B}$. That is, let us find the *flows* corresponding to the basic arcs in the rooted spanning tree. Mathematically, this involves the computation of $\mathbf{x}_B = \boldsymbol{\beta} = \mathbf{B}^{-1}\mathbf{b}$. However, because $\mathbf{B}$ has such a special structure, we will avoid the actual computation and storage of $\mathbf{B}^{-1}$. As we will see, explicit knowledge of the basis inverse is not necessary to efficiently carry out the steps of the network simplex algorithm. This is due to the lower triangular form of basis matrix **B** and its graphical representation as a rooted spanning tree. Note that $\mathbf{x}_B = \mathbf{B}^{-1}\mathbf{b}$ can be thought of as the solution of the linear system

$$\mathbf{Bx}_B = \mathbf{b} \tag{9.23}$$

But because **B** is lower triangular, the solution of this system can be found quite readily by front solving. That is, we first solve for the first element of $\mathbf{x}_B$ by using the first equation of $\mathbf{Bx}_B = \mathbf{b}$. This value is then substituted into the remaining equations and the second equation can be solved to yield the value of the second element of $\mathbf{x}_B$. This process is then repeated until all elements of $\mathbf{x}_B$ are found. Because the lower triangular representation of **B** was formed by working from the ends of the tree toward the root node, we can find this basic solution graphically utilizing the rooted spanning tree representation and working our way from the ends toward the root.

For example, consider the network flow problem given in Figure 9.10. Figure 9.11 depicts a rooted spanning tree for this network that corresponds to the lower triangular basis matrix

$$\mathbf{B} = \begin{array}{c} \\ 2 \\ 1 \\ 3 \\ 4 \\ 5 \\ 6 \end{array} \begin{array}{c} (1,2) \ (1,3) \ (3,5) \ (5,4) \ (5,6) \ \text{root arc} \\ \left(\begin{array}{cccccc} -1 & 0 & 0 & 0 & 0 & 0 \\ 1 & 1 & 0 & 0 & 0 & 0 \\ 0 & -1 & 1 & 0 & 0 & 0 \\ 0 & 0 & 0 & -1 & 0 & 0 \\ 0 & 0 & -1 & 1 & 1 & 0 \\ 0 & 0 & 0 & 0 & -1 & 1 \end{array} \right) \end{array}$$

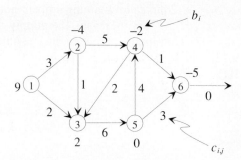

Figure 9.10 Network flow problem.

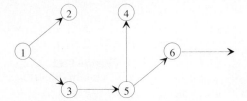

Figure 9.11 A rooted spanning tree basis.

Now consider the linear system $\mathbf{Bx}_B = \mathbf{b}$:

$$
\begin{pmatrix}
-1 & 0 & 0 & 0 & 0 & 0 \\
1 & 1 & 0 & 0 & 0 & 0 \\
0 & -1 & 1 & 0 & 0 & 0 \\
0 & 0 & 0 & -1 & 0 & 0 \\
0 & 0 & -1 & 1 & 1 & 0 \\
0 & 0 & 0 & 0 & -1 & 1
\end{pmatrix}
\begin{pmatrix}
x_{1,2} \\
x_{1,3} \\
x_{3,5} \\
x_{5,4} \\
x_{5,6} \\
x_6
\end{pmatrix}
=
\begin{pmatrix}
-4 \\
9 \\
2 \\
-2 \\
0 \\
-5
\end{pmatrix}
$$

We can easily front solve this system of equations. That is, because the first equation only involves $x_{1,2}$, we easily find that $x_{1,2} = 4$. Next, substituting $x_{1,2} = 4$ into the second equation and solving for $x_{1,3}$ yields $x_{1,3} = 5$. By continuing in this manner, the basic flows are found as follows:

$$x_{1,2} = 4$$

$$x_{1,3} = b_1 - x_{1,2} = 9 - 4 = 5$$

$$x_{3,5} = b_3 + x_{1,3} = 2 + 5 = 7$$

$$x_{5,4} = -b_4 = 2$$

$$x_{5,6} = b_5 + x_{3,5} - x_{5,4} = 0 + 7 - 2 = 5$$

$$x_6 = b_6 + x_{5,6} = -5 + 5 = 0$$

However, recall that when front solving the system $\mathbf{Bx}_B = \mathbf{b}$, we are actually working from the ends of the tree toward the root. Thus, it is quite straightforward to determine the flows without actually writing the system of equations $\mathbf{Bx}_B = \mathbf{b}$. For example, in Figure 9.12, we see that because $b_2 = -4$ and node 4 has only one incident arc, then the flow on this arc must be equal to 4 to satisfy the demand at node 2 (that is, $x_{1,2} = 4$). But if $x_{1,2} = 4$, then there are only 5 units remaining at node 1 because $b_1 = 9$. Thus, all remaining units at node 1 must flow along arc (1, 3) and we have $x_{1,3} = 5$. This increases the supply at node 3 to 7 because $b_3 = 2$ and forces $x_{3,5} = 7$. Continuing in this manner results in the primal solution given in Figure 9.12, which, of course, is precisely the same solution derived previously using $\mathbf{Bx}_B = \mathbf{b}$.

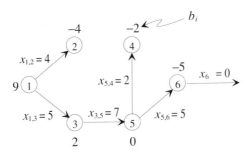

Figure 9.12 Computing the primal solution (flows) x_{ij}.

Checking for Optimality and Choosing the Entering Arc

The next task that must be performed is the check for optimality. This involves the computation of $z_{i,j} - c_{i,j}$ for each nonbasic variable $x_{i,j}$. That is, we need to compute

$$z_{i,j} - c_{i,j} = \boldsymbol{\pi}\mathbf{a}_{i,j} - c_{i,j} \tag{9.24}$$

where

$$\boldsymbol{\pi} = \mathbf{c}_B\mathbf{B}^{-1} \tag{9.25}$$

However, because $\mathbf{a}_{i,j} = \mathbf{e}_i - \mathbf{e}_j$, (9.24) simplifies to

$$z_{i,j} - c_{i,j} = \boldsymbol{\pi}(\mathbf{e}_i - \mathbf{e}_j) - c_{i,j} = \pi_i - \pi_j - c_{ij} \tag{9.26}$$

and $z_{i,j} - c_{i,j}$ can be computed quite easily if the values of $\boldsymbol{\pi} = (\pi_1, \pi_2, \ldots, \pi_m)$ are known.

Each constraint corresponds to a node of the network, so the dual variables $\boldsymbol{\pi} = (\pi_1, \pi_2, \ldots, \pi_m)$ correspond to nodes $1, 2, \ldots, m$, respectively. For this reason, the dual variables are sometimes called *node potentials*. Note, however, that $\boldsymbol{\pi} = \mathbf{c}_B\mathbf{B}^{-1}$ is simply the solution of the triangular linear system

$$\pi\mathbf{B} = \mathbf{c}_B \tag{9.27}$$

and can be found by back solving this system. Graphically, this corresponds to working from the root to the ends of the rooted spanning tree.

Also, because each column of $\mathbf{B}$ contains precisely one 1 and one -1, each equation comprising system $\pi\mathbf{B} = \mathbf{c}_B$ is of the form

$$\pi_i - \pi_j = c_{ij} \tag{9.28}$$

where node i represents the tail node, and node j is the head node. Noting the form of $z_{i,j} - c_{i,j}$ given by (9.26), we see that (9.28) is equivalent to setting $z_{i,j} - c_{i,j}$ for the basic variable $x_{i,j}$ equal to zero.

By rearranging (9.28), note that given the node potential π_i at the tail node i of arc (i, j) of the rooted spanning tree, the potential at head node j can be easily computed using

$$\pi_j = \pi_i - c_{ij} \tag{9.29}$$

That is, *the potential decreases by the cost of the arc when moving in the direction of the arc from the root.* Similarly, given π_j, the potential at the head node j, the node potential at tail node i can be found using

$$\pi_i = \pi_j + c_{ij} \tag{9.30}$$

and *the potential increases when moving opposite the direction of the arc.*

Expanding the linear system $\pi\mathbf{B} = \mathbf{c}_B$ for the rooted spanning tree of Figure 9.11 results in

$$\pi_1 - \pi_2 = c_{1,2} = 3 \tag{9.31}$$

$$\pi_1 - \pi_3 = c_{1,3} = 2 \tag{9.32}$$

$$\pi_3 - \pi_5 = c_{3,5} = 6 \tag{9.33}$$

$$\pi_5 - \pi_4 = c_{5,4} = 4 \tag{9.34}$$

$$\pi_5 - \pi_6 = c_{5,6} = 3 \tag{9.35}$$

$$\pi_6 = c_6 = 0 \tag{9.36}$$

Note that the cost associated with the root arc, c_6, was set equal to zero in (9.36). This is actually an arbitrary choice and is only chosen for convenience. The flow on the root arc is always zero, so c_6 can be chosen arbitrarily without changing the value of the objective. For simplicity, we will always assume that the cost associated with the root arc is zero; however, any value will work equally as well.

The solution to the linear system $\pi \mathbf{B} = \mathbf{c}_B$ in (9.31–9.36) is readily obtained by back solving. That is, we begin by utilizing (9.36) to obtain $\pi_6 = 0$. This, in turn, yields the value of π_5 by substituting into (9.35). By continuing in this manner, the dual solution obtained is

$$\pi_6 = 0$$

$$\pi_5 = \pi_6 + 3 = 0 + 3 = 3$$

$$\pi_4 = \pi_5 - 4 = 3 - 4 = -1$$

$$\pi_3 = \pi_5 + 6 = 3 + 6 = 9$$

$$\pi_1 = \pi_3 + 2 = 9 + 2 = 11$$

$$\pi_2 = \pi_1 - 3 = 11 - 3 = 8$$

Of course, these simple computations can be performed directly, without explicitly considering the linear system $\pi \mathbf{B} = \mathbf{c}_B$. This is done by utilizing the graph along with the relationships defined in (9.29) and (9.30) and working from the root toward the ends of the tree in Figure 9.13. We begin by setting $\pi_6 = c_6 = 0$. By moving from node 6 to node 5, the potential increases by $c_{35} = 3$, and thus $\pi_5 = 3$. Continuing in this manner yields the solution provided in Figure 9.13.

Once the dual solution π is computed in this straightforward manner,

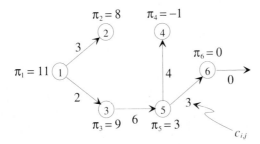

Figure 9.13 Computing the dual solution (node potentials) π_i.

$z_{i,j} - c_{i,j}$ for the nonbasic variables can be computed using

$$z_{i,j} - c_{i,j} = \pi_i - \pi_j - c_{i,j} \tag{9.37}$$

Finally, if $z_{i,j} - c_{i,j} \leq 0$, for all arcs (i, j), then the current basic set of flows is optimal. Otherwise, we choose that $x_{i,j}$ with the most positive $z_{i,j} - c_{i,j}$ as the entering variable (arc). We will refer to the entering variable as $x_{k,l}$. Arc (k, l) will enter the basis tree and a departing arc (variable) must then be selected.

Consider again the rooted spanning tree and the dual solution provided by Figure 9.13. Note from Figure 9.10 that there are four nonbasic arcs, $x_{2,3}$, $x_{2,4}$, $x_{4,3}$, and $x_{4,6}$. Utilizing (9.37), we can then compute the $z_{i,j} - c_{i,j}$ for each of these nonbasic arcs as follows:

$$z_{2,3} - c_{2,3} = \pi_2 - \pi_3 - c_{2,3} = 8 - 9 - 1 = -2$$

$$z_{2,4} - c_{2,4} = \pi_2 - \pi_4 - c_{2,4} = 8 - (-1) - 5 = 4$$

$$z_{4,3} - c_{4,3} = \pi_4 - \pi_3 - c_{4,3} = -1 - 3 - 2 = -5$$

$$z_{4,6} - c_{4,6} = \pi_4 - \pi_6 - c_{4,6} = -1 - 0 - 1 = -2$$

Observing that $z_{2,4} - c_{2,4} > 0$, we see that the current basic feasible solution is not optimal. Thus, $x_{2,4}$ is chosen as the entering variable, or, equivalently, arc $(2, 4)$ is chosen as the entering arc.

Determining the Departing Arc

Recall that the standard simplex method uses the minimum-ratio test to identify the value of the entering variable and at the same time identify the blocking or departing variable. The departing variable is that basic variable that is driven to zero first by increasing the entering variable from its present value of zero. Mathematically, the computation of the minimum ratio involves two key components, the current basic solution vector, $\mathbf{x}_B = \boldsymbol{\beta} = \mathbf{B}^{-1}\mathbf{b}$, and the updated column, $\boldsymbol{\alpha}_{k,l} = \mathbf{B}^{-1}\mathbf{a}_{k,l}$, associated with the entering variable $x_{k,l}$. The vector $\mathbf{x}_B$ has already been computed and we could, of course, determine $\boldsymbol{\alpha}_{k,l}$ by front solving the system $\mathbf{B}\boldsymbol{\alpha}_{k,l} = \mathbf{a}_{k,l}$. However, there is an interesting graphical method for determining $\boldsymbol{\alpha}_{k,l}$. This method is based on the observation that $\boldsymbol{\alpha}_{k,l}$ actually specifies how to write column $\mathbf{a}_{k,l}$ as a linear combination of the columns of $\mathbf{B}$. (This concept was discussed in detail in Chapter 4.)

We have already seen that $\mathbf{x}_B$ can be computed directly from the rooted spanning tree basis by working from the ends of the tree toward the root. How then does one compute the column vector $\boldsymbol{\alpha}_{k,l} = \mathbf{B}^{-1}\mathbf{a}_{k,l}$, and is there a graphical method for determining the minimum ratio? This is best answered by way of an example, so consider the basic feasible solution represented by the rooted spanning in Figure 9.12, and consider the entering variable $x_{k,l} = x_{2,4}$ selected previously.

Note that the addition of arc $(2, 4)$ creates exactly one cycle in the resulting graph (see Figure 9.14). This, of course, will always be the case, because there is a unique chain between every pair of nodes in a tree and the addition of an arc will

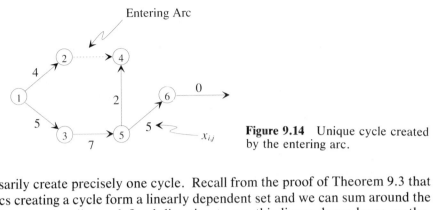

Figure 9.14 Unique cycle created by the entering arc.

necessarily create precisely one cycle. Recall from the proof of Theorem 9.3 that the arcs creating a cycle form a linearly dependent set and we can sum around the cycle in an arbitrarily predefined direction to see this linear dependence mathematically. By using the graph of Figure 9.14, this would result in

$$\mathbf{a}_{2,4} - \mathbf{a}_{5,4} - \mathbf{a}_{3,5} - \mathbf{a}_{1,3} + \mathbf{a}_{1,2} = \mathbf{0} \tag{9.38}$$

Solving for $\mathbf{a}_{k,l} = \mathbf{a}_{2,4}$ in (9.38), we see that the column associated with the entering arc can be written in terms of the basic columns as

$$\mathbf{a}_{2,4} = -\mathbf{a}_{1,2} + \mathbf{a}_{1,3} + \mathbf{a}_{3,5} + \mathbf{a}_{5,4} \tag{9.39}$$

Noting that $\mathbf{B} = (\mathbf{a}_{1,2}, \mathbf{a}_{1,3}, \mathbf{a}_{3,5}, \mathbf{a}_{5,4}, \mathbf{a}_{5,6}, \mathbf{a}_6)$ and because $\boldsymbol{\alpha}_{2,4}$ specifies how to write $\mathbf{a}_{2,4}$ as a linear combination of the basic columns, it follows immediately that

$$\boldsymbol{\alpha}_{2,4} = \begin{pmatrix} -1 \\ 1 \\ 1 \\ 1 \\ 0 \\ 0 \end{pmatrix}$$

Note that all of the entries in $\boldsymbol{\alpha}_{2,4}$ are either 0, $+1$, or -1. This will always be the case because of the network structure of the problem.

Therefore, in a manner analogous to (9.21), we have

$$\mathbf{x}_B - x_{2,4}\boldsymbol{\alpha}_{2,4} = \begin{pmatrix} x_{1,2} \\ x_{1,3} \\ x_{3,5} \\ x_{5,4} \\ x_{5,6} \\ x_6 \end{pmatrix} - x_{2,4} \begin{pmatrix} -1 \\ 1 \\ 1 \\ 1 \\ 0 \\ 0 \end{pmatrix} = \begin{pmatrix} 4 \\ 5 \\ 7 \\ 2 \\ 5 \\ 0 \end{pmatrix} - x_{2,4} \begin{pmatrix} -1 \\ 1 \\ 1 \\ 1 \\ 0 \\ 0 \end{pmatrix} = \begin{pmatrix} 4 + x_{2,4} \\ 5 - x_{2,4} \\ 7 - x_{2,4} \\ 2 - x_{2,4} \\ 5 \\ 0 \end{pmatrix} \geq \mathbf{0}$$

$$\tag{9.40}$$

Clearly, (9.40) results in $x_{2,4} = $ minimum $\{5, 7, 2\} = 2$, and the departing variable is $x_{5,4}$. Although this overall process may at first seem complicated, all the preceding computations can be performed quite readily in a graphical manner.

Instead of actually computing $\alpha_{k,l} = \alpha_{2,4}$ and performing the minimum ratio as we did earlier, consider inducing a flow of $x_{2,4} = \Delta$ around the created cycle in the direction of arc $(k, l) = (2, 4)$. Then, the flow on those arcs that are in the same direction as $x_{2,4}$ will increase in flow by Δ, whereas the flow on the arcs oriented in the opposite direction will decrease by Δ. The flows on basic arcs not involved in the cycle will not change. From Figure 9.15, we see that this results in

$$x_{1,2} = 4 + \Delta$$

$$x_{1,3} = 5 - \Delta$$

$$x_{3,5} = 7 - \Delta$$

$$x_{5,4} = 2 - \Delta$$

$$x_{5,6} = 5$$

$$x_6 = 0$$

which are precisely the same conditions as in (9.40) that were derived using $\alpha_{2,4}$.

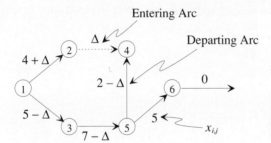

Figure 9.15 Determining the departing arc.

Thus, the minimum-ratio test can be performed by simply inducing a flow of $x_{k,l} = \Delta$ on the created cycle, examining the arcs on which the flow is decreasing, and choosing that arc among these with the minimum flow. The flows on the cycle are then updated by the computed amount Δ, and the departing arc is removed to form a new rooted spanning tree basis. Figure 9.16 illustrates the updated solu-

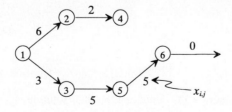

Figure 9.16 The updated primal solution (flows).

tion for the previous example. The entire process is then repeated by recomputing the corresponding dual solution and checking the optimality conditions.

Updating the Dual Solution

As we have seen, recomputing the primal solution involves the very simple process of updating the flows around the unique cycle created by the entering arc. Before formally stating the network simplex method, let us investigate updating the dual solution. Is there also a straightforward way of updating the dual solution without starting over from scratch and recomputing all the node potentials by starting again at the root? Which dual variables actually change and by how much?

To answer these questions, consider the schematic representation of a basic rooted spanning tree given in Figure 9.17. Only the entering and departing arcs are explicitly represented with all other arcs assumed to be in one of subtrees T_p or T_q. There are actually two cases to consider. Figure 9.17(a) illustrates the case when the departing arc is directed toward the root node, whereas Figure 9.17(b) depicts the case when the departing arc is directed away from the root. For our discussion, let us concentrate on the case of Figure 9.17(a).

Consider removing the departing arc (p, q) before the entering arc is added to the tree. Clearly, the rooted spanning tree T is divided into two disjoint sub-

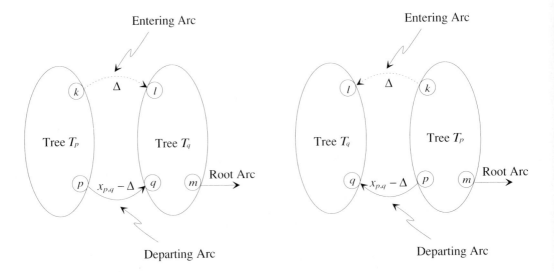

(a) Departing Arc Directed Toward the Root   (b) Departing Arc Directed Away From the Root

Figure 9.17   Updating the dual solution.

trees, T_p and T_q, which contain nodes p and q, respectively. Note that because the root node is contained in subtree T_q, the values of all of the dual variables associated with nodes in T_q will remain the same. This is because the dual solution is computed by working from the root to the ends of the tree. And because all nodes in T_q remained connected to the root node by the same chains, the dual solutions associated with these nodes will remain the same.

Thus, it is only necessary to update the dual solution of those nodes that are disconnected from the root when the departing arc is removed. These, of course are the nodes that are contained in the subtree T_p. However, all arcs contained in T_p remain basic and $z_{i,j} - c_{i,j} = \pi_i - \pi_j - c_{i,j} = 0$ for all basic arcs. Therefore, the values of all dual variables for nodes in T_p must change by precisely the same amount if $z_{i,j} - c_{i,j}$ is to remain 0 for each of these arcs. Let us then compute the change associated with node k.

For simplicity, denote the dual solution prior to the pivot by $\pi = (\pi_1, \pi_2, \ldots, \pi_m)$ and denote the dual solution after the pivot by $\pi' = (\pi'_1, \pi'_2, \ldots, \pi'_m)$. Then, as previously noted

$$\pi'_i = \pi_i, \qquad \text{for all nodes } i \in T_q \tag{9.41}$$

Arc (k, l) is the entering arc, so we have

$$\pi_k - \pi_l - c_{k,l} = z_{k,l} - c_{k,l} > 0 \tag{9.42}$$

And because arc (k, l) is a basic arc after the pivot operation, it follows that

$$\pi'_k - \pi'_l - c_{k,l} = 0 \tag{9.43}$$

But, from (9.41), $\pi'_l = \pi_l$ because $l \in T_q$. Therefore, subtracting (9.43) from (9.42) yields

$$\pi_k - \pi'_k = (z_{k,l} - c_{k,l}) \tag{9.44}$$

which then gives

$$\pi'_k = \pi_k - (z_{k,l} - c_{k,l}) \tag{9.45}$$

Thus, the dual solution associated with node k decreases by $(z_{k,l} - c_{k,l})$, and we have established that

$$\pi'_i = \pi_i - (z_{k,l} - c_{k,l}), \qquad \text{for all nodes } i \in T_q \tag{9.46}$$

That is, in the case when the departing arc is directed toward the root, the value of the dual variable for each node in T_p *decreases* by $(z_{k,l} - c_{k,l})$, whereas the value of the dual variable for each node in T_q remains the same.

By a parallel argument, it can be shown that when departing arc (p, q) is directed away from the root, as in Figure 9.17(b), the value of the dual variable for each node in T_q *increases* by $(z_{k,l} - c_{k,l})$, whereas the value of the dual variable for each node in T_p remains the same. The details are left to the reader.

The network simplex can now be summarized as follows.

The network simplex algorithm (minimization problem)

STEP 1. *Determine the primal and dual solutions.* Given a feasible basis represented by a rooted spanning tree, compute the primal solution $\mathbf{x}_B$ (i.e., the flows) by starting at the ends of the tree and working toward the root using the values of the b_i. Compute the dual solution $\boldsymbol{\pi}$ (i.e., the node potentials) by starting at the root node and working toward the ends of the tree using the relationships defined in (9.29) and (9.30).

STEP 2. *Check for optimality.* Compute $z_{i,j} - c_{i,j}$ each nonbasic arc using

$$z_{i,j} - c_{i,j} = \pi_i - \pi_j - c_{i,j}$$

If $z_{i,j} - c_{i,j} \leq 0$, for all i, j, then the current solution is optimal; stop. Otherwise, choose as the entering arc that arc with the most positive $z_{i,j} - c_{i,j}$. Call the entering arc (k, l) and call the corresponding flow $x_{k,l}$.

STEP 3. *Determine the departing variable.* Add the entering arc (k, l) to the rooted spanning tree and induce a flow of $x_{k,l} = \Delta$ around the resulting cycle in the direction of arc (k, l). Then each arc in the cycle is either increasing in flow by Δ or decreasing in flow by Δ. Let D represent the set of arcs that are decreasing in flow. Compute

$$\Delta = \begin{cases} \infty, & \text{if } D = \phi \\ \displaystyle\min_{(i,j) \in D} \{x_{i,j}\}, & \text{otherwise} \end{cases}$$

If $\Delta = \infty$, then there is no blocking variable and the objective is unbounded. Otherwise, let $(p, q) \in D$ be the arc associated with this minimum. That is, $x_{p,q}$ is the departing variable.

STEP 4. *Pivot and update.* Update the flows along the created cycle by $+\Delta$ or $-\Delta$, adding the entering arc (k, l) to the rooted spanning tree and removing the departing arc (p, q). Update the dual solution and return to Step 2.

Example 9.3: The Network Simplex Method

Consider the simple network flow problem given in Figure 9.18 and assume that a starting basic feasible solution is given by the rooted spanning tree in Figure 9.19.

Then the steps of the network simplex method for this problem are summarized in Figure 9.20. Note that the basic solution represented in iteration 4 is

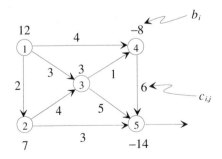

Figure 9.18 Network for Example 9.3.

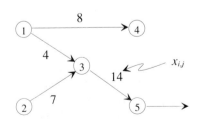

Figure 9.19 Starting solution for Example 9.3.

| Iteration | Primal Solution (Flows) | Dual Solution (Node Potentials) | Reduced Costs $(z_{i,j} - c_{i,j})$ | Pivot |

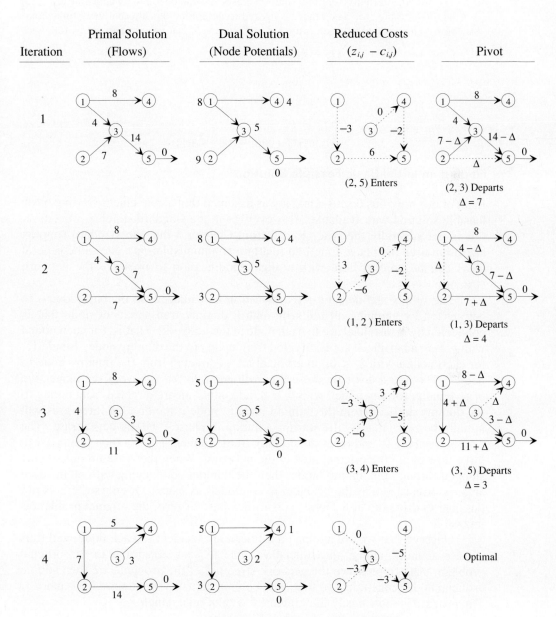

Figure 9.20   Solution of Example 9.3.

an optimal solution because $z_{i,j} - c_{i,j} \leq 0$, for all arcs (i, j). However, because $z_{1,3} - c_{1,3} = 0$, an alternative optimal solution exists and can be found by choosing arc $(1, 3)$ as the entering arc. The reader is invited to determine this alternative optimal solution. The optimal solution provided in Figure 9.20 can be summarized as follows:

$$x_{1,2}^* = 7$$

$$x_{1,4}^* = 5$$

$$x_{2,5}^* = 14$$

$$x_{3,4}^* = 3$$

$$z^* = 2(7) + 4(5) + 1(3) + 3(14) = 79$$

Finding an Initial Basic Feasible Solution

In all of our previous discussions, it was assumed that a convenient starting basic feasible solution was available. However, as in the standard simplex procedure, this is not generally the case. Recall from Chapter 4 that the standard simplex algorithm uses the phase I method to attain an initial solution. The same general ideas can be applied here with a slight modification to preserve the network structure.

The supply and demand are assumed to be balanced, so the basic idea is to get started by shipping all units through a dummy transshipment node that is connected by an artificial arc to each node in the network. That is, for each node i with $b_i > 0$, an artificial arc is directed from node i to the dummy node. Similarly, for each node i with $b_i < 0$, an artificial arc is directed from the dummy node to node i. The initial flows on these arcs will equal to the values of the respective b_i's. Finally, for any node i with $b_i = 0$, add either an arc directed from node i to the dummy node or from the dummy node to node i; the flow on these arcs will initially be zero, that is, the starting solution in phase I will be degenerate. The cost of shipping to and from the dummy transshipment node is then penalized to force flow along the original arcs in the network. When there are no units being shipped through the dummy node, then the dummy node, along with all incident arcs, is discarded and the problem is continued as usual. Of course, if it is not possible to drive the flow along each artificial arc to zero, the original problem is infeasible.

Observe that we are adding an artificial arc incident to each node, and thus we are actually adding an artificial variable to each constraint in the original problem. And to preserve the network structure of the problem, we then add an artificial node (constraint) so that each column contains the required $+1$ and -1. The process is most easily illustrated by way of an example.

Example 9.4: Finding an Initial Basic Feasible Solution

Consider the network flow problem depicted in Figure 9.21. Although the problem is somewhat trivial, let us assume that we first must determine an initial basic feasible solution to this problem. We begin by setting up the phase I problem as in Figure

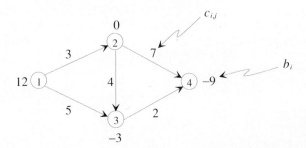

Figure 9.21 Network flow problem for Example 9.4.

9.22(a) with each original node being connected by an artificial arc to dummy node 5. Note the directions of the artificial arcs as compared to the values of the b_i's. Also observe that the phase I cost associated with each artificial arc is 1 and the phase I cost associated with each original arc is 0. This is the same strategy that was used in the phase I method discussed in Chapter 4. Equivalently, each artificial arc could be penalized with a large positive cost, M, and the original costs could be used for the remaining arcs. The Big-M method then could be applied to achieve the same results as the two-phase procedure. However, to avoid the task of choosing an appropriate value for M, we will emphasize the use of the two-phase method.

The starting basis for the phase I problem is given in Figure 9.22(b). Note that the flow on each arc is precisely equal to the original values of the b_i's. Obviously, because $b_2 = 0$, the flow on artificial arc (2, 5) is 0 and the starting basic solution is degenerate. The steps of the network simplex method required to solve the phase I problem are summarized in Figure 9.23.

Note, from Figure 9.23, that the phase I problem yields an optimal solution in which the flows on all artificial arcs are 0. Thus, we begin phase II by discarding all

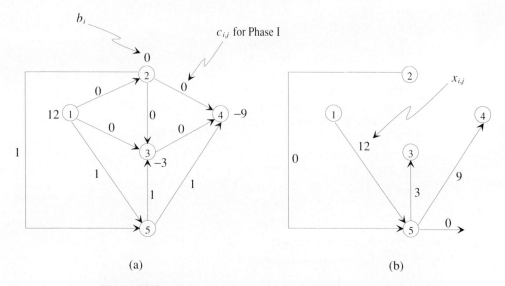

(a)   (b)

Figure 9.22 Phase I problem and the initial basic feasible solution.

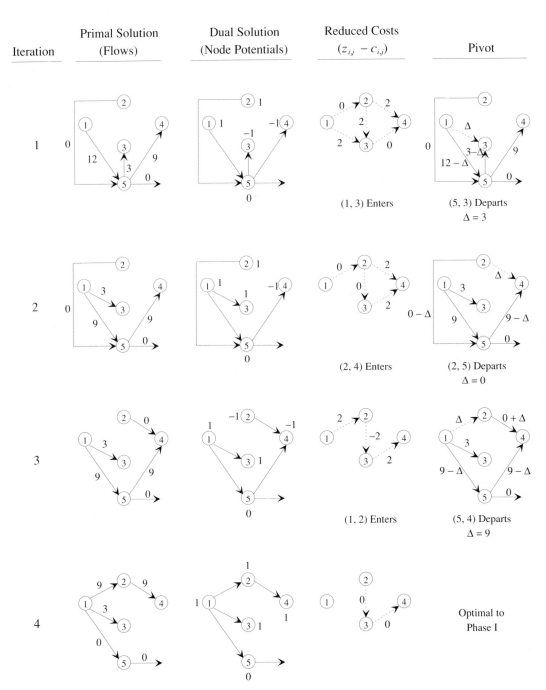

Iteration	Primal Solution (Flows)	Dual Solution (Node Potentials)	Reduced Costs $(z_{i,j} - c_{i,j})$	Pivot

Figure 9.23 Phase I solution of Example 9.4.

artificial arcs and using the original costs. The resulting starting basis that would be used to begin phase II is given in Figure 9.24. The reader is invited to complete the details of phase II.

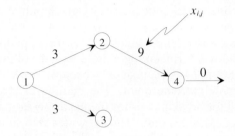

Figure 9.24 Phase II starting basic feasible solution for Example 9.4.

DEGENERACY AND CYCLING

Recall, from Chapter 4, that cycling is an infinite loop resulting from repeatedly iterating through a sequence of degenerate basic feasible solutions corresponding to the same extreme point. As was the case with the standard simplex method, degeneracy and cycling can be a problem when implementing the network simplex method. In fact, Cunningham (1979) gives an example of cycling occurring in a graph with three nodes. Consequently, several methods have been developed to prevent cycling. One of the most elegant of these methods is the method of strongly feasible trees. A *strongly feasible tree* is simply a tree in which all degenerate basic arcs (i.e., basic arcs with zero flow) are directed toward the root. An arc (i, j) is directed toward the root if node j lies on the unique chain connecting node i and the root node. Figure 9.25 provides an example of a strongly feasible tree. Note that there are three degenerate arcs, $(2, 3)$, $(4, 1)$, and $(5, 6)$. Note also that each of these degenerate arcs is directed toward the root node. For example, $(2, 3)$ is directed toward the root because node 3 lies on the unique chain connecting node 2 and root node 9.

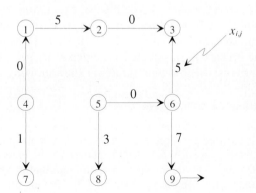

Figure 9.25 An example of a strongly feasible tree.

Strongly Feasible Trees and Finite Convergence

Let arc (k, l) be the entering arc with $x_{k,l} = \Delta$. If $\Delta > 0$, then the pivot is nondegenerate and results in a strict improvement (decrease) in the objective function. In fact, the objective function decreases by an amount equal to $\Delta(z_{k,l} - c_{k,l})$. If $\Delta = 0$, then the pivot is degenerate and results in no change in the objective function. There are only a finite number of extreme-point solutions, so the network simplex method will terminate in a finite number of iterations in the absence of degeneracy. This follows immediately because each nondegenerate pivot results in visiting a new extreme point. The method of strongly feasible trees is essentially a method for breaking ties for the departing arc, so that cycling cannot occur at a degenerate extreme point. The basic idea is to simply maintain a strongly feasible tree as the basis at each iteration of the network simplex method.

Clearly, if we are to maintain a strongly feasible tree at each iteration, then we must begin with a strongly feasible tree. Because an initial solution is generally determined using phase I of the two-phase procedure, it is actually a trivial process to begin with a strongly feasible tree. Recall that the phase I procedure begins by adding an artificial node that is connected to each node of the original network by an artificial arc. Consider an original node i. If $b_i > 0$, the corresponding artificial arc will be directed toward the artificial node and have a flow of b_i. Similarly, if $b_i < 0$, a flow of b_i will be directed along the artificial arc toward node i. However, if $b_i = 0$, we have a choice for the original orientation of the associated artificial arc. Thus, we will always direct the artificial arcs with zero flow toward the root. An example of this can be seen in Figure 9.22(b). Note that $(2, 5)$ is a degenerate artificial arc and was chosen so that it was directed toward the root. All other arcs are nondegenerate and their directions are, of course, determined automatically to ensure a feasible solution. Thus, getting started with a strongly feasible tree is a simple process. How, then, do we maintain a strongly feasible tree at each iteration?

Again, let (k, l) represent the entering arc. In the current rooted spanning tree basis, trace the unique chain from node k to the root node. Also trace the unique chain from node l to the root node. Let node h be the first node common to both chains. Adding arc (k, l) to the current tree creates exactly one cycle. Now, consider tracing this unique cycle in the direction of (k, l), starting and ending at node h. A strongly feasible tree can be maintained by breaking ties for the minimum ratio by choosing the *last* arc in this cycle as the departing arc, which yields the minimum ratio. The process is now illustrated via the following example.

Example 9.5: Maintaining Strongly Feasible Trees

Consider again the basic feasible solution represented by the strongly feasible tree in Figure 9.25. Recall that this tree has three degenerate arcs, each of which is directed toward the root node. Now, suppose that $(7, 8)$ is chosen as the entering arc as in Figure 9.26(a). Note that there is a tie for the minimum ratio, that is, arcs $(2, 3)$ and $(4, 1)$ are both candidates for the departing arc with $\Delta = 0$. Tracing the unique chain

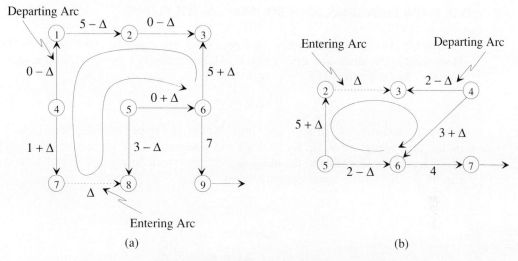

Figure 9.26 Maintaining strongly feasible trees.

from nodes 7 and 8 to the root node results in the first common node $h = 6$. Beginning at node 6, we now trace the unique cycle in the direction $(7, 8)$. Thus, the last arc encountered in this cycle that satisfies the minimum ratio is $(4, 1)$. Therefore, $(4, 1)$ is chosen as the departing arc. In the next basis, there will still be three degenerate arcs, $(7, 8)$, $(2, 3)$, and $(5, 6)$, all of which will be directed toward the root. As a final note, observe that if $(2, 3)$ had been chosen as the departing arc, then the next basic tree would have had degenerate arc $(4, 1)$ directed *away* from the root and would no longer be a strongly feasible tree.

Another example of maintaining a strongly feasible tree is presented in Figure 9.26(b). Note that, in this case, $\Delta = 2$ and arcs $(4, 3)$ and $(5, 6)$ are candidates for the departing arc. Unlike the previous case, this is a nondegenerate pivot. However, the pivot will lead to a degenerate extreme point because there is a tie for the minimum-ratio test. The first common node in this case is node 6. Beginning at node 6 and tracing the unique cycle in the direction of the entering arc $(2, 3)$, we see that $(4, 3)$ is chosen as the departing arc. Note that $(5, 6)$ will be a degenerate arc in the next basic feasible solution and will be directed toward the root.

It is thus a simple process to start with and maintain strongly feasible trees during the course of the network simplex method. Why then does this ensure finite convergence? Because all degenerate arcs are directed toward the root, the departing arc in each degenerate pivot will be directed toward the root. But, we have already shown that updating the dual variables under the condition that the departing arc is directed toward the root node results in a strict decrease in a subset of the dual variables, whereas the remaining dual variables do not change. Therefore, the sum of the dual variables strictly decreases for each degenerate pivot, and it is impossible to repeat a basic tree during a sequence of degenerate pivots.

NETWORK FLOW PROBLEMS WITH BOUNDS ON THE FLOWS

Frequently, in developing a network flow model, arc capacities or bounds on the flows along the arcs are encountered. Mathematically, these additional constraints can be written as

$$l_{i,j} \leq x_{i,j} \leq u_{i,j}$$

where the quantities $l_{i,j}$ and $u_{i,j}$ are simply lower and upper bounds, respectively, on the flow along arc (i, j). By letting $\mathbf{l}$ and $\mathbf{u}$ represent the vectors of lower and upper bounds, respectively, the minimum-cost network flow problem with bounds can be written in the usual compact form:

$$\text{minimize } z = \mathbf{cx}$$

$$\text{subject to:}$$

$$\mathbf{Ax} = \mathbf{b}$$

$$\mathbf{l} \leq \mathbf{x} \leq \mathbf{u}$$

As we saw in Chapter 5, the basic idea in solving a linear programming problem with bounded variables is to handle the simple bounds on the variables in an implicit manner, thus maintaining a standard $m \times m$ basis matrix or, in this case, a standard rooted spanning tree. Recall that in the bounded-variables simplex method, a nonbasic variable will represent a variable that is either fixed at its lower bound or upper bound. Thus, vector $\mathbf{x}$ will be partitioned into the basic variables $\mathbf{x}_B$, the nonbasic variables at their lower bound $\mathbf{x}_{N_l}$, and the nonbasic variables at their upper bound $\mathbf{x}_{N_u}$. Partitioning the problem data into the appropriate components [i.e., $\mathbf{A} = (\mathbf{B} : \mathbf{N}_l : \mathbf{N}_u)$ and $\mathbf{c} = (\mathbf{c}_B, \mathbf{c}_{N_l}, \mathbf{c}_{N_u})$] results in the canonical system

$$z = \mathbf{c}_B \mathbf{x}_B + \mathbf{c}_{N_l} \mathbf{x}_{N_l} + \mathbf{c}_{N_u} \mathbf{x}_{N_u} \tag{9.47}$$

$$\mathbf{B} \mathbf{x}_B + \mathbf{N}_l \mathbf{x}_{N_l} + \mathbf{N}_u \mathbf{x}_{N_u} = \mathbf{b} \tag{9.48}$$

Following the development of Chapter 5, and letting J_l, J_u denote the index sets of the variables that are nonbasic at their lower bounds and upper bounds, respectively, we now solve for z and $\mathbf{x}_B$ in terms of the nonbasic variables to get

$$z = \mathbf{c}_B \mathbf{B}^{-1} \mathbf{b} - \sum_{(i,j) \in J_l} (z_{i,j} - c_{i,j}) x_{i,j} - \sum_{(i,j) \in J_u} (z_{i,j} - c_{i,j}) x_{i,j} \tag{9.49}$$

$$\mathbf{x}_B = \mathbf{B}^{-1} \mathbf{b} - \sum_{(i,j) \in J_l} \boldsymbol{\alpha}_{i,j} x_{i,j} - \sum_{(i,j) \in J_u} \boldsymbol{\alpha}_{i,j} x_{i,j} \tag{9.50}$$

By noting the form of (9.49), the optimality conditions for this minimization problem can be summarized as follows.

Optimality conditions (bounded-variables minimization problem)

$$z_{i,j} - c_{i,j} \leq 0, \qquad \text{for all } (i, j) \in J_l$$

and

$$z_{i,j} - c_{i,j} \geq 0, \qquad \text{for all } (i, j) \in J_u$$

Determining the Entering Variable

If the current basic feasible solution is not optimal, then we either choose to increase the flow on some arc where $x_{k,l}$ is currently nonbasic at its lower bound and $z_{k,l} - c_{k,l} > 0$, or we decrease the flow on an arc where $x_{k,l}$ is currently nonbasic at its upper bound and $z_{k,l} - c_{k,l} < 0$. As usual, we will choose to enter that nonbasic variable that forces the greatest rate of change of the objective. The entering variable $x_{k,l}$ will be that variable with $z_{k,l} - c_{k,l}$ defined by

$$z_{k,l} - c_{k,l} = \text{maximum}\left\{\underset{(i,j) \in J_l}{\text{maximum}} (z_{i,j} - c_{i,j}), \underset{(i,j) \in J_u}{\text{maximum}} - (z_{i,j} - c_{i,j})\right\} \qquad (9.51)$$

There are two cases to consider when determining the departing variable. These cases correspond to whether the flow on the entering arc is being increased from its lower bound or being decreased from its upper bound.

Increasing the Flow on a Nonbasic Arc from Its Lower Bound

Suppose that $z_{k,l} - c_{k,l} > 0$ and $x_{k,l}$ is currently nonbasic at its lower bound (i.e., $x_{k,l} = l_{k,l}$). Then the solution can be improved by increasing $x_{k,l}$. As with the standard network simplex method, let us add arc (k, l) to the current basic spanning tree and induce a flow change of $+\Delta$ around the created cycle in the *same* direction as arc (k, l) (see Figure 9.27). Thus, the flow on arc (k, l) becomes $x_{k,l} = l_{k,l} + \Delta$. Denote by D the set of basic arcs in this cycle that are decreasing in flow as a result of this induced flow and let I be the set of basic arcs in the cycle that are increasing in flow.

Entering Arc:
Flow Increasing

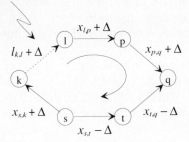

Figure 9.27 Increasing the flow on the entering arc.

Then, to maintain a feasible set of flows, Δ must satisfy the following:

$$\Delta \leq u_{k,l} - l_{k,l}$$

$$\Delta \leq x_{i,j} - l_{i,j}, \qquad \text{for all } (i, j) \in D$$

$$\Delta \leq u_{i,j} - x_{i,j}, \qquad \text{for all } (i, j) \in I$$

Thus, to achieve the maximum improvement possible in the objective value while maintaining feasibility, Δ should be chosen using

$$\Delta = \text{minimum } \{\delta_1, \delta_2, u_{k,l} - l_{k,l}\} \qquad (9.52)$$

where

$$\delta_1 = \begin{cases} \infty, & \text{if } D = \phi \\ \underset{(i, j) \in D}{\text{minimum}} \{x_{i,j} - l_{i,j}\}, & \text{otherwise} \end{cases} \qquad (9.53)$$

$$\delta_2 = \begin{cases} \infty, & \text{if } I = \phi \\ \underset{(i, j) \in I}{\text{minimum}} \{u_{i,j} - x_{i,j}\}, & \text{otherwise} \end{cases} \qquad (9.54)$$

Note that if $\Delta = \delta_1 = x_{p,q} - l_{p,q}$, then the departing variable is $x_{p,q}$, which becomes nonbasic at its lower bound. Similarly, if $\Delta = \delta_2 = u_{p,q} - x_{p,q}$, then $x_{p,q}$ departs at its upper bound. Finally, if $\Delta = u_{k,l} - l_{k,l}$, then the entering variable $x_{k,l}$ blocks itself and $x_{k,l}$ moves from nonbasic at its lower bound to nonbasic at its upper bound. In this last case, the basis matrix remains the same; the only things that change are the flows on the basic arcs and the objective value. Of course, if the earlier computations result in $\Delta = \infty$, then the flow on arc (k, l) can be increased without bound and no finite optimal solution exists.

Decreasing the Flow on a Nonbasic Arc from Its Upper Bound

Now consider the case when $z_{k,l} - c_{k,l} < 0$ and $x_{k,l}$ is currently nonbasic at its upper bound (i.e., $x_{k,l} = u_{k,l}$). Then the solution can be improved by decreasing the flow on arc (k, l). As in the previous case, let us add arc (k, l) to the current basic spanning tree. However, in this case, we will induce a flow change of $+\Delta$ around the created cycle in the *opposite* direction of arc (k, l), as in Figure 9.28. This will effectively decrease the flow on arc (k, l) and this flow will be given by $x_{k,l} = u_{k,l} - \Delta$.

Again denote by D the set of basic arcs in this cycle that are decreasing in flow as a result of this induced flow, and let I be the set of basic arcs in the cycle that are increasing in flow. Then Δ will again be defined precisely as in (9.52–9.54) and the network simplex method with bounded variables can be summarized as follows.

Entering Arc:

Flow Decreasing

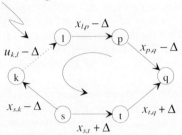

Figure 9.28 Decreasing the flow on the entering arc.

The network simplex algorithm with bounds on the flows (minimization problem)

STEP 1. *Determine the primal and dual solutions.* Given a feasible basis represented by a rooted spanning tree, compute the primal solution $\mathbf{x}_B$ (i.e., the flows) by starting at the ends of the tree and working toward the root using the values of the b_i and the nonbasic variable values. Compute the dual solution $\boldsymbol{\pi}$ (i.e., the node potentials) by starting at the root node and working toward the ends of the tree using the relationships defined in (9.29) and (9.30).

STEP 2. *Check for optimality.* Compute $z_{i,j} - c_{i,j}$ for each nonbasic arc using

$$z_{i,j} - c_{i,j} = \pi_i - \pi_j - c_{i,j}$$

If $z_{i,j} - c_{i,j} \le 0$, for all $(i, j) \in J_l$, and $z_{i,j} - c_{i,j} \ge 0$, for all $(i, j) \in J_u$, then the current basic feasible solution is optimal; stop. Otherwise, select as the entering arc that arc (k, l) with

$$z_{k,l} - c_{k,l} = \text{maximum}\{\underset{(i, j) \in J_l}{\text{maximum}}(z_{i,j} - c_{i,j}), \underset{(i, j) \in J_u}{\text{maximum}} -(z_{i,j} - c_{i,j})\}$$

Ties in the selection of $x_{k,l}$ may be broken arbitrarily. If $x_{k,l}$ is currently at its lower bound [i.e., $(k, l) \in J_l$], then go to Step 3. If $x_{k,l}$ is currently at its upper bound [i.e., $(k, l) \in J_u$], then go to Step 4.

STEP 3. *Increase $x_{k,l}$ from its current value of $l_{k,l}$.* Add the entering arc (k, l) to the rooted spanning tree and induce a flow of $+\Delta$ around the resulting cycle in the *same* direction as arc (k, l). Then each arc in the cycle is either increasing in flow by Δ or decreasing in flow by Δ. Let D represent the set of arcs that are decreasing in flow and let I represent the set of arcs that are increasing in flow. Compute Δ using (9.52–9.54).

(a) If $\Delta = \infty$, then the problem has an unbounded objective value; stop.

(b) If $\Delta = u_{k,l} - l_{k,l}$, then $x_{k,l}$ becomes nonbasic at its upper bound. Update the flows along the cycle by $+\Delta$ or $-\Delta$. The basis tree does not change. Return to Step 2.

(c) If $\Delta = \delta_1 = x_{p,q} - l_{p,q}$, then the departing variable $x_{p,q}$ becomes nonbasic at its lower bound. If $\Delta = \delta_2 = u_{p,q} - x_{p,q}$, then the departing variable $x_{p,q}$ becomes nonbasic at its upper bound. Update the flows along the cycle by $+\Delta$ or $-\Delta$, adding the entering arc (k, l) to the rooted spanning tree and removing the departing arc (p, q). Update the dual solution and return to Step 2.

STEP 4. *Decrease $x_{k,l}$ from its current value of $u_{k,l}$.* Add the entering arc (k, l) to the rooted spanning tree and induce a flow of $+\Delta$ around the resulting cycle in the *opposite* direction of arc (k, l). Then each arc in the cycle is either increasing in flow by Δ or decreasing in flow by Δ. Let D represent the set of arcs that are decreasing in flow and let I represent the set of arcs that are increasing in flow. Compute Δ using (9.52–9.54).

(a) If $\Delta = \infty$, then the problem has an unbounded objective value; stop.

(b) If $\Delta = u_{k,l} - l_{k,l}$, then $x_{k,l}$ becomes nonbasic at its lower bound. Update the flows along the cycle by $+\Delta$ or $-\Delta$. The basis tree does not change. Return to Step 2.

(c) If $\Delta = \delta_1 = x_{p,q} - l_{p,q}$, then the departing variable $x_{p,q}$ becomes nonbasic at its lower bound. If $\Delta = \delta_2 = u_{p,q} - x_{p,q}$, then the departing variable $x_{p,q}$ becomes nonbasic at its upper bound. Update the flows along the cycle by $+\Delta$ or $-\Delta$, adding the entering arc (k, l) to the rooted spanning tree and removing the departing arc (p, q). Update the dual solution and return to Step 2.

Example 9.6: Network Flows with Bounds

Consider the network flow problem given in Figure 9.29(a). For illustration purposes, assume that a starting basic feasible solution is given by the rooted spanning tree in Figure 9.29(b). An initial basic feasible solution could also be found using the phase I procedure discussed in the previous section. Note that $x_{1,2}$ and $x_{3,4}$ are nonbasic at positive values. (The notation $\overset{f}{\mapsto}$ is used to denote a nonbasic arc with a positive flow of f.) Whereas $x_{3,4}$ is nonbasic at its lower bound, $x_{1,2}$ is nonbasic at its upper bound. The flow on arc $(3, 2)$ is also nonbasic at its lower bound; however, because its lower bound is 0, it does not show graphically. The solution of this problem is outlined in Figure 9.30. The optimal solution given in iteration 4 can be

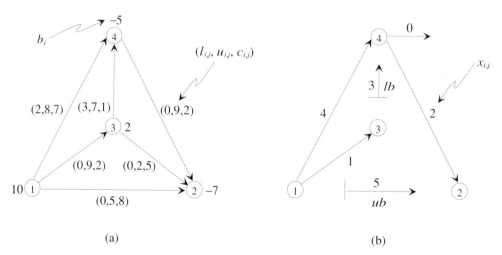

(a)   (b)

Figure 9.29   (a) Network for Example 9.6 and (b) initial basic feasible solution.

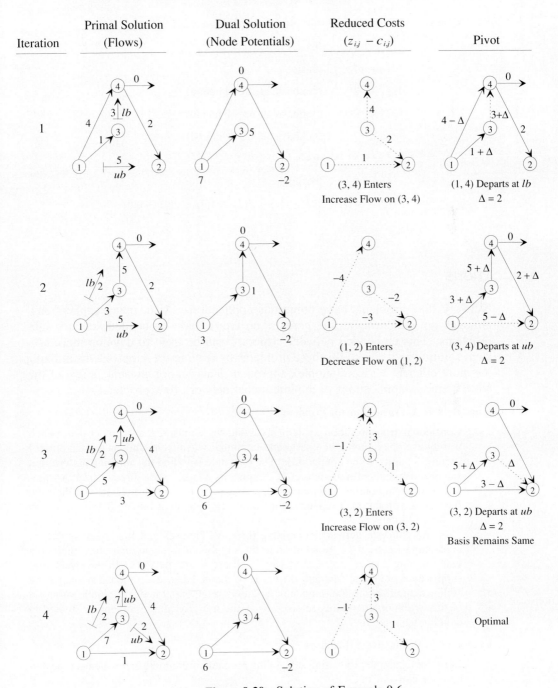

Figure 9.30   Solution of Example 9.6.

summarized as follows:

$$x^*_{1,2} = 1 \quad \text{(basic)}$$

$$x^*_{1,3} = 7 \quad \text{(basic)}$$

$$x^*_{1,4} = 2 \quad \text{(nonbasic at lower bound)}$$

$$x^*_{3,2} = 2 \quad \text{(nonbasic at upper bound)}$$

$$x^*_{3,4} = 7 \quad \text{(nonbasic at lower bound)}$$

$$x^*_{4,2} = 4 \quad \text{(basic)}$$

$$x^*_4 = 0 \quad \text{(basic)}$$

$$z^* = 8(1) + 2(7) + 7(2) + 5(2) + 1(7) + 2(4) + 0(0) = 61$$

APPLICATIONS

Network flow problems have numerous applications. Also, many applications, although not exhibiting a pure network structure, have embedded network sub-problems. This embedded network structure can be used to develop more efficient solution procedures for these problems by solving the subproblems utilizing the more efficient network simplex approach. This section presents a few of the more common applications of minimum-cost network flow problems.

Example 9.7: A Transshipment Problem

Consider a transportation system involving warehouses and retail outlets. Each warehouse i has s_i units of a particular commodity available, and each retail outlet j requires d_j units of the commodity. The objective is to find the shipping pattern that satisfies demands at minimum cost. Shipments can be made directly from a warehouse to a retail outlet or shipments can be made through intermediate warehouses. It is even possible to exchange shipments between retail outlets if this results in reduced shipping costs.

An example network containing three warehouses and four retail outlets is depicted in Figure 9.31. In addition to the supply and demand quantities, there is a unit shipping cost $c_{i,j}$ associated with each arc in the network. In the event that supply does not equal demand, the network can be balanced in the usual manner by adding a dummy warehouse node or a dummy retail outlet node. A simple variation of this problem, commonly referred to as the transportation problem, is studied in detail in Chapter 10.

Example 9.8: Production Scheduling

Consider again the production scheduling problem of Example 2.4, which is repeated here for convenience.

IMC, Inc., needs to schedule the monthly production of a certain item for the next 4 months. The unit production cost is estimated to be $12 for the first 2 months,

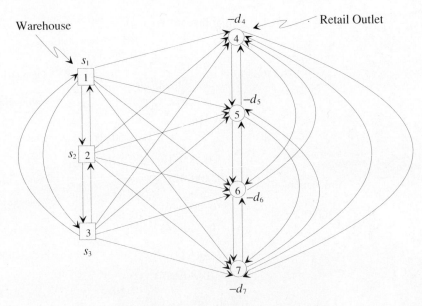

Warehouse

$-d_4$ Retail Outlet

Figure 9.31   A transshipment problem.

and $14 for the last 2 months. The monthly demands are 400, 750, 950, and 900 units. IMC, Inc., can produce a maximum of 800 units each month. In addition, the company can employ overtime during the second and third months, which increases monthly production by an additional 200 units. However, the cost of production increases by $4 per unit. Excess production can be stored at a cost of $3 per unit per month, but a maximum of 50 units can be stored in any month. By assuming that beginning and ending inventory levels are zero, how should the production be scheduled so as to minimize the total costs?

Although a linear programming formulation of this problem was developed in Chapter 2, a minimum-cost network flow representation of this problem is shown in Figure 9.32. Node s represents the supply or production node. Note that node s has been assigned a supply of 3000 units because that is the sum of the quantities demanded in the 4-month planning period. Flows along the arcs from s to each node m_i represent the amounts produced during regular time and overtime. The flow from node i to m_1 represents the initial inventory, which for this example is zero. Similarly, the flow from node m_4 to node f represents the final inventory. Flow along the arc from m_i to d_i represents the transfer of the finished products to meet the demands. Finally, flows along the arcs from m_i to m_{i+1} represent the inventory held from month i to month $i + 1$.

Example 9.9: A Leasing Problem

Consider the problem of leasing warehouse space over the next 3-month period. The requirements for each month i, $i = 1, 2, 3$, are known and are given by r_1, r_2, and r_3 square feet, respectively. There are several short-term lease options available and the problem is to choose the leasing policy that meets the requirements at minimum cost. For example, one policy would be to lease r_i square feet during month i for 1

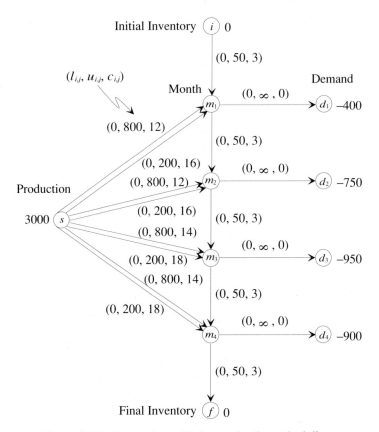

Figure 9.32 Network model for production scheduling.

month for each $i = 1, 2, 3$. However, because longer-term leases usually receive a lower leasing rate, it may be more economical to lease the maximum amount of space required for the entire 3-month period at the beginning of month 1. Obviously, there are many alternative leasing patterns in which different-length leases can be combined to meet the space requirements.

Let $x_{i,j}$ be the number of square feet leased at the beginning of month i for a period of j months, and let $c_{i,j}$ be the associated unit cost. Then a linear programming problem for determining the least-cost leasing policy can be written as follows:

$$\text{minimize } z = c_{1,1}x_{1,1} + c_{1,2}x_{1,2} + c_{1,3}x_{1,3} + c_{2,1}x_{2,1} + c_{2,2}x_{2,2} + c_{3,1}x_{3,1} \tag{9.55}$$

subject to

$$x_{1,1} + x_{1,2} + x_{1,3} \qquad\qquad\qquad \geq r_1 \tag{9.56}$$

$$x_{1,2} + x_{1,3} + x_{2,1} + x_{2,2} \qquad\quad \geq r_2 \tag{9.57}$$

$$x_{1,3} + \qquad\quad + x_{2,2} + x_{3,1} \geq r_3 \tag{9.58}$$

$$x_{i,j} \geq 0 \tag{9.59}$$

Note that constraints (9.56–9.58) do not exhibit a network structure. However, as we will see, it is possible to establish a network representation of this problem with a few simple manipulations. Let us begin by changing the constraints to equalities by subtracting the respective surplus variables, s_1, s_2, $s_3 \geq 0$. This results in the linear system:

$$x_{1,1} + x_{1,2} + x_{1,3} \qquad\qquad\qquad -s_1 \qquad\qquad = r_1 \qquad (9.60)$$

$$x_{1,2} + x_{1,3} + x_{2,1} + x_{2,2} \qquad\qquad -s_2 \qquad = r_2 \qquad (9.61)$$

$$x_{1,3} + \qquad + x_{2,2} + x_{3,1} \qquad\qquad -s_3 = r_3 \qquad (9.62)$$

Now, form a new linear system by subtracting (9.60) from (9.61), (9.61) from (9.62), and adding the redundant equation derived by multiplying equation (9.62) by -1. We then obtain

$$x_{1,1} + x_{1,2} + x_{1,3} \qquad\qquad\qquad -s_1 \qquad\qquad = r_1 \qquad (9.63)$$

$$-x_{1,1} \qquad\qquad + x_{2,1} + x_{2,2} \qquad + s_1 - s_2 \qquad = r_2 - r_1 \qquad (9.64)$$

$$-x_{1,2} \qquad - x_{2,1} \qquad + x_{3,1} \qquad + s_2 - s_3 = r_3 - r_2 \qquad (9.65)$$

$$-x_{1,3} \qquad - x_{2,2} - x_{3,1} \qquad\qquad + s_3 = -r_3 \qquad (9.66)$$

Note that this new constraint set has a network structure because each column contains exactly one $+1$ and one -1. Also observe that this constraint set represents a balanced network flow problem on four nodes because the sum of the right-hand-side values is zero. (That is, "supply" equals "demand.") To avoid confusion, let us denote the nodes corresponding to equations (9.63) to (9.66) by a, b, c, and d, respectively. Then the network flow representation of this problem is illustrated in Figure 9.33.

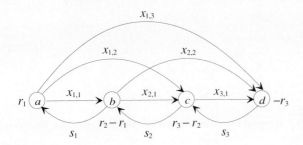

Figure 9.33 Network model for the leasing problem.

SUMMARY

In this chapter, an important class of specially structured linear programming problems was introduced. These problems can be represented in a convenient graphical manner and are referred to as network flow problems. Due to their special network structure, a very efficient implementation of the simplex algorithm can be used to derive an optimal solution with only a fraction of the effort required by the standard simplex method.

EXERCISES

9.1. Consider the following minimum cost network flow problem.

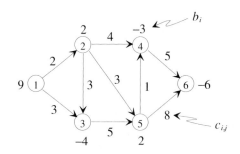

(a) Consider the rooted spanning tree consisting of arcs (1, 2), (2, 3), (2, 4), (5, 4), and (4, 6) with root node 6. Write the corresponding basis matrix in lower triangular form.

(b) Determine the primal solution (flows) and the dual solution (node potentials) corresponding to the rooted spanning tree basis in part (a).

(c) Beginning with the starting solution found in part (b), solve this problem by the network simplex method. Is the optimal solution unique? Explain.

9.2. Consider the following minimum-cost network flow problem.

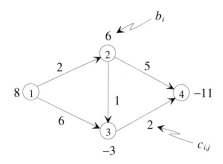

(a) Write the linear programming problem corresponding to this network flow problem.

(b) Using $x_{1,2}$, $x_{1,3}$, and $x_{2,4}$ as part of a starting basis, solve this problem by the network simplex method.

(c) For what range of values for $c_{1,3}$ will the current basis remain optimal?

(d) Suppose that $c_{2,4}$ changes from its current value of 5 to 2. Use sensitivity analysis to find the new optimal solution.

(e) Suppose that an additional arc from node 1 to node 4 is added with cost $c_{1,4} = 4$. Find the new optimal solution.

9.3. Solve the minimum-cost network flow problem given in Exercise 9.2 by the two-phase method.

9.4. Solve the following minimum-cost network flow problem by the two-phase method.

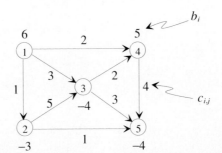

9.5. A manufacturer has three warehouses that must deliver a commodity to four customers. Warehouses 1, 2, and 3 have 300, 400, and 200 units available, respectively,
and Customers 1, 2, 3, and 4 have placed orders for 190, 260, 140, and 310 units,
respectively. The per unit costs (in dollars) of shipping from each warehouse to each
customer are given in Table 9.1. The manufacturer needs to determine the minimum-cost shipping schedule that satisifes all the demands.

TABLE 9.1

Warehouse	Customer			
	1	2	3	4
1	2	3	4	1
2	3	2	6	2
3	4	4	3	5

(a) Give a minimum cost network flow representation of this problem.
(b) By exploiting the special structure of the network in part (a), suggest a simple
method for finding a starting basic feasible solution.
(c) Beginning with the starting solution found in part (b), solve this problem by the
network simplex method.

9.6. Using $x_{1,2}$, $x_{2,3}$, $x_{2,5}$, $x_{2,6}$, and $x_{5,4}$ as part of a starting basis, solve the following
minimum-cost network flow problem.

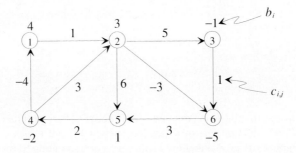

9.7. Solve the following problem by the network simplex method. (Observe that
$\sum_{i=1}^{4} b_i \neq 0$.)

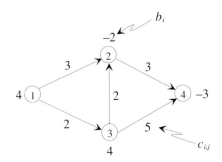

9.8 Consider the following network.

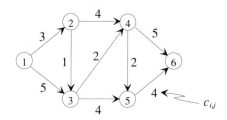

(a) Use the network simplex method to find the least-cost (shortest) path from node 1 to node 6. (Hint: Let $b_1 = 1$, $b_6 = -1$, and $b_i = 0$, for $i \neq 1, 6$.)

(b) Suppose that you wanted to find the shortest path from *each* node to node 6. Set up a minimum-cost network flow problem, the solution of which will yield the desired paths.

9.9. Consider the following linear programming problem. Give a balanced-network representation of this problem. (Note that not all constraints are equalities.)

$$\text{minimize } 6y_1 + 3y_2 + 5y_3 - 2y_4 + 4y_5 - y_6 + y_7$$

subject to:

$$
\begin{array}{rcl}
y_1 \quad\quad\quad - y_4 + y_5 \quad\quad\quad & \geq & 7 \\
-y_1 - y_2 + y_3 \quad\quad\quad\quad\quad\quad & = & -5 \\
\quad\quad\quad\quad\quad - y_5 + y_6 \quad\quad & = & 4 \\
+ y_2 \quad\quad + y_4 \quad\quad - y_6 + y_7 & \leq & 3 \\
- y_3 \quad\quad\quad\quad\quad - y_7 & \leq & -6 \\
y_j \geq 0, & & \text{for all } j
\end{array}
$$

9.10. Consider the following minimum-cost network flow problem. Suppose that $x_{1,2}$, $x_{2,5}$, $x_{3,2}$, and $x_{4,5}$ are part of a starting basis with root arc x_5 corresponding to node 5. Compute the primal and dual solutions corresponding to this basis. Now, imagine a simplex tableau representing this basic feasible solution. Complete the column of this tableau corresponding to the nonbasic variable $x_{2,4}$. Also give the row of this tableau corresponding to the basic variable $x_{2,5}$.

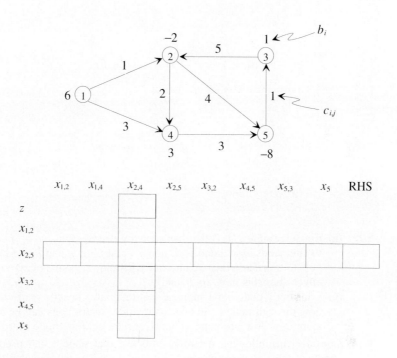

	$x_{1,2}$	$x_{1,4}$	$x_{2,4}$	$x_{2,5}$	$x_{3,2}$	$x_{4,5}$	$x_{5,3}$	x_5	RHS
z									
$x_{1,2}$									
$x_{2,5}$									
$x_{3,2}$									
$x_{4,5}$									
x_5									

9.11. Consider the following network. Let $x_{4,1}$, $x_{4,3}$, $x_{3,2}$, and $x_{5,4}$ be part of a starting basis with root node 5, where $x_{1,2}$ is nonbasic at its upper bound, and all other variables are nonbasic at their lower bounds (i.e., zero). Solve this problem by the network simplex method for bounded variables.

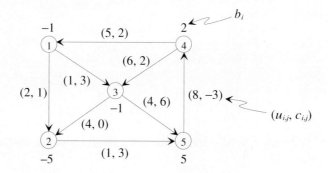

9.12. Solve the following minimum-cost network flow problem.

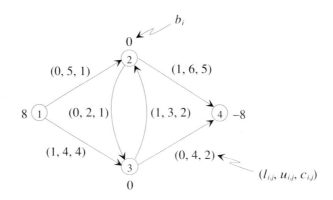

9.13. (a) Consider an arc (i, j) that has capacity restrictions $(l_{i,j}, u_{i,j})$. Suppose also that the demands associated with nodes i and j are b_i and b_j, respectively. Indicate how to transform this arc into one with a lower bound of zero on the flow. What effect does this have on b_i and b_j?

(b) Consider a node i and suppose that there are capacity restrictions (l_i, u_i) on the flow through node i. Indicate how these restrictions can be handled in a minimum-cost flow problem. (Hint: Consider splitting node i into two nodes.)

9.14. Consider the following network. Suppose that node 1 can produce an unlimited supply of units at a cost of α per unit, and node 2 can produce up to s_2 units at a cost of β per unit. Node 4 has a demand of d_4 units, and node 5 requires d_5 units. In addition, at most t_3 units can flow through node 3 at a cost of λ per unit. If $c_{i,j}$ is the cost per unit flow along each existing arc (i, j), set up a network that can be used to solve the problem of minimizing total cost.

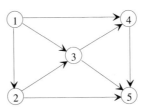

9.15. Solve the following minimum-cost network flow problem, where the lower bounds $l_{i,j} = 0$ for all arcs.

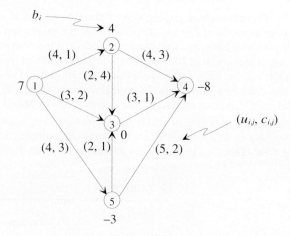

9.16. A company needs to schedule the production of a certain item for the next four periods. Due to seasonal fluctuations in the price of raw materials, the unit production cost is estimated to be $40, $35, $37, and $44 for periods 1 through 4, respectively. The finished product is shipped to two outlets where it is sold to the consumer. The estimated demands at each outlet during each planning period is given in Table 9.2. The per unit shipping cost from the production center to each outlet is anticipated to vary from period to period and is summarized in Table 9.3.

TABLE 9.2

| Period | Demands | |
	Outlet 1	Outlet 2
1	100	150
2	150	250
3	300	320
4	250	450

TABLE 9.3

| Period | Unit shipping costs | |
	Production to outlet 1	Production to outlet 2
1	$6	$4
2	6	5
3	7	6
4	8	5

The production facility can produce a maximum of 600 units each period. Excess production can be stored at either the production facility or at the outlets. The cost of storing one unit from one period to the next at the production facility is $2, but a maximum of 100 units can be stored in any period. The unit storage cost at Outlet 1 is $4, with a storage capacity of 30 units. The corresponding figures for Outlet 2 are $3 and 40 units. The objective is to satisfy all the demands at minimum total cost. Set up a minimum-cost network flow problem for solving this problem.

9.17. Consider the problem of making a one-to-one assignment of four workers to four jobs. Due to differing skills, each worker can complete each job at a different rate and cost. The total per job costs are summarized in Table 9.4.

TABLE 9.4

Worker	Job 1	Job 2	Job 3	Job 4
1	11	12	14	9
2	16	10	17	16
3	14	14	17	15
4	10	13	11	8

(a) Formulate a minimum-cost network flow problem for solving this problem. (Hint: Let $b_i = 1$ for each worker, $i = 1, \ldots, 4$, and $b_j = -1$ for each job, $j = 1, \ldots, 4$.)

(b) Note that each basic feasible solution for this problem will have seven basic arcs and the root arc. However, because only four assignments (i.e., $x_{i,j} = 1$) are possible, three of the basic variables must be zero in every basic feasible solution, that is, every basic feasible solution is degenerate of order 3. Determine a starting solution by denoting job node 4 as the root node, letting $x_{1,1} = x_{2,2} = x_{3,3} = x_{4,4} = 1$ and choosing three zero-flow arcs to form a *strongly feasible tree*.

(c) Starting with the basic feasible solution of part (b), solve this network flow problem maintaining a strongly feasible tree at each iteration.

9.18. Show that the following tree with two roots cannot be part of a network basis.

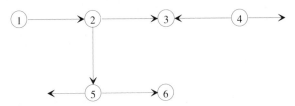

9.19. Two paths in a network are said to be *arc disjoint* if they contain no common arcs. Consider a connected directed graph with m nodes. Develop a minimum-cost network flow problem that will determine the number of arc disjoint paths between node 1 and node m. (Hint: Add an arc from node m to node 1 and assign appropriate costs and bounds to all the arcs in the resulting network.) Illustrate with an example.

CHAPTER 10

THE TRANSPORTATION
AND ASSIGNMENT PROBLEMS

CHAPTER OVERVIEW

In this chapter, we introduce two special network models, the transportation and assignment problems. As we will see, both of these problems are actually special cases of the minimum-cost network flow problem studied in Chapter 9. However, because of their own unique characteristics and applications, much research has been devoted to developing solution procedures tailored especially for these problems.

We begin by introducing the transportation problem. The name transportation problem is more or less a result of tradition and/or convenience, because the problem itself has many other applications in addition to that of the transportation of goods. Although the problem exhibits essentially the same mathematical structure as the minimum-cost network flow problem, the underlying network is a *bipartite graph*, that is, the nodes of the network model can be divided into two disjoint sets. Although the transportation problem can be solved as a standard linear programming problem as well as a minimum-cost network flow problem, the special bipartite structure allows it to be solved in a very efficient manner using a convenient tabular format. After some basic properties of the transportation model are discussed, a special implementation of the simplex algorithm is presented for solving this class of problems.

Following the transportation problem, we discuss the linear assignment problem, which is actually a special subclass of the transportation problem. As such, it could be treated by any of the approaches for solving the transportation problem. However, because of other complications, most notably degeneracy, the assignment problem can be more efficiently solved by other methods that have been especially tailored to the problem structure. The chapter concludes by presenting one of these methods for solving the assignment problem known as the *Hungarian algorithm*.

THE TRANSPORTATION PROBLEM

Figure 10.1 provides a convenient illustration of a general transportation problem. The underlying graph is a direct graph, however, the nodes of this network model can be partitioned into two sets. This partitioning is noted by sets S and D in Figure 10.1. Unlike the general directed networks discussed in Chapter 9, observe that all arcs in the transportation network are directed *from* nodes in S *to* nodes in D. There are no arcs directed from D to S. Also, there are no arcs connecting two nodes in S or two nodes in D. A graph of this type is called a *bipartite graph*. Each node in S is usually called a *source node*, and can be thought of as a node or terminal at which a supply of a particular commodity exists. The nodes in D are referred to as *destination nodes*. Destination nodes are those nodes that demand the goods stored at the source nodes. Thus, a convenient analogy is that source nodes represent warehouses and destination nodes represent customers.

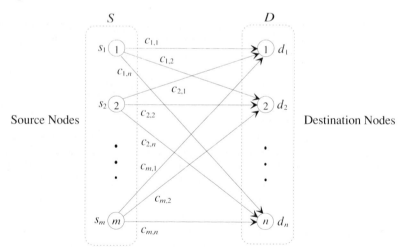

Figure 10.1   A general transportation problem.

Associated with each source node i, $i = 1, \ldots, m$, is a number s_i that represents the supply of goods available at node i. Similarly, a quantity d_j is associated with destination node j, $j = 1, \ldots, n$, and represents the quantity demanded at node j. The arcs connecting the source nodes to the destination nodes represent transportation routes between each individual source and destination. The quantity $c_{i,j}$ corresponding to each arc represents the *per unit* cost of shipping between the associated source and destination nodes.

The objective of the transportation problem is to find the minimal cost pattern of shipment, where the decision variables are given by

$$x_{i,j} = \text{quantity shipped from source node } i \text{ to destination node } j$$

The constraints are associated with the amounts available at each source and demanded at each destination. We will, in general, assume that there is an arc connecting each supply node to each demand node. If in reality flow on one or more of the arcs is not possible, then the $c_{i,j}$ values associated with these arcs may be set to an arbitrarily large value to make it prohibitively costly for any flow to actually occur.

As in Chapter 9, we will assume that the total supply equals the total demand, that is,

$$\sum_{i=1}^{m} s_i = \sum_{j=1}^{n} d_j$$

If this is not the case, supply and demand may be balanced in the usual manner by adding a dummy destination node or a dummy source node. That is, if $\sum_{i=1}^{m} s_i > \sum_{j=1}^{n} d_j$ (i.e., supply exceeds demand), we will add a dummy destination node with a demand of $(\sum_{i=1}^{m} s_i - \sum_{j=1}^{n} d_j)$ units. Physically, the allocation of goods from a source to a dummy destination simply means that these goods stay at the source node. Consequently, the cost of "shipping" from a source node to a dummy destination should reflect the actual costs associated with holding the goods at the source. Typically, in textbook examples, this cost is given as zero; however, in reality, it may reflect actual costs, such as inventory costs.

In a similar manner, if demand exceeds supply (i.e., $\sum_{j=1}^{n} d_j > \sum_{i=1}^{m} s_i$), a new dummy source node must be added to the problem. Physically, the allocation of goods from a dummy source to a given destination means that this destination node will be *short* the units shown allocated. Consequently, the cost of "shipping" from a dummy source node to a destination node should reflect the actual cost associated with a shortage at the respective destination. Again, the cost is typically given as zero. The process of balancing supply and demand will be discussed again later via an example.

Because we are assuming that total supply equals total demand, all supplies must be shipped and all demands must be satisfied. Note that the quantity $\sum_{j=1}^{n} x_{i,j} = x_{i,1} + x_{i,2} + \cdots + x_{i,n}$ is the total quantity *shipped from* node i, whereas $\sum_{i=1}^{m} x_{i,j} = x_{1,j} + x_{2,j} + \cdots + x_{m,j}$ is the total quantity *received at* node j. Thus, the resulting mathematical model can be written in summation notation as

$$\text{(TP)} \quad \text{minimize } z = \sum_{i=1}^{m} \sum_{j=1}^{n} c_{i,j} x_{i,j} \tag{10.1}$$

subject to:

$$\sum_{j=1}^{n} x_{i,j} = s_i, \qquad \text{for } i = 1, \ldots, m \qquad \text{(Supply Constraints)} \qquad (10.2)$$

$$\sum_{i=1}^{m} x_{i,j} = d_j, \qquad \text{for } j = 1, \ldots, n \qquad \text{(Demand Constraints)} \qquad (10.3)$$

$$x_{i,j} \geq 0, \qquad \text{for } i = 1, \ldots, m, j = 1, \ldots, n \qquad (10.4)$$

Note that the foregoing mathematical model for the transportation problem contains two main sets of constraints (ignoring, for the moment, the nonnegativity restrictions on $x_{i,j}$): one set of m constraints associated with the source nodes and one set of n constraints associated with the destination nodes. Also because there are m sources and n destinations, there will be mn variables in the problem each corresponding to an arc from a source node to a destination node. Thus, the transportation model is an equality constrained model with mn nonnegative variables and $m + n$ constraints.

Because of the bipartite nature of the network, the transportation model depicted in Figure 10.1 may also be represented in the convenient tabular form in Table 10.1. Note that the tableau has m rows, one for each source node, and n columns corresponding to the n destination nodes. Each cell corresponds to an arc and thus variable $x_{i,j}$ is associated with cell (i, j) in the tableau. Observe that the constraints are completely defined by this tableau because the sum of the $x_{i,j}$ in each row i must equal s_i, and the sum of the $x_{i,j}$ in each column j must equal d_j,

TABLE 10.1 TABLEAU REPRESENTATION OF THE TRANSPORTATION PROBLEM

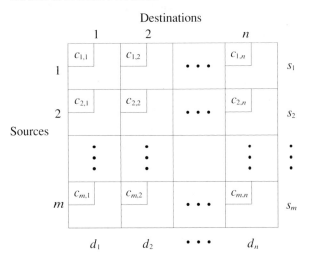

which are precisely the supply and demand constraints defined in (10.2) and (10.3), respectively.

So that we may more easily investigate the mathematical structure associated with this network model, consider the following simple example.

Example 10.1: A Transportation Problem

To help fix ideas, consider the transportation network with $m = 2$ source nodes and $n = 3$ destination nodes in Figure 10.2.

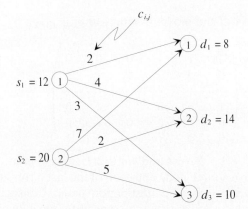

Figure 10.2 Transportation network of Example 10.1.

First, note that total supply equals total demand because $\sum_{i=1}^{2} s_i = \sum_{j=1}^{3} d_j = 32$. By utilizing the basic structure provided in (10.1–10.4), the mathematical model for this problem can be written as follows with the corresponding tableau representation provided in Table 10.2.

$$\text{minimize } 2x_{1,1} + 4x_{1,2} + 3x_{1,3} + 7x_{2,1} + 2x_{2,2} + 5x_{2,3} \qquad (10.5)$$

TABLE 10.2 TRANSPORTATION TABLEAU FOR EXAMPLE 10.1

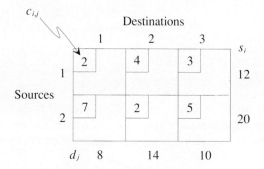

subject to:

$$x_{1,1} + x_{1,2} + x_{1,3} \hspace{2.5cm} = 12 \Big\}$$
$$x_{2,1} + x_{2,2} + x_{2,3} = 20 \Big\} \quad \text{(Supply Constraints)} \qquad (10.6)$$

$$x_{1,1} + \hspace{1.5cm} x_{2,1} \hspace{2cm} = 8 \Big\}$$
$$x_{1,2} + \hspace{1.5cm} x_{2,2} \hspace{1.5cm} = 14 \Big\} \quad \text{(Demand Constraints)} \qquad (10.7)$$
$$x_{1,3} + \hspace{2cm} x_{2,3} = 10 \Big\}$$

$$x_{i,j} \geq 0, \qquad \text{for } i = 1, 2; j = 1, 2, 3 \qquad (10.8)$$

The coefficient matrix of the model in (10.5–10.8) is given by

$$\begin{array}{c} node \\ 1 \\ 2 \\ \hline 1 \\ 2 \\ 3 \end{array}
\begin{array}{cccccc} x_{1,1} & x_{1,2} & x_{1,3} & x_{2,1} & x_{2,2} & x_{2,3} \\ \end{array}
\left(\begin{array}{ccc|ccc}
1 & 1 & 1 & 0 & 0 & 0 \\
0 & 0 & 0 & 1 & 1 & 1 \\
\hline
1 & 0 & 0 & 1 & 0 & 0 \\
0 & 1 & 0 & 0 & 1 & 0 \\
0 & 0 & 1 & 0 & 0 & 1
\end{array}\right)$$

This matrix has $m + n = 2 + 3 = 5$ rows and $mn = 2(3) = 6$ columns. Each of the first m rows corresponds to a source node and each of the last n rows (i.e., rows $m + 1$ through $m + n$) corresponds to a destination node. Consequently, each column, which corresponds to an arc of the transportation network, contains exactly two $+1$'s, with the remaining elements being 0. In general, variable $x_{i,j}$, which corresponds to arc (i, j), has $+1$ in supply constraint i and $+1$ in demand constraint j.

This at first may seem like a slightly different structure than that encountered in Chapter 9. However, it is easily seen to be equivalent by simply multiplying all demand constraints by -1. Because all constraints are equality constraints, this is quite straightforward and the coefficient matrix of the previous model becomes

$$\mathbf{A}^\circ = \begin{array}{c} node \\ 1 \\ 2 \\ \hline 1 \\ 2 \\ 3 \end{array}
\left(\begin{array}{ccc|ccc}
1 & 1 & 1 & 0 & 0 & 0 \\
0 & 0 & 0 & 1 & 1 & 1 \\
\hline
-1 & 0 & 0 & -1 & 0 & 0 \\
0 & -1 & 0 & 0 & -1 & 0 \\
0 & 0 & -1 & 0 & 0 & -1
\end{array}\right) \qquad (10.9)$$

Matrix $\mathbf{A}^0$, in this form, is often referred to as a *node–arc incidence matrix*. In general, modifying the transportation problem by multiplying the demand constraints by -1 results in the following mathematical model:

$$\text{minimize } z = \sum_{i=1}^{m} \sum_{j=1}^{n} c_{i,j} x_{i,j} \qquad (10.10)$$

subject to:

$$\sum_{j=1}^{n} x_{i,j} = s_i, \qquad \text{for } i = 1, \ldots, m \qquad \text{(Supply Constraints)} \qquad (10.11)$$

$$-\sum_{i=1}^{m} x_{i,j} = -d_j, \qquad \text{for } j = 1, \ldots, n \qquad \text{(Demand Constraints)} \qquad (10.12)$$

$$x_{i,j} \geq 0, \qquad \text{for } i = 1, \ldots, m; j = 1, \ldots, n \qquad (10.13)$$

To remain consistent with the development of Chapter 9, we will use the model of (10.10–10.13) when developing a solution procedure for the transportation problem. This, of course, is not necessary and an entirely equivalent solution procedure could be developed by directly using the model in (10.1–10.4). In the equivalent formulation (10.10–10.13), each column contains precisely one $+1$ and one -1, and the transportation problem has exactly the same mathematical structure as the minimum-cost network flow problem.

Letting $\mathbf{a}_{i,j}$ denote the column associated with $x_{i,j}$ in (10.11–10.12), observe that $\mathbf{a}_{i,j}$ can be written as

$$\mathbf{a}_{i,j} = \mathbf{e}_i - \mathbf{e}_{m+j} \qquad (10.14)$$

where the notation $\mathbf{e}_k$ corresponds to a vector of all zeros except for $+1$ in the kth position. For example, from (10.9), we see that

$$\mathbf{a}_{1,2} = \begin{pmatrix} 1 \\ 0 \\ 0 \\ -1 \\ 0 \end{pmatrix} = \begin{pmatrix} 1 \\ 0 \\ 0 \\ 0 \\ 0 \end{pmatrix} - \begin{pmatrix} 0 \\ 0 \\ 0 \\ 1 \\ 0 \end{pmatrix} = \mathbf{e}_1 - \mathbf{e}_{m+2}$$

where $m = 2$.

Thus, the transportation problem could be solved by the network simplex algorithm presented in Chapter 9. However, let us instead develop a variant of this algorithm that is especially suited to the bipartite network structure of the transportation problem. This new algorithm will be referred to as the *transportation simplex method* and will be applied using the transportation tableau depicted in Table 10.1. We begin the development of a solution procedure by investigating various properties of the transportation problem.

Properties of the Coefficient Matrix

As demonstrated in the previous section, the coefficient matrix $\mathbf{A}^0$ of the transportation problem (10.10–10.13) shares the same mathematical structure as that of the network flow problem of Chapter 9. Thus, the coefficient matrix also shares the property of *unimodularity* that guarantees that every basic feasible solution to the transportation will be all integer provided that the s_i and d_j values are integer.

By noting the form of $\mathbf{A}^0$ in (10.9), it is clear that the rows of $\mathbf{A}^0$ are linearly dependent because summing the rows results in the zero vector. In fact, because $\mathbf{A}^0$ has $m + n$ rows, it follows from Theorems 9.1 and 9.3 that the rank of $\mathbf{A}^0$ is $m + n - 1$. Therefore, as in the network simplex method, we will need to add a linearly independent column to $\mathbf{A}^0$ to attain a coefficient matrix with full row rank so that the simplex method can be applied. As in Chapter 9, this linearly independent column actually corresponds to an arc that is referred to as the root arc. Without loss of generality, we will assume that this artificial root arc is attached to demand node n as in Figure 10.3 (any node will work equally well), with the corresponding column vector

$$\mathbf{e}_{m+n} = \begin{pmatrix} 0 \\ \vdots \\ 0 \\ 1 \end{pmatrix}$$

Let $\mathbf{A} = (\mathbf{A}^0, \mathbf{e}_{m+n})$. Then the model to be solved by the simplex algorithm can be written

$$\text{minimize } \mathbf{cx} \tag{10.15}$$

subject to:

$$\mathbf{Ax} = \mathbf{b} = \begin{pmatrix} \mathbf{s} \\ -\mathbf{d} \end{pmatrix}$$

$$\mathbf{x} \geq \mathbf{0}$$

where

$$\mathbf{x} = \begin{pmatrix} x_{1,1} \\ \vdots \\ x_{m,n} \\ x_{\text{root}} \end{pmatrix}$$

$$\mathbf{s} = \begin{pmatrix} s_1 \\ \vdots \\ s_m \end{pmatrix}$$

$$\mathbf{d} = \begin{pmatrix} d_1 \\ \vdots \\ d_n \end{pmatrix}$$

Because $\text{rank}(\mathbf{A}) = m + n$, the artificial variable corresponding to the root arc will necessarily be a member of every basis. However, because supply equals demand, the flow on the root arc is always zero (i.e., $x_{\text{root}} = 0$) and its only role is in maintaining a basis matrix $\mathbf{B}$ of full row rank.

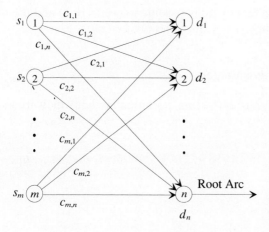

Figure 10.3 Transportation network with root arc.

Consequently, each basic feasible solution to the transportation problem will consist of the root arc along with $m + n - 1$ basic cells (arcs) in the transportation tableau. Because we will be using the transportation tableau, the root arc will not actually appear explicitly during the solution process. However, as in Chapter 9, it plays a fundamental role in maintaining a basis, which corresponds graphically to a rooted spanning tree.

Thus, each basic feasible solution will consist of $m + n - 1$ basic (nonempty) cells in the transportation tableau. This is a very important concept, and, as we will see, is fundamental to the operation of the transportation simplex method.

Prior to investigating methods for actually determining a basic feasible solution, let us look at some other basic properties that enable the transportation problem to be solved more efficiently than a general linear programming problem.

Feasibility of the Model

Let $s = \sum_{i=1}^{m} s_i = \sum_{j=1}^{n} d_j$ and consider the solution defined by

$$x_{i,j} = \frac{s_i d_j}{s}, \qquad \text{for } i = 1, \ldots, m; j = 1, \ldots, n \qquad (10.16)$$

Clearly, $x_{i,j} \geq 0$ if $s_i \geq 0$ and $d_j \geq 0$. Now substituting the expression for $x_{i,j}$ into (10.2) and (10.3) yields

$$\sum_{j=1}^{n} \frac{s_i d_j}{s} = \frac{s_i}{s} \sum_{j=1}^{n} d_j = \frac{s_i}{s} s = s_i \qquad (10.17)$$

$$\sum_{i=1}^{m} \frac{s_i d_j}{s} = \frac{d_j}{s} \sum_{i=1}^{m} s_i = \frac{d_j}{s} s = d_j \qquad (10.18)$$

Thus, the solution defined by (10.16) is feasible to (10.2–10.4) and the transportation problem will always have a feasible solution if $s_i \geq 0$, for $i = 1, \ldots, m$, and $d_j \geq 0$, for $j = 1, \ldots, n$.

Finiteness of the Objective Value

Consider again the constraints of the transportation model, which are written here in expanded form:

$$\sum_{j=1}^{n} x_{i,j} = x_{i,1} + x_{i,2} + \cdots + x_{i,n} = s_i, \qquad \text{for } i = 1, \ldots, m \qquad (10.19)$$

$$\sum_{i=1}^{m} x_{i,j} = x_{1,j} + x_{2,j} + \cdots + x_{m,j} = d_j, \qquad \text{for } j = 1, \ldots, n \qquad (10.20)$$

$$x_{i,j} \geq 0, \qquad \text{for } i = 1, \ldots, m; j = 1, \ldots, n \qquad (10.21)$$

Noting that $x_{i,j} \geq 0$, for all i, j, it follows from (10.19) and (10.20), respectively, that

$$0 \leq x_{i,j} \leq s_i, \qquad \text{for } j = 1, \ldots, n \qquad (10.22)$$

and

$$0 \leq x_{i,j} \leq d_j, \qquad \text{for } i = 1, \ldots, m \qquad (10.23)$$

Combining these results yields

$$0 \leq x_{i,j} \leq \text{minimum}\{s_i, d_j\}, \text{ for } i=1, \ldots, m; j = 1, \ldots, n \qquad (10.24)$$

Therefore, each $x_{i,j}$ is bounded below by zero and bounded above by minimum$\{s_i, d_j\}$. Consequently, the objective value of the transportation problem cannot be unbounded regardless of the values of the $c_{i,j}$.

Thus, we have shown that the transportation problem always has a finite optimal solution. Therefore, in developing the transportation simplex method, it is not necessary to check for feasibility or unboundedness.

Finding an Initial Basic Feasible Solution

Recall that a basic feasible solution has $m + n - 1$ basic (nonempty) cells in the transportation tableau. However, we cannot choose just any $m + n - 1$ cells and expect them to form a feasible basis. Care must be taken to ensure that the basis cells chosen correspond to a feasible rooted spanning tree, as in Chapter 9. To better understand this correspondence, consider the balanced transportation tableaux and corresponding graphical representations in Figure 10.4.

First, observe that $m = 3$ and $n = 3$, and therefore there will be $m + n - 1 = 5$ basic cells plus the root. Note that the solution in Figure 10.4(a) is a feasible solution because all flows are nonnegative and all supply and demand constraints

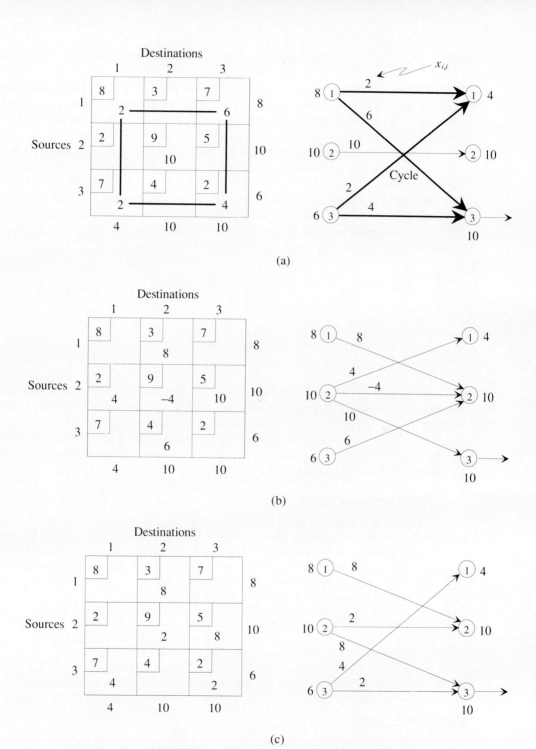

Figure 10.4 (a) a feasible solution that is not a basic solution, (b) a basic solution that is not a feasible solution, and (c) a basic feasible solution.

are satisfied. However, it is not a *basic* feasible solution even though there are five nonempty cells. This is because it does not correspond to a rooted spanning tree. This can be seen graphically by the highlighted cycle that is also noted on the tableau.

In contrast, the solution depicted in Figure 10.4(b) is a basic solution because it corresponds to a rooted spanning tree; however, it is not feasible because $x_{2,2} = -4 < 0$.

Finally, the solution given in Figure 10.4(c) is an example of a basic feasible solution because it not only corresponds to a rooted spanning tree, but also all flows are nonnegative and all constraints are satisfied. This basic feasible solution can be summarized as follows:

$$x_{1,2} = 8$$

$$x_{2,2} = 2$$

$$x_{2,3} = 8$$

$$x_{3,1} = 4$$

$$x_{3,3} = 2$$

As illustrated earlier, finding an initial basic feasible solution is not always an obvious task, and it is generally necessary to employ a systematic procedure that ensures that a feasible rooted spanning tree is constructed for the underlying bipartite graph. There are several methods for finding such a solution and we will present two of the most common approaches. The first of these is often referred to as the *northwest corner rule*.

The Northwest Corner Rule

STEP 0. Begin with a balanced transportation tableau.

STEP 1. Examine the cell in the northwest corner of the transportation tableau (i.e., the cell in the uppermost left-hand corner). Call this cell (i, j).

STEP 2. Assign the maximum number of units possible to cell (i, j) without violating supply constraint i and demand constraint j. Reduce the supply associated with row i and the demand associated with column j by the assigned amount.

STEP 3. If all units have been allocated, stop. Otherwise, continue as follows:

(a) If the supply in row i is completely satisfied, eliminate row i.

(b) If the demand in column j is completely satisfied, eliminate column j.

(c) If, however, *both* the supply in row i and the demand in row j are completely satisfied, eliminate *either* row i or column j, but *do not eliminate both*. (The next assignment will be 0 units and the resulting solution will be degenerate.)

Return to Step 1.

Example 10.2: The Northwest Corner Rule

Consider the transportation model given in tableau form in Table 10.3.

Observe that $\sum_{i=1}^{3} s_i = 200$, whereas $\sum_{j=1}^{3} d_j = 130$, and thus, supply exceeds demand by 70 units. We must start by creating a balanced system. Because supply exceeds demand, this is accomplished by adding a dummy destination node with a

TABLE 10.3 TRANSPORTATION TABLEAU OF EXAMPLE 10.2

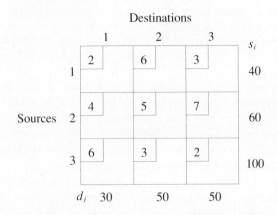

demand of 70 units. Each destination node corresponds to a column of the transportation tableau and, therefore, we are adding an additional column to the transportation tableau. This column is labeled as destination 4 in Table 10.4. Note that the demand associated with this column is 70 units, which corresponds precisely to the excess supply. Also note that we have assumed that the cost of "shipping" goods from a given source node to the dummy demand node is zero, because we are actually holding the goods at the source node.

TABLE 10.4 THE NORTHWEST CORNER RULE

Destinations

Sources	1	2	3	4	s_i
1	2 (30)	6 (10)	3	0	40 10 0
2	4	5 (40)	7 (20)	0	60 20 0
3	6	3	2 (30)	0 (70)	100 70 0
d_j	30 0	50 40 0	50 30 0	70 0	

Consider again the balanced transportation tableau in Table 10.4. The northwest corner rule begins in cell (1, 1). The maximum amount that can be allocated to cell (1, 1) without violating supply and demand is 30 units. Adjusting the supplies as a result of this allocation results in the supply of row 1 decreasing to 10 and the demand of column 1 decreasing to 0. Because the demand of destination 1 is satisfied, we eliminate column 1 from further consideration and move right to cell (1, 2). The maximum amount that can be allocated to cell (1, 2) is 10 units, which results in the supply at source 1 being fully utilized. Moving to cell (2, 2) next and then continuing in this manner results in the basic feasible solution derived in Table 10.4. Note that, as expected, there are $m + n - 1 = 3 + 4 - 1 = 6$ basic (nonempty) cells.

An improved algorithm for finding an initial basic feasible solution is the Vogel approximation method (VAM). VAM is slightly more involved than the northwest corner rule, but it is usually worth the extra effort in that, normally, a far better initial solution is obtained. In fact, the only thing that would lead the northwest corner rule to a good initial solution is sheer luck.

The reason for the improved performance of VAM is that it pays attention, in its development of a basic feasible solution, to the cost information contained within the transportation tableau. It does this through the establishment of "penalty numbers," which indicate the possible cost penalty associated with *not* assigning an allocation to the lowest cost cell in a given row or column. A version of the Vogel approximation method is given in what follows and then demonstrated using the example previously solved by the northwest corner rule.

The Vogel Approximation Method

STEP 0. Begin with a balanced transportation tableau.
STEP 1. Determine the VAM penalty numbers for each row and column as follows:
    (a) The penalty number for each row i is the absolute value of the difference between the cost of the lowest-cost cell and the cost of the next-lowest-cost cell in row i.
    (b) The penalty number for each column j is the absolute value of the difference between the cost of the lowest-cost cell and the cost of the next-lowest-cost cell in column j.
STEP 2. Select the row or column having the largest penalty number. In the event of ties, break the ties arbitrarily.
STEP 3. Assign the maximum number of units possible to the lowest-cost cell in the selected row or column. Call this cell (i, j). Reduce the supply associated with row i and the demand associated with column j by the assigned amount.
STEP 4. (a) If the supply in row i is completely satisfied, eliminate row i.
    (b) If the demand in column j is completely satisfied, eliminate column j.
    (c) If, however, *both* the supply in row i and the demand in row j are completely satisfied, eliminate *either* row i or column j, but *do not eliminate both*.
STEP 5. If there is only one row or one column remaining, assign all remaining units to the cells in the remaining row/column (make certain that *each and every* cell in the remaining row/column receives an assignment even if that assignment is 0 units); stop. Otherwise, return to Step 1.

Various refinements to the algorithm are possible, particularly in additional rules for the breaking of ties (as cited in Step 2). For example, in the event of a tie for the maximum penalty number, we could (and we will in our example) break the tie in favor of the row or column having the smallest associated $c_{i,j}$ value.

Example 10.3: The Vogel Approximation Method

Consider again the balanced transportation tableau of Example 10.2. Table 10.5 details the steps of the Vogel approximation method applied to find a basic feasible solution to this problem. In Table 10.5(a), the row and column penalties are computed as described in Step 1 of VAM. The maximum penalty is 4 (circled) and is associated with source row 2 of the tableau. Therefore, row 2 is chosen and the maximum amount possible is assigned to the lowest-cost cell in row 2. This results in $x_{2,4} = 60$ and the supply in row 2 is fully utilized. Row 2 is then eliminated and the tableau is updated as in Table 10.5(b).

Continuing with the steps of the algorithm results in the remaining tableau in Table 10.5. Note that in Table 10.5(c), there was a tie for the maximum penalty between source 1 and destination 2. Choosing source row 1 would result in an allocation to cell $(1, 4)$, whereas choosing destination column 2 would result in an allocation to cell $(3, 2)$. Because $c_{1,4} = 0$ and $c_{3,2} = 3$, the tie is broken by choosing cell $(1, 4)$, the lowest cost cell of the two. In assigning 10 units to cell $(1, 4)$, both the supply in source row 1 is completely utilized and the demand in destination column 4 is completely satisfied. However, by following Step 4(c) of the algorithm, source row 1 was chosen to be eliminated in this case, while retaining column 4.

Also note that when there was one row left in Table 10.5(d), it was necessary to assign a value of 0 to cell $(3, 4)$ because every cell must receive an allocation in the final row/column. This, of course, results in a degenerate solution, but note that again there are $m + n - 1 = 6$ basic (nonempty) cells. The basic feasible solution derived is summarized in Table 10.6. The reader is invited to graphically confirm that this basic feasible solution is, in fact, a rooted spanning tree.

As a final note, observe that the cost associated with the basic feasible solution of Table 10.6 is

$$z = 2(30) + 0(10) + 0(60) + 3(50) + 2(50) + 0(0) = 310$$

whereas the solution derived using the northwest corner rule in Example 10.2 yields

$$z = 2(30) + 6(10) + 5(40) + 7(20) + 2(30) + 0(70) = 520$$

Therefore, VAM generated a much better starting solution than the northwest corner rule in this case.

Optimality Conditions

Consider a basic feasible solution to the transportation problem defined by basis matrix $\mathbf{B}$. Let $z_{i,j} - c_{i,j}$ represent the reduced cost corresponding to variable $x_{i,j}$. Then, because the objective of the transportation problem is one of minimization, the optimality conditions are given by

$$z_{i,j} - c_{i,j} = \mathbf{c}_B \mathbf{B}^{-1} \mathbf{a}_{i,j} - c_{i,j} \le 0, \qquad \text{for all } (i, j) \qquad (10.25)$$

Now, consider the vector of dual variables for the transportation problem. Clearly, because there are $m + n$ constraints, there will be $m + n$ dual variables. Denote these dual variables by $\boldsymbol{\pi} = (\boldsymbol{\lambda}, \boldsymbol{\mu}) = (\lambda_1, \lambda_2, \ldots, \lambda_m, \mu_1, \mu_2, \ldots, \mu_n)$, where λ_i corresponds to supply constraint i, and μ_j corresponds to demand con-

TABLE 10.5 THE VOGEL APPROXIMATION METHOD (VAM)

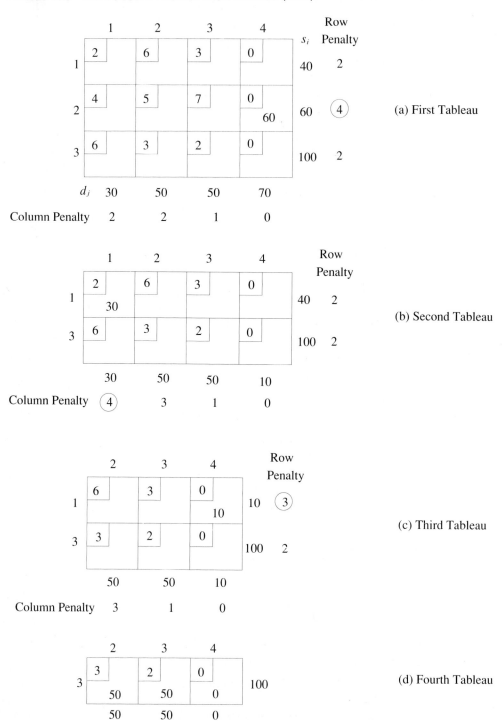

(a) First Tableau

(b) Second Tableau

(c) Third Tableau

(d) Fourth Tableau

TABLE 10.6 FINAL SOLUTION BY VAM

		Destinations		
	1	2	3	4
1	2 30	6	3	0 10
Sources **2**	4	5	7	0 60
3	6	3 50	2 50	0 0

straint j. With respect to the transportation tableau, λ_i corresponds to row i, and μ_j corresponds to column j.

Now, recalling that $\boldsymbol{\pi} = \mathbf{c}_B \mathbf{B}^{-1}$ and that $\mathbf{a}_{i,j} = \mathbf{e}_i - \mathbf{e}_{m+j}$, we can derive a simplified expression for the reduced cost:

$$
\begin{aligned}
z_{i,j} - c_{i,j} &= \mathbf{c}_B \mathbf{B}^{-1} \mathbf{a}_{i,j} - c_{i,j} \\
&= \boldsymbol{\pi} \mathbf{a}_{i,j} - c_{i,j} \\
&= (\boldsymbol{\lambda}, \boldsymbol{\mu})(\mathbf{e}_i - \mathbf{e}_{m+j}) - c_{i,j} \\
&= \lambda_i - \mu_j - c_{i,j}
\end{aligned}
\tag{10.26}
$$

Note that this expression for $z_{i,j} - c_{i,j}$ is of exactly the same form as that derived in Chapter 9.

Therefore, if the dual solution is known, it is quite straightforward to check for optimality using the optimality conditions:

$$
z_{i,j} - c_{i,j} = \lambda_i - \mu_j - c_{i,j} \leq 0, \qquad \text{for each cell } (i, j)
\tag{10.27}
$$

However, to check for optimality in this manner, we must first compute the values of the dual variables.

Determining the Dual Solution

From the previous section, we know that checking for optimality requires the computation of

$$
z_{i,j} - c_{i,j} = \lambda_i - \mu_j - c_{i,j}
\tag{10.28}
$$

for each *nonbasic cell* (i, j), which in turn, requires the computation of the current dual solution. How then does one efficiently compute the values of the dual variables?

Recall that for each *basic cell*,

$$
z_{i,j} - c_{i,j} = \lambda_i - \mu_j - c_{i,j} = 0
\tag{10.29}
$$

and it follows that

$$\lambda_i - \mu_j = c_{i,j}, \qquad \text{for each } \textit{basic cell } (i, j) \tag{10.30}$$

Because there are $m + n - 1$ basic cells and $m + n$ dual variables, (10.30) defines a system of $m + n - 1$ equations in $m + n$ unknowns. Recalling from Chapter 9 that the dual variable associated with the root node will be zero, we may set $\mu_n = 0$ (any other dual variable will work equally as well). We then solve for the remaining $m + n - 1$ dual variables using the $m + n - 1$ equations defined by (10.30). This, of course, is equivalent to solving for the dual solution by starting at the root node (demand node n) and working towards the ends of the tree, as we did in Chapter 9. However, because the graph is bipartite and is represented by the transportation tableau, it is more convenient to solve for the dual solution as described before.

Let us illustrate this technique by solving for the dual solution associated with the basic solution provided by Table 10.4, which is reproduced in Table 10.7 for convenience.

TABLE 10.7 COMPUTING THE DUAL SOLUTION

	1	2	3	4	
1	2 30	6 10	3	0	$\lambda_1 = 6$
2	4	5 40	7 20	0	$\lambda_2 = 5$
3	6	3	2 30	0. 70	$\lambda_3 = 0$
	$\mu_1 = 4$	$\mu_2 = 0$	$\mu_3 = -2$	$\mu_4 = 0$	

First, observe that cells $(1, 1)$, $(1, 2)$, $(2, 2)$, $(2, 3)$, $(3, 3)$, and $(3, 4)$ are the basic cells. By using (10.30), these basic cells define the system of equations:

$$\lambda_1 - \mu_1 = c_{1,1} \tag{10.31}$$

$$\lambda_1 - \mu_2 = c_{1,2} \tag{10.32}$$

$$\lambda_2 - \mu_2 = c_{2,2} \tag{10.33}$$

$$\lambda_2 - \mu_3 = c_{2,3} \tag{10.34}$$

$$\lambda_3 - \mu_3 = c_{3,3} \tag{10.35}$$

$$\lambda_3 - \mu_4 = c_{3,4} \tag{10.36}$$

Setting $\mu_4 = 0$, it is a trivial process to solve for the remaining dual variables using Equations (10.31–10.36). In fact, it is a quite straightforward to compute these values directly from the tableau as in Table 10.7.

Checking for Optimality

Checking for optimality is really quite simple once the dual solution is computed; the quantity

$$z_{i,j} - c_{i,j} = \lambda_i - \mu_j - c_{i,j} \qquad (10.37)$$

is computed for each nonbasic cell. The current basic feasible solution will be optimal if

$$z_{i,j} - c_{i,j} \leq 0, \qquad \text{for all cells } (i, j) \qquad (10.38)$$

For the transportation tableau of Table 10.7, the nonbasic cells are (1, 3), (1, 4), (2, 1), (2, 4), (3, 1), and (3, 2). Checking for optimality results in

$$z_{1,3} - c_{1,3} = \lambda_1 - \mu_3 - c_{1,3} = 6 - (-2) - 3 = 5$$

$$z_{1,4} - c_{1,4} = \lambda_1 - \mu_4 - c_{1,4} = 6 - 0 - 0 = 6$$

$$z_{2,1} - c_{2,1} = \lambda_2 - \mu_1 - c_{2,1} = 5 - 4 - 4 = -3$$

$$z_{2,4} - c_{2,4} = \lambda_2 - \mu_4 - c_{2,4} = 5 - 0 - 0 = 5$$

$$z_{3,1} - c_{3,1} = \lambda_3 - \mu_1 - c_{3,1} = 0 - 4 - 6 = -10$$

$$z_{3,2} - c_{3,2} = \lambda_3 - \mu_2 - c_{3,2} = 0 - 0 - 3 = -3$$

These values are typically computed directly on the tableau and recorded in the upper right-hand corner of each nonbasic cell, as in Table 10.8. Observe that the current basic solution is not optimal and we would chosen cell (1, 4) as the entering cell because variable $x_{1,4}$ corresponds to the most positive $z_{i,j} - c_{i,j}$. We will generally refer to the entering cell as cell (k, l).

TABLE 10.8 DETERMINING $z_{i,j} - c_{i,j}$ FOR THE NONBASIC CELLS

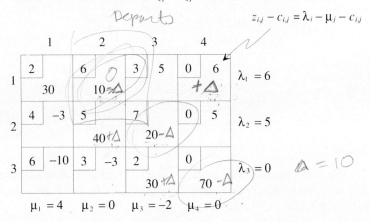

Determining the Departing Variable

To determine the departing variable, we will, as we did in Chapter 9, utilize the unique cycle created by the entering cell (arc). We begin by inducing a flow of $+\Delta$ in the entering cell, which is denoted by cell (k, l). Because all other nonbasic variables must remain fixed at zero, the flow in some basic cell in row k must be *reduced* by Δ so that row k of the tableau stills sums to s_k. Call the basic cell in row k, which is reduced by Δ, cell (k, t). But because the flow in cell (k, t) has been reduced by Δ, another basic cell in column t must be *increased* by Δ to maintain the sum of column t at d_t. This process is repeated until the unique Δ-cycle is found that begins and ends at cell (k, l).

Once the Δ-cycle is found, the determination of the departing cell is found by simply examining the cells that are *decreasing* by Δ and choosing that cell that will be driven to zero first by increasing Δ. This process is now illustrated using the entering cell $(k\ l) = (1, 4)$ determined previously in Table 10.8.

We begin by entering a flow of $+\Delta$ in cell $(1, 4)$, as in Table 10.9. Because we have increased the flow in row 1 by an amount Δ, we must reduce the flow of some basic cell in row 1 by Δ. Similarly, the flow of some basic cell in column 4 must be reduced by Δ. Obviously, because cell $(3, 4)$ is the only basic cell in column 4, $x_{3,4}$ will be reduced by Δ. Note that this is indicated by the entry of $(70 - \Delta)$ in cell $(3, 4)$. Now, because the flow in cell $(3, 4)$ has decreased by Δ, the flow in another basic cell in row 3 must increase by Δ. This results in $(30 + \Delta)$ in cell $(3, 3)$. Continuing in this manner results in the *unique* Δ-cycle in Table 10.9. Note that not all basic cells are involved in the cycle. In this case, the flow in basic cell $(1, 1)$ does not change. The flow in this cell could not possibly be modified because there is no other basic cell in column 1.

TABLE 10.9 DETERMINING THE DEPARTING VARIABLE

$\Delta = \min\{10, 20, 70\} = 10$

Because the flows in cells $(1, 2)$, $(2, 3)$, and $(3, 4)$ are decreasing by Δ, it is quite easy to compute Δ as $\Delta = \text{minimum}\{10, 20, 70\} = 10$. The departing cell is then cell $(1, 2)$.

The simplex method as applied to the transportation tableau can now be summarized as follows.

The transportation simplex algorithm (minimization problem)

STEP 0. *Determine a basic feasible solution.* Find an initial basic feasible solution using the northwest corner rule or VAM.

STEP 1. *Determine the dual solution.* Given a basic feasible solution represented by $m + n - 1$ basic (nonempty) cells, compute the dual solution $\pi = (\lambda, \mu) = (\lambda_1, \ldots, \lambda_m, \mu_1, \ldots, \mu_n)$ by using the *basic cells* and the fact that $z_{i,j} - c_{i,j} = \lambda_1 - \mu_j - c_{i,j} = 0$ for each of these basic cells. This results in a system of $m + n$ equations in $m + n - 1$ unknowns. Arbitrarily set one dual variable equal to 0 (e.g., $\mu_n = 0$) and then solve for the remaining λ_i and μ_j using

$$\lambda_i - \mu_j = c_{i,j}$$

STEP 2. *Check for optimality.* Compute $z_{i,j} - c_{i,j}$ for each nonbasic (empty) cell using

$$z_{i,j} - c_{i,j} = \lambda_i - \mu_j - c_{i,j}$$

If $z_{i,j} - c_{i,j} \leq 0$, for all i, j, then the current solution is optimal; stop. Otherwise, choose as the entering cell that cell with the most positive $z_{i,j} - c_{i,j}$. Call the entering cell (k, l) and call the corresponding flow $x_{k,l}$.

STEP 3. *Determine the departing variable.* Assign a value of $+\Delta$ units to the entering cell. By using the entering cell and the basic cells, construct the unique Δ-cycle through the tableau, beginning and ending in the entering cell. Then each cell in the cycle is either increasing by Δ or decreasing by Δ. Let D represent the set of cells that are decreasing by Δ. Compute

$$\Delta = \underset{(i,j)\in D}{\text{minimum}} \{x_{i,j}\}$$

Let $(p, q) \in D$ be the cell associated with this minimum. That is, $x_{p,q}$ is the departing variable.

STEP 4. *Pivot and update.* Update the flows along the created cycle by $+\Delta$ or $-\Delta$. The entering cell (k, l) now has a value of $+\Delta$ and the departing cell (p, q) becomes empty (nonbasic).

Return to Step 1.

Example 10.4: The Transportation Simplex Method

A widget manufacturer has three nationwide distribution centers that are to distribute widgets to four retail stores. The distribution centers have 100, 300, and 200 units available, and the demands at the retail stores are projected to be 80, 100, 200, and 220 units. Given the cost matrix (in hundreds of dollars) in Table 10.10, find the transportation schedule that minimizes the cost.

We begin the solution process by setting up the *balanced* transportation tableau in Table 10.11. And for illustrative purposes, we will begin the execution of the transportation simplex algorithm with the basic feasible solution provided in the table. The iterations of the algorithm are summarized in Tables 10.12 to 10.14, with Table 10.14 providing the optimal solution.

TABLE 10.10 UNIT TRANSPORTATION COSTS (IN HUNDREDS OF DOLLARS)

Distribution centers	Retail stores			
	1	2	3	4
1	3	2	6	6
2	4	5	4	3
3	1	6	9	6

TABLE 10.11 TRANSPORTATION TABLEAU AND INITIAL SOLUTION

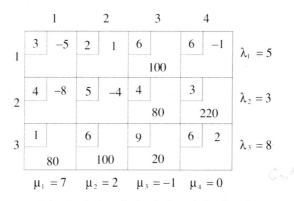

TABLE 10.12 ITERATION 1

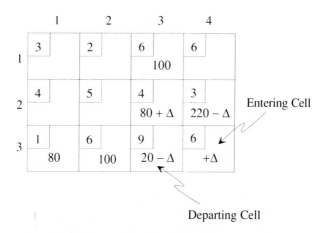

(a) Primal solution, dual solution, and reduced costs

(b) Pivot, $\Delta = \min\{20, 220\} = 20$

382

TABLE 10.13 ITERATION 2

	1	2	3	4	
1	3 −3	2 3	6 100	6 −1	$\lambda_1 = 5$
2	4 −6	5 −2	4 100	3 200	$\lambda_2 = 3$
3	1 80	6 100	9 −2	6 20	$\lambda_3 = 6$
	$\mu_1 = 5$	$\mu_2 = 0$	$\mu_3 = -1$	$\mu_4 = 0$	

(a) Primal solution, dual solution, and reduced costs

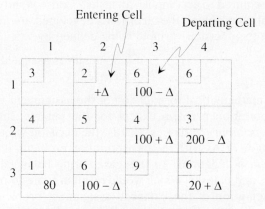

(b) Pivot, $\Delta = \min\{100, 100, 200\} = 100$

TABLE 10.14 OPTIMAL SOLUTION

	1	2	3	4	
1	3 −6	2 100	6 −3	6 −4	$\lambda_1 = 2$
2	4 −6	5 −2	4 200	3 100	$\lambda_2 = 3$
3	1 80	6 0	9 −2	6 120	$\lambda_3 = 6$
	$\mu_1 = 5$	$\mu_2 = 0$	$\mu_3 = -1$	$\mu_4 = 0$	

Note that the optimal solution provided by Table 10.14 is degenerate because basic variable $x_{3,2} = 0$. This is a direct result of the tie for the departing cell that occurred in Table 10.13(b). The tie was broken arbitrarily by choosing cell $(1, 3)$ as the departing cell and maintaining cell $(3, 2)$ as a basic cell at value zero.

THE ASSIGNMENT PROBLEM

The assignment problem gets its name from a particular application in which we wish to assign "individuals" to "tasks" (or tasks to machines, and so on). It is assumed that each individual must be assigned to *only one task*, and each task is assigned to *only one individual*. Various personnel assignment problems *may* fit the assumptions and structure of such a model. The objective of the assignment problem is to minimize the total cost of the assignments made, where this cost may be in terms of dollars required to perform the tasks or, perhaps, the sum of the times required to accomplish all tasks. Consequently, as was the case with the transportation problem, the algorithms developed are normally of the minimizing type, with the decision variable being defined by

$$x_{i,j} = \begin{cases} 1, & \text{if individual } i \text{ is assigned to task } j \\ 0, & \text{otherwise} \end{cases}$$

The mathematical formulation of the assignment problem is very similar to the transportation problem except for the notable difference that the decision variables are binary variables, that is, the decision variables can only assume the values of 0 or 1. As with the transportation problem, it is generally assumed that the problem is balanced. In this case, being balanced means that there are m individuals and m tasks, which would result in a one-to-one assignment. A graphical representation of the balanced assignment problem is illustrated in Figure 10.5. If the problem is not balanced, then additional artificial individuals or tasks are created so that a one-to-one assignment is possible. This is essentially the

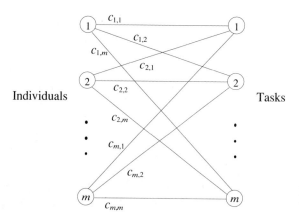

Individuals   Tasks

Figure 10.5 Network representation of the assignment problem.

same idea that was used to balance the transportation problem earlier in this chapter.

The *integer programming* formulation for the balanced assignment problem can then be written as follows:

$$\text{(AP)} \quad \text{minimize } z = \sum_{i=1}^{m} \sum_{j=1}^{m} c_{i,j} x_{i,j} \tag{10.39}$$

subject to:

$$\sum_{j=1}^{m} x_{i,j} = 1, \qquad \text{for } i = 1, \ldots, m \tag{10.40}$$

$$\sum_{i=1}^{m} x_{i,j} = 1, \qquad \text{for } j = 1, \ldots, m \tag{10.41}$$

$$x_{i,j} = 0, 1, \quad \text{for } i = 1, \ldots, m; j = 1, \ldots, m \tag{10.42}$$

Obviously, constraints (10.40) and (10.41) are of precisely the same form as those of the transportation problem, except that each $s_i = 1$ and each $d_j = 1$. Constraint (10.40) specifies that each individual is assigned to exactly one task, and (10.41) ensures that each task is assigned to precisely one individual.

Because the constraint matrix has the same form as the transportation problem, the assignment problem shares the property of unimodularity, which guarantees that every basic feasible solution is all integer. Therefore, the integer restrictions (10.42) are satisfied automatically and the assignment problem can be written as the equivalent *linear programming problem*:

$$\text{(ALP)} \quad \text{minimize } z = \sum_{i=1}^{m} \sum_{j=1}^{m} c_{i,j} x_{i,j} \tag{10.43}$$

subject to:

$$\sum_{j=1}^{m} x_{i,j} = 1, \qquad \text{for } i = 1, \ldots, m \tag{10.44}$$

$$\sum_{i=1}^{m} x_{i,j} = 1, \qquad \text{for } j = 1, \ldots, m \tag{10.45}$$

$$x_{i,j} \geq 0, \qquad \text{for } i = 1, \ldots, m; j = 1, \ldots, m \tag{10.46}$$

Notice that it is not necessary to write the variable restrictions (10.46) as $0 \leq x_{i,j} \leq 1$, because the nonnegativity restrictions together with constraints (10.44) and (10.45) implicitly force $x_{i,j}$ to be less than or equal to 1.

The linear programming formulation of the assignment problem makes it quite obvious that it is a special subclass of the transportation problem. As such, it can be solved by any of the methods that were applicable to the transportation problem. However, basic feasible solutions of the linear programming formulation of the assignment problem have an additional property that reduces the efficiency of the simplex algorithm when applied to this problem.

Observe that the mathematical model of (10.43–10.46) has $2m$ constraints and m^2 variables. Because it shares a common network structure with the transportation problem, the rows of the coefficient matrix associated with (10.44–10.45) are linearly independent and the rank of this matrix will be $2m - 1$. Therefore, each basic feasible solution of the assignment problem will have $2m - 1$ basic variables and the root arc. But because we are constructing a one-to-one assignment between m individuals and m tasks, only m of the $x_{i,j}$ can take on the value of 1. Consequently, each basic feasible solution consists of $2m - 1$ basic variables, m of which have the value 1 with the remaining $m - 1$ basic variables having the value 0. Thus, every basic feasible solution to the assignment problem is a degenerate solution with $m - 1$ degenerate basic variables.

Recall, from Chapter 4, that during a degenerate pivot, the solution actually remains at the same extreme point and only the basis matrix changes. Consequently, the objective does not improve during a degenerate pivot. Because there are always $m - 1$ degenerate basic variables, there is a high likelihood of degenerate pivots occurring. In fact, in practice, over 90% of the pivots that occur are degenerate pivots, dramatically decreasing the efficiency of the simplex algorithm.

In an attempt to overcome this problem, several alternative algorithms have been developed for the assignment problem. One such algorithm, which was developed by Kuhn (1955), is called the *Hungarian algorithm*. The algorithm falls into a class of algorithm called *primal-dual algorithms*, and its operation is quite different from the standard simplex method. As the name primal-dual suggests, the algorithm relies on both the primal and dual problems during the solution process. Let us then begin our development of the Hungarian algorithm by writing the dual problem corresponding to problem (ALP).

The Dual Problem and Complementary Slackness

Let $\pi = (\lambda, \mu) = (\lambda_1, \ldots, \lambda_m, \mu_1, \ldots, \mu_m)$ be the vector of dual variables corresponding to the constraints of (ALP), where λ corresponds to constraint (10.44), and μ corresponds to constraint (10.45). Clearly, the dual variables are unrestricted because all the constraints of the primal problem are equalities. The dual problem of (ALP) can then be written as follows:

$$\text{(DALP)} \quad \text{maximize} \sum_{i=1}^{m} \lambda_i + \sum_{j=1}^{m} \mu_j \tag{10.47}$$

subject to:

$$\lambda_i + \mu_j \leq c_{i,j}, \quad \text{for } i = 1, \ldots, m; j = 1, \ldots, m \tag{10.48}$$

$$\lambda, \mu \text{ unrestricted} \tag{10.49}$$

The dual constraint $\lambda_i + \mu_j \leq c_{i,j}$ corresponds to primal variable $x_{i,j}$ and as a result the complementary slackness conditions are given by

$$x_{i,j}(c_{i,j} - \lambda_i - \mu_j) = 0, \qquad \text{for } i = 1, \ldots, m; j = 1, \ldots, m \qquad (10.50)$$

Notice that the quantity $(c_{i,j} - \lambda_i - \mu_j)$ is simply the value of the dual slack variable in constraint (10.48). Therefore, the complementary slackness conditions simply state that if the dual constraint is nonbinding (i.e., $\lambda_i + \mu_j < c_{i,j}$), then the corresponding $x_{i,j} = 0$. Similarly, if $x_{i,j} = 1$, then the corresponding dual constraint is binding (i.e., $\lambda_i + \mu_j = c_{i,j}$).

Thus, given a feasible solution to the dual, the complementary slackness conditions could be used to identify the set of primal variables that must be equal to zero, and the set of primal variables which *may* be equal to one. For notational convenience, define the index sets

$$P = \{(i, j): c_{i,j} - \lambda_i - \mu_j > 0\} \qquad (10.51)$$

$$Q = \{(i, j): c_{i,j} - \lambda_i - \mu_j = 0\} \qquad (10.52)$$

for a given dual solution, λ, μ. Then, by complementary slackness,

$$x_{i,j} = 0, \qquad \text{for all } (i, j) \in P \qquad (10.53)$$

$$x_{i,j} = 0 \text{ or } 1, \qquad \text{for all } (i, j) \in Q \qquad (10.54)$$

A Basic Primal-Dual Strategy

Before presenting the details of the Hungarian algorithm, a basic primal-dual strategy can be described as follows:

1. Choose a feasible solution to the dual problem. (Note that this solution need not be a *basic* feasible solution.)
2. Using the complementary slackness conditions, identify the sets P and Q defined in (10.51) and (10.52), respectively.
3. Set $x_{i,j} = 0$, for all $(i, j) \in P$, and try to find a feasible solution to the primal constraint set using *only* $x_{i,j}$, for $(i, j) \in Q$. If such a primal solution is found, then it is optimal; stop. Otherwise, modify the dual solution and return to Step 2.

Note that the determination of a primal feasible solution in Step 3 depends on the set Q. Obviously, if Q is empty, then determining a primal feasible solution is not possible. Logically, it would be advantageous for Q to contain as many elements as possible, thus making it more likely that a feasible solution to the primal can be found.

If a feasible solution is found in Step 3, then this solution represents an optimal solution because all three components of the optimality conditions have been satisfied. That is, primal feasibility, dual feasibility, and complementary slackness are all satisfied by the current solution.

If a primal feasible solution cannot be found in Step 3, then the index set Q does not contain an adequate set of indices. The dual solution needs to be modified in some way. The purpose of modifying the dual solution is to render a different set of dual constraints binding and thus change the elements of the set Q. By introducing at least one new element into the set Q, it may be possible to find a primal feasible solution.

Choosing an Initial Dual Feasible Solution

Let us now begin to examine some of the details of the basic algorithmic process. For simplicity throughout, we will use the assignment matrix illustrated in Table 10.15 to describe the assignment problem. Note that this simple matrix completely describes the problem because the supply and demands associated with each row and column are one.

TABLE 10.15 ASSIGNMENT MATRIX

Tasks

		1	2		m
	1	$c_{1,1}$	$c_{1,2}$	$\bullet\bullet\bullet$	$c_{1,m}$
	2	$c_{2,1}$	$c_{2,2}$	$\bullet\bullet\bullet$	$c_{2,m}$
Individuals	3	$\vdots$	$\vdots$		$\vdots$
	m	$c_{m,1}$	$c_{m,2}$	$\bullet\bullet\bullet$	$c_{m,m}$

Now consider the process of choosing a feasible dual solution. Consider the dual solution defined by

$$\lambda_i = \underset{j}{\text{minimum}}\ \{c_{i,j}\} \tag{10.55}$$

$$\mu_j = \underset{i}{\text{minimum}}\ \{c_{i,j} - \lambda_i\} \tag{10.56}$$

Equation (10.55) simply specifies that the dual variable λ_i is chosen as the smallest element in row i of the assignment matrix. For each i, we then subtract λ_i from all the entries in row i. This will generate at least one zero element in every row of the matrix. The dual variable μ_j is then chosen as the smallest element in column j. Subtracting μ_j from each of the entries in column j results in what will be referred to as the *reduced matrix*, which has as entries $c_{i,j} - \lambda_i - \mu_j$. Note that by construction, the reduced matrix has at least one zero in each row and at least one zero in each column and all the entries are nonnegative.

Thus, this process produces a dual feasible solution because $c_{i,j} - \lambda_i - \mu_j \geq 0$, for all (i, j). And because each row and each column contains at least one zero, we see that at least m of the dual constraints will be binding. Thus, using this method for determining a dual feasible solution will result in the index set Q containing at least m elements. Choosing the dual solution and constructing the reduced matrix is now illustrated via the following example.

Example 10.5: Determining the Reduced Matrix

Consider the assignment problem with the cost matrix provided in Table 10.16(a). The dual solution determined using (10.55) and (10.56) is shown in Tables 10.16(a) and 10.16(b), with the computed reduced matrix depicted in Table 10.16(c). Note that as expected, the reduced matrix contains at least one zero in every row and column. In this case, there are six zeros contained in the reduced matrix, each corresponding to a binding dual constraint. Thus, for the reduced matrix of Table 10.16(c), the index set $Q = \{(1, 1), (1, 3), (2, 2), (2, 4), (3, 2), (4, 1)\}$.

TABLE 10.16   DETERMINING THE REDUCED MATRIX

$c_{i,j}$

2	10	3	17	$\lambda_1 = 2$
5	3	9	10	$\lambda_2 = 3$
8	2	5	14	$\lambda_3 = 2$
3	5	10	16	$\lambda_4 = 3$

(a) Original matrix

$c_{i,j} - \lambda_i$

0	8	1	15
2	0	6	7
6	0	3	12
0	2	7	13

$\mu_1 = 0 \quad \mu_2 = 0 \quad \mu_3 = 1 \quad \mu_4 = 7$

(b) Matrix after subtracting λ_i

$c_{i,j} - \lambda_i - \mu_j$

0	8	0	8
2	0	5	0
6	0	2	5
0	2	6	6

(c) Reduced matrix

Identifying an Assignment Corresponding to a Reduced Matrix

Now that the reduced matrix has been determined, the next step is to attempt to find a corresponding feasible primal solution. Recall that only those $x_{i,j}$ corresponding to binding dual constraints (i.e., $c_{i,j} - \lambda_i - \mu_j = 0$) may be set equal to one. These primal variables correspond precisely to the cells that contain zero in the reduced matrix; the index set of these variables is $Q = \{(1, 1), (1, 3), (2, 2), (2, 4), (3, 2), (4, 1)\}$. All other primal variables must remain fixed at zero. That is, $x_{i,j} = 0$, for all $(i, j) \in P = \{(1, 2), (1, 4), (2, 1), (2, 3), (3, 1), (3, 3), (3, 4), (4, 2), (4, 3), (4, 4)\}$.

A feasible primal solution is one in which each individual is assigned to exactly one task and each task to precisely one individual. This would correspond to making an assignment in exactly one cell in each row and exactly one cell in each column. Thus, if a row or column of the reduced matrix contains a single zero, then an assignment *must* be made in that cell. We could then begin our attempt at finding a primal feasible solution by choosing one such cell and forcing an assignment in that cell. For example, in Table 10.17, we could choose cell (3, 2) because it contains the only zero in row 3. (Note that several other initial choices are possible. For example, cell (1, 3) contains the only zero in column 3.) We symbolize the fact that we are setting $x_{3,2} = 1$ by placing a box around the zero in cell (3, 2), as in Table 10.17. However, in doing so, no other zero in column 2 may be used for an assignment, and we cross out all other zeros in column 2. This process is now repeated by choosing and making an assignment in another zero cell, which is the only remaining uncrossed zero cell in its row/column. Continuing in this manner results in the solution defined in Table 10.17.

TABLE 10.17 AN OPTIMAL ASSIGNMENT

$\cancel{0}$	8	$\boxed{0}$	8
2	$\cancel{0}$	5	$\boxed{0}$
6	$\boxed{0}$	2	5
$\boxed{0}$	2	6	6

If, after all zero cells are either assigned or crossed out, *exactly m assignments have been made*, the resulting solution is optimal because the solution satisfies dual feasibility, complementary slackness, and primal feasibility. If, however, fewer than m assignments were possible, then it was not possible to find a complementary primal feasible solution corresponding to the chosen dual solution. We must therefore modify the dual solution in an effort to introduce at least one additional zero into the reduced matrix. The details of this modification process are discussed in the following section.

Observe, from Table 10.17, that, for Example 10.5, it is possible to determine a feasible primal solution using only the primal variables corresponding to the zero cells. This assignment is

$$x_{1,3} = 1$$

$$x_{2,4} = 1$$

$$x_{3,2} = 1$$

$$x_{4,1} = 1$$

and is an optimal solution to the given assignment problem.

Modifying the Dual Solution and the Reduced Matrix

Suppose that when utilizing a particular reduced matrix, only k assignments were possible, where $k < m$. Then boxing k of the zeros in the reduced matrix resulted in all other zeros being crossed out. For example, consider the original matrix and the corresponding reduced matrix given in Table 10.18. Note that only $k = 2$ assignments are possible in Table 10.18(b).

TABLE 10.18 ORIGINAL AND REDUCED MATRIX

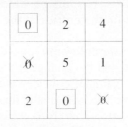

(a) Original matrix   (b) Reduced matrix

Because boxing k zeros resulted in all other zeros being crossed out, it follows that it is possible to cover all of the zeros in the reduced matrix with k horizontal and/or vertical lines. This is illustrated for the example problem in Table 10.19(a). There are several systematic ways of performing the operation of covering all of the zeros. One simple procedure for drawing such lines can be

TABLE 10.19 COVERING
ALL ZEROS WITH $k =$
2 LINES

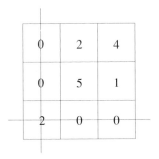

described as follows:

1. Count the number of uncovered zeros in each row and column.
2. Draw a line through the row or column with the most uncovered zeros.
3. Repeat Steps 1 and 2 until all zeros are covered.

The basic idea in modifying the dual solution is to preserve as many zeros in the reduced matrix as possible while introducing at least one new zero (corresponding to a new binding dual constraint). Toward this end, consider the following modifications of the dual solution:

$$\text{1. Replace } \lambda_i \text{ by } \lambda_i + c_0 \text{ for all } \textit{uncovered rows} \tag{10.57}$$

$$\text{2. Replace } \mu_j \text{ by } \mu_j - c_0 \text{ for all } \textit{covered columns} \tag{10.58}$$

where c_0 is some positive constant. What effect does this dual variable change have on the entries in the reduced matrix?

Covering the zeros in the reduced matrix by k horizontal/vertical lines creates four different types of cells. These four types of cells are cataloged in Table 10.20 and the change in the entry $(c_{i,j} - \lambda_i - \mu_j)$ is computed for each cell type. For example, consider a cell (i, j) in an uncovered row and uncovered column. Because cell (i, j) is in an uncovered row, λ_i increases by c_0, whereas μ_j remains the same. Therefore, we can compute the change in the quantity $(c_{i,j} - \lambda_i - \mu_j)$ by replacing λ_i by $\lambda_i + c_0$:

$$c_{i,j} - (\lambda_i + c_0) - \mu_j = (c_{i,j} - \lambda_i - \mu_j) - c_0$$

Thus, the entries in cells that are uncovered decrease by c_0. Similarly, as noted in Table 10.20, the entries in cells that are covered by both a horizontal and vertical line increase by c_0. Finally, all entries covered by a single horizontal or vertical line remain the same.

Thus, the operation of updating the dual solution in the manner described in (10.57) and (10.58) can be performed directly on the reduced matrix without

TABLE 10.20   EFFECT ON THE ENTRIES IN THE REDUCED MATRIX

Cell type	Change in $(c_{i,j} - \lambda_i - \mu_j)$
Uncovered row Uncovered column	$c_{i,j} - (\lambda_i + c_0) - \mu_j = (c_{i,j} - \lambda_i - \mu_j) - c_0$ Therefore, $(c_{i,j} - \lambda_i - \mu_j)$ decreases by c_0
Covered row Uncovered column	$c_{i,j} - \lambda_i - \mu_j = c_{i,j} - \lambda_i - \mu_j$ Therefore, $(c_{i,j} - \lambda_i - \mu_j)$ does not change
Uncovered row Covered column	$c_{i,j} - (\lambda_i + c_0) - (\mu_j - c_0) = c_{i,j} - \lambda_i - \mu_j$ Therefore, $(c_{i,j} - \lambda_i - \mu_j)$ does not change
Covered row Covered column	$c_{i,j} - \lambda_i - (\mu_j - c_0) = (c_{i,j} - \lambda_i - \mu_j) + c_0$ Therefore, $(c_{i,j} - \lambda_i - \mu_j)$ increases by c_0

explicitly considering the dual solution. We must, however, choose c_0 so that the dual solution remains feasible (i.e., $c_{i,j} - \lambda_i - \mu_j \geq 0$) and so that the reduced matrix gains at least one zero.

Note that the only entries in the reduced matrix that are decreasing are the entries that are uncovered. Because all of these entries are decreasing by c_0, choosing c_0 as the *minimum uncovered element* will not only maintain dual feasibility, but it will also result in at least one new zero entry in the reduced matrix. For the reduced matrix of Table 10.19, we see that the minimum uncovered element is 1 and therefore $c_0 = 1$. Using the modifications summarized in Table 10.20 results in the new reduced matrix in Table 10.21. Note that an additional zero was introduced in cell (2, 3) and an optimal assignment is now possible.

TABLE 10.21   MODIFIED
MATRIX AND OPTIMAL
ASSIGNMENT

[0]	1	3
0̸	4	[0]
3	[0]	0̸

We are now ready to summarize all of the steps of the Hungarian algorithm. Note that, although as we have seen, the algorithmic process depends on the dual solution, there is no explicit mention of the dual solution in the algorithm. This is due to the simple procedure required to derive an initial dual feasible solution, and because the reduced matrix can be updated without explicitly updating the dual solution.

The Hungarian algorithm

STEP **0.** Begin with a balanced assignment matrix.

STEP **1.** *Form the reduced matrix.*

    (a) Select the smallest element in each row and subtract that element from each element in that row. This will generate at least one zero in every row of the matrix.

    (b) Select the smallest element in each column of the resulting matrix and subtract that element from each element in that column. This will form the reduced matrix and generate at least one zero element in every column of the matrix.

STEP **2.** *Identify the assignments.* By complementary slackness, assignments may be made only on the cells of the matrix that contain zeros.

    (a) Make an assignment in a zero cell in any row that has only one zero in it. This assignment is noted by placing a box around the zero. Cross out all other zeros in the column containing the box.

    (b) Make an assignment in a zero cell in any column that has only one zero in it by placing a box around the zero. Cross out all other zeros in the row containing the box.

    (c) Repeat Steps 2(a) and 2(b) until no further assignments are possible. In the event that some zeros remain uncrossed, but no row/column contains a single uncrossed zero, then make an assignment in an arbitrary zero cell and cross out all other zeros in the row and column containing the assigned (boxed) cell.

    (d) If m assignments were possible (i.e., every row/column contains a box), stop; the boxed cells represent an optimal assignment. If only k assignment were possible, where $k < m$, go to Step 3.

STEP **3.** *Cover all zeros.* If Steps 1 and 2 have not led to a feasible solution, it is necessary to attempt to generate additional zeros in the matrix by first drawing the *minimum* number of lines (either horizontal or vertical, or some mixture) that will cover (pass through) *all* the zeros in the matrix. If k assignments were possible, then all of the zeros will be covered with k lines. A simple procedure to aid in drawing such lines follows:

    (a) Count the number of uncovered zeros in each row and column.

    (b) Draw a line through the row or column with the most uncovered zeros.

    (c) Repeat Steps 3(a) and 3(b) until all zeros are covered.

STEP **4.** *Modify the reduced matrix.* Let c_0 be the value of the smallest uncovered element. Subtract c_0 from all *uncovered elements*. Add c_0 to all *elements that are covered by two lines*.

      Return to Step 2.

As a final example of the algorithmic process, consider the following assignment problem.

Example 10.6: The Hungarian Algorithm

Find the minimal-cost assignment based on the cost matrix given in Table 10.22(a).

The initial reduced matrix is given in Table 10.22(b), and the steps of the Hungarian algorithm are summarized in Table 10.22(c), 10.22(d), and 10.22(e). Note that the optimal solution is to assign individual 1 to task 2, individual 2 to task 1, individual 3 to task 4, and individual 4 to task 3, at a cost of $z = 49$.

TABLE 10.22 SOLUTION TO EXAMPLE 10.6

10	12	11	14
7	13	12	16
12	19	20	18
8	18	12	17

(a) Original matrix

0	0	0	0
0	4	4	5
0	5	7	2
0	8	3	5

(b) Reduced matrix

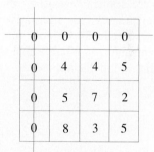

k = 2 assignments possible

$c_0 = 2$

(c) Iteration 1

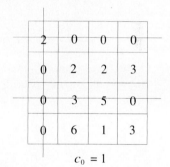

k = 3 assignments possible

$c_0 = 1$

(d) Iteration 2

3	0	0̸	0̸
0	1	1	2
1	3	5	0
0̸	5	0	2

Optimal Solution

$x_{1,2} = 1$

$x_{2,1} = 1$

$x_{3,4} = 1$

$x_{4,3} = 1$

$z = 12 + 7 + 18 + 12 = 49$

(e) Iteration 3

SUMMARY

In this chapter, we introduced the special linear programming problems known as the transportation problem and the assignment problem. The transportation problem was solved by the transportation simplex method, which is essentially the same as the network simplex method of Chapter 9. However, the bipartite structure of the transportation network allows the problem to be solved in a convenient tabular manner.

The assignment problem exhibited the same mathematical structure as the transportation problem, and, as such, could be solved by any of the methods that were applicable to the transportation problem. However, due to problems with degeneracy, an alternative algorithm for solving the assignment problem was presented. This algorithm, known as the Hungarian algorithm, is a primal-dual type algorithm that is able to solve the assignment problem very efficiently using a simple tabular approach.

EXERCISES

10.1. Consider the unit cost matrix for a transportation problem shown in Table 10.23.

TABLE 10.23

Source	Destination 1	2	3	Supply
1	4	3	5	10
2	6	8	9	20
3	2	5	4	25
Demand	15	35	5	

(a) Find an initial basic feasible solution using the northwest corner rule.

(b) Beginning with the initial solution found in part (a), use the transportation simplex method to find the minimum-cost shipping pattern.

(c) Is the optimal solution found in part (b) unique? Explain?

10.2. Consider again Exercise 10.1 and suppose that the entries in the given table are unit profits. Use the transportation simplex method to find the shipping pattern that maximizes profit.

10.3. Consider the following transportation tableau.

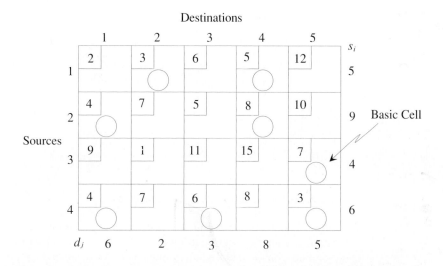

(a) Find the solution corresponding to the indicated basic cells.

(b) Is the solution found in part (a) optimal? If not, find the optimal solution.

10.4. Consider the cost matrix in Table 10.24 for a transportation problem in which the objective is to minimize cost.

TABLE 10.24

Source	Destination			Supply
	1	2	3	
1	8	5	4	50
2	6	8	9	20
Demand	10	20	40	

(a) Write the linear programming formulation for this problem.

(b) Set up the transportation tableau and use the northwest corner rule to find an initial basic feasible solution.

(c) Beginning with the initial solution found in part (b), solve the problem using the transportation simplex method. Give an optimal primal solution and an optimal dual solution.

(d) Write the dual of the linear program formulated in part (a). Verify that the dual solution found in part (c) is feasible to the dual problem.

10.5. Consider the following minimal cost transportation problem. Table 10.25 gives the unit transportation costs, $c_{i,j}$, along with the supplies and demands.

TABLE 10.25

Source	Destination 1	2	3	4	Supply
1	6	9	4	7	20
2	3	2	5	6	40
3	4	8	2	3	30
Demand	10	25	15	25	

(a) Find an initial solution using the northwest corner rule. Compute the cost associated with this solution.

(b) Find an initial solution using the Vogel approximation method. Compare the cost of this solution with that found by the northwest corner rule in part (a).

(c) Beginning with the initial solution found in part (b), solve the problem using the transportation simplex method.

(d) For what range of values for $c_{1,1}$ will the current basis remain optimal?

(e) Suppose that $c_{1,4}$ changes from its current value of 7 to 4. Use sensitivity analysis to determine the new optimal solution.

10.6. A widget manufacturer operates five production facilities, and must purchase the required raw material from one of three vendors. Vendors 1, 2, and 3 can supply at most 360, 270, and 450 pounds of raw material, respectively. In order to satisfy the current demand for widgets, Production Facilities 1, 2, 3, 4, and 5 require 180, 90, 230, 210, and 140 pounds of raw material, respectively. The unit cost of shipping 1 pound of raw material from each vendor to each production facility is given in Table 10.26.

TABLE 10.26

Vendor	Production facility 1	2	3	4	5
1	12	14	13	16	18
2	15	19	14	15	16
3	20	17	16	12	17

(a) Determine the optimal shipping pattern.

(b) For what range of values for $c_{1,5}$ will the current basis remain optimal?

(c) If the unit cost $c_{2,1}$ changes from its current value of 15 to 12, use sensitivity analysis to determine the new optimal solution.

10.7. A company has three plants, all of which make the same product. The product is made to order and the company has received four orders. Customers A, B, C, and D have placed orders for 600, 1000, 1400, and 700 units, respectively. The production cost and available capacity at the different plants is summarized in Table 10.27.

TABLE 10.27

Plant	Unit production cost	Available capacity (units)
1	$26	1200
2	$31	1600
3	$29	1400

The cost of shipping one unit from each plant to each customer is given in the Table 10.28. Assuming that orders can be split among the plants, determine the minimum-cost production–distribution schedule.

TABLE 10.28

	Customer			
Plant	A	B	C	D
1	$5.00	$7.00	$4.00	$6.00
2	3.00	6.00	7.00	4.00
3	4.00	5.00	4.00	8.00

10.8. A firm has three plants that all produce the same product. Their monthly capacities and per unit production costs are given in Table 10.29. Four warehouses are supplied from the three plants. The monthly demands of Warehouses A, B, C, and D are 150, 210, 100, and 90 units per month, respectively. Unit shipping costs from plant to warehouse are summarized in Table 10.30. What is the optimal production and routing scheme?

TABLE 10.29

Plant	Regular capacity	Cost per unit	Overtime capacity	Cost per unit
1	140	$3.00	60	$5.00
2	110	5.00	40	7.50
3	200	4.25	80	7.00

TABLE 10.30

	Warehouse			
Plant	A	B	C	D
1	$2.00	$1.00	$2.50	$4.00
2	4.00	3.00	5.00	3.00
3	1.00	2.00	4.00	3.00

10.9. Consider the cost matrix shown in Table 10.31. The objective is to find the minimum cost one-to-one assignment of workers to jobs.

TABLE 10.31

	Job		
Worker	1	2	3
1	6	3	2
2	2	5	4
3	7	4	5

(a) Set up this problem as a transportation problem, find an initial solution using the northwest corner rule, and then solve using the transportation simplex method.

(b) Set up this problem as an assignment problem and solve using the Hungarian algorithm.

10.10. Consider the assignment problem shown in Table 10.32.

TABLE 10.32

	Individual		
Task	1	2	3
1	17	18	16
2	14	19	17
3	15	19	18

(a) Write the linear programming formulation of this problem.

(b) Write the dual of the linear program formulated in part (a). List the complementary slackness conditions.

(c) Using the Hungarian algorithm, find an optimal primal and an optimal dual solution.

(d) Show that the optimal primal and dual solutions found in part (c) satisfy primal feasibility, dual feasibility, and complementary slackness.

10.11. Solve the assignment problem associated with the cost matrix shown in Table 10.33.

TABLE 10.33

	Job			
Worker	1	2	3	4
1	6	2	4	6
2	6	3	2	3
3	4	6	7	7
4	4	2	4	4

10.12. Table 10.34 gives the cost of machining each of four jobs on each of four machines. Assuming that only one job is to be assigned to a machine, use the Hungarian algorithm to find the assignment that minimizes total machining cost.

TABLE 10.34

Job	Machine				
	1	2	3	4	5
1	28	42	35	40	49
2	21	14	35	36	42
3	42	21	14	32	21
4	42	14	35	30	42

10.13. A quality check shows that the number of defects in six components produced by the Mho Electronics Company are a function of the workers assigned to produce these components. The average number of defective components, per week, expected for each of six workers is given in Table 10.35. How should workers be assigned to tasks so as to minimize the total expected number of defects.

TABLE 10.35

Worker	Component					
	1	2	3	4	5	6
1	12	41	17	15	18	13
2	14	20	17	15	53	15
3	22	23	25	40	27	28
4	18	50	20	19	27	25
5	7	8	20	10	10	30
6	23	25	25	30	25	43

10.14. Nuclear Technology, Inc., intends to subcontract three special valves for its nuclear power plants. Four bids have been received, and, according to government restrictions, no subcontractor may be permitted to produce more than one valve type. The bids, in terms of thousands of dollars per valve type, and the potential subcontractor information are given in Table 10.36. Use the Hungarian algorithm to determine the optimal assignments.

TABLE 10.36

Valve type	Contractor			
	1	2	3	4
A	12	13	20	18
B	8	20	10	18
C	20	22	30	25

CHAPTER **11**

INTEGER PROGRAMMING

CHAPTER OVERVIEW

The previous chapters were almost exclusively devoted to linear models in which the decision variables were assumed to be continuous. That is, it was assumed that the divisibility assumption holds and noninteger values for the decision variables are acceptable. One notable exception is the linear assignment problem addressed in Chapter 10. Recall that the mathematical formulation of the assignment problem required that the variables be binary (zero–one). However, due to the special network structure and the property of unimodularity, the assignment problem could be equivalently formulated as a linear programming problem.

In this chapter, we introduce integer linear programming models, that is, linear programming problems in which some or all of the variables are restricted to be integers. Because of the ease with which we solved linear programming models, the reader may, at first, erroneously assume that integer linear programming problems can be approached in a similar straightforward manner. However, as we will see, integer programming problems are combinatorial problems, and, in general, much more difficult to solve than linear programming problems. In fact, there is no single algorithm that can be applied to all integer linear programming problems as the simplex algorithm was used to effectively solve any linear programming problem.

The chapter begins by examining the graphical solution of simple integer programs. This is followed by a discussion of various techniques for formulating integer programming models. Finally, several algorithmic strategies are pre-sented for solving integer programming problems. These include branch-and-bound enumeration, implicit enumeration, and cutting plane methods. It should be noted, however, that even these methods are only applicable to problems of moderate size. General large-scale integer programming problems can still only be solved approximately by heuristic methods. Some heuristic methods of solu-tion are discussed in Chapter 12.

INTRODUCTION

As we have seen throughout this text, a linear programming problem is a mathe-matical model in which one attempts to optimize a linear objective function sub-ject to a set of linear constraints. However, in numerous applications, it may be necessary to specify that certain variables can only assume integer values. For example, a decision variable may be used to model the number of vehicles or workers. Clearly, in this case, noninteger solutions would be unacceptable. In addition, many modeling situations involve yes–no type decisions. For example: Should a certain factory be built? Should a certain investment be made? Situa-tions such as these can be modeled by introducing binary (zero–one) decision variables to mathematically handle the yes–no decisions. Various modeling tech-niques used in integer programming formulations will be discussed in more detail later.

An *integer linear programming problem* is simply a linear program in which some or all of the variables are restricted to integer values. Typically, we will refer to these models simply as integer programming models, because the term linear is seldom used except to contrast the models with integer nonlinear pro-gramming models. If *all* the variables must assume only integer values, then the problem is called a *pure* integer programming model, whereas if some of the variables are restricted to integer values and others remain continuous variables, then the problem is referred to as a *mixed* integer programming model. By using matrix notation, a mixed integer programming problem may be written in the general form:

$$\text{(MIP)} \quad \text{maximize } \mathbf{cx} + \mathbf{dy}$$

subject to:

$$\mathbf{Ax} + \mathbf{Dy} \leq \mathbf{b} \tag{11.1}$$

$$\mathbf{x} \geq \mathbf{0}, \ \mathbf{x} \text{ integer-valued}$$

$$\mathbf{y} \geq \mathbf{0}$$

where

$$\mathbf{A} = m \times n \text{ matrix}$$

$$\mathbf{D} = m \times p \text{ matrix}$$

$$\mathbf{c} = 1 \times n \text{ vector}$$

$$\mathbf{d} = 1 \times p \text{ vector}$$

$$\mathbf{b} = m \times 1 \text{ vector}$$

$$\mathbf{x} = n \times 1 \text{ vector of integer variables}$$

$$\mathbf{y} = p \times 1 \text{ vector of continuous variables}$$

Similarly, a pure integer programming problem can be written as

(IP)   maximize $\mathbf{cx}$

subject to:

$$\mathbf{Ax} \leq \mathbf{b}$$

$$\mathbf{x} \geq \mathbf{0}, \mathbf{x} \text{ integer-valued}$$

If, in addition, $\mathbf{x}$ is restricted to be binary, that is, each $x_j = 0$ or 1, then problem (IP) is called a zero–one (or binary) integer programming (BIP) problem.

The linear programming problem derived by omitting all the integer restrictions on the variables is called the *linear programming relaxation (LP relaxation)*. For example, the LP relaxation associated with problem (MIP) is simply

maximize $\mathbf{cx} + \mathbf{dy}$

subject to:

$$\mathbf{Ax} + \mathbf{Dy} \leq \mathbf{b} \tag{11.2}$$

$$\mathbf{x} \geq \mathbf{0}$$

$$\mathbf{y} \geq \mathbf{0}$$

The feasible region associated with an integer program is always a subset of the feasible region associated with its LP relaxation. This is true because the feasible region of the integer program is derived by adding restrictions (i.e., the integer restrictions) to the feasible region of the LP relaxation. Thus, when solving a maximization problem, the optimal objective value of the integer program will always be less than or equal to the optimal objective value of the LP relaxation. That is, for a maximization integer program, the LP relaxation provides an upper bound for the optimal objective value. Similarly, the LP relaxation provides a lower bound in the case of a minimization integer program. The LP relaxation is used extensively in constructing solution algorithms for integer programming problems.

There are actually a multitude of algorithms for addressing integer programming problems. However, most techniques can be classified as either enumeration techniques or cutting-plane methods. Enumeration techniques are designed to exploit the fact that the feasible region of a bounded integer program always contains a finite set of feasible points. In the crudest form, an enumeration technique could consist of total enumeration, that is, enumerating every feasible integer point. However, more refined techniques, such as branch-and-bound enumeration and implicit enumeration, attempt to enumerate only a small subset of the feasible integer points, while concluding that the remaining points are inferior to those examined.

Recall that if an optimal solution to a linear programming problem exists, the simplex algorithm always finds an extreme-point optimum. This is the basic motivation for cutting-plane methods. Constraints (or cutting planes) are successively added to the LP relaxation of an integer programming problem in such a way that the current noninteger optimal extreme point is cut away or made infeasible. However, this is done so that all integer points remain feasible. By proceeding in this manner, a new convex set is constructed that eventually has an integer point as an optimal extreme point. Thus, an optimal integer solution can be found by solving a sequence of linear programs.

Enumeration and cutting-plane methods will be discussed in more detail later. But, first, to help fix ideas and to begin to see the differences between linear programming problems and integer programming problems, let us discuss the graphical solution of two-dimensional integer programming models.

GRAPHICAL SOLUTION OF TWO-DIMENSIONAL INTEGER PROGRAMS

The mechanics of the graphical approach for solving integer programming problems parallel those used to solve linear programming problems in Chapter 3. Basically, one identifies the set of feasible points, and then, using the level curves of the objective function, the optimal integer solution is found. The process is illustrated using the following pure integer programming model with two variables.

Example 11.1: Graphical Solution

Consider the following pure integer programming problem:

$$\text{maximize } z = -x_1 + 4x_2 \tag{11.3}$$

subject to:

$$-10x_1 + 20x_2 \leq 22 \tag{11.4}$$

$$5x_1 + 10x_2 \leq 49 \tag{11.5}$$

$$8x_1 - x_2 \leq 36 \tag{11.6}$$

$$\mathbf{x} \geq \mathbf{0}, \mathbf{x} \text{ integer} \tag{11.7}$$

First, we identify the feasible region of the model, as shown in Figure 11.1. We begin in the usual manner by finding the intersection of the half-planes defined by the linear constraints. Note that although these linear constraints form a convex set, the addition of the integer restrictions results in a discrete set of integer points within this

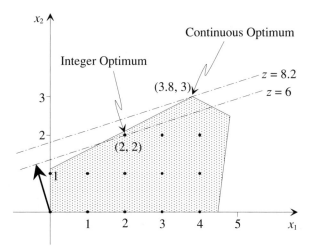

Figure 11.1   Graphical solution of Example 11.1.

convex set. This discrete set of points, which represents the feasible region of the integer programming problem, is obviously a nonconvex set.

The next step is to determine the integer point(s) in the feasible region that yield the maximum value of the objective function. As usual, the direction of improving z can be identified by examining the gradient of the objective function, $z = f(x_1, x_2) = -x_1 + 4x_2$. For our example, this results in

$$\nabla f(x_1, x_2) = \begin{pmatrix} \dfrac{\partial z}{\partial x_1} \\ \dfrac{\partial z}{\partial x_2} \end{pmatrix} = \begin{pmatrix} -1 \\ 4 \end{pmatrix}$$

This gradient vector is illustrated graphically in Figure 11.1. The level curves of the objective are normal to this vector. Sliding the level curves in the direction of increasing z, we see that the optimal integer solution is

$$x_1^* = 2$$
$$x_2^* = 2$$
$$z^* = 6$$

whereas, the optimal solution of the LP relaxation is

$$x_1 = 3.8$$
$$x_2 = 3$$
$$z = 8.2$$

Clearly, the optimal objective of the LP relaxation provides an upper bound for the optimal objective of the integer program.

Notice that rounding the linear programming solution results in the solution (4, 3), which is not only infeasible, but also nowhere near the optimal integer solution of (2, 2). In general, rounding the linear programming solution will not result in even a feasible solution.

COMPUTATIONAL COMPLEXITY

As we saw in Example 11.1, integer constraints serve to further restrict the solution space of a problem. Moreover, if the feasible region of a particular integer programming problem is bounded, then there will always be only a finite number of feasible solutions. Thus, the reader may, at first, try to conclude that because the solution space is smaller, the integer programming problem will actually be easier to solve. This, however, is not generally the case with integer programming problems. Problems in which the variables take on strictly integer values are problems in combinatorics, and are typically orders of magnitude more difficult to solve than problems with continuous variables.

Recall from Chapter 7 that the theory of computational complexity allows us to classify algorithms in terms of their computational efficiency. An algorithm whose worst-case performance (running time) is bounded by a polynomial function of the problem size is referred to as a polynomial algorithm. On the other hand, the worst-case performance of an exponential algorithm grows at least as fast as an exponential function of the problem size. Polynomial algorithms are considered to be ''good'' algorithms, whereas algorithms that experience exponential growth are not.

Classifying algorithms in this way, however, is not always a true indication of the algorithm's performance in practice. For example, the simplex algorithm has been shown to have exponential worst-case behavior, but it performs extremely well in practice. Its poor, theoretical performance seems reserved for pathological examples constructed solely for that purpose. On the other hand, Khachian's ellipsoid algorithm discussed in Chapter 7 exhibits polynomial worst-case behavior, but has been shown to perform poorly in practice.

Problems are classified in a similar way. A problem is classified as polynomial if it can be solved by a polynomial algorithm. Such problems are considered tractable and are generally thought of as easy to solve. Since the introduction of Khachian's algorithm, the class of polynomial problems, commonly referred to as class P, has included linear programming problems. Unfortunately, almost all discrete optimization problems do not share this solution characteristic. Problems for which no known polynomial algorithm exists are said to be in the class NP. Almost all integer programming problems are members of the class NP. There are, however, a few notable exceptions. For example, as we saw in Chapter 10, the linear assignment problem could be expressed as an equivalent linear programming problem because of its special structure and is therefore polynomial. Other examples of polynomial solvable integer programming problems are matching problems and spanning-tree problems. The reader interested in computational complexity is referred to Garey and Johnson (1979).

Because there are typically only a finite number of feasible points in an integer programming problem, a naive, brute-force approach to their solution would be to attempt to evaluate every possible combination of integer values for the problem variables. The hopelessness of such an approach can be readily demonstrated for problems of even quite small sizes. For example, if we have a problem involving just 10 variables and if, in turn, each variable could take on only the integer values 0 through 9, it can be seen that the total number of solutions to be examined is 10^{10}. Obviously, more intelligent approaches are necessary.

There are actually numerous approaches to solving integer programming problems, ranging from brute-force enumeration through highly esoteric methods. The majority of these algorithms can be broadly classified as either "exact" algorithms or "heuristic" algorithms. Exact methods of solution are those that, rather obviously, promise to yield the exact, optimal solution to a problem in combinatorics and employ various techniques so as to reduce the number of solutions to be searched. These exact methods include branch-and-bound enumeration, implicit enumeration, and some cutting-plane methods; these methods are discussed further in subsequent sections of this chapter. However, because of the combinatorial nature of integer programming problems, exact solutions cannot be found presently for many moderate- to large-size problems. Heuristic methods are then used to address these problems.

A heuristic method is one that has no formal mathematical basis, is developed more or less through intuition, and cannot guarantee an exact optimal solution. A "good" heuristic, however, can normally find "good" solutions (often near optimal) in a minimal amount of time. Several heuristic methods for solving combinatorial optimization problems are discussed in Chapter 12.

FORMULATING INTEGER PROGRAMMING PROBLEMS

As mentioned earlier, some decision variables may naturally assume only integer values because the variables denote some discrete physical quantity in the problem. For example, an integer variable may denote the number of workers to hire, the number of facilities to build, or the number of cars to manufacture. The integer variables represent an actual physical quantity in the model, so integer programming models with such variables are often referred to as direct integer programs.

There are also many integer programs that are known as coded integer programs. These generally involve yes–no type decisions that are modeled using binary variables. The zero–one values of the binary variables are used to mathematically characterize the yes–no decisions in the model. Finally, some integer programs are transformed integer programs. They are mathematical programs that have been transformed into integer programs to make them more mathematically tractable.

The remainder of this section illustrates a variety of techniques for modeling integer programming problems. We begin our discussion with the knapsack problem, one of the most useful, but also one of the simplest of integer models.

The Knapsack Problem

Suppose that a hiker must select from among several items those items that will give him maximum utility. However, the hiker has only a knapsack in which to carry the items. Obviously, the knapsack has only a limited amount of space. Thus, the problem is to choose those items that will fit in the knapsack and at the same time maximize utility.

We can model this problem mathematically by first numbering the items from 1 through n, and then for each $i = 1, \ldots, n$, defining the binary variable

$$x_i = \begin{cases} 1, & \text{if item } i \text{ is selected} \\ 0, & \text{otherwise} \end{cases}$$

Now let c_j denote the utility of item j and let a_j represent the amount of space consumed by item j. Then the *knapsack problem* can be formulated as follows, where b denotes the size of the knapsack.

$$\text{maximize} \sum_{j=1}^{n} c_j x_j \tag{11.8}$$

subject to:

$$\sum_{j=1}^{n} a_j x_j \leq b \tag{11.9}$$

$$\mathbf{x} \text{ binary} \tag{11.10}$$

Observe that the objective function maximizes utility. Also note that the knapsack problem has a single constraint that models the capacity restriction resulting from the knapsack.

The knapsack problem can also be viewed as a problem in investment selection. In this context, c_j denotes the profit expected from investment j, a_j represents the capital investment required for investment j, and b is the total capital available to invest. The binary variable x_j corresponds to the yes–no decision of whether to invest in investment j. If $x_j = 1$, then an investment of a_j is made in investment j. Thus, the objective of the knapsack formulation, in this case, is to maximize profit, and constraint (11.9) is a budgetary constraint that controls the total amount invested.

The knapsack problem is important not only as an application, but knapsack problems also occur as subproblems in the context of other applications. For a comprehensive study of the knapsack problem along with extensions, applications, and special solution algorithms, the interested reader is referred to Martello and Toth (1990).

The knapsack problem is an example of a pure zero–one (binary) integer programming problem. Another example of a pure zero–one integer program is the set-covering problem, which is discussed next.

The Set-Covering Problem

Suppose that there are n potential sites for new service facilities and the cost associated with erecting a facility at site j is c_j. The proposed facilities are to service (or cover) m areas. For example, the facilities may represent fire stations and the areas may be sections of a city. The problem is then to find the least-cost set of facilities that is able to service (or cover) all areas.

To formulate the integer programming model, we first define the decision variables,

$$x_j = \begin{cases} 1, & \text{if facility } j \text{ is opened} \\ 0, & \text{otherwise} \end{cases}$$

Also, define coefficient matrix $\mathbf{A}$ such that entry $a_{i,j}$ is 1 if facility j is capable of covering area i and 0 otherwise. Finally, let $\mathbf{1} = (1, 1, \ldots, 1)^t \in E_m$. Then, the *set-covering problem* can be written in matrix notation as

$$\text{minimize } \mathbf{cx} \tag{11.11}$$

subject to:

$$\mathbf{Ax} \geq \mathbf{1} \tag{11.12}$$

$$\mathbf{x} \text{ binary} \tag{11.13}$$

Observe that each constraint comprising (11.12) is of the form

$$\sum_{j=1}^{n} a_{i,j} x_j \geq 1 \tag{11.14}$$

with each $a_{i,j}$ being fixed at either 0 or 1. Thus, constraint i forces at least one x_j to take on the value 1, and consequently cover area i.

In the event that each area must be covered by exactly one facility, then the problem is referred to as the *set-partitioning problem* and can be formulated in a similar manner as

$$\text{minimize } \mathbf{cx} \tag{11.15}$$

subject to:

$$\mathbf{Ax} = \mathbf{1}$$

$$\mathbf{x} \text{ binary}$$

The Fixed-Charge Problem

Consider the situation in which there are m potential sites to construct new facilities, each of which is capable of producing a certain product. If constructed, each facility would serve as a source that would help supply the product to n destinations in order to satisfy the demands of the customers. If source i is constructed, then a one-time fixed charge of f_i is incurred. Let b_i denote the production capacity of source i if it is constructed, and let d_j represent the quantity demanded at destination j. Also suppose that $c_{i,j}$ denotes the unit production/shipping cost from source i to destination j. Then the *fixed-charge location problem* may be formulated as follows:

$$\text{minimize} \sum_{i=1}^{m} \left(f_i y_i + \sum_{j=1}^{n} c_{i,j} x_{i,j} \right) \tag{11.16}$$

subject to:

$$\sum_{i=1}^{m} x_{i,j} = d_j, \qquad \text{for } j = 1, \ldots, n \tag{11.17}$$

$$\sum_{j=1}^{n} x_{i,j} \le b_i y_i, \qquad \text{for } i = 1, \ldots, m \tag{11.18}$$

$$\mathbf{x} \ge \mathbf{0} \tag{11.19}$$

$$\mathbf{y} \text{ binary} \tag{11.20}$$

where the decision variables are

$$y_i = \begin{cases} 1, & \text{if facility } i \text{ is constructed} \\ 0, & \text{otherwise} \end{cases}$$

and

$$x_{i,j} = \text{units shipped from source } i \text{ to destination } j$$

Constraints (11.17) are the demand constraints corresponding to the n destinations, and constraints (11.18) are the capacity restrictions. Now consider a particular facility site k. Note, from (11.18), that the capacity of facility k is either 0 or b_k. That is, if $y_k = 0$, then facility k is not opened and the corresponding constraint (11.18) becomes

$$\sum_{j=1}^{n} x_{k,j} \le 0$$

Obviously, because $x_{k,j} \ge 0$, this constraint forces $x_{k,j} = 0$ for all j.

Similarly, if $y_k = 1$, then the capacity constraint for facility k becomes

$$\sum_{j=1}^{n} x_{k,j} \leq b_k$$

and, at the same time, a cost of f_k is incurred in the objective function. In this case, the total contribution of facility k to the objective function is

$$f_k + \sum_{j=1}^{n} c_{k,j} x_{k,j}$$

Note that for facility k, a fixed cost of f_k is incurred as well as a cost that is proportional to the production at facility k.

The fixed-charge problem is an example of a mixed integer program and the basic modeling approach can also be used to model other one-time costs such as setup costs in scheduling models.

The Traveling Salesman Problem

The *traveling salesman problem* is a classical combinatorial problem that has received much attention in the literature. In fact, it is still an area of active research and has a multitude of applications, with two of the most important being vehicle routing and job sequencing. The basic problem can be stated simply as follows: Starting from his home city, a salesman is to visit each city on a given list exactly once and return to his home city. Given that the distance between any two cities is known, the objective of the traveling salesman problem is to determine the order in which to visit the cities so that the total distance traveled is a minimum.

There are several mathematical formulations of the traveling salesman problem, and the one we have chosen to present is due to Miller, Tucker, and Zemlin (1960). We begin the formulation process by first numbering the cities from 1 through n, with city 1 being designated as the home city. By denoting the distance from city i to city j by $c_{i,j}$, the decision variables are defined by

$$x_{i,j} = \begin{cases} 1, & \text{if city } j \text{ is visited immediately following city } i \\ 0, & \text{otherwise} \end{cases}$$

Then an integer programming formulation of the traveling salesman problem can be written as follows:

$$\text{minimize} \sum_{i=1}^{n} \sum_{j=1}^{n} c_{i,j} x_{i,j} \tag{11.21}$$

subject to:

$$\sum_{j=1}^{n} x_{i,j} = 1, \qquad \text{for } i = 1, \ldots, n \tag{11.22}$$

$$\sum_{j=1}^{n} x_{i,j} = 1, \qquad \text{for } j = 1, \ldots, n \qquad (11.23)$$

$$t_i - t_j + nx_{i,j} \le n - 1, \qquad \text{for } i, j = 2, \ldots, n \qquad (11.24)$$

$$\mathbf{x} \ge \mathbf{0} \qquad (11.25)$$

where the t_i are arbitrary real numbers.

Constraints (11.22) ensure that on the traveling salesman tour, each city i is followed by exactly one city j. Similarly, constraints (11.23) specify that a unique city i is visited immediately before city j. The remaining constraints (11.24) are referred to as *subtour elimination* constraints. Their purpose is to ensure that a single tour results rather than a number of disjoint subtours. Because city 1 is the home city, the constraints operate by ensuring that every tour contains city 1.

To better understand how constraints (11.24) eliminate subtours, consider a problem with $n = 5$ cities and consider the solution given by $x_{1,2} = x_{2,1} = x_{3,4} = x_{4,5} = x_{5,3} = 1$. Then, using the definition of $x_{i,j}$, we can visualize this solution, as shown in Figure 11.2. Note that this solution clearly satisfies constraints (11.22) and (11.23); however, the subtour containing cities 3, 4, and 5 does not contain city 1, the home city. Let us examine the constraints of the form (11.24) corresponding to the arcs in this subtour. These can be written as

$$t_3 - t_4 + 5x_{3,4} \le 4 \qquad (11.26)$$

$$t_4 - t_5 + 5x_{4,5} \le 4 \qquad (11.27)$$

$$t_5 - t_3 + 5x_{5,3} \le 4 \qquad (11.28)$$

where t_3, t_4, and t_5 are arbitrary, fixed real numbers.

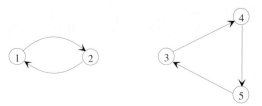

Figure 11.2 Subtours.

Now, summing constraints (11.26–11.28) yields

$$5(x_{3,4} + x_{4,5} + x_{5,3}) \le 4(3) \qquad (11.29)$$

Finally, noting that $x_{3,4} = x_{4,5} = x_{5,3} = 1$, we have the obviously false condition that $5(3) \le 4(3)$. Thus, constraints (11.24) are violated by this solution. In this same manner, constraints (11.24) eliminate all tours not containing city 1 and hence eliminate all subtours.

As a final note, observe that if constraints (11.24) are eliminated from the formulation of the traveling salesman problem, then the resulting relaxation is the linear assignment problem studied in Chapter 10. For this reason, many of the algorithms that have been developed to address the traveling salesman problem utilize the assignment problem in the solution process. The reader interested in reading further about applications and special solution algorithms for the traveling salesman problem is referred to Lawler et al. (1985).

Either–Or Constraints

There are many instances when at least one of two constraints must be satisfied, but it may not be possible to satisfy both constraints. For example, consider the problem of scheduling jobs on a single machine. Let x_i denote the start time of job i and let t_i represent the machine time required for job i. Then the completion time of job i is $(x_i + t_i)$.

Only one machine is available, so it is not possible for two jobs to be scheduled during the same time interval. Therefore, for any two jobs j and k, it must be true that either

$$x_j + t_j \leq x_k \tag{11.30}$$

or

$$x_k + t_k \leq x_j \tag{11.31}$$

Constraint (11.30) specifies that the start time of job k is after the completion time of job j. Similarly, (11.31) stipulates that job j starts no sooner than the completion time of job k. Obviously, both of these constraints cannot be satisfied simultaneously, but at least one must be satisfied to ensure a feasible schedule. Rewriting the constraints in standard form by placing the variables on the left and the constants on the right, we have that either

$$x_j - x_k \leq t_j \tag{11.32}$$

or

$$x_k - x_j \leq t_k \tag{11.33}$$

We would like to model this either–or condition (disjunction) while preserving the linear programming structure of the problem. Let M be an arbitrarily large positive number and consider the following constraint pairs:

$$x_j - x_k \leq t_j \tag{11.34}$$

$$x_k - x_j \leq t_k + M \tag{11.35}$$

or

$$x_j - x_k \leq t_j + M \tag{11.36}$$

$$x_k - x_j \leq t_k \tag{11.37}$$

Note that adding M to the right-hand side of constraint (11.35), in effect, simply relaxes the constraint. That is, because the right-hand side is arbitrarily large, x_j and x_k are not restricted by (11.35). Similarly, constraint (11.36) also does not restrict x_j and x_k. Thus, (11.34–11.37) are equivalent to (11.32–11.33).

Finally, let us introduce the binary variable y and rewrite (11.34–11.37) as follows:

$$x_j - x_k \leq t_j + My \tag{11.38}$$

$$x_k - x_j \leq t_k + M(1 - y) \tag{11.39}$$

Observe that if $y = 1$, then (11.38–11.39) are precisely equivalent to (11.36–11.37), whereas if $y = 0$, (11.38–11.39) are the same as (11.34–11.35). Thus, the introduction of a single binary variable was all that was necessary to derive an equivalent linear representation of an either–or condition.

If–then conditions can be handled in exactly the same manner as either–or constraints. For example, suppose that if the constraint $g(\mathbf{x}) < b$ is satisfied, then it must also be true that the condition $h(\mathbf{x}) \leq d$ is satisfied. However, if $g(\mathbf{x}) < b$ is not satisfied, then $h(\mathbf{x}) \leq d$ may or may not be satisfied. To clearly see how this if–then condition is converted into an either–or condition, let P represent the event that $g(\mathbf{x}) < b$ is satisfied and, similarly, let Q represent $h(\mathbf{x}) \leq d$. Then this implication can be written as the simple logic expression $P \Rightarrow Q$. But by using elementary logic, $P \Rightarrow Q$ is equivalent to [(not P) or Q]. Thus, this implication is equivalent to an either–or condition. Noting that (not P) is given by $g(\mathbf{x}) \geq b$, the if–then condition can be written in the form

$$g(\mathbf{x}) \geq b - My \tag{11.40}$$

$$h(\mathbf{x}) \leq d + M(1 - y) \tag{11.41}$$

where, again, y is a binary variable, and M is an arbitrarily large positive number.

p Out of *m* Constraints

Enforcing the either–or constraints in the previous section can also be thought of as satisfying at least one of two conditions. This idea can be generalized quite easily to the problem of satisfying at least p out of m constraints. Suppose that p of the following m constraints must be satisfied.

$$\begin{pmatrix} \mathbf{a}_1\mathbf{x} \leq b_1 \\ \mathbf{a}_2\mathbf{x} \leq b_2 \\ \vdots \\ \mathbf{a}_m\mathbf{x} \leq b_m \end{pmatrix} \tag{11.42}$$

Then, following the logic of the previous section, we can write an equivalent linear system as follows:

$$\mathbf{a}_i\mathbf{x} \leq b_i + M(1 - y_i), \qquad \text{for } i = 1, \ldots, m \tag{11.43}$$

$$\sum_{i=1}^{m} y_i \geq p \tag{11.44}$$

where

$$y_i = \begin{cases} 1, & \text{if constraint } i \text{ is satisfied} \\ 0, & \text{otherwise} \end{cases}$$

The problem of satisfying *exactly p* of *m* constraints can also be modeled by simply changing constraint (11.44) to

$$\sum_{i=1}^{m} y_i = p \tag{11.45}$$

Representing General Integer Variables Using Zero-One Variables

Consider a bounded integer variable x_j, where $0 \le x_j \le u_j$. Then, assuming that the upper bound u_j is integer, the variable x_j can take on $u_j + 1$ values. That is, $x_j = 0, 1, 2, \ldots, u_j - 1, u_j$. Now let r be the *smallest* positive integer such that $2^{r+1} > u_j$. For example, if $u_j = 13$, then $r = 3$ because $2^{r+1} = 2^{3+1} = 2^4 = 16 > 13$.

Because every positive integer is easily represented in binary form, we can express x_j as the sum of binary variables as follows:

$$x_j = y_{j,0}(2^0) + y_{j,1}(2^1) + \cdots + y_{j,r-1}(2^{r-1}) + y_{j,r}(2^r) \tag{11.46}$$

$$y_{j,i} \text{ binary} \tag{11.47}$$

For example, if $0 \le x_j \le 13$, then $r = 3$ and

$$x_j = y_{j,0}(2^0) + y_{j,1}(2^1) + y_{j,2}(2^2) + y_{j,3}(2^3) = y_{j,0} + 2y_{j,1} + 4y_{j,2} + 8y_{j,3} \tag{11.48}$$

$$y_{j,0}, y_{j,1}, y_{j,2}, y_{j,3} \text{ binary} \tag{11.49}$$

We can then substitute expression (11.46) for every instance of x_j in the integer programming model. In this manner, it is possible to convert every integer programming problem with bounded variables into an equivalent zero–one integer programming problem. Because zero–one problems are typically easier to solve than general integer programs, the reader may be tempted to simply convert all bounded variables problems in this manner. However, this results in a problem with many variables, and usually any advantage gained by utilizing only zero–one variables is offset by the increase in combinatorial complexity.

Transforming a Piecewise Linear Function

In many applications, mathematical modeling requires the use of *piecewise linear functions*. A graphical example of a continuous piecewise linear function is given in Figure 11.3. Note that the function is actually a nonlinear function composed of a number of linear segments. The points, t_i, where the slope of the function changes, are referred to as *breakpoints*.

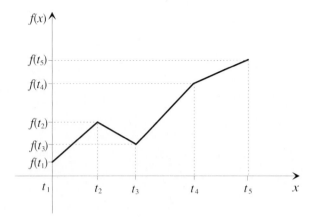

Figure 11.3   A piecewise linear function.

To see more clearly how such a function could originate, consider the following scenario. Suppose that at most 800 pounds of a particular commodity may be purchased. The cost of the commodity is $10 per pound for the first 200 pounds, $8 per pound for the next 300 pounds, and $5 per pound for the next 300 pounds. Let x denote the number of pounds purchased and let $f(x)$ represent the total cost associated with the purchase of x pounds.

To represent the function $f(x)$ as a piecewise linear function, let us first write a mathematical expression for $f(x)$. Clearly, $f(x) = 10x$, for $0 \leq x \leq 200$. Now, consider the interval $200 \leq x \leq 500$. Note that the first 200 pounds cost $10 per pound, thus if $x \in [200, 500]$, we will always incur a fixed cost of $200(\$10) = \$2000$. Also each additional pound between 200 and 500 pounds costs $8; therefore, we can write $f(x) = 2000 + 8(x - 200) = 400 + 8x$, for $200 \leq x \leq 500$. In a similar manner, the cost associated with the interval $500 \leq x \leq 800$ is $f(x) = 4400 + 5(x - 500) = 1900 + 5x$. We can thus summarize function $f(x)$ as follows:

$$f(x) = \begin{cases} 10x, & \text{for } 0 \leq x \leq 200 \\ 400 + 8x, & \text{for } 200 \leq x \leq 500 \\ 1900 + 5x, & \text{for } 500 \leq x \leq 800 \end{cases} \qquad (11.50)$$

The mathematical representation provided by (11.50) is also represented graphically in Figure 11.4. Notice that the breakpoints of the function occur at $t_1 = 0$, $t_2 = 200$, $t_3 = 500$, and $t_4 = 800$.

It is very difficult to use this nonlinear representation of $f(x)$ in the context of an optimization problem; therefore, we would like to express it in an alternative manner that could be used more readily in an optimization framework. In particular, we would like to transform this nonlinear function so that it could be used in an integer linear programming model. Toward this end, consider two consecutive breakpoints, t_k and t_{k+1}. If $x \in [t_k, t_{k+1}]$ then, by using the concept of a convex

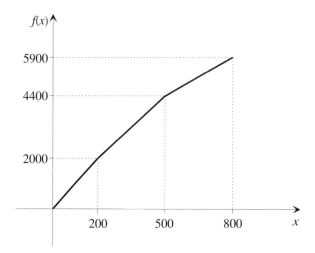

Figure 11.4   Graphical representation of $f(x)$.

combination, there exists $\delta_k \in [0, 1]$ such that

$$x = \delta_k t_k + (1 - \delta_k)t_{k+1} \tag{11.51}$$

Now, because $f(x)$ is linear on the interval $[t_k, t_{k+1}]$, it also follows that

$$f(x) = \delta_k f(t_k) + (1 - \delta_k)f(t_{k+1}) \tag{11.52}$$

Generalizing this idea, we can write

$$x = \delta_1 t_1 + \delta_2 t_2 + \cdots + \delta_{r+1} t_{r+1} \tag{11.53}$$

$$f(x) = \delta_1 f(t_1) + \delta_2 f(t_2) + \cdots + \delta_{r+1} f(t_{r+1}) \tag{11.54}$$

where

$$\delta_1 + \delta_2 + \cdots + \delta_{r+1} = 1 \tag{11.55}$$

$$\delta_i \leq 0, \qquad \text{for } i = 1, \ldots, r + 1 \tag{11.56}$$

$$\text{at most two adjacent } \delta_i \text{ are positive} \tag{11.57}$$

Condition (11.57) is necessary to ensure that each piecewise linear segment is traced out by the mathematical representation. We will rewrite this in a more mathematical way by introducing a binary variable y_j for each linear segment of the piecewise linear function. The complete model will then be written

$$x = \delta_1 t_1 + \delta_2 t_2 + \cdots + \delta_{r+1} t_{r+1} \tag{11.58}$$

$$f(x) = \delta_1 f(t_1) + \delta_2 f(t_2) + \cdots + \delta_{r+1} f(t_{r+1}) \tag{11.59}$$

where

$$\delta_1 + \delta_2 + \cdots + \delta_{r+1} = 1 \tag{11.60}$$

$$\delta_1 \leq y_1 \tag{11.61}$$

$$\delta_2 \leq y_1 + y_2 \tag{11.62}$$

$$\delta_3 \leq y_2 + y_3 \tag{11.63}$$

$$\vdots$$

$$\delta_r \leq y_{r-1} + y_r \tag{11.64}$$

$$\delta_{r+1} \leq y_r \tag{11.65}$$

$$y_1 + y_2 + y_3 + \cdots + y_r = 1 \tag{11.66}$$

$$\boldsymbol{\delta} \geq \mathbf{0} \tag{11.67}$$

$$\mathbf{y} \text{ binary} \tag{11.68}$$

Notice that a particularly y_k controls the values of δ_k and δ_{k+1}. That is, if $y_k = 0$, then the constraints (11.61–11.65) force δ_k and δ_{k+1} to be zero. Similarly, if $y_k = 1$, then δ_k, $\delta_{k+1} \in [0, 1]$. Finally, because constraint (11.66) specifies that precisely one y_i will have the value 1, exactly two adjacent δ_i are allowed to be nonzero in any solution of (11.60–11.68). In fact, the y_i that assumes the value of 1 corresponds precisely to the linear segment that is being used.

By using the mathematical representation in (11.58–11.68), the piecewise linear function of Figure 11.4 can be written as

$$x = 0\delta_1 + 200\delta_2 + 500\delta_3 + 800\delta_4$$

$$f(x) = f(0)\delta_1 + f(200)\delta_2 + f(500)\delta_3 + f(800)\delta_4$$

$$= 0\delta_1 + 2000\delta_2 + 4400\delta_3 + 5900\delta_4$$

where

$$\delta_1 + \delta_2 + \delta_3 + \delta_4 = 1$$

$$\delta_1 \leq y_1$$

$$\delta_2 \leq y_1 + y_2$$

$$\delta_3 \leq y_2 + y_3$$

$$\delta_4 \leq y_3$$

$$y_1 + y_2 + y_3 = 1$$

$$\boldsymbol{\delta} \geq \mathbf{0}$$

$$\mathbf{y} \text{ binary}$$

To utilize this technique in the context of a mathematical programming problem, the expressions in (11.58) and (11.59) would be substituted for all occurrences of x and $f(x)$, respectively. The remaining conditions (11.60–11.68) would be added to the problem as constraints. Notice that the variables $\mathbf{x}$ would be eliminated from the problem and the new problem would have as variables vectors δ and $\mathbf{y}$. After solving, the $\mathbf{x}$ solution could be recovered by using (11.58).

BRANCH-AND-BOUND ENUMERATION

The branch-and-bound method is a quasi-enumerative approach to problem solving that has been applied to a wide variety of combinatorial problems. It is fairly efficient for modest-size problems, and the general methodology forms an important part of the set of (exact) solution methods for the general class of integer linear programming problems.

The basic idea of branch-and-bound is to partition a given problem into a number of subproblems. This process of partitioning is usually called *branching* and its purpose is to establish subproblems that are easier to solve than the original problem because of their smaller size or more amenable structure. Branching is generally represented in terms of a tree structure, as in Figure 11.5, where each node i of the search tree represents a subproblem P_i.

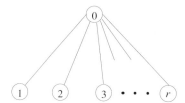

Figure 11.5   Branching.

As we will see, this search tree may have many levels, with the nodes at the bottom of the branchings (i.e., nodes with only one incident arc) being referred to as *pendant nodes*. The solution process involves a systematic evaluation of the pendant nodes of the search tree. The evaluation process consists of three key components: *branching, computing bounds,* and *fathoming*. Each of these is now discussed in more detail.

Branching

Clearly, if we are to derive the optimal solution to a given problem P_0, then the set of subproblems of P_0 must represent all of P_0. For simplicity, let $\{P_i\}$ represent the set of feasible integer solutions to a problem P_i. Then, if P_0 is partitioned into $P_1, P_2, \ldots, P_r$, it must be true that

$$\{P_0\} = \{P_1\} \cup \{P_2\} \cup \cdots \cup \{P_r\} \tag{11.69}$$

Also, it is generally more efficient to also choose subproblems $P_1, P_2, \ldots, P_r$ such that

$$\{P_i\} \cap \{P_j\} = \phi, \qquad \text{for all } i \neq j \tag{11.70}$$

This is especially true when it is necessary to enumerate all solutions to a problem, because some solutions would be enumerated multiple times if the feasible regions of some subproblems have a nonempty intersection.

To help understand the branching process, consider an integer program P_0 with n variables, and suppose that a particular variable, say, x_k, can take on the integer values 0, 1, 2, or 3. Then two branchings that satisfy the conditions (11.69) and (11.70) are illustrated in Figure 11.6. Of course, these are only two possibili-

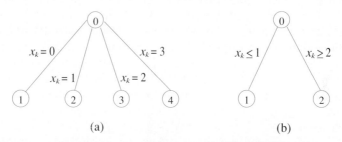

Figure 11.6 Different branching strategies.

ties, and several others are also possible. Note that in Figure 11.6(a), four subproblems are created, each of which corresponds to fixing the variable x_k at one of its possible values. Because x_k is now fixed in value, each of the subproblems involves only $n - 1$ variables.

Another possibility for a branching that satisfies conditions (11.69) and (11.70) is illustrated in Figure 11.6(b). Notice that in this case, the additional constraints $x_k \leq 1$ and $x_k \geq 2$ are used to create a disjoint partitioning. In effect, these conditions are specifying that $x_k = 0, 1$ in subproblem P_1 and $x_k = 2, 3$ in subproblem P_2.

Note, further, that during the branching process, we are essentially adding restrictions to a particular problem to form the resulting subproblems. Consequently, the feasible region of a subproblem is a subset of the feasible region of the parent problem. Thus, in the case of a maximization problem, the optimal objective value associated with a subproblem is always less than or equal to the optimal objective value associated with the parent problem. Therefore, as we descend in the search tree, the optimal objective values associated with each subproblem decrease for a maximization problem (and increase for a minimization problem).

Computing Bounds

Suppose that we know a feasible integer solution to a particular maximization integer program. Then the objective value provided by this solution is a lower bound for the optimal objective value of the integer program. That is, because we know the objective value of a particular feasible integer solution, we are assured of obtaining an optimal objective value at least that good. We will designate this lower bound by z_L. If several feasible integer solutions are known, then z_L will correspond to the largest known objective value. That is, z_L is the *greatest lower bound* and the integer solution corresponding to this value is called the *incumbent solution*, because it is the best known integer solution. (Of course, if no feasible solution is known as yet, we may set $z_L = -\infty$.)

The purpose of computing upper bounds (in a maximization problem) is to determine how good the optimal solution at a node can be without actually solving the integer program at that node. This is usually done by solving the LP relaxation. Consider an integer subproblem P_i associated with some pendant node i. Let z denote the optimal objective value associated with subproblem P_i. That is, z corresponds to an optimal integer solution of P_i. Because z may be difficult to determine, we are interested in finding an upper bound for z that can be readily computed. Consider solving the LP relaxation of subproblem P_i, and let $\bar{z}$ denote the optimal objective value of the LP relaxation. Then, clearly, $\bar{z} \geq z$ because the feasible region of the integer program is a subset of the feasible region of the LP relaxation.

Now suppose that $\bar{z} \leq z_L$. Then, $z \leq \bar{z} \leq z_L$ and subproblem P_i does not need to be considered further because it will never yield a solution any better than the current best integer solution. (This process of eliminating a subproblem P_i from further consideration is referred to as *fathoming* and is discussed further in the following section.) However, if $\bar{z} > z_L$, then a conclusion cannot be reached and further branching is needed.

Fathoming

During the branch-and-bound process, an attempt is made to resolve each of the subproblems corresponding to the pendant nodes of the search tree. Once all of the subproblems associated with the pendant nodes are resolved, then the problem is solved. A subproblem can be eliminated from further consideration in one of three ways:

1. The subproblem yields an optimal integer solution. In this case, we update z_L and the incumbent solution if necessary and continue the node-selection process.
2. It can be shown that the optimal solution value of the subproblem is no better than the best integer solution found thus far. This is usually done by computing a bound on the optimal integer objective value by solving the LP relaxation. This bound is then compared with the objective value of the incumbent solution.
3. The subproblem is infeasible.

A node (subproblem) that has been removed from further consideration in one of these ways is said to be *fathomed* and no further branching on that node is necessary. If, however, it is not possible to fathom a given pendant node, then the subproblem associated with that node is again partitioned into smaller subproblems by branching in some prescribed manner. The process is then repeated until all pendant nodes have been fathomed.

Search Strategies

Branching also involves choosing the next subproblem (pendant node) to examine. There are several branching strategies for choosing the next pendant node, with the most common being *depth-first search* and *best-bound search*. In each of these strategies, the pendant nodes are placed in a list according to some measure of importance. If the current node under examination is fathomed, then the next node in the list is selected. If the examination of the current node is complete but we were not able to fathom the current node, then the current subproblem is partitioned into additional subproblems that are then added to the list according to the branching strategy being used. A new node is then selected and the process is repeated until the list of available pendant nodes is empty.

The solution of the linear relaxation at each node generates a bound on the optimal integer objective value that can be derived from that node. In the best-bound search, the next subproblem chosen is simply the one with the best such bound. That is, for a maximization problem, we would branch next on the node with the largest upper bound. The rationale for branching in this way is to attempt to generate a good integer solution early in the branching process. This incumbent solution then could be used to fathom nodes with smaller upper bounds, thus expediting the search procedure. However, in computer implementations, it has the disadvantage of generally requiring more memory than a depth-first search.

Depth-first search is also called last-in-first-out (LIFO). Last-in-first-out refers to the strategy for placing nodes in and selecting nodes from the list of pendant nodes. Using the depth-first strategy, we always choose the subproblem (node) that was placed in the list most recently. We essentially work down one side of the search tree first and then backtrack once a node is fathomed. Because we are branching on the most recently created node, the process of reoptimization can be used more efficiently. This is a result of subproblems being created by adding restrictions to the parent problem. In large-scale problems, it is obviously more efficient to solve subproblems using reoptimization techniques (as in sensitivity analysis) rather than solving each subproblem from scratch.

Example 11.2: Branch-and-Bound Enumeration

Let us now illustrate the branch-and-bound approach by solving a simple integer programming problem in which we can view the process graphically. In this example, we will compute bounds using the LP relaxations and base our branching strategy on the best-bound approach. The depth-first approach will be illustrated subsequently in conjunction with implicit enumeration.

To better explain the branching process which will be used, let the notation $\lfloor t \rfloor$ represent the greatest integer less than or equal to a real number t. For example,

$$\lfloor 4.2 \rfloor = 4$$

$$\lfloor -2.1 \rfloor = -3$$

Now, suppose that we first solve the linear relaxation. If this yields an-all-integer solution, then the problem is solved. Otherwise, some integer variable will be selected, say, $x_k = \beta_k$, where β_k is the current *noninteger* value of x_k. A partitioning of the problem will then be created by utilizing the conditions

$$x_k \le \lfloor \beta_k \rfloor \tag{11.71}$$

$$x_k \ge \lfloor \beta_k \rfloor + 1 \tag{11.72}$$

This branching process is illustrated graphically in Figure 11.7. Obviously, this creates a disjoint partitioning and all integer solutions of the parent problem are included in one of these two subproblems. For example, if $x_k = 5.82$, then the next branching will be defined by $x_k \le \lfloor 5.82 \rfloor = 5$, and $x_k \ge \lfloor 5.82 \rfloor + 1 = 5 + 1 = 6$.

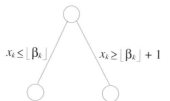

Figure 11.7   Branching.

Now, consider the following integer programming problem, which was also solved graphically in Example 11.1.

$$\text{maximize } z = -x_1 + 4x_2 \tag{11.73}$$

subject to:

$$-10x_1 + 20x_2 \le 22 \tag{11.74}$$

$$5x_1 + 10x_2 \le 49 \tag{11.75}$$

$$8x_1 - x_2 \le 36 \tag{11.76}$$

$$\mathbf{x} \ge \mathbf{0}, \mathbf{x} \text{ integer} \tag{11.77}$$

Let us denote this original integer program by P_0 and recall, from Example 11.1, that the optimal solution is given by $x_1^* = 2$, $x_2^* = 2$, and $z^* = 6$. We will now use branch-and-bound to derive this integer optimum. We begin by relaxing the integer restrictions of problem P_0 and finding the optimal solution of the LP relaxation. This is displayed graphically in Figure 11.8. Because no integer solution is currently known, we set $z_L = -\infty$.

Note, from Figure 11.8, that the optimal solution to the initial LP relaxation is $x_1 = 3.8$, $x_2 = 3$, and $\bar{z} = 8.2$. The corresponding optimal tableau is given in Table 11.1. Clearly, the solution of the LP relaxation is not optimal to the integer program because x_1 is not integer-valued. Therefore, we will choose the variable x_1 to create our first branching. In large problems, there may be many integer variables with noninteger values from which to choose. Typically, an integer variable is chosen from among those with noninteger values based on some other criterion. For example, the variable with the greatest economic importance may be selected. That is, the variable with the largest impact on the objective value.

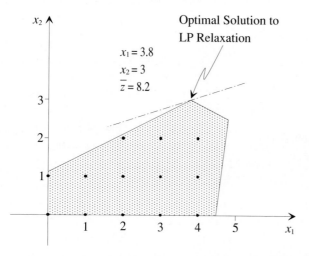

Figure 11.8 Graph for Example 11.2.

Based on our selection of $x_1 = 3.8$, the first pair of branches are defined by $x_1 \leq 3$ and $x_1 \geq 4$, as shown in Figure 11.9. The two new subproblems created by adding these constraints to problem P_0 are depicted as subproblems P_1 and P_2 in Figure 11.10. Clearly, all feasible integer solutions are contained in one of these two subproblems.

Graphically solving the LP relaxation of subproblem P_1 yields $x_1 = 3$, $x_2 = 2.6$, and $\bar{z} = 7.4$. Similarly, the LP relaxation of subproblem P_2 results in $x_1 = 4$, $x_2 = 2.9$, and $\bar{z} = 7.6$. Of course, in practice, we would not be solving these subproblems graphically; we would be utilizing reoptimization techniques. For example, to solve the LP relaxation of subproblem P_1, we would append the constraint $x_1 \leq 3$ to the tableau in Table 11.1 and then reoptimize using the dual simplex method. This is illustrated in more detail in conjunction with cutting-plane methods in Example 11.4. But here we continue with a simple graphical approach.

Note that in both cases, the subproblems were feasible with noninteger optimal solutions. Also, in both cases, $\bar{z} > z_L = -\infty$. Therefore, neither of the nodes associated with these subproblems may be fathomed. Because the objective value of the LP relaxation of subproblem 2 was the greater of the two (i.e., $7.6 > 7.4$), the best-bound approach specifies that we first examine subproblem P_2.

TABLE 11.1

	z	x_1	x_2	x_3	x_4	x_5	RHS
z	1	0	0	0.06	0.1	0	8.2
x_2	0	0	1	0.025	0.05	0	3
x_1	0	1	0	-0.05	0.1	0	3.8
x_5	0	0	0	0.425	-0.75	1	8.6

Figure 11.9   Branching on variable x_1.

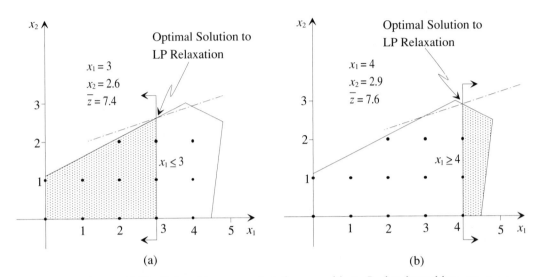

Figure 11.10   Subproblems created from problem P_0 by branching on x_1: (a) subproblem P_1 and (b) subproblem P_2.

Because $x_2 = 2.9$ as a result of the LP relaxation of subproblem 2, we create two new subproblems based on the restrictions $x_2 \leq 2$ and $x_2 \geq 3$. The branching process is illustrated in Figure 11.11, and the two subproblems are displayed graphically as subproblems P_3 and P_4 in Figure 11.12. Note that subproblem P_4 is infeasible, and thus node 4 may be fathomed.

Observe also that the LP relaxation of subproblem P_3 results in an integer solution, and we have $x_1 = 4$, $x_2 = 2$, and $\bar{z} = z = 4$. Therefore, node 3 can also be fathomed and we now have a candidate integer solution. Because this is the best

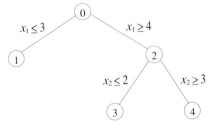

Figure 11.11   Branching on variable x_2 at node 2.

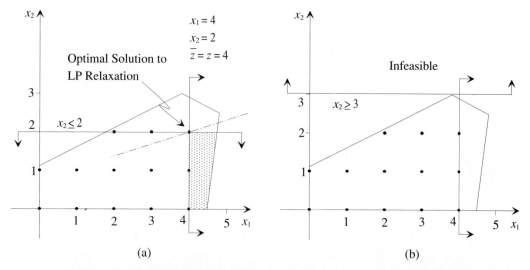

Figure 11.12 Subproblems created from subproblem P_2 by branching on x_2: (a) subproblem P_3 and (b) subproblem P_4.

integer solution found thus far, it is called the incument solution, and the lower bound is updated as $z_L = 4$.

The only remaining unfathomed pendant node is node 1 (i.e., subproblem P_1). Noting that $\bar{z} = 7.4 > z_L = 4$, we see that we cannot fathom node 1 based on its upper bound. The fractional variable $x_2 = 2.6$ results in nodes 5 and 6, corresponding to subproblems 5 and 6, which are based on the additional restrictions $x_2 \leq 2$ and $x_2 \geq 3$, respectively (see Figures 11.13 and 11.14). Note that node 6 can be fathomed due to infeasibility, and node 5 results in a noninteger optimal solution of $x_1 = 1.8$, $x_2 = 2$, and $\bar{z} = 6.2$. Because $\bar{z} = 6.2 > z_L = 4$, we continue branching on node 5.

Adding the restrictions $x_1 \leq 1$ and $x_1 \geq 2$ based on the noninteger variables $x_1 = 1.8$ results in branching on node 5, as shown in Figure 11.15. The corresponding subproblems P_7 and P_8 are given in Figure 11.16. Solving the LP relaxation of subproblem P_7 yields a noninteger optimum, whereas subproblem P_8 results in an integer optimal solution of $x_1 = 2$, $x_2 = 2$, and $\bar{z} = z = 6$. We, therefore, can fathom node 8 and because $z = 6 > z_L = 4$, this new integer solution becomes the incumbent solution with $z_L = 6$.

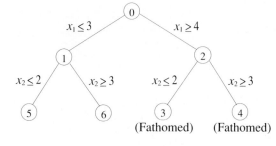

Figure 11.13 Branching on variable x_2 at node 1.

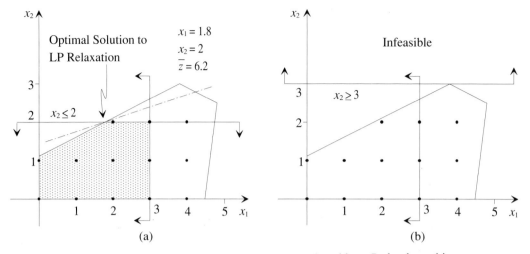

Figure 11.14 Subproblems created from subproblem P_1 by branching on x_2: (a) subproblem P_5 and (b) subproblem P_6.

Figure 11.15 Branching on variable x_1 at node 5.

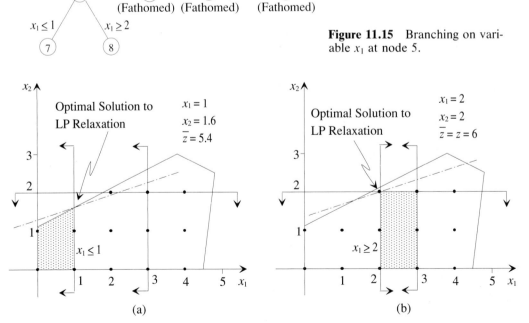

Figure 11.16 Subproblems created from subproblem P_5 by branching on x_1: (a) subproblem P_7 and (b) subproblem P_8.

The only remaining pendant node is node 7. Note, however, that the upper bound associated with node 7 is $\bar{z} = 5.4$, and the greatest lower bound is $z_L = 6$. Thus, node 7 can never yield a solution as good as the current incumbent solution. Therefore, node 7 is fathomed and we stop with the incumbent solution as the optimal solution. Of course, this is exactly the same solution that was found graphically in Example 11.1. A complete summary of the solutions and the branching process is given in Figure 11.17.

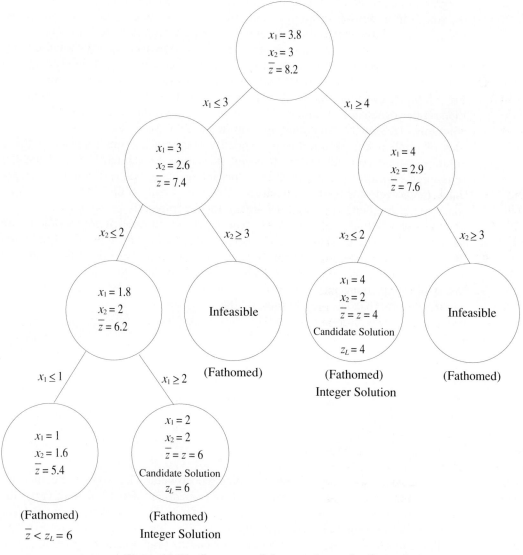

Figure 11.17 Summary of the search tree for Example 11.2.

Branch-and-bound can be used in much the same way to solve mixed integer programming problems. The only significant difference is that branching only occurs with respect to variables that are restricted to be integer.

IMPLICIT ENUMERATION

Implicit enumeration is a technique that is usually applied to zero–one integer programming problems. It is similar to branch-and-bound enumeration; however, the rules for branching, bounding, and fathoming have been simplified and refined because each integer variable can only take on the values of zero or one.

In general, zero–one integer programming problems only have a finite number of feasible points. Of course, this finite number may be extremely large. If there are n zero–one variables, then there are at most 2^n feasible integer points. For example, even in a moderate-size problem with $n = 100$, an upper bound on the number of feasible points is 2^{100}, which is on the order of 10^{30} feasible integer solutions. Obviously, total enumeration in such a case is impossible. The basic idea of implicit enumeration is to explicitly enumerate a small subset of these solutions while concluding that it is not necessary to explicitly investigate the remaining solutions because they are either infeasible or will result in an objective value that is inferior to the best integer solution already found. Those solutions that are not explicitly investigated are said to be implicitly enumerated.

As with branch-and-bound, the search process in implicit enumeration may be viewed graphically as a search tree. In this case, however, the branching process has been simplified because each variable can only assume the values of zero or one. We move from one node of the search tree to a node on the next lower level by fixing an additional variable to either zero or one, as shown in Figure 11.18. That is, each branch of the tree corresponds to some variable x_i fixed so that $x_i = 0$ or $x_i = 1$.

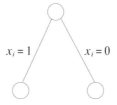

$x_i = 1$ $x_i = 0$

Figure 11.18   Branching with binary variables.

At a particular node, those variables whose values have been specified by the branching process are referred to as *fixed variables*. The remaining variables whose values have not yet been specified at that node are called *free variables*. Such a nodal solution in which some of the variables are fixed at either zero or one is called a *partial solution*. Note that by design, the branching process results in a partial solution being augmented at each branch. That is, an additional variable is fixed at each branching. The idea is to systematically construct a feasible integer solution.

We will consider a *forward step* in the branching process as fixing some free variable at the value one. *Backtracking* is tracing back to the origin until you encounter the first node with only one descending branch. If such a node is located, then a new subproblem is formed by creating the remaining branch at that node. The process is continued until all pendant nodes are fathomed and each nonpendant node has exactly two branches. To illustrate the branching process, consider the complete enumeration of a problem with three binary variables. Figure 11.19 depicts a search tree for this complete enumeration. Each node corresponds to a subproblem with the nodes being *numbered in the order of their creation*. Note that we work down one side of the tree in a depth-first fashion. Also, observe that the completed search tree has $2^3 = 8$ pendant nodes. In general, the number of pendant nodes in a *complete* search tree will be 2^n, where n is the number of binary variables.

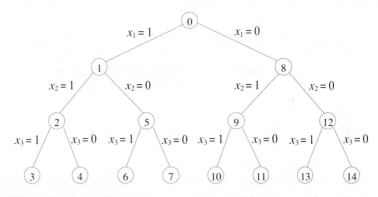

Figure 11.19 Complete binary search tree with nodes numbered in the order of creation.

Give a node and a partial solution, a *completion* of the partial solution is a solution in which values are specified for all of the remaining free variables. A partial solution is *fathomed* by demonstrating that there are no improving feasible completions or by finding the best feasible completion.

For ease of implementation, we will assume that the zero–one integer programming problem to be solved is in the following form, where $\mathbf{c} \geq \mathbf{0}$.

$$\text{minimize } \mathbf{cx}$$

$$\text{subject to:}$$

$$\mathbf{Ax} \leq \mathbf{b} \tag{11.78}$$

$$\mathbf{x} \text{ binary}$$

It is actually a simple process to transform a zero–one programming problem into this form. First, all equality constraints are replaced by a pair of inequalities. If the original problem is one of maximization, then the objective is multiplied by (-1) and changed to minimization. Finally, if a variable x_k appears in the objective function with a negative coefficient (i.e., $c_k < 0$), then x_k is replaced with $(1 - x'_k)$, where x'_k is also a binary variable. Thus, the objective term corresponding to x_k becomes $c_k(1 - x'_k) = c_k - c_k x'_k$. The objective coefficient of x'_k is now positive because $-c_k > 0$. Note that the objective function has also been modified by a constant equal to c_k. Once the problem is solved, the original solution can be recovered by noting that $x_k = 1 - x'_k$.

Note that if the right-hand-side vector $\mathbf{b}$ is nonnegative, then the trivial solution defined by $\mathbf{x} = \mathbf{0}$ will be the optimal solution because $\mathbf{c} \geq \mathbf{0}$. This is the basic motivation for the previous formulation. We begin our discussion of the basic steps of implicit enumeration by considering a solution called the zero completion.

The Zero Completion at a Node

Given a partial solution corresponding to a node of the search tree, we would like to find a good completion of the partial solution. Because the objective coefficients of the free variables at the node are nonnegative, assigning each free variable the value zero will always result in the minimum possible objective value. Such a solution is called the *zero completion* at node k.

Note that it is a trivial process to check the feasibility of the zero completion. One only needs to check if the right-hand-side constants are nonnegative. If the zero completion is feasible to the constraint set, then it obviously must be the optimal solution at node k and the node is fathomed. In addition, the objective value of this *feasible* zero completion provides an upper bound for the objective value of the original problem. We will denote by z_U the smallest such upper bound that has been found, and, as before, the corresponding solution will be called the incumbent solution.

If the zero completion is an infeasible solution, then additional branching may be necessary. However, even if the zero completion is infeasible, the objective value associated with the zero completion, denoted by $\underline{z}$, provides a lower bound on the best objective value that could be obtained from node k. This bound can then be used in the same way as in the traditional branch-and-bound method. If the lower bound is greater than the objective value, z_U, associated with the best feasible integer solution (incumbent solution) found thus far, then the node may be fathomed.

The Infeasibility Test

The infeasibility test attempts to establish that there are no feasible completions possible at a given node by examination of the constraint set. If it is possible to establish that there are no feasible completions at a node, then it is not possible to augment the current partial solution and obtain a feasible solution. If this is the case, then obviously the node can be fathomed.

The test is performed on one constraint at a time and as soon as an infeasibility is detected, the test is terminated. The technique is most easily illustrated by way of an example.

Consider the constraint

$$4x_1 - 2x_2 + 3x_3 + x_4 - x_5 + 3x_6 \leq 2 \tag{11.79}$$

and suppose that the current partial solution corresponds to the fixed variables $x_1 = 1$ and $x_3 = 1$. Then, setting $x_1 = 1$ and $x_3 = 1$ in constraint (11.79) yields

$$-2x_2 + x_4 - x_5 + 3x_6 \leq 2 - 4(1) - 3(1) + 0 = -5 \tag{11.80}$$

Denoting the slack variable in constraint (11.80) by $s \geq 0$ and solving for s yields

$$s = -5 + 2x_2 - x_4 + x_5 - 3x_6 \tag{11.81}$$

Now consider computing the maximum value of s. We can do this in a manner similar to the one that was used to compute the objective value of the zero completion. If the coefficient of some x_k is positive in the expression for s, then we set $x_k = 1$. Similarly, $x_k = 0$ if it has a coefficient that is negative. Clearly, for expression (11.81), this results in

$$s_{\max} = -5 + 2(1) - (0) + (1) - 3(0) = -2 < 0 \tag{11.82}$$

Because the maximum value of s is negative, it is not possible to find a feasible completion at this node. The node, therefore, may be fathomed and the search continued at another node. If, however, we had found that $s_{\max}$ was nonnegative, then we would proceed to check the feasibility of the next constraint. If each constraint, in turn, does not indicate infeasibility, then the infeasibility check is inconclusive and further branching may be needed.

The basic steps of implicit enumeration are now illustrated by the following example.

Example 11.3: Implicit Enumeration

Consider the following zero–one integer programming problem that has already been placed in the required form of (11.78).

$$\text{minimize } z = 4x_1 + 5x_2 + 6x_3 + 2x_4 + 3x_5 \tag{11.83}$$

subject to:

$$-4x_1 - 2x_2 + 3x_3 - 2x_4 + x_5 \leq -1 \tag{11.84}$$

$$-x_1 - 5x_2 - 2x_3 + 2x_4 - 2x_5 \leq -5 \tag{11.85}$$

$$\mathbf{x} \text{ binary} \tag{11.86}$$

Subproblem P_0

We begin with all variables free, and because no feasible solution is known yet, we set $z_U = +\infty$. Subproblem P_0 corresponds precisely to the original problem formulation given in (11.83–11.86). The two major tests are now performed on subproblem P_0.

Zero Completion Test

Substituting $x_i = 0$ for all free variables, we see that the zero completion is infeasible to both constraints (11.84) and (11.85). The trivial lower bound for node 0 generated by the zero completion is $\underline{z} = 0$. Because $\underline{z} = 0 < z_U = +\infty$, we continue to investigate problem P_0.

Infeasibility Test

Letting s_1 and s_2 denote the slack variables in constraints (11.84) and (11.85), respectively, we find that

$$s_1 = -1 + 4x_1 + 2x_2 - 3x_3 + 2x_4 - x_5$$

$$s_2 = -5 + x_1 + 5x_2 + 2x_3 - 2x_4 + 2x_5$$

Now computing the maximum value of s_1 and s_2 yields

$$s_{1\,max} = -1 + 4(1) + 2(1) - 3(0) + 2(1) - (0) = 7 > 0$$

$$s_{2\,max} = -5 + (1) + 5(1) + 2(1) - 2(0) + 2(1) = 5 > 0$$

Because $s_{1\,max}$ and $s_{2\,max}$ are both positive, the infeasibility test is inconclusive and we continue by branching on node 0. We will assume that problem P_0 is augmented by fixing $x_1 = 1$. This results in subproblem P_1 at node 1 of the search tree in Figure 11.20.

$x_1 = 1$

Figure 11.20   Search tree with new node 1.

Subproblem $P_1(x_1 = 1)$

$$\text{minimize } z = 4 + 5x_2 + 6x_3 + 2x_4 + 3x_5 \tag{11.87}$$

subject to:

$$-2x_2 + 3x_3 - 2x_4 + x_5 \le 3 \tag{11.88}$$

$$-5x_2 - 2x_3 + 2x_4 - 2x_5 \le -4 \tag{11.89}$$

$$\mathbf{x} \text{ binary} \tag{11.90}$$

Zero Completion Test

The zero completion is infeasible to constraint (11.89). The trivial lower bound generated by the zero completion is $\underline{z} = 4$. Because $\underline{z} = 4 < z_U = +\infty$, we continue our investigation of node 1.

Infeasibility Test

Determining the expression for the slack variables in (11.88) and (11.89), we get

$$s_1 = 3 + 2x_2 - 3x_3 + 2x_4 - x_5$$

$$s_2 = -4 + 5x_2 + 2x_3 - 2x_4 + 2x_5$$

which yields

$$s_{1\ max} = 3 + 2(1) - 3(0) + 2(1) - (0) = 7 > 0$$

$$s_{2\ max} = -4 + 5(1) + 2(1) - 2(0) + 2(1) = 5 > 0$$

Therefore, again, the infeasibility test is inconclusive and we continue by branching on node 1. Augmenting subproblem P_1 by fixing $x_2 = 1$ results in subproblem P_2 at node 2 of the search tree. (See Figure 11.21.)

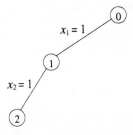

Figure 11.21 Search tree with new node 2.

Subproblem $P_2(x_1 = 1, x_2 = 1)$

$$\text{minimize } z = 9 + 6x_3 + 2x_4 + 3x_5 \qquad (11.91)$$

subject to:

$$3x_3 - 2x_4 + x_5 \le 5 \qquad (11.92)$$

$$-2x_3 + 2x_4 - 2x_5 \le 1 \qquad (11.93)$$

$$\mathbf{x} \text{ binary} \qquad (11.94)$$

Zero Completion Test

The zero completion is feasible to subproblem P_2, and therefore a candidate integer solution is

$$x_1 = 1$$

$$x_2 = 1$$

$$x_3 = 0$$

$$x_4 = 0$$

$$x_5 = 0$$

$$z = 9$$

Because a feasible integer solution has been found, we update the upper bound on the objective by setting $z_U = 9$. This solution is now the incumbent solution. Node 2 is fathomed and we backtrack to node 1 and branch on $x_2 = 0$ to form subproblem P_3, as shown in Figure 11.22.

Subproblem $P_3(x_1 = 1, x_2 = 0)$

$$\text{minimize } z = 4 + 6x_3 + 2x_4 + 3x_5 \qquad (11.95)$$

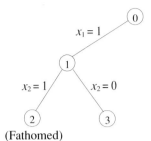

Figure 11.22 Search tree with new node 3.

subject to:

$$3x_3 - 2x_4 + x_5 \leq 3 \tag{11.96}$$

$$-2x_3 + 2x_4 - 2x_5 \leq -4 \tag{11.97}$$

$$\mathbf{x} \text{ binary} \tag{11.98}$$

Zero Completion Test

The zero completion is infeasible due to constraint (11.97). The trivial lower bound generated by the zero completion is $\underline{z} = 4$ and we see that $\underline{z} < z_U = 9$. Therefore, we continue with the infeasibility test.

Infeasibility Test

$$s_1 = 3 - 3x_3 + 2x_4 - x_5$$

$$s_2 = -4 + 2x_3 - 2x_4 + 2x_5$$

and

$$s_{1 \, max} = 3 - 3(0) + 2(1) - (0) = 5 > 0$$

$$s_{2 \, max} = -4 + 2(1) - 2(0) + 2(1) = 0$$

Thus, the infeasibility test is inconclusive and we continue by branching on node 3. Augmenting subproblem P_3 by fixing $x_3 = 1$ results in subproblem P_4 at node 4 of the search tree depicted in Figure 11.23.

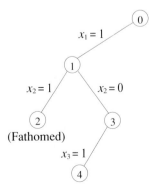

Figure 11.23 Search tree with new node 4.

Subproblem $P_4(x_1 = 1, x_2 = 0, x_3 = 1)$

$$\text{minimize } z = 10 + 2x_4 + 3x_5 \tag{11.99}$$

subject to:

$$-2x_4 + x_5 \leq 0 \tag{11.100}$$

$$2x_4 - 2x_5 \leq -2 \tag{11.101}$$

$$\mathbf{x} \text{ binary} \tag{11.102}$$

Zero Completion Test

Note that the zero completion is infeasible to constraint (11.101); however, the lower bound generated is $\underline{z} = 10$. Because $\underline{z} = 10 > z_U = 9$, then node 4 can be fathomed because additional branching can never produce an integer solution as good as the incumbent solution found in subproblem 2. We then backtrack to node 3 and create a new branch by fixing $x_3 = 0$. This results in subproblem P_5, which corresponds to node 5 in Figure 11.24.

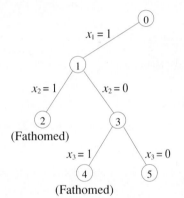

Figure 11.24 Search tree with new node 5.

Subproblem $P_5(x_1 = 1, x_2 = 0, x_3 = 0)$

$$\text{minimize } z = 4 + 2x_4 + 3x_5 \tag{11.103}$$

subject to:

$$-2x_4 + x_5 \leq 3 \tag{11.104}$$

$$2x_4 - 2x_5 \leq -4 \tag{11.105}$$

$$\mathbf{x} \text{ binary} \tag{11.106}$$

Zero Completion Test

The zero completion is infeasible due to constraint (11.105) and the lower bound is $\underline{z} = 4 < z_U = 9$.

Infeasibility Test

$$s_1 = 3 + 2x_4 + x_5$$

$$s_2 = -4 - 2x_4 + 2x_5$$

and

$$s_{1\,max} = 3 + 2(1) - 2(0) = 5 > 0$$

$$s_{2\,max} = -4 - 2(0) + 2(1) = -2 < 0$$

Thus, no feasible completion is possible for subproblem P_5. Node 5 is fathomed and backtracking results in branching on $x_1 = 0$ at node 0. This results in the formation of node 6 in Figure 11.25.

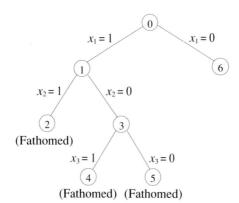

Figure 11.25 Search tree with new node 6.

Subproblem $P_6(x_1 = 0)$

$$\text{minimize } z = 5x_2 + 6x_3 + 2x_4 + 3x_5 \tag{11.107}$$

subject to:

$$-2x_2 + 3x_3 - 2x_4 + x_5 \leq -1 \tag{11.108}$$

$$-5x_2 - 2x_3 + 2x_4 - 2x_5 \leq -5 \tag{11.109}$$

$$\mathbf{x} \text{ binary} \tag{11.110}$$

Zero Completion Test

The zero completion is infeasible to both constraints (11.108) and (11.109) and the lower bound is $\underline{z} = 0 < z_U = 9$.

Infeasibility Test

$$s_1 = -1 + 2x_2 - 3x_3 + 2x_4 - x_5$$

$$s_2 = -5 + 5x_2 + 2x_3 - 2x_4 + 2x_5$$

and

$$s_{1\,max} = -1 + 2(1) - 3(0) + 2(1) - (0) = 3 > 0$$

$$s_{2\,max} = -5 + 5(1) + 2(1) - 2(0) + 2(1) = 4 > 0$$

Therefore, the infeasibility test is inconclusive and we continue by branching on node 6. Augmenting the current partial solution by fixing $x_2 = 1$ results in subproblem P_7 (see Figure 11.26).

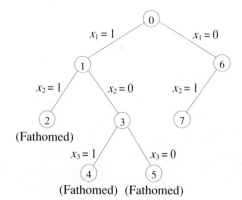

Figure 11.26 Search tree with new node 7.

Subproblem $P_7(x_1 = 0,\ x_2 = 1)$

$$\text{minimize } z = 5 + 6x_3 + 2x_4 + 3x_5 \qquad (11.111)$$

subject to:

$$3x_3 - 2x_4 + x_5 \leq 1 \qquad (11.112)$$

$$2x_3 + 2x_4 - 2x_5 \leq 0 \qquad (11.113)$$

$$\mathbf{x} \text{ binary} \qquad (11.114)$$

Zero Completion Test

The zero completion is feasible and yields the candidate integer solution:

$$x_1 = 0$$
$$x_2 = 1$$
$$x_3 = 0$$
$$x_4 = 0$$
$$x_5 = 0$$
$$z = 5$$

Note that the objective value is less than z_U; therefore, we update by now setting $z_U = 5$ and this solution becomes the new incumbent solution. Node 7 is fathomed and we backtrack to node 6 and branch on $x_2 = 0$ to form problem P_8, as shown in Figure 11.27.

Subproblem $P_8(x_1 = 0,\ x_2 = 0)$

$$\text{minimize } z = 6x_3 + 2x_4 + 3x_5 \qquad (11.115)$$

subject to:

$$3x_3 - 2x_4 + x_5 \leq -1 \qquad (11.116)$$

$$-2x_3 + 2x_4 - 2x_5 \leq -5 \qquad (11.117)$$

$$\mathbf{x} \text{ binary} \qquad (11.118)$$

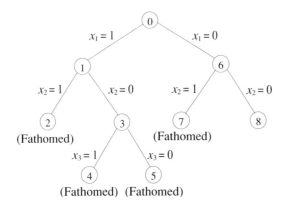

Figure 11.27 Search tree with new node 8.

Zero Completion Test
The zero completion is infeasible to both constraints (11.116) and (11.117) and the lower bound is $\underline{z} = 0$.

Infeasibility Test

$$s_1 = -1 - 3x_3 + 2x_4 - x_5$$

$$s_2 = -5 + 2x_3 - 2x_4 + 2x_5$$

and

$$s_{1 \text{ max}} = -1 - 3(0) + 2(1) - (0) = 1 > 0$$

$$s_{2 \text{ max}} = -5 + 2(1) - 2(0) + 2(1) = -1 < 0$$

Therefore, there are no feasible completions of the partial solution at node 8. Backtracking we find that the search tree is complete. Therefore, the optimal integer solution is given by the incumbent solution found at node 7.

Note that in the solution process, we generated a very small portion of the complete search tree. Because there are $n = 5$ variables, the complete tree would have five levels and $2^5 = 32$ pendant nodes.

Refinements

Much research has been done on methods to increase the efficiency of the implicit enumeration process. As a result, there are many refinements to the basic enumeration algorithm that was presented in Example 11.3. Two of the simplest tests for increasing the efficiency of the algorithm are the *cancellation-zero test* and the *cancellation-one test*.

Give a partial solution, the cancellation-zero test is designed to answer the question: Should any variable be fixed necessarily at zero? Similarly, the cancellation-one test attempts to determine if any variable should be necessarily fixed at one. This is done by examining each constraint individually along with the corresponding value of s_{max} computed in the infeasibility test.

For example, consider the constraint defined by

$$-4x_1 - 2x_2 + 9x_3 + 7x_4 - 3x_5 - 10x_6 + 3x_7 + x_8 \leq -13 \qquad (11.119)$$

and assume that the current subproblem is associated with the partial solution $x_2 = 1$ and $x_8 = 0$. Then constraint (11.119) simplifies to

$$-4x_1 + 9x_3 + 7x_4 - 3x_5 - 10x_6 + 3x_7 \leq -11 \qquad (11.120)$$

and we can write an expression for the slack variable s as

$$s = -11 + 4x_1 - 9x_3 - 7x_4 + 3x_5 + 10x_6 - 3x_7 \qquad (11.121)$$

Computing the maximum value of s yields

$$s_{max} = -11 + 4 - 0 - 0 + 3 + 10 - 0 = 6 \qquad (11.122)$$

From (11.122), we see that the infeasibility test is inclusive. However, consider variable x_3. Note that if $x_3 = 1$, it is impossible to satisfy constraint (11.119) because $s_{max} = 6$. This can be easily seen by setting $x_3 = 1$ in the computation of s_{max} and noting that this results in $s_{max} < 0$. Therefore, if a feasible completion is to be found for this partial solution, then we must have $x_3 = 0$. This is the cancellation-zero test and it can be easily performed by comparing the value of the positive coefficients in the constraint with the value of s_{max}. If a positive coefficient of a variable x_j is greater than s_{max}, then $x_j = 0$. Therefore, for constraint (11.119), we see that any feasible completion must have $x_3 = 0$ because $9 > s_{max}$ and $x_4 = 0$ because $7 > s_{max}$.

Similarly, consider the variable x_6. Note that in the computation of s_{max}, if it were not for setting $x_6 = 1$, then s_{max} would be negative. That is, a feasible completion is not possible unless $x_6 = 1$. This, of course, is the cancellation-one test, and its implementation is just as straightforward as the cancellation-zero test. If the absolute value of a negative coefficient of a variable x_j is greater than s_{max}, then $x_j = 1$. Thus, for constraint (11.119), the cancellation-one test results in $x_6 = 1$ because $|-10| > s_{max}$.

Therefore, any feasible completion of the partial solution with $x_2 = 1$ and $x_8 = 0$ must have $x_3 = x_4 = 0$ and $x_6 = 1$. This, obviously, will accelerate the enumeration process.

Among the other refinements for implicit enumeration are surrogate constraints, preferred sets, special branching rules, and various bookkeeping schemes. The interested reader is referred to, for example, Taha (1975) and Salkin and Mathur (1989).

CUTTING-PLANE METHODS

Consider the integer points contained in the polyhedral set depicted graphically in Figure 11.28(a). Now, suppose that it was possible to determine the convex hull of these feasible integer points, as shown in Figure 11.28(b). (The *convex hull* of a

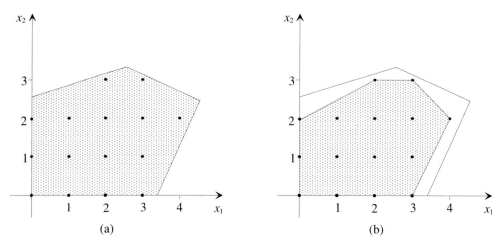

Figure 11.28   (a) Feasible integer points and (b) convex hull of feasible integer points.

set of points refers to the smallest convex set containing the points.) Then the optimal solution of the integer programming problem could be determined by solving a single linear programming problem in which the convex hull is used as the feasible region. This is because the extreme points of the convex hull correspond to integer solutions. This is the basic motivation behind cutting-plane methods.

In reality, however, it is very difficult (if not impossible) to actually compute the convex hull of a set of feasible integer points. Thus, the goal of a cutting-plane method is to iteratively construct the convex hull in the vicinity of the optimal integer solution. This is done in a systematic manner by introducing additional constraints (cutting planes) that cut off portions of the feasible region without excluding any feasible integer points. The purpose of this construction is to eventually force the optimal integer solution to be an extreme point of the constructed feasible region. Once this is accomplished, the optimal integer solution can be found by solving the corresponding LP relaxation. The general idea of a cutting-plane algorithm is portrayed in Figure 11.29.

The polyhedral region depicted in Figure 11.29(a) corresponds to the LP relaxation, and the integer points within this region are the feasible integer points. Noting the levels curves of the maximization objective, we can easy identify the continuous and integer optima, as shown in Figure 11.29(a). Now, consider adding constraints (cutting planes) 1 and 2 to the LP relaxation, as shown in Figure 11.29(b). Observe that a portion of the feasible region associated with the LP relaxation is cut off, but all integer points remain feasible. In fact, after adding these two cuts, we see that the optimal solution to the LP relaxation also satisfies the integer restrictions, and hence is also the optimal integer solution. The general process is, however, not always as simple as illustrated in Figure 11.29. First,

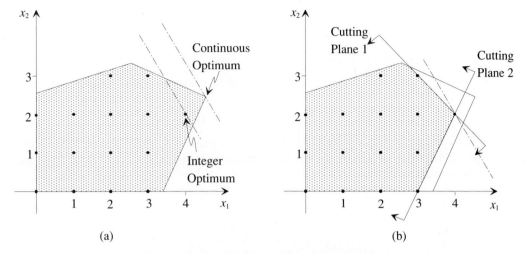

Figure 11.29 Cutting-plane algorithm.

note that there are infinitely many ways to introduce a cut that will remove the current continuous optimal solution without excluding any feasible integer points. Also, in reality, it may require many cuts until the optimal integer point becomes an extreme point of the restructured problem.

There are a number of methods for constructing cutting planes. Some of these methods are applicable to pure integer programming problems, whereas others are designed for mixed integer programs. In this section, we discuss a basic cutting-plane method for pure integer programs. The method described was originally proposed by Gomory (1960) and the cutting planes generated are called *dual fractional cuts*. The adjective "dual" refers to the fact that dual feasibility is maintained throughout the algorithm with a primal feasible solution being determined only at termination.

Dual Fractional Cuts for Pure Integer Programs

Consider the pure integer programming problem

$$\text{(IP)} \quad \text{maximize } \mathbf{cx} \tag{11.123}$$

$$\text{subject to:}$$

$$\mathbf{Ax} = \mathbf{b}$$

$$\mathbf{x} \geq \mathbf{0}, \mathbf{x} \text{ integer-valued}$$

where all the entries in $\mathbf{A}$, $\mathbf{b}$, and $\mathbf{c}$ are integer.
The LP relaxation associated with (IP) is

$$\text{(LP)} \quad \text{maximize } \mathbf{cx} \tag{11.124}$$

subject to:

$$\mathbf{Ax} = \mathbf{b}$$

$$\mathbf{x} \geq \mathbf{0}$$

Recall from Chapter 4 that a basic feasible solution to (LP) is defined by the relationship

$$\mathbf{x}_B + \mathbf{B}^{-1}\mathbf{N}\mathbf{x}_N = \mathbf{B}^{-1}\mathbf{b} \tag{11.125}$$

where the coefficient matrix has been partitioned as $\mathbf{A} = (\mathbf{B} : \mathbf{N})$, and vector $\mathbf{x}$ has been partitioned into the basic and nonbasic components, $\mathbf{x}_B = \mathbf{B}^{-1}\mathbf{b} \geq \mathbf{0}$ and $\mathbf{x}_N = \mathbf{0}$, respectively.

By letting J denote the index set of the nonbasic variables, (11.125) can be rewritten as

$$\mathbf{x}_B + \sum_{j \in J} (\mathbf{B}^{-1}\mathbf{a}_j)x_j = \mathbf{B}^{-1}\mathbf{b} \tag{11.126}$$

or, in the equivalent form,

$$\mathbf{x}_B + \sum_{j \in J} \boldsymbol{\alpha}_j x_j = \boldsymbol{\beta} \tag{11.127}$$

where, as usual,

$$\boldsymbol{\alpha}_j = \mathbf{B}^{-1}\mathbf{a}_j \tag{11.128}$$

and

$$\boldsymbol{\beta} = \mathbf{B}^{-1}\mathbf{b} \tag{11.129}$$

Recall, also, that the objective function can be similarly written as

$$z + \sum_{j \in J} (z_j - c_j)x_j = \mathbf{c}_B \mathbf{B}^{-1}\mathbf{b} = \mathbf{c}_B \boldsymbol{\beta} \tag{11.130}$$

Then (11.126) and (11.130) can be summarized in the simplex tableau in the usual manner, as shown in Table 11.2.

TABLE 11.2

	z	$x_{B,1}$	$\cdots$	$x_{B,r}$	$\cdots$	$x_{B,m}$	$\cdots$	x_k	$\cdots$	x_j	$\cdots$	RHS
z	1	0	$\cdots$	0	$\cdots$	0	$\cdots$	$z_k - c_k$	$\cdots$	$z_j - c_j$	$\cdots$	$\mathbf{c}_B \boldsymbol{\beta}$
$x_{B,1}$	0	1	$\cdots$	0	$\cdots$	0	$\cdots$	$\alpha_{1,k}$	$\cdots$	$\alpha_{1,j}$	$\cdots$	β_1
$\vdots$	$\vdots$	$\vdots$		$\vdots$		$\vdots$		$\vdots$		$\vdots$		$\vdots$
$x_{B,r}$	0	0	$\cdots$	1	$\cdots$	0	$\cdots$	$\alpha_{r,k}$	$\cdots$	$\alpha_{r,j}$	$\cdots$	β_r
$\vdots$	$\vdots$	$\vdots$		$\vdots$		$\vdots$		$\vdots$		$\vdots$		$\vdots$
$x_{B,m}$	0	0	$\cdots$	0	$\cdots$	1	$\cdots$	$\alpha_{m,k}$	$\cdots$	$\alpha_{m,j}$	$\cdots$	β_m

Suppose that the tableau in Table 11.2 is optimal to (LP). If $\boldsymbol{\beta} = \mathbf{B}^{-1}\mathbf{b}$ is all-integer, then clearly the optimal solution to the LP relaxation is also optimal to (IP). Otherwise, the vector $\boldsymbol{\beta}$ has at least one fractional element, say, β_r. We will now proceed to develop a cutting plane based on row r of the tableau. Row r will be referred to as the *source row*.

We begin by writing row r of Table 11.2 in algebraic form. This results in

$$x_{B,r} + \sum_{j \in J} \alpha_{r,j} x_j = \beta_r \tag{11.131}$$

Again, let the notation $\lfloor t \rfloor$ represent the greatest integer less than or equal to t and define f_r and $f_{r,j}$ by

$$f_r = \beta_r - \lfloor \beta_r \rfloor \tag{11.132}$$

and

$$f_{r,j} = \alpha_{r,j} - \lfloor \alpha_{r,j} \rfloor, \qquad \text{for all } j \in J \tag{11.133}$$

Then, clearly,

$$0 < f_r < 1 \tag{11.134}$$

and

$$0 \le f_{r,j} < 1, \qquad \text{for all } j \in J \tag{11.135}$$

Now, rewrite (11.132) and (11.133) as

$$\beta_r = f_r + \lfloor \beta_r \rfloor \tag{11.136}$$

and

$$\alpha_{r,j} = f_{r,j} + \lfloor \alpha_{r,j} \rfloor \qquad \text{for all } j \in J \tag{11.137}$$

Substituting these expressions into (11.131) yields

$$x_{B,r} + \sum_{j \in J} (\lfloor \alpha_{r,j} \rfloor + f_{r,j}) x_j = \lfloor \beta_r \rfloor + f_r \tag{11.138}$$

which can be rewritten as

$$x_{B,r} + \sum_{j \in J} \lfloor \alpha_{r,j} \rfloor x_j + \sum_{j \in J} f_{r,j} x_j = \lfloor \beta_r \rfloor + f_r \tag{11.139}$$

Finally, rearranging terms in (11.139), we get

$$\sum_{j \in J} f_{r,j} x_j - f_r = \lfloor \beta_r \rfloor - x_{B,r} - \sum_{j \in J} \lfloor \alpha_{r,j} \rfloor x_j \tag{11.140}$$

Note that the right-hand side of (11.140) is always integer-valued for all feasible integer solutions. Therefore, it follows that the left-hand side of (11.140) is also integer-valued for all feasible integer solutions. That is, $\sum_{j \in J} f_{r,j} x_j$ differs from f_r by an integer. Note also that $\sum_{j \in J} f_{r,j} x_j$ is nonnegative for any nonnegative solution. Therefore, $\sum_{j \in J} f_{r,j} x_j$ is equal to either $f_r, f_r + 1, f_r + 2, \ldots$, and it follows that

$$\sum_{j \in J} f_{r,j} x_j \geq f_r \tag{11.141}$$

The inequality in (11.141) is referred to as a Gomory (1960) cut. By construction, all feasible integer solutions satisfy (11.141). Also, because currently $x_j = 0$, for all $j \in J$, and $f_r > 0$, the current linear programming solution is not feasible to this new constraint. Thus, adding constraint (11.141) to the LP relaxation will cut off the current noninteger optimum.

Now multiplying (11.141) by (-1) and adding the slack variable $x_{n+1} \geq 0$, we get

$$-\sum_{j \in J} f_{r,j} x_j + x_{n+1} = -f_r \tag{11.142}$$

Note, further, that by the preceding argument, the slack variable x_{n+1} is integer for all integer solutions. Thus, the problem remains a pure integer programming problem even after introducing this new variable. Once constraint (11.142) is added to the LP relaxation, the problem is reoptimized using the dual simplex method.

We now illustrate the cutting-plane algorithm with the following example.

Example 11.4: A Cutting-Plane Method

Consider the following pure integer programming problem:

$$\text{maximize } z = 18x_1 + 12x_2 \tag{11.143}$$

subject to:

$$2x_1 - x_2 + x_3 = 5 \tag{11.144}$$

$$2x_1 + 3x_2 + x_4 = 13 \tag{11.145}$$

$$\mathbf{x} \geq \mathbf{0}, \ \mathbf{x} \text{ integer} \tag{11.146}$$

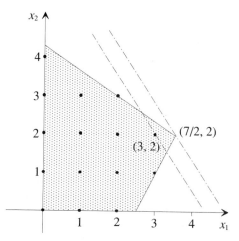

Figure 11.30 Graph for Example 11.4.

Noting that x_3 and x_4 are nothing more than the slack variables in constraints (11.144) and (11.145), respectively, the feasible region can be viewed graphically in x_1–x_2 space, as shown in Figure 11.30. The feasible region of the integer program consists of the integer points within the polyhedron defining the feasible region of the corresponding LP relaxation. Clearly, the optimal solution of the LP relaxation occurs at $x_1 = \frac{7}{2}$ and $x_2 = 2$, whereas the optimal solution of the integer program is $x_1 = 3$ and $x_2 = 2$. Let us now derive the optimal integer solution by generating a sequence of cutting planes.

We begin the solution process by using the simplex method to determine the optimal solution to the LP relaxation. The final simplex tableau giving this optimal linear programming solution is presented in Table 11.3.

TABLE 11.3

	z	x_1	x_2	x_3	x_4	RHS
z	1	0	0	$\frac{15}{4}$	$\frac{21}{4}$	87
x_1	0	1	0	$\frac{3}{8}$	$\frac{1}{8}$	$\frac{7}{2}$
x_2	0	0	1	$-\frac{1}{4}$	$\frac{1}{4}$	2

Note that as expected, from Figure 11.30, $x_{B,1} = x_1$ has a fractional value. Observing that the nonbasic variables are x_3 and x_4, we can use row 1 as the source row to generate the Gomory cut

$$(\tfrac{3}{8})x_3 + (\tfrac{1}{8})x_4 \geq \tfrac{1}{2} \tag{11.147}$$

Now multiplying (11.147) by (-1) and adding the slack variable x_5 results in

$$-(\tfrac{3}{8})x_3 - (\tfrac{1}{8})x_4 + x_5 = -\tfrac{1}{2} \tag{11.148}$$

Appending this constraint to the tableau of Table 11.3 results in Table 11.4. Note that primal feasibility has been lost and therefore the problem is reoptimized by using the dual simplex method. This results in the new optimal tableau of Table 11.5

TABLE 11.4

	z	x_1	x_2	x_3	x_4	x_5	RHS
z	1	0	0	$\frac{15}{4}$	$\frac{21}{4}$	0	87
x_1	0	1	0	$\frac{3}{8}$	$\frac{1}{8}$	0	$\frac{7}{2}$
x_2	0	0	1	$-\frac{1}{4}$	$\frac{1}{4}$	0	2
x_5	0	0	0	$-\frac{3}{8}$	$-\frac{1}{8}$	1	$-\frac{1}{2}$

TABLE 11.5

	z	x_1	x_2	x_3	x_4	x_5	RHS
z	1	0	0	0	4	10	82
x_1	0	1	0	0	0	1	3
x_2	0	0	1	0	$\frac{1}{3}$	$-\frac{2}{3}$	$\frac{7}{3}$
x_3	0	0	0	1	$\frac{1}{3}$	$-\frac{8}{3}$	$\frac{4}{3}$

Again, notice that the solution is not all integer. Selecting row 2 as the source row yields the cut

$$(\tfrac{1}{3})x_4 + (\tfrac{1}{3})x_5 \geq \tfrac{1}{3} \tag{11.149}$$

which can be written in the form

$$-(\tfrac{1}{3})x_4 - (\tfrac{1}{3})x_5 + x_6 = -\tfrac{1}{3} \tag{11.150}$$

by introducing another nonnegative integer slack variable. This cut is added to the previous tableau to form the tableau in Table 11.6.

TABLE 11.6

	z	x_1	x_2	x_3	x_4	x_5	x_6	RHS
z	1	0	0	0	4	10	0	82
x_1	0	1	0	0	0	1	0	3
x_2	0	0	1	0	$\frac{1}{3}$	$-\frac{2}{3}$	0	$\frac{7}{3}$
x_3	0	0	0	1	$\frac{1}{3}$	$-\frac{8}{3}$	0	$\frac{4}{3}$
x_6	0	0	0	0	$-\frac{1}{3}$	$-\frac{1}{3}$	1	$-\frac{1}{3}$

Reoptimizing the linear relaxation by utilizing the dual simplex method results in the optimal solution given in Table 11.7. Note that in this case, the solution is all-integer and is thus the optimal solution to the original problem given in (11.143–11.146).

TABLE 11.7

	z	x_1	x_2	x_3	x_4	x_5	x_6	RHS
z	1	0	0	0	0	6	12	78
x_1	0	1	0	0	0	1	0	3
x_2	0	0	1	0	0	-1	1	2
x_3	0	0	0	1	0	-3	1	3
x_4	0	0	0	0	1	1	-3	1

Before concluding this example, let us view graphically in x_1–x_2 space the cuts that were derived. In order to do this, we must find expressions for the two cuts in terms of x_1 and x_2.

The first cutting plane that was derived is

$$(\tfrac{3}{8})x_3 + (\tfrac{1}{8})x_4 \geq \tfrac{1}{2} \tag{11.151}$$

However, from (11.146) and (11.145), we have that

$$x_3 = 5 - 2x_1 + x_2 \tag{11.152}$$

and

$$x_4 = 13 - 2x_1 - 3x_2 \tag{11.153}$$

Now, substituting these expressions for x_3 and x_4 into (11.151) yields

$$(\tfrac{3}{8})(5 - 2x_1 + x_2) + (\tfrac{1}{8})(13 - 2x_1 - 3x_2) \geq \tfrac{1}{2} \tag{11.154}$$

which simplifies to

$$x_1 \leq 3 \tag{11.155}$$

Thus, the first cutting plane in x_1–x_2 space is simply $x_1 \leq 3$.

Now, consider the second cutting plane, which was defined by the inequality

$$(\tfrac{1}{3})x_4 + (\tfrac{1}{3})x_5 \geq \tfrac{1}{3} \tag{11.156}$$

From (11.148), we know that

$$x_5 = -\tfrac{1}{2} + (\tfrac{3}{8})x_3 + (\tfrac{1}{8})x_4 \tag{11.157}$$

By utilizing (11.152) and (11.153), it is easy to see that (11.157) simplifies to

$$x_5 = -\tfrac{1}{2} + (\tfrac{3}{8})(5 - 2x_1 + x_2) + (\tfrac{1}{8})(13 - 2x_1 - 3x_2) = 3 - x_1 \tag{11.158}$$

Finally, by substituting (11.153) and (11.158) into (11.156), the second cut simplifies to

$$x_1 + x_2 \leq 5 \tag{11.159}$$

Figure 11.31 shows the effect the two cuts have on the original feasible region of the LP relaxation. Note that the optimal integer point is now an extreme point of the feasible region, and, in fact, the two cuts have provided an exact local approximation of the convex hull of the feasible integer points.

The algorithm illustrated in the previous example is actually a very rudimentary cutting-plane algorithm, but it serves to illustrate the general principles of cutting planes. The algorithm is guaranteed to converge finitely under certain conditions, namely:

1. Always choose the topmost fractional row of the tableau as the source row. This includes the z-row of the tableau.
2. Add cuts to the bottom of the tableau, and whenever the slack variable from a cut becomes basic, throw out that row.
3. Use the lexicographic dual simplex method to update the tableau.

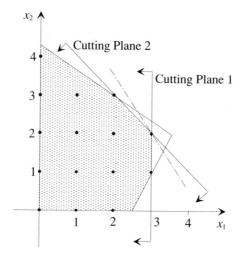

Figure 11.31 Cutting planes of Example 11.4.

However, the performance of the algorithm can be very slow in practice. For this reason, many enhancements and alternative cutting-plane techniques have been developed. The interested reader is referred to Taha (1975) and Salkin and Mathur (1989).

SUMMARY

In this chapter, we introduced integer linear programming models. After a brief discussion of computational complexity, several modeling techniques were presented for developing integer programming models. It was seen that integer programming problems are combinatorial in nature and are much more difficult to solve than linear programming problems in which the variables are continuous.

There are a number of methods for solving integer programming problems. However, exact solutions can usually be found only for moderate-size problems unless the problem exhibits some special characteristics. The basic solution methods presented in this chapter include branch-and-bound enumeration, implicit enumeration, and cutting-plane methods, as well as the graphical solution of simple integer programming problems.

EXERCISES

11.1. A company is planning the weekly production of three products, each of which requires two machining operations to produce. Each unit of Product A requires 12 minutes on Machine 1 and 9 minutes on Machine 2. Similarly, Product B needs 10 minutes on Machine 1 and 10 minutes on Machine 2, whereas the corresponding values for Product C are 13 and 11. Machines 1 and 2 are each available for 40 hours during the week. If any units of Product A are manufactured a one-time setup and fixturing cost of $10 and $13 are incurred on Machines 1 and 2, respectively. The setup costs for Product B are $11 on Machine 1 and $9 on Machine 2. The corresponding costs for Product C are $9 and $15. The unit profits of Products A, B, and C are $8, $7, and $9, respectively. Assuming that the company can sell all the units that it can produce, formulate an integer programming model to maximize profit.

11.2. A company is considering several investment opportunities, each of which differs in the initial capital required. The accounting department has done a thorough analysis of each of these investments and has estimated the long-term profit of each. The initial capital required and estimated profits (in millions of dollars) are summarized in Table 11.8. Investments 1 and 4 are considered high-risk investments and management has decided to invest in at most one of these. In addition, Investment 6 is contingent upon also investing in Investment 3. If $120 million of initial capital is available, formulate an integer programming model to determine the optimal investment strategy.

TABLE 11.8

Investment	Initial capital	Estimated profit
1	26	18
2	34	12
3	18	7
4	45	24
5	31	11
6	39	15
7	23	9
8	13	6

11.3. **(a)** Show how integer programming techniques can be used to ensure that an integer variable x can only assume the values 3, 5, 13, or 21.

(b) Show how integer programming techniques can be used to reduce the following constraint to linear integer form.

$$|a_1x_1 + \cdots + a_2x_2| \geq b, \qquad \text{where } b > 0$$

11.4. Suppose that $-u_1 \leq f_1(\mathbf{x}) \leq u_1$ and $-u_2 \leq f_2(\mathbf{x}) \leq u_2$, where u_1 and u_2 are constants. Show how to use integer programming techniques to model the following.

(a) Either $f_1(\mathbf{x}) \geq 0$ or $f_2(\mathbf{x}) \geq 0$.

(b) If $f_1(\mathbf{x}) > 0$, then $f_2(\mathbf{x}) \geq 0$.

(c) Either $f_1(\mathbf{x}) \geq 0$ or $f_2(\mathbf{x}) \geq 0$, but both are not positive.

11.5. Consider the following constraint set. At least two of the following must hold:

$$x_1 + 2x_2 \leq 12$$

$$3x_1 + 2x_2 \leq 12$$

$$-x_1 + 3x_2 \leq 3$$

(a) Sketch this feasible region. Is the feasible region convex?

(b) Use integer programming techniques to write an equivalent linear integer model.

11.6. A city is trying to establish a municipal emergency ambulance service that can adequately service all sections of the city. The planning committee has established eight potential sites for the emergency service facilities. However, due to time and distance requirements, each potential site can provide coverage to only a subset of the city's six sections. Table 11.9 summarizes the sections for which each site provides coverage. Formulate an integer programming model for determining the fewest number of facilities that will provide all sections of the city with adequate coverage.

TABLE 11.9

Site	Sections of city covered
1	A, B, E
2	C, D
3	B, C, D
4	A, D, F
5	B, F
6	A, D, E, F
7	A, C, E
8	B, D, F

11.7. Formulate the following problem as a mixed integer programming problem.

$$\text{maximize } z = f_1(x_1) + f_2(x_2)$$

subject to:

at least two of the following hold: $x_1 + 3x_2 \leq 12$
$$2x_1 + x_2 \leq 16$$
$$x_1 + x_2 \leq 9$$

$$x_1, x_2 \geq 0$$

where

$$f_1(x_1) = \begin{cases} 10 + 2x_1, & \text{if } 0 \leq x_1 \leq 5 \\ 15 + x_1, & \text{if } x_1 \geq 5 \end{cases}$$

$$f_2(x_2) = \begin{cases} 8 + x_2, & \text{if } 0 \leq x_2 \leq 2 \\ 4 + 3x_2, & \text{if } x_2 \geq 2 \end{cases}$$

11.8. A company manufactures two products, A and B, which are produced using two raw materials. Currently, both products are in high demand and it is anticipated that all that is manufactured can be sold. The selling price for one unit of Product A is \$3.20 and the selling price for Product B is \$2.70. Product A requires 4.2 pounds of Raw Material 1 and 1.7 pounds of Raw Material 2, and the corresponding figures for Product B are 3.9 pounds and 1.1 pounds. Raw Materials 1 and 2 are purchased from a local vendor and the purchase prices are determined according to the schedule shown in Table 11.10. Prior to sale, Products A and B must also undergo final inspection and packaging. Final inspection and packaging of each unit of Product A requires 1.2 hours, and Product B requires 0.9 hour per unit. If there are 480 hours available for final inspection and packaging, formulate an integer programming problem for determining the product mix that maximizes profit.

TABLE 11.10

Raw material 1		Raw material 2	
First 500 lb	\$0.30/lb	First 200 lb	\$0.40/lb
Next 1000 lb	\$0.25/lb	Next 600 lb	\$0.35/lb
Next 1000 lb	\$0.20/lb		
Maximum Available	2500 lb	Maximum Available	800 lb

11.9. ABC, Inc., is considering several investment options. Each option has a minimum investment required as well as a maximum investment allowed. These restrictions along with the expected return are summarized in the Table 11.11. (Figures are in millions of dollars.)

TABLE 11.11

Option	Minimum investment	Maximum investment	Expected return (%)
1	3	27	13
2	2	12	9
3	9	35	17
4	1	15	10
5	12	46	22
6	4	18	12

Because of the high-risk nature of Option 5, company policy requires that the total amount invested in Option 5 be no more than the combined amount invested in Options 2, 4, and 6. In addition, if an investment is made in Option 3, it is required that at least a minimum investment be made in Option 6. ABC has \$80 million to invest and obviously wants to maximize its total expected return on investment. Formulate an integer programming model to determine which options to invest in and how much should be invested.

11.10. Consider the following integer programming problem.

$$\text{maximize } z = -3x_1 + 4x_2$$

subject to:

$$6x_1 - 4x_2 \geq 15$$

$$x_1 + x_2 \geq 5$$

$$4x_1 + 2x_2 \leq 31$$

$$x_1, x_2 \geq 0$$

$$x_1, x_2 \text{ integers}$$

(a) Solve this problem graphically.

(b) Solve the LP relaxation graphically. Round the optimal solution to the LP relaxation in every possible way (i.e., round each noninteger value either up or down). Check these points for feasibility and compute the objective values of those that are feasible. Compare these with the solution found in part (a).

(c) Solve this problem by the branch-and-bound procedure described in Example 11.2. Solve the LP relaxation of each subproblem graphically.

11.11. Consider the following problem.

$$\text{maximize } z = 2x_1 + x_2$$

subject to:

$$2x_1 - 2x_2 \leq 3$$

$$-2x_1 + x_2 \leq 2$$

$$2x_1 + 2x_2 \leq 13$$

$$x_1, x_2 \geq 0$$

$$x_1, x_2 \text{ integers}$$

(a) Solve this problem graphically.

(b) Solve this problem by the branch-and-bound procedure described in Example 11.2. Solve the LP relaxation of each subproblem graphically.

11.12. Solve the following problem by the implicit enumeration procedure described in Example 11.3.

$$\text{minimize } z = 3x_1 + 2x_2 + 5x_3 + x_4$$

subject to:

$$-2x_1 + x_2 - x_3 - 2x_4 \leq -2$$

$$-x_1 - 5x_2 - 2x_3 + 3x_4 \leq -3$$

$$x_j \text{ binary, for all } j$$

11.13. Consider the following zero–one integer programming problem.

$$\text{maximize } z = -4x_1 + 3x_2 - 2x_3 + 7x_4$$

subject to:

$$-x_1 + 2x_2 + x_3 + 4x_4 \geq 6$$
$$-2x_1 + 2x_2 + x_3 + x_4 \leq 2$$

$$x_j \text{ binary, for all } j$$

(a) Transform this problem into the form minimize $\mathbf{cx}$ subject to $\mathbf{Ax} \leq \mathbf{b}$, $\mathbf{x}$ binary, where $\mathbf{c} \geq \mathbf{0}$.

(b) Solve the problem formulated in part (a) by implicit enumeration. Give the solution to the original problem.

11.14. Solve the following problem by implicit enumeration.

$$\text{minimize } z = 3x_1 + x_2 + 2x_3 - x_4$$

subject to:

$$2x_1 - 3x_2 - 4x_3 + 5x_4 \leq -1$$

$$x_1 + 2x_2 - x_3 + x_4 \geq 1$$

$$4x_1 + x_2 + x_3 + 2x_4 \geq 3$$

$$x_j \text{ binary, for all } j$$

11.15. Solve the following problem by implicit enumeration.

$$\text{minimize } z = 4x_1 + 5x_2 - 6x_3 + 2x_4 - 3x_5$$

subject to:

$$4x_1 + 2x_2 + 3x_3 + 2x_4 + x_5 \leq 5$$

$$x_1 + 5x_2 - 2x_3 - 2x_4 - 2x_5 \geq 1$$

$$x_j \text{ binary, for all } j$$

11.16. Consider the following problem.

$$\text{maximize } z = 4x_1 + 8x_2$$

subject to:

$$2x_1 + 2x_2 \leq 19$$

$$-2x_1 + 2x_2 \leq 3$$

$$x_1, x_2 \geq 0$$

$$x_1, x_2 \text{ integers}$$

(a) Solve this problem graphically.

(b) Solve this problem by the branch-and-bound procedure. Solve the LP relaxation of each subproblem graphically.

(c) Solve this problem by the cutting-plane method described in Example 11.4. Derive an expression for each cut in terms of x_1 and x_2. Illustrate the progress of the algorithm graphically in x_1–x_2 space.

11.17. Solve the following knapsack problem using the cutting-plane method.

$$\text{maximize } z = 10x_1 + 24x_2 + 10x_3 + 2x_4$$

subject to:

$$2x_1 + 4x_2 + 3x_3 + x_4 \leq 23$$

$$x_j \geq 0, \qquad \text{for all } j$$

$$x_j \text{ integer}, \qquad \text{for all } j$$

11.18. Consider the following problem.

$$\text{maximize } z = 4x_1 + 4x_2$$

subject to:

$$2x_1 \leq 3$$

$$2x_1 + 2x_2 \geq 5$$

$$2x_2 \leq 3$$

$$x_1, x_2 \geq 0$$

$$x_1, x_2 \text{ integer}$$

(a) Solve this problem by the cutting-plane method.
(b) Illustrate the algorithm graphically.

11.19. Consider the following problem.

$$\text{maximize } z = 6x_1 + 24x_2$$

subject to:

$$-2x_1 + 2x_2 \leq 5$$

$$x_1 + 2x_2 \leq 8$$

$$x_1, x_2 \geq 0$$

$$x_1, x_2 \text{ integer}$$

(a) Solve this problem by the cutting-plane method.
(b) Derive an expression for each cut in terms of x_1 and x_2. Illustrate the progress of the algorithm graphically in x_1–x_2 space.

CHAPTER 12

HEURISTIC PROGRAMMING—AND AI

CHAPTER OVERVIEW

In the preceding chapters, our attention has focused almost exclusively on *algorithms* for the *optimization* of linear models—or, as in Part 2, on linear models having integer and/or discrete variables. The preceding three chapters on methods of network and integer programming are illustrations of this focus as there we describe algorithms that—by definition—guarantee convergence (i.e., if given enough time and computational resources) to the best possible answer. Unfortunately, although the simplex method (and alternatives such as the interior point method of Karmarkar) is an efficient tool for the solution of large-scale, linear, *continuous* models, algorithms for networks and more general integer programs are not nearly so robust or efficient. There are, of course, exceptions—such as the application of the network simplex to *certain* classes of models that may be represented by *specific* types of networks. However, in general, algorithms for integer models are very often ineffective on real-world problems.

There are, however, other approaches that may be used to find a solution to models involving integer and/or discrete variables. And, in many cases, such approaches have proven (through experience) capable of providing *acceptable* solutions to truly massive-size problems. The key word, of course, is "acceptable." As soon as we are willing to consider pragmatic, rather than optimal methods, then we open the door to an entire "new world" of problem solving—and we move from the use of algorithms to the employment of heuristic procedures, or heuristic programming (Ignizio, 1980). Further, by means of what is often little more than a change in terminology, we may also extend the notion of heuristic programming to encompass a set of techniques that are typically associated with the discipline of artificial intelligence (AI).

In this chapter, we discuss heuristics and heuristic programming, provide several examples of heuristic methods for integer programming problems, describe heuristic programming's evolution/transformation into a tool now associated with AI, and describe just when such *satisficing* approaches are appropriate. Although the presentation herein will focus on the application of the heuristic programming approach to but one type of problem (i.e., problems of the *construction* type, and more specifically those of integer programming), it may be noted that the heuristic programming methodology is just as appropriate for problems of *association* (e.g., prediction and classification). Because it would be impossible to cover all existing heuristic programming methods and applications within one chapter (or even one book), we shall illustrate the fundamental notions associated with this extraordinarily versatile concept via the presentation of several representative heuristic programming methods. First, however, we shall deal with certain basic and underlying concepts.

TERMINOLOGY AND DEFINITIONS

In Chapter 8, we remarked on one proposal for the classification of problem types (and we advise the reader to refer to that chapter for additional details). Specifically, we noted that it would appear—to at least some investigators—that problems may be categorized as those of (1) *association*, (2) *construction*, or (3) of problems that are simply beyond our means to address in a scientific manner. You may further recall that problems of construction (e.g., scheduling, planning, layout, deployment, clustering, and configuration) could be further characterized by their methods of solution. Specifically, (nontrivial) problems of the construction type would all appear to require solution techniques that employ a combination of *feedback* and *comparison*. For example, the simplex algorithm—which is, of course, an algorithm used to *construct* a solution for this problem type—requires feedback in the form of such things as information as to reduced costs; and comparisons are made among such values in the determination of the generation of a new program (basic feasible solution).

When we deal with problems of integer programming, we are dealing with problems of the construction type. Thus, any methods—be they exact or heuristic—for the solution of such problems will involve feedback and comparison. As just one example, the branch-and-bound algorithm (as described in preceding chapters) requires feedback in the form of node bounds, which are then compared in the determination of which node to explore next. However, although algorithms guarantee (given the time and computational resources) convergence to an optimal solution, heuristic methods provide—at best—*acceptable* solutions.

In order to clarify our discussion, and also to provide a basis upon which we may more effectively address the material to follow, let us first define certain terms and notions.

- *Algorithm.*   An algorithm is a method for solving a problem, using operations from a given set of basic operations, which produces the answer in a finite number of such operations. These basic operations are, typically, those of elementary mathematics (e.g., addition, subtraction, multiplication, and division). Further, note carefully that *an algorithm is guaranteed to converge to the optimal solution.* The simplex method is just one example of an algorithm.
- *Heuristic* (or heuristic rule).   These are rules of thumb that are developed through intuition, experience, and judgment. Typically, they are used when we do not completely understand the elements and interrelationships within a given system, but have some confidence as to the response of a system for a given input—or, alternatively, when the problem is of such size and complexity as to preclude the use of exact methods. The result of the application of a heuristic cannot guarantee the optimal solution. It may also be noted that most heuristic rules can be, if so desired, expressed as if–then statements, or *production rules*. For example, *if* it is cloudy, *then* it will likely rain.*
- *Heuristic programming.*   When one or more heuristics are combined with a procedure for deriving a solution from the associated rules, we have a heuristic program. Heuristic programming involves finding a solution to a problem using operations from a given set of basic operations, where such a solution is produced in a finite number of steps. However, unlike an algorithm, heuristic programming cannot guarantee the optimal solution. In a typical heuristic program, the heuristic rules *commingle* with procedural statements. Heuristic programming is a concept that is broad enough to encompass the solution of problems of *either* association or construction.
- *Satisficing.*   Whenever one uses heuristics or heuristic programming (or—as we shall see—expert systems, neural networks, genetic "algorithms," tabu search, etc.), one is implicitly accepting the notion of satisficing. Satisficing, in turn, is a concept proposed by Simon (1957) for use in the explanation of how individuals and organizations *actually* arrive at decisions. Specifically, they typically do not seek the generally elusive and utopian optimal solution; rather they seek an *acceptable* solution—one that satisfies their aspirations.
- *Expert Systems.*   Expert systems is a model and associated procedure that exhibits, within a specific domain, a degree of expertise in problem solving

---

* The reader may note that this particular heuristic rule is one that is applied to a problem of *association*, and more specifically that of pattern classification. That is, the observance of clouds is associated with the possibility of rain—through simple substitution.

that is comparable to that of a human expert. In turn, the model involved in expert systems consists of two primary and distinct parts: the knowledge base (a collection of heuristic rules) and the knowledge processor (or inference engine)—and it is this separation of knowledge from procedures that often is the only significant distinction between a heuristic program and an expert system (Ignizio, 1991). As with heuristic programming, the expert systems approach may be applied to either problems of association or those of construction.

We advise the reader to pay particular attention to the last definition. Note carefully that (1) expert systems does not necessarily involve the use of a computer (although, like most procedures, the use of a computer serves to greatly reduce the burden of problem solving), (2) expert systems and heuristic programming have, at the very least, a great deal in common, and (3) both approaches represent *philosophies* of problem solving rather than specific procedures for specific problem types.

ATTITUDES, OPINIONS, AND AI

As discussed before, heuristics and heuristic programming do not guarantee convergence to an optimal solution (i.e., as would an algorithm). Instead, they—at best—provide *acceptable* results in actual practice, and are based upon such fuzzy and unsettling notions as intuition, experience, and judgment. For this reason, a large portion of the scientific community—particularly mathematicians—have been extremely reluctant to accept heuristics as either tools for problems solving or even as topics for polite conversation. Considering the vague nature of heuristics, and its lack of any guarantee as to solution quality, this is hardly a surprising reaction. What is surprising, however, is to see so many recent instances of the same individuals readily embracing such techniques as expert systems, neural networks, and genetic "algorithms"—themselves strictly heuristic methods.

The operations research (OR) community, especially that within the U.S. academic sector, has made its distaste for heuristic methods particularly clear. As such, it is extremely rare to encounter a paper on such methods in the *Journal of Operations Research* (generally considered to be the flagship journal of the OR community). Considering the history of operations research (or, as it is generally known elsewhere, operational research), this aversion toward heuristic methods is somewhat surprising. When one reads the original OR case studies, describing the highly successful OR efforts of World War II, it would certainly appear that heuristic methods—in conjunction with some rather basic data collection and statistics—played a truly major role in the solution of a wide variety of critical, real-world problems (Waddington, 1973; Christopherson and Baughan, 1992).

In 1958, there appeared—in the *Journal of Operations Research*—a particularly ironically titled paper: "Heuristic Problem Solving: The Next Advance in Operations Research" (Simon and Newell, 1958). (This paper, along with the book by Waddington, should be *required reading* by anyone within the OR—or management science, decision science, systems engineering,—sector.) In their paper, Simon and Newell described the use of heuristic methods (i.e., *heuristic programming*) for the solution of those problems beyond the capabilities of algorithmic approaches. However, upon closer reading, and in light of more recent developments, one may recognize that what they proposed was, *in essence*, the *expert systems* concept.

The reaction of the OR community in the United States toward the Simon and Newell article was—for the most part—one of indifference. And this attitude did not change until heuristic programming was given a new name and a new identity. Today, there is considerable interest in the OR community—worldwide—in methods of "artificial intelligence," including such highly publicized "AI tools" as expert systems, genetic "algorithms," simulated annealing, and neural networks. *However, it should be noted once again that all of these methods involve heuristics and cannot—in general—guarantee optimality, and that they rely (implicitly) on the notion of satisficing.*

Thus, it is intriguing to note the sudden fascination on the part of the OR community with these "AI tools." Whereas methods that are titled heuristic programming are still ignored, those *same* methods—under the title of, say, expert systems—are of immense interest. For those who might doubt such assertions, we need only point to the numerous instances of the "rebranding" of heuristic methods that have appeared in the AI literature. This is particularly true in the expert systems area, where we see methods originally developed as heuristic programming that are now called expert systems—sometimes simply as a result of the transformation of the original FORTRAN code into an "AI language," such as PROLOG or LISP, and other times by nothing more than a change in name. We would advise the reader to examine carefully two papers in particular—both dealing with the same problem (i.e., the loading of military cargo aircraft) and both using (*at the time the articles appeared*) virtually identical procedures (and heuristic rules). The first paper concerns a *heuristic method* termed DMES (Cochard and Yost, 1985) and the second describes an *expert system* designated as AALPS (Anderson and Ortiz, 1987). A more detailed discussion of these two papers appears in Chapter 8 of the text on expert systems by Ignizio (1991).

In this chapter, we shall, for the most part, use the term *heuristic programming* when describing the methods under consideration. But, as is hopefully clear from the preceding discussion, with but a few changes (primarily that of the separation of the heuristic rules from the control statements and a slight change in terminology), these methods may also be considered expert systems. However, before proceeding to such examples, let us first examine in more detail the very notion of heuristics.

FUNDAMENTAL HEURISTIC TYPES

We may find, in the literature (see, in particular, such journals as the *European Journal of Operational Research, Journal of the* [British] *Operational Research Society, Computers and Operations Research, Interfaces, Transactions of IIE,* and *IEEE Journal on Systems, Man, and Cybernetics*) numerous instances of the development and employment of heuristic programming. From such articles, it might appear that each different problem requires the development of an entirely new heuristic procedure (and entirely new set of heuristic rules). However, from the evidence that we have examined over nearly three decades of study, we have concluded that there is, more likely, but a few fundamental heuristics—and that virtually all heuristic programming methods that have been developed are but *combinations* of a few of these fundamental types.

In particular, there are eight heuristics (and about half of these would appear to be but combinations of the others) that appear to have been (and continue to be) the most widely employed in practice. These are listed in what follows. In order to illustrate the philosophy espoused by each rule, let us assume—except in the case of the steepest-ascent heuristic—that the rules are to be applied to a problem of combinatorial nature, as may be represented by a 0–1 programming model and wherein the objective function is to be maximized.

- *Add, or greedy, heuristic.*   Such a rule emphasizes immediate gain, with little if any regard to future decisions or consequences (and has been termed, by some, as the *Wall Street rule*). For example, given a 0–1 model, we start our search with the null set (i.e., all variables set to zero). We then proceed to examine, one at a time, each variable and set that variable to a value of one if it improves the solution, or keep it at a value of zero otherwise. The process continues until any further addition would cause the solution to become infeasible. (In practice, the greedy, myopic nature of the add heuristic is sometimes tempered by the inclusion of some consideration for future flexibility—or regret—and instances of this may be noted in the examples that follow.)

- *Drop, or survival of the fittest, heuristic.*   This heuristic is the complement (or perhaps "dual") of the add heuristic. Here, we start the search with all variables set to a value of one (and thus an infeasible solution), and then drop off (i.e., set to zero), one at a time, the variable that contributes least to the solution. This procedure is continued until feasibility is reached.

- *Exchange, or mutation, heuristic.*   The exchange heuristic operates a bit like a somewhat more sophisticated combination of the add and drop heuristics. The premise is that some variables are in solution (a value of one) and others are out of solution (a value of zero), and certain exchanges of one or more "in solution" variables for one or more "out of solution" variables are then examined. The exchange is made if the solution may be improved, and is not made otherwise. The process continues until a complete cycle of exchanges have been examined with no improvement in the solution.

- *Steepest-ascent heuristic.* The steepest-ascent heuristic might be considered just another version of the add, or greedy, heuristic as it certainly espouses the greedy philosophy (i.e., movement in the direction of immediate gain). In practice, it most often is found as a combination of three heuristics: one to determine a starting point, another to probe the "neighborhood" of that point for a direction that permits improvement of the solution, and the third to determine just how far to move out in the direction of improvement. The steepest-ascent heuristic forms the basis for the vast majority of search techniques used in nonlinear programming (i.e., with continuous valued variables).

- *Permutation search, or reshuffle, heuristic.* Permutation search might be considered to be the discrete analogy of the steepest ascent heuristic. In problems involving continuous variables, the search neighborhood is well defined (i.e., that region physically "close" to a point under consideration); but this is not the case when dealing with integer variables, and the various combinations and permutations that correspond to their representation. Permutation search defines a neighborhood to be that set of permutations that may be formed through a specified set of allowable exchange operations on the permutation associated with the starting permutation (or permutation presently under consideration). This concept will be further clarified in a section to follow.

- *Partitioning, or decomposition, heuristic.* The notion of partitioning, or problem decomposition, is an appealing one and an approach that is frequently employed to reduce large problems to smaller, more manageable problems. Karp (Karp, 1975; Lewis and Papadimitrion, 1978) has, for example, employed partitioning to reduce the traveling salesperson problem to a cluster of subtours, each of which is relatively easy to solve. Most typically, the partitioning heuristic used is some combination of add, drop, and exchange heuristics. The techniques of cluster analysis are all representative of partitioning heuristics.

- *Stochastic augmentation* (or the "kick in the pants") *heuristic.* A typical real-world problem is characterized by numerous *local* optimal solutions. The heuristic rules just discussed will all likely terminate once such a local optimal is reached, and the value of that local optimal solution may be considerably less than the global optimal—or some other local optimal. As such, practitioners of the heuristic approach often include stochastic adjuncts in the heuristic programming method—in an attempt to "kick" the solution out of the local optimal. One commonly employed procedure is to randomly select a set of starting points, another is to employ a probability distribution for the determination of a given move (e.g., exchange, permutation). This latter concept permits moves to be made that actually *worsen* the solution—in the hope that they may lead away from a local optimal solution to some better solution.

- *Reduced bookkeeping/storage heuristic.* In problems of a combinatorial nature, it is possible—unless means are taken to mitigate the situation—to investigate the same solution over and over. We might consider storing the solutions previously investigated (and, perhaps, the value of the solution at those points); however, this would likely require excessive computational requirements. As such, one heuristic that has found frequent use is to simply store a limited list of these previous solutions and whenever a new solution is to be added to that list, we remove the oldest solution on the list. Another use of this heuristic is that of moving away from local optimal solutions (as was the case for the stochastic augmentation rule listed before). In this instance, once we have arrived at a local optimal solution (or suspect this to be the case), we attempt to make a move that is "far enough" away so as to permit movement to other local/global optima. We try to accomplish by not moving to any points on our list of recently visited solutions.

Ignizio (1980) goes into further detail with regard to most of these fundamental, or base, heuristics and discusses additional examples of their use. Those interested in such details are directed to the subject reference. Here, we shall simply indicate just where such heuristics exist within the examples described in the following sections.

THE ADD/DROP HEURISTIC PROGRAMMING METHOD FOR CLUSTER ANALYSIS/SITE LOCATION

In Chapter 8, we discussed the mathematical formulation of a specific type of clustering problem. Such a problem could also be, as was discussed, described as a *site-location* problem, a *deployment* problem, or a *partial covering* problem (1968, 1971). In essence, we have a set of candidate *server* sites (i.e., potential cluster locations), a set of *customers* (objects to be clustered; i.e., to be assigned to a site), and the desire to locate no more than a certain number of servers so as to minimize the sum of the distances from all customers to that site physically nearest the customer (i.e., the measure of cluster similarity). Alternatively, it may be that there is some measure other than physical distance; that is, some measure that represents the ability of a given server to "cover" a specific customer should that customer be assigned to that server. In this latter instance, we would attempt to select no more than a certain number of servers so as to maximize the resultant sum of server/customer coverage. The mathematical model for such a problem may be expressed as follows.

Find **x** and **y** so as to

$$\text{maximize} \sum_{i=1}^{m} \sum_{j=1}^{n} c_{ij} x_{ij} \tag{12.1}$$

subject to:

$$\sum_{j=1}^{n} y_j \leq K \tag{12.2}$$

$$\sum_{j=1}^{n} x_{ij} = 1, \qquad \text{for all } i \tag{12.3}$$

$$(x_{i,j} - y_j) \leq 0, \qquad \text{for all } i \text{ and } j \tag{12.4}$$

where

$x_{ij} = 1$ if customer i is assigned to site j; 0 otherwise

$y_j = 1$ if site j is selected; 0 otherwise

$K = $ maximum number of sites that may be selected

$c_{ij} = $ coverage provided to customer i by site j

The size of such problems should be clear from the formulation just provided. That is:

$$\text{number of } variables = n(m + 1) \tag{12.5}$$

$$\text{number of } constraints = m(n + 1) + 1 \tag{12.6}$$

where

$n = $ number of sites under consideration

$m = $ number of customers

During the 1960s, and extending into the early 1970s, Ignizio developed a heuristic programming method—and various extensions of that method—for the solution of such problems. Over the years, the concept has found wide application in a host of areas (e.g., location of fire stations, location of solid-waste collection centers, warehouse siting, printed circuit board layout), but its first real-world application—in the late 1960s—was to that of the deployment of air defense systems. Here, the "servers" were the possible locations for the components of the system (i.e., missile launchers and radars,) and the "customers" were regions of airspace (i.e., to be covered so as to minimize the likelihood of enemy aircraft or missile penetration). The approach was first used to evaluate a number of competing air defense missile systems. Each candidate was deployed by the heuristic method and its defensive performance then evaluated by means of simulated attacks. As a result of this effort, the air defense system selected was the Patriot system of Gulf War fame.

During the evaluation of the candidate air defense systems, it was required to routinely solve problems involving the siting of up to three (i.e., $K \leq 3$) sites from among 306,252 candidate sites so as to provide coverage to 25,521 sectors of airspace (the reader is advised to substitute these values into Equations (12.5) and (12.6) so as to gain some appreciation of just how large are such problems). By using a computer that would today be considered a grindingly slow antique (the UNIVAC 1107), such problems were solved in under 3 minutes of computation time (Ignizio, 1971).

This particular heuristic is composed of nothing more than a combination of add and drop heuristics. For sake of discussion, let us assume that we wish to locate no more than K warehouses to a set of n possible sites ($n \gg K$). We begin with no sites selected (i.e., the null set). We then select the one site that is physically closest to all of the customers (ties are broken arbitrarily). Next, we determine the site that, *in combination with the site already selected*, would serve to minimize total site to customer distances (i.e., where it is assumed that each customer will be served by the site to which it is nearest). We then determine that site that, *in combination with those already selected*, serves to minimize total site to customer distances. Note that, thus far, we have simply used the add, or greedy, heuristic. However, once we have selected three (or more) sites, we include a drop step *after every add step*. That is, we evaluate the contributions of all sites that are presently in the solution. If the site with the *least* contribution (i.e., having the least impact on the resultant sum of distances if it were dropped from solution) is *not* the most recent site selected, then it is dropped from the solution and we proceed to an add step. If the site having the least contribution was, however, the most recent site selected, we retain the present solution and proceed to an add step. These add/drop steps are then repeated until either: (1) K warehouses have been selected or (2) there are no sites left that will serve to improve the solution. The specific steps of the add/drop heuristic programming method follow.

Add/drop heuristic for clustering/siting

1. *Initialization.* Establish a matrix with the candidate sites/servers as column headings and the customers as row headings. The cells of the matrix represent some measure of coverage as would be provided if the respective candidate site served the associated customer. Because the method is one of maximization, *the cell elements should represent measures that are to be maximized* (if, however, the measures are to be minimized, then we simply subtract all cell elements from the largest cell element). Once the matrix has been represented for the maximization case, we call it matrix **A** with column vectors $\mathbf{a}_j$ and cell elements $a_{i,j}$. Set θ is initially the null set.

2. *Selection of Site 1.* For each column vector $\mathbf{a}_j$, sum the elements of that

column. Let that sum be denoted as T_j. Select the column vector $\mathbf{a}_j$ having the maximum T_j as the first site selected and designate it as $\mathbf{a}^*$. Place the index j of max T_j into the set θ. Ties may be broken arbitrarily.

3. *Selection of an Additional Site.* For each column $\mathbf{a}_j$, where $j \notin \theta$, calculate S_j, where

$$S_j = \sum_{j=1}^{m} \max[(a_{i,j} - a^*_{i,j}), 0]), \qquad \text{for all } i \qquad (12.7)$$

If any $S_j > 0$, then select max S_j. Ties may be broken arbitrarily. Add the index associated with the max S_j to the set θ, where the most recent index appears to the farthest right side of the set, and then proceed to step 4. However, if all $S_j \leq 0$, terminate the procedure and go to step 9.

4. *Formation of the Best Combination.* Remove the column vector associated with the most recent index assigned to set θ. Replace the old $\mathbf{a}^*$ with a new column vector $\mathbf{a}^*$, where

$$a^*_{i,j} = \max_{j \in \theta} \{a_{i,j}\}, \qquad \text{for all } i \qquad (12.8)$$

If the number of indices in $\theta = 2$, then repeat steps 3 and 4. Otherwise, proceed to step 5.

5. *Drop Impact Evaluation.* Let $\mathbf{R}$ represent the matrix having only the columns $\mathbf{a}^*$ and $\mathbf{a}_j$, where $j \in \theta$ (i.e., $\mathbf{R}$ contains only the columns of $\mathbf{A}$ associated with those sites thus far selected, plus column $\mathbf{a}^*$). For each column "t" of $\mathbf{R}$, except that of $\mathbf{a}^*$, compute E_t (the elimination, or drop, effect of site t):

$$E_t = \sum_{i=1}^{m} \max[\{a_{i,s}\} - a^*_{i,j}], \qquad \text{for all } s \in \theta \text{ and } s \neq t \qquad (12.9)$$

6. *Drop Step.* If the maximum E_t is associated with the last site in θ, then proceed to step 8. Otherwise, drop the column of $\mathbf{R}$ associated with the site having the maximum value of E_t and remove the associated index from θ.

7. *Formation of the Best Combination After a Drop.* Construct a new column vector $\mathbf{a}^*$, where $a^*_{i,j}$ is computed via (12.8). Then return to step 3.

8. *Termination Check.* If the number of indices in $\theta = K$, then proceed to step 9. Otherwise, return to step 3.

9. *Assignment.* From the matrix $\mathbf{R}$, find the max $\{a_{i,j}\}$, for all i and $j \in \theta$. Assign customer i to site j for those i and j pairs corresponding to each max $\{a_{i,j}\}$.

Example 12.1: Site Location via the Add/Drop Heuristic

Let us demonstrate this heuristic program on the warehouse location example depicted in Table 12.1. Here, the elements of the table are the customer to site distances. *Because the heuristic programming method employed is one of maximiza-*

TABLE 12.1   INITIAL MATRIX

| | Candidate site locations | | | | |
Customer	1	2	3	4	5
I	0	100	250	150	400
II	40	50	500	300	30
III	400	200	50	350	160
IV	100	150	60	10	250

tion, and because the problem depicted is one requiring minimization, we need to first preprocess the distance matrix of Table 12.1. That is, we find the maximum distance in that table (500 units, the distance from customer II to site 3) and simply subtract all matrix elements from that value, leading to Table 12.2—the table upon which we shall perform the operations of the heuristic program. Further, let us assume that the total number of sites to be selected must be 3 or less (i.e., $K \leq 3$).

TABLE 12.2   PROCESSED MATRIX/ADD STEP

| | Candidate site locations | | | | |
Customer	1	2	3	4	5
I	500	400	250	350	100
II	460	450	0	200	470
III	100	300	450	150	340
IV	400	350	440	490	250
T_j	1460	**1500**	1440	1190	1160
θ			{2}		

Notice that we have added two new rows (the bottom two rows) to the processed matrix of Table 12.2. The row headed by T_j is simply the sum of all the values in each column. The maximum value from among these is associated with the single best site. This is site 2, and this selection is noted by the fact that the total for this site is in boldface. Beneath this row is the row headed by θ, and it is used to represent that set of sites presently in solution—along with the order in which they have been selected (i.e., the sites are listed, left to right, according to order of entry in the solution). Thus, site 2 is presently in solution.

We now proceed with the next add step. Here, however, we evaluate the *increase* that may be achieved by any given site when it is *combined* with the site already in solution. *And realize that, with the processed matrix, we now assign customers to the sites associated with the maximum cell entries.* We may then proceed with this step, as is summarized in Table 12.3. Note that site 3 should be combined with the site already selected (i.e., site 2) because the S_j row for site 3 has the largest value. This value (i.e., 240 units) was, in turn, calculated by step 3 of the procedure, as previously listed.

TABLE 12.3   PROCESSED MATRIX/SECOND
ADD STEP

Customer	Candidate site locations				
	1	a*	3	4	5
I	500	400	250	350	100
II	460	450	0	200	470
III	100	300	450	150	340
IV	400	350	440	490	250
S_j	160	—	**240**	140	60
θ			{2, 3}		

Continuing with the steps of the procedure, we develop Tables 12.4 through 12.7.

TABLE 12.4   PROCEDURE/THIRD ADD STEP

Customer	Candidate site locations			
	1	a* (sites 2 & 3)	4	5
I	500	400	350	100
II	460	450	200	470
III	100	450	150	340
IV	400	440	490	250
S_j	**110**	—	50	20
θ		{2, 3, 1}		

TABLE 12.5
PROCEDURE/DROP STEP

Columns of matrix **R**			
2	**3**	**1**	a*
400	250	500	500
450	0	460	460
300	450	100	450
350	440	400	440
0	−190	−110	—
θ		{3, 1}	

TABLE 12.6   PROCEDURE/ADD STEP

Customer	Candidate site locations			
	a* (sites 3 & 1)	2	4	5
I	500	400	350	100
II	460	450	200	470
III	450	300	150	340
IV	440	350	490	250
S_j	—	0	**50**	10
θ		{3, 1, 4}		

TABLE 12.7
PROCEDURE/DROP STEP &
TERMINATION

Columns of matrix **R**			
3	**1**	**4**	**a***
250	[500]	350	500
0	[460]	200	460
[450]	100	150	450
440	400	[490]	490
−300	−310	**−50**	—
θ		{3, 1, 4}	

As we may note from Table 12.7, the solution derived is to select sites 3, 1, and 4, where site 3 covers customer III, site 1 covers customers I and II, and site 4 covers customer IV (note that the site-to-customer assignment is denoted by the elements in brackets in Table 12.7). However, although we have certainly derived a solution, the question remains as to just how good is this solution? One way to check this out is to find all possible combinations of three sites from among the five possible locations, and thus determine the best possible combination. With but five sites, the number of combinations to be evaluated is just 10. However, with a modest-size problem of, say, 30 sites to be selected among 100 possible alternatives, exhaustive enumeration would require the evaluation of about 2.95×10^{25} combinations. Clearly, other methods for the assessment of the efficiency of the approach are required. This very important matter will be addressed in a later section. However, at this point, the reader should be cautioned as to the acceptance of heuristics based upon just a few examples—and perhaps just might begin to wonder just how many examples are "enough."

Observations and Extensions

It just so happened that the add/drop heuristic produced an optimal solution for this example. However, that was simply our good fortune, as a heuristic method cannot guarantee optimality (if it did, it would no longer be a heuristic).

Rather obviously, the particular example addressed is a very special case of the site-location problem. That is, it is implicitly assumed that the costs associated with the selection of each site are equal; that a site can cover any number of customers; and so forth. In most real-life situations, such factors as costs and coverage constraints may have to be considered. However, we can deal with most of these by means of various extensions to the add/drop method listed before. As just one example, we could break any ties in the selection of sites by choosing the site with the least cost. Or we might weight each distance in the matrix by some value associated with the cost of the site (and/or cost of serving that customer). In this manner, sites with high initial costs are penalized and we would hope that this might bias the ultimate selection to those sites with the lowest costs.

Realize, however, that we are simply combining heuristics with existing heuristics. And each of those additional heuristics just described fall into one of the eight types listed earlier. In fact, both the tie-breaking and cell-weighting heuristics are simply further examples of greedy, or add, heuristics.

As one final note, those readers with a background in pattern recognition (or, more specifically, in *feature selection*) might recognize the add/drop heuristic presented here to be virtually identical (i.e., in terms of the philosophy of approach) to the so-called *full stepwise method of feature selection* (James, 1985). In the feature-selection problem, we seek to determine those attributes that serve to be the most important ones in pattern recognition.

THE EXCHANGE HEURISTIC PROGRAMMING METHOD FOR CLUSTER ANALYSIS/SITE LOCATION

Another, and oftentimes effective heuristic approach to the clustering/site-location problem (as well as other types of problems, such as those of set covering) of the previous section is that achieved by means of the use of the exchange heuristic. This method generally requires some starting solution and the establishment of the set of exchanges that is to be permitted. We shall let λ equal the number of variables *in solution* (i.e., have values of one) that are to be exchanged for K variables that are *out of solution* (i.e., have values of zero).

Of course, some limits must be placed on λ and K or else we could wind up using exhaustive enumeration—or worse. Based on the experience of numerous investigators, the λ/K exchanges that are most typically employed are simply

$$2/1 \quad 2/2 \quad 1/1 \quad 1/2$$

We shall employ the same set of exchanges, and employ them in precisely the order listed. And *note carefully that as soon as we have an improvement during any λ/K exchange, we proceed to the next type of λ/K exchange* (this will be illustrated in the example). The procedure starts with some initial solution and cycles through the four exchanges listed until no improvement is noted *in a complete cycle*. To demonstrate the method, let us return to the site-location problem of Table 12.1, and listed as Table 12.8.

TABLE 12.8   INITIAL MATRIX

Customer	Candidate site locations				
	1	2	3	4	5
I	0	100	250	150	400
II	40	50	500	300	30
III	400	200	50	350	160
IV	100	150	60	10	250

Example 12.2: Site Location via Exchange Heuristic

We now demonstrate the application of the exchange heuristic, using the set of exchanges listed earlier, to the site-location/clustering problem previously solved by the add/drop heuristic.

For sake of discussion, let us assume that our starting solution is given as $x = (0, 1, 0, 1, 1)$; i.e., site 2, 4, and 5 are selected. With such a solution the sum of the distances between each customer and its nearest site is 300 units. (The solution derived by the add/drop heuristic was that of sites 3, 1, and 4—with a sum of distances of 100—and thus we are beginning with a very poor solution.) The results of the application of the exchange heuristic procedure, employing the cycle of exchanges of 2/1, 2/2, 1/1, and 1/2, are listed in what follows.

1. *Initialization.* The initial solution vector is $(0, 1, 0, 1, 1)$ with a value of 300. Proceed to the 2/1 exchanges.

2. *2/1 Exchanges ($\lambda = 2$, $K = 1$).* We attempt to exchange two of the sites selected for one site not selected.
   If sites 2 and 4 are exchanged for site 1, the sum of the distances is 290 units. Because we have improvement at this step, we move to the next type of λ/K exchanges (i.e., 2/2). At this point, our new best solution is $x = (1, 0, 0, 0, 1)$.

3. *2/2 Exchanges ($\lambda = 2$, $K = 2$).* We attempt to exchange two of the sites selected for two not selected.
   If sites 1 and 5 are exchanged for 2 and 3, the sum of distances is 260 units. We have improvement and thus we do not examine any more 2/2 exchanges at this time. We move to the next set of exchanges (i.e., 1/1) with a new best solution of $x = (0, 1, 1, 0, 0)$.

4. *1/1 Exchanges (λ = 1, K = 1).* We now attempt to exchange one of the sites selected for one not selected.

If site 2 is exchanged for site 1, the sum of the distances is 150 units. Again, we have (by chance only) immediate improvement and move to the next set of exchanges (1/2) with a new best solution of $\mathbf{x} = (1, 0, 1, 0, 0)$.

5. *1/2 Exchanges (λ = 1, K = 2).* Now we attempt exchanges of one site in solution for two that are not.

If site 1 is exchanged for sites 2 and 4, the sum of distances is 210 units. This is worse than the present best solution, so we attempt another 1/2 exchange.

If site 1 is exchanged for sites 2 and 5, the sum of distances is 240 units.
If site 1 is exchanged for sites 4 and 5, the sum is 240 units.
If site 3 is exchanged for sites 2 and 4, the sum is 250.
If site 3 is exchanged for sites 2 and 5, the sum is 290.
If site 3 is exchanged for sites 4 and 5, the sum is 200.

At this point, we have examined all 1/2 exchanges and completed one cycle. Because improvement was found during this cycle (actually, it was found on three occasions), we must recycle using the present best solution of $\mathbf{x} = (1, 0, 1, 0, 0)$—with an associated distance of 150 units.

6. *Second Cycle.* We next examine exchanges of two sites in solution for one that is not in the solution. Evaluating such exchanges, we will find no improvement over the present best solution and thus we move to the 2/2 exchanges. Again, no improvement will be found. Thus, we move to the 1/1 exchanges and we will still find no improvement. Finally, we examine the 2/1 exchanges and again find no improvement. Thus, the final solution achieved by the exchange heuristic is to select sites 1 and 3, resulting in a sum of distances of 150 units.

At this point, the reader may be a bit alarmed. The exchange heuristic did not achieve a solution nearly as good as the add/drop method, and it certainly did not produce an optimal solution—even for this very small problem. However, such a reaction is inappropriate and certainly premature. There is no way to judge a heuristic on the basis of just one problem; and one should never judge a heuristic on the basis of its performance on *small* problems. In fact, had our starting solution been some other selection of sites, we may well have achieved optimality in just one cycle (e.g., try starting the procedure with the solution $\mathbf{x} = 0, 1, 0, 0, 1$).

We have had considerable success by means of combining the add/drop heuristic with the exchange heuristic (Ignizio and Harnett, 1974). Here, we use the add/drop procedure to generate an initial solution that is then delivered to the λ/K exchange heuristic. The reader may consult the cited reference for the results achieved by such a combination.

Observations and Extensions

It should be obvious to the reader that when employing the λ/K exchange heuristic, there is always the possibility of evaluating the same solution vector over and over. Of course, such duplication of effort is wasteful. However, if one is to completely avoid this he or she would be forced to store all previously examined solution vectors and compare the latest one derived by the λ/K exchange with those on this list. If a match occurs, then we need not evaluate that solution. Otherwise, an evaluation is necessary and the new solution would be appended to the list of previously generated solution vectors. And it should be immediately obvious that in the case of problems of more realistic sizes, such a list would be of enormous length—and both the amount of storage for that list, as well as the matching process, could soon exceed one's computational resource capacities.

However, we may still use this notion by means of including the reduced bookkeeping/storage heuristic listed among our set of eight fundamental heuristic types. Thus, instead of attempting to store all solutions generated, we simply limit the length of the list and, once it is full, add the new solution to the bottom of the list and delete the solution on the top of the list. The logic employed here is that it is considered more likely to develop solutions that are similar to those on the existing list than those that have been deleted from the list—and considerable empirical evidence tends to verify this notion. It may also be noted that such a list embodies both the add and drop heuristics—adding recent moves and dropping old ones.

The idea of using such a limited listing of previous solutions is hardly new, having been employed for decades by practitioners of heuristic programming. However, this notion has recently been associated with a concept denoted as the *tabu search* (actually, it is just one of the two main notions that provide the foundation of the tabu search). Here, such lists are termed *tabu*—as it is "tabu" (i.e., not permitted) to move to a solution on the list. Glover (Glover, 1986; Glover, Taillard, and de Werra, forthcoming) has proposed this concept as an aid in the application of virtually any heuristic search method. We will provide a very brief overview of this concept later in the chapter.

A HEURISTIC PROGRAM FOR SCHEDULING AND DEPLOYMENT PROBLEMS

One of the fundamental heuristic types mentioned earlier was that of permutation search. In our practice, we have found this heuristic—as well as variations of the approach—to be particularly effective in the solution of certain types of problems. It is also exceptionally easy to convert to computer code and is computationally efficient for extremely large problems. Rather obviously from the name of the approach, the technique may be applied to problems in which we seek the best, or least an acceptable, permutation. One such problem type is that of scheduling, where one seeks the permutation of jobs so as to optimize some measure of performance. Typically, that measure used is the makespan of the schedule; and that is the measure we shall employ for the sake of discussion. First, however, let us explain permutation search and list the steps of the procedure (Nicholson, 1971).

Given, say, four objects, one permutation of these objects may be listed as

$$[P^0] = [1, 2, 3, 4]$$

where $[P^0]$ is the initial (0th) permutation. The number of objects in the permutation will be denoted by N; thus in this permutation, $N = 4$. If this permutation corresponded to, say, a schedule, then it would infer that job 1 is followed by job 2, job 2 by job 3, and job 3 by job 4. We may define the neighborhood of a given permutation by first establishing a set of permissible exchanges. To keep the discussion simple, let us focus solely on what are known as "one-for-one adjacent exchanges." That is, we may exchange one element in the permutation for its adjacent element. As such, it should be easy to see that there are always $N - 1$ permutations within the neighborhood of any given permutation. For example, the three permutations in the neighborhood of our initial permutation just listed are

$$[P^1] = [2, 1, 3, 4]$$

$$[P^2] = [1, 3, 2, 4]$$

$$[P^3] = [1, 2, 4, 3]$$

Obviously, this notion could be extended to any variety of exchanges such as two-for-two adjacent exchanges, three-for-three adjacent exchanges, and so on. Fortunately, considerable empirical evidence indicates that simple one-for-one adjacent exchanges do a remarkably good job—and that is the type of exchange to which we shall restrict our discussion.

The steps of our particular version of the permutation search procedure may be then summarized as follows.

Permutation search heuristic with one-for-one adjacent exchanges

1. *Initialization.* Obtain, by whatever means, some initial starting permutation. Designate this as permutation $[P^0]$ and list it as the present base solution.

2. *Neighborhood Search.* Investigate, one at a time, the permutations in the neighborhood of the present base solution. As soon as a permutation is found that improves the best solution thus far, make this permutation the new base solution. Proceed to step 3.

3. *Termination Check.* If, during step 2, no improvement could be found, then terminate the procedure (in practice, we often try other starting permutations). If, however, improvement was found, then repeat step 2.

In order to better appreciate the simplicity of the procedure, we shall demonstrate its application by means of a scheduling example. Here, we assume that we seek the schedule that will minimize the production makespan.

Sequence Position (p)	Job (n)	t(p,1) ⟋ T(p,1)	t(p,2) ⟋ T(p,2)
1	2	4 ⟋ 4	2 ⟋ 6
2	1	2 ⟋ 6	3 ⟋ 9
3	4	2 ⟋ 8	6 ⟋ 15
4	3	5 ⟋ 13	2 ⟋ 17

Figure 12.1 Makespan table for the permutation [2, 1, 4, 3].

Example 12.3: Flowshop Scheduling via Permutation Search

The makespan of a schedule is the elapsed time between the start of the first job scheduled and that of the finish time for the last job finished. To facilitate discussion, we shall employ the use of a makespan table, as demonstrated in Figure 12.1. Here, we assume a flowshop situation (the method, however, is not limited to any one special type of scheduling problem). In a flowshop, all "jobs" flow in the same sequence, through the same set of "machines," and interruption of a job is not permitted. Thus, the schedule being evaluated in Figure 12.1 is that of the job sequence [2, 1, 4, 3] through two machines (A and B), and the processing times for each job/machine pairing is listed in Table 12.9.

TABLE 12.9 JOB-MACHINE PROCESSING TIMES

Job	Time on machine A	Time on machine B
1	2	3
2	4	2
3	5	2
4	2	6

We shall let

$t(p, m)$ = processing time, on machine m, of job p in processing sequence P [e.g., $t(3, 2) = 6$ in Figure 12.1]

$T(p, m)$ = completion time, on machine m of job p in processing sequence P [e.g., $T(3, 2) = 15$ in Figure 12.1]

The establishment of the makespan table then follows these rules:

1. Let N = number of jobs to be scheduled and M = number of machines. Develop a table with N internal rows and M internal columns. Add two columns to the table to depict the job-processing sequence under consideration (e.g., see the previous makespan table) and a top row to head the columns.
2. For each row associated with each job n, enter the process times of that job in the *upper* half of the cells for that row.
3. $T(1, 1) = t(1, 1)$
4. For row $p = 1$ and $m > 1$:

$$T(p, m) = T(p, m - 1) + t(p, m)$$

5. For column $m = 1$ and $p > 1$:

$$T(p, m) = T(p - 1, m) + t(p, m)$$

6. For all other elements in the table (i.e., $p \neq 1$ and $m \neq 1$):

$$T(p, m) = t(p, m) + \max\{T(p, m - 1), T(p - 1, m)\}$$

Once this table has been established for any given schedule (i.e., permutation), the makespan of that schedule may be read directly from the table—and is the value listed in the last cell of the matrix (i.e., the cell in the last row, last column). In Figure 12.1, the makespan of the schedule is thus 17 units. In the discussion to follow, we shall assume that such a makespan table has been used to compute the makespan associated with each schedule (i.e., permutation) under investigation.

For a two-machine flowshop problem as depicted before, there are exact procedures that may be employed to determine the schedule providing the minimum makespan. However, in the general case, such problems are combinatorially explosive and heuristic methods are justified. Thus, for sake of illustration, we demonstrate the employment of a permutation search in the solution of the problem listed before. The results of the permutation search are summarized in Table 12.10. We

TABLE 12.10 PERMUTATION SEARCH APPLIED TO FLOWSHOP EXAMPLE

Iteration	Permutation	Makespan	Best permutation thus far
0	$[1, 2, 3, 4] = [P^0]$	19*	$[P^0]$
1	$[2, 1, 3, 4] = [P^1]$	19	$[P^0]$
2	$[1, 3, 2, 4] = [P^2]$	19	$[P^0]$
3	$[1, 2, 4, 3] = [P^3]$	16*	$[P^3]$
4	$[2, 1, 4, 3] = [P^4]$	17	$[P^3]$
5	$[1, 4, 2, 3] = [P^5]$	15*	$[P^5]$
6	$[4, 1, 2, 3] = [P^6]$	15	$[P^5]$
7	$[1, 2, 4, 3] = [P^7]$	16	$[P^5]$
8	$[1, 4, 3, 2] = [P^8]$	15	$[P^5]$

have assumed a starting solution of [1, 2, 3, 4]. The appearance of an asterisk in the third column is used to indicate the generation of an improved solution.

Thus, our best schedule is $[P^5] = [1, 4, 2, 3]$ (and also schedules $[P^6]$ and $[P^8]$). We should note that when starting the search with a relatively poor solution, the procedure typically takes longer to converge. Notice, from iterations 5 and 6, that whenever we obtain an improved solution, we have returned to the cycle of exchanges beginning with the exchange of the first two elements of the permutation.

Extensions to Problems of Deployment

Permutation search may be employed in the solution of any problem involving permutations, even many of those that at first glance might not seem to fit the requirement. One such problem is that of deployment; and here we describe the use of permutation search on a problem of the deployment (or placement) of acoustic transducers in acoustic arrays. This example is one encountered in an actual design effort for the U.S. Navy (Ignizio, Wiemann, and Hughes, 1987). However, it is not necessary for the reader to have any background whatsoever in acoustics or array design; and it should be noted that the general approach used may be applied to a variety of other types of deployment problems.

A transducer is a physical device that transmits or receives energy, or power, from one system to another. For example, in the field of electromagnetics, an antenna is a transducer; in acoustics, we may have analogous transducers—ceramic elements that convert vibrations into acoustic energy. The acoustic transducer that most readers will be familiar with is that of the loudspeaker for one's stereo system.

Often, the transmission of power is achieved by means of an ensemble of transducers, denoted as an *array*. However, the design of transducer arrays—of any type—involves certain decisions such as: (1) the location of each transducer, or element of the array, (2) the amplitude of the energy delivered to each element, and (3) the phase of the energy delivered to each element. In the case of an acoustic array for, say, ships or torpedoes, the designer seeks the values of these variables so as to optimize (or achieve) certain measures of performance. Such measures include (1) the shape of the beam of energy delivered by the array, and (2) the sensitivity of the array to extraneous signals outside of the "main beam" of the system. Typically, one cites the width of the beam (at the −3-dB level) and the maximum level of any sidelobes that may be developed.

Early array design was often a case of trial and error or brute-force simulation. Today, arrays are typically designed by means of the establishment of a mathematical model (Ignizio et al., 1987), and the solutions are derived by means of mathematical programming methods wherever possible. However, the model most often used is the so-called error-free model. Consequently, the design derived by the mathematical programming solution method is only approximated by the parameters of the actual system that is manufactured. In the case of small arrays, such variations are usually tolerable. However, in the case of the large arrays now being built, slight variations (from the paper solution) often result in major distortions of the beam shape desired.

For sake of illustration, we will cite the case of the design of a planar array (i.e., all elements arrayed on a flat surface) for torpedoes. The size of this array was relatively modest, eight rows of eight transducers per row (where the three elements in each corner of the array were omitted, leaving 52 elements in the array). The Navy wanted to build such an array in which the main beam of energy (at the −3-dB level) was 20 degrees in width and with no sidelobes higher than −40 dB below the main beam. On paper, the solution was ultimately derived. However, once the actual transducers were fabricated (using extremely tight, and costly, tolerances), it was discovered that the array had sidelobes as high as −31 dB rather than −40 dB. The approach used in an attempt to overcome the problem was to construct a large number of transducers and pick the 52 of these that were the closest to being identical in performance. However, little if any improvement was noted. The next step taken was to randomly place each of the 52 transducers into their slots, in the hope that any differences in phasing would be canceled out. However, and after the generation of numerous random deployments, the best deployment had sidelobes of −33 dB, but a 2-dB improvement.

The approach taken by Ignizio et al. to overcome this problem consisted of two phases. First, a relatively well-known heuristic procedure (the Kendig error-terms heuristic) was employed in order to achieve a relatively good starting solution.* This starting solution consisted of a deployment of 52 transducers in their slots and about 100 more transducers on the "workbench" (i.e., extra elements that had been fabricated). However, rather than employing transducers that had been manufactured to extremely tight tolerances, the transducers used were fabricated to much looser tolerances (and much, much lower production costs).

The second phase of the procedure was to employ permutation search. To illustrate how this was accomplished, let us now assume a smaller problem with just four transducer slots and eight transducers. Further, let us assume that the initial deployment of transducers is given as

    transducer 3 to slot 1
    transducer 5 to slot 2
    transducer 8 to slot 3
    transducer 2 to slot 4

with the remaining transducers (numbers 1, 4, 6, and 7) on the workbench. We next form a permutation such as this:

    1   <u>3</u>   4   <u>5</u>   6   <u>8</u>   7   <u>2</u>

---

* Actually, an expert system was developed to replicate the steps of the Kendig procedure. The results of this phase were then delivered to the permutation search heuristic.

Notice carefully that those transducers that have actually been assigned to a given slot are underlined, and that we have interspersed deployed elements with workbench elements. This has been done so as to ensure that the transducers on the workbench will have a good opportunity to be exchanged with those in the slots.

One permutation of this deployment within the neighborhood of one-for-one adjacent exchanges is then

$$3 \quad \underline{1} \quad 4 \quad \underline{5} \quad 6 \quad \underline{8} \quad 7 \quad \underline{2}$$

Here, we have exchanged transducer 1 (from the workbench) for transducer 3 (previously in slot 1). Once this exchange has been made, we then test the resulting deployment and determine if an improvement has been found. If not, we proceed to the next exchange (element 4 for element 3). The procedure then continues until no further improvement is possible. The simplicity of the concept is obvious.

So, how did the permutation search perform on the 8 × 8 array problem? Those results, denoted as the *hybrid system*, are summarized in Table 12.11 and compared to those achieved by random assignment and the Kendig method alone.

TABLE 12.11 RESULTS ON AN 8 × 8 ARRAY

Deployment method used	Highest sidelobe level (dB below main beam)
Random assignment	−31 to −33
Kendig method (alone)	−35
Hybrid system	−39

As is clear from Table 12.11, the hybrid system—composed of the Kendig method in phase 1 and the permutation search in phase 2—was able to achieve results within 1 dB of the theoretical limit. What is not noted in this table, however, is the differences in tolerances used for the elements of these two phases. Specifically, the tolerances were set to ±9% in amplitude and ±11 degrees in phase, as compared to the much tighter tolerances of ±1% in amplitude and ±0.6 degrees in phase as used by the Navy in its random assignments. This was done primarily because of the costs associated with attempting to hold to the tighter tolerances. Even with this relaxation, we were able to obtain substantially *improved* performances at even more substantially *reduced* costs.

A HEURISTIC PROGRAM FOR MINIMAL CONFLICT SCHEDULING

The eight-queen problem is one that is encountered throughout the literature in both heuristic programming and artificial intelligence. The objective in such a problem is to place as many queens as possible on a chessboard. However, no pair of queens may be in conflict (i.e., no queen may be placed in the same row, or same column, or on the same diagonal as associated with another queen). It may be shown, either heuristically or by means of exact approaches, that a maximum of eight queens may be placed on the board, in several possible configurations, just one of which is demonstrated in Figure 12.2.

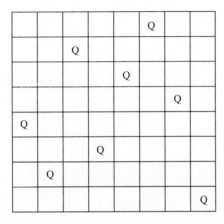

Figure 12.2 Eight-queen solution.

   Although the eight-queen problem may be viewed as more or less an interesting puzzle, there are important real-world problems that may be viewed as generalizations of such a situation. Specifically, there are certain problems in which one hopes to schedule *as many jobs as possible*, during some given time period. And the only thing that may stand in the way of scheduling all the jobs is certain pairwise conflicts between jobs. To provide a real-world example of such a problem, consider the task of scheduling the interceptions of incoming missiles. We shall assume that each incoming missile, denoted as a target, may be intercepted at a few specific points during its flight. Interceptors (i.e., surface-to-air missiles) may be launched at various times so as to intercept the target at its intercept points. However, there is the possibility that one interceptor might interfere with, or even destroy another if a certain combination of launch times and targets is selected. Another twist to this problem is given when one not only wishes to shoot down as many targets as possible, but also wishes to maximize the number of shots taken at the targets.

   Whereas this particular example is that of a military problem, other problems having virtually identical structure have also been identified (Ignizio, 1978; Ignizio-Burke, 1985). As just one illustration, consider the situation in which a number of jobs is to be scheduled and where each job has several different possible start times. Once a job is started, it must not be interrupted. Further, each job consumes—at different times during its processing—different types of resources. However, should one job require a certain resource during the same time that some other job requires that same resource, a conflict will occur. Finally, we wish to schedule the maximum number of jobs that may be performed without conflict.

Such a problem requires a somewhat unusual mathematical formulation, one that is associated with multiple objective problems and that will be described in considerably more detail in Part 3. Here, we simply note that the formulation for this problem may be given as follows (Ignizio-Burke, 1985).

Find **x** so as to *lexicographically* maximize:

$$\mathbf{u} = \left\{ \left(\sum_{i=1}^{m} S_i \right), \left(\sum_{j=1}^{n_i} \sum_{i=1}^{m} x_{ij} \right) \right\} \tag{12.10}$$

subject to:

$$S_i - \sum_{j=1}^{n_i} x_{ij} \le 0, \qquad \text{for all } i \tag{12.11}$$

$$x_{ij} + c_{(ij)(ks)} x_{ks} \le 1, \qquad \text{for all } [(ij, ks)/k \le i \text{ and } s < j \text{ if } k = s] \tag{12.12}$$

where

m = total number of targets (e.g., missiles, jobs)

n_i = number of intercept points for target i

S_i = 1 if a shot taken at target i; 0, otherwise

x_{ij} = 1 if an interceptor for target i has been fired at time j, where $j \in n_i$

$c_{(ij)(ks)}$ = 1 if an interceptor for target i at time j conflicts with an interceptor for target k at time s; 0, otherwise

Recognize that the "objective function" here is to be *lexicographically* maximized. That is, our first priority is to maximize the first term—which corresponds to maximizing the number of targets fired upon. The second term, and second priority, is to maximize the number of interceptors fired. (The concept of lexicographic maximization and minimization will be more fully explained in Part 3.) The first set of constraints serves to define the value of S_i. That is, if *any* shots are fired at target i, S_i must take on a value of one. The second set of constraints indicates the conflicts between interceptors in pairs and thus maintains a firing schedule so that no interceptor-to-interceptor conflicts are encountered.

A conflict scheduling problem may also be represented by a symmetric, square matrix; and this depiction is much easier to comprehend than that of the associated mathematical model. For example, consider the conflict matrix of Figure 12.3. Here, we have three enemy targets and a representation of (1) the times—designated as job-start times—at which each target may be intercepted, and (2) the conflicts between interceptor pairs. For example, it is possible to intercept target 1 at two different times, target 2 at three different times, and target 3 at two different times. Further, if one attempts to intercept target 1 at time 1 *and* target 2 at time 1 (or time 3), a conflict will occur, as indicated by 1's in the associated matrix cells. Note carefully that time 1 for target 1 may not be the

		Target 1		Target 2			Target 3	
Time		1	2	1	2	3	1	2
Target 1	1	0	0	1	0	1	0	0
	2	0	0	0	1	0	0	1
Target 2	1	1	0	0	1	0	1	0
	2	0	1	1	0	0	0	0
	3	1	0	0	0	0	0	0
Target 3	1	0	0	1	0	0	0	1
	2	0	1	0	0	0	1	0

Figure 12.3 Conflict matrix.

same clock time as time 1 for targets 2 or 3. We hope then to find some interceptor to target schedule that will (1) maximize the different number of targets fired upon and (2) maximize the number of shots fired (i.e., attempts made on all targets).

Whereas small conflict-scheduling problems may be approached by exact methods, those of more realistic size will require a heuristic approach. Here, we demonstrate a heuristic programming method that invokes the add, or greedy, heuristic. However, our greed is tempered by the recognition that we would like to select—from among those candidates that are presently most attractive—that one that will hopefully minimize any regret we may have in future selections. This heuristic procedure follows the steps that follow. While Ignizio-Burke terms this the column-sum procedure, we have denoted it as the add heuristic for conflict scheduling in order to emphasize the fact that the add heuristic (tempered by the minimization of regret in future choices) is employed.

Add heuristic for conflict scheduling

1. *Initialization.*   Establish the conflict matrix associated with the given problem. Initially, all columns (i.e., job-start times) are *eligible* and the list for all candidate targets is *empty* (indicating that no shots/attempts have yet been taken at that target/job). All rows and columns of the conflict matrix are initially unchecked.
2. *Column Sum.*   Sum each eligible column (i.e., the columns associated with the possible job-start times) of the conflict matrix—but do not include, in this sum, any numbers associated with either a row or column that has been checked.
3. *Candidate List.*   List the target, or targets, having the fewest previous assignments and store these in the set of candidate targets.
4. *Minimal Column Sum.*   For all targets in the set of candidate targets, determine the associated job-start time(s) having the minimal column sum. Store these in the list of candidate job-start times.

5. *Selection.* Select the job-start time having the minimum sum in the list of candidate job-start times as the job-start time to be next assigned. If there are no ties, go to step 7. Otherwise proceed to step 6.

6. *Tie Breaker.* For the tied job-start times, compute the difference between the minimum sum and the next-to-minimum sum for the respective targets. Find the job-start time whose differences is the *maximum* for all tied job-start times. Assign the next job-start time to that job-start time having the maximum difference.

7. *Matrix Reduction.* Check all columns and rows of the conflict matrix that are associated with 1's in the row and column of the latest assigned job-start time. In addition, check the row and column associated with the latest assigned job-start time.

8. *Ineligible Job-Start Times.* All columns (i.e., job-start times) with checks are now ineligible.

9. *Assignment List.* Increase, by one, the number of assignments of the target associated with the most recently selected job-start time.

10. *Termination Check.* If all columns in the conflict matrix have been either eliminated (checked) or scheduled, terminate the process. Otherwise, return to step 2.

Example 12.4: Illustration of Procedure

We now demonstrate the application of this heuristic procedure for the conflict matrix illustrated in Figure 12.4. This figure depicts both the problem under investigation (i.e., three targets, each with two possible job-start times) and also summarizes the result of summing all eligible job-start columns. As may be seen, there are four tied columns, those having column sums of a value of 3 (and depicted in boldface).

Because there are four ties for the column sum, we must, via step 6 of the procedure, determine the maximum differences for these tied elements. Thus, for target 1/job-start time 1, we note that the difference is $4 - 3 = 1$ (i.e., the difference between the column sum for target 1/job-start time 2 and that for target 1/job-start

		Target 1		Target 2		Target 3	
Time		1	2	1	2	1	2
Target 1	1	0	1	1	0	0	1
	2	1	0	1	1	1	0
Target 2	1	1	1	0	0	0	1
	2	0	1	0	0	1	1
Target 3	1	0	1	0	1	0	1
	2	1	0	1	1	1	0
Column Sums		**3**	4	**3**	**3**	**3**	4

Figure 12.4 Conflict matrix and column sums.

time 1). For target 2/job-start time 1, the difference is $3 - 3 = 0$, as is the difference for target 2/job-start time 2. For target 3/job-start time 1, the difference is $4 - 3 = 1$.

There is a tie for the maximum differences (i.e., between target 1/job-start time 1 and target 3/job-start time 1), so we will arbitrarily break that tie in favor of job-start time 1/target 1. In order to readily identify this assignment, we have placed a box—in Figure 12.5—around the element associated with target 1/job-start time 1. Once this assignment has been made, we must check all rows and columns associated with this assignment (row 1 and column 1 of the matrix) as well as all rows (2, 3, 6) and columns (2, 3, 6) associated with values of 1 in the row or column associated with the selected job-start time. This results in the matrix of Figure 12.5. Also depicted in this latest matrix are the column sums for the remaining eligible job-start times.

		Target 1		Target 2		Target 3		
Time		1	2	1	2	1	2	
Target 1	1	[0]	1	1	0	0	1	√
	2	1	0	1	1	1	0	√
Target 2	1	1	1	0	0	0	1	√
	2	0	1	0	0	1	1	
Target 3	1	0	1	0	1	0	1	
	2	1	0	1	1	1	0	√
Column Sums		–	–	–	1	1	–	
		√	√	√			√	

Figure 12.5 Conflict matrix and column sums.

Examining the column sums for Figure 12.5, we find a tie between target 2/job-start time 2 and target 3/job-start time 1. There are no maximum differences to calculate (i.e., step 6 of the procedure), so we will simply break the tie in favor of target 2/job-start time 2. We may then check row 4 and column 4 (that row and column associated with the assignment), and we also check row 5 and column 5 because there is a conflict for that assignment. At this point, we have checked all rows and columns of the matrix (see Figure 12.6) and thus we terminate the proce-

		Target 1		Target 2		Target 3		
Time		1	2	1	2	1	2	
Target 1	1	[0]	1	1	0	0	1	√
	2	1	0	1	1	1	0	√
Target 2	1	1	1	0	0	0	1	√
	2	0	1	0	[0]	1	1	√
Target 3	1	0	1	0	1	0	1	√
	2	1	0	1	1	1	0	√
Column Sums		–	–	–	–	–	–	
		√	√	√	√	√	√	

Figure 12.6 Conflict matrix and column sums.

dure. Our final solution then is to assign an interceptor to target 1 at intercept time 1, and assign another interceptor to target 2 at intercept time 2. We may also notice that we were unable to fire at all three targets.

GENETIC "ALGORITHMS"

Under our definition of an algorithm, the methodology that is popularly known as genetic algorithms is a misnomer as, in practice, solution optimality cannot be guaranteed. *Thus, the approach should be considered as simply another heuristic programming method.* However, this in no way should diminish our interest in the concept, as it has proven effective in actual problem solving in a number of instances (Goldberg, 1989). Further, and unlike some of the specially tailored heuristic programs presented previously, the use genetic algorithms is a very general approach to problem solving. Some of the more interesting characteristics of the approach include the following:

- The approach invokes a coding of the parameter set rather than working with the parameters themselves (e.g., in a manner such as that employed to represent a schedule, or deployment, in permutation search).
- Search is conducted from a population of points, rather than from just a single point.
- Probabilistic rather than deterministic rules are employed.

Goldberg (1989) defines genetic algorithms to be "search algorithms based on the mechanics of natural selection and natural genetics." Of course, the search procedures, or heuristic methods already presented, could be viewed in precisely the same light as they employ heuristics (e.g., the drop heuristic, or survival of the fittest; the add heuristic, or path of least resistance) that may also be associated, philosophically, with natural selection and natural genetics—in fact, it would be difficult to find a fundamental (and successful) heuristic that could not be associated with such concepts. Goldberg then goes on to note that "they [genetic algorithms] combine survival of the fittest among string structures with a structured yet randomized information exchange to form a search algorithm with some of the innovative flair of human search. In every generation, a new set of artificial creatures (strings) is created using bits and pieces of the fittest of the old: an occasional new part is tried for good measure."

Again, from our point of view, genetic algorithms are not algorithms, but are, rather, heuristic procedures. However, the name is now so widely employed and accepted that we shall also employ such terminology and thus, from here on, we will omit the quotes about the word algorithm when referring to the procedure. The three heuristics that form the typical genetic algorithm are

1. Reproduction
2. Crossover
3. Mutation

These heuristics are composed of combinations from the set of fundamental heuristics cited earlier. Thus, the reproduction heuristic is a combination of the add, or greedy, heuristic, in conjunction with a stochastic adjunct (i.e., the better solutions are more likely to be replicated in future trials). The crossover heuristic is a combination of the exchange heuristic and a stochastic adjunct (i.e., exchanges are selected at random). The last heuristic, mutation, is—as employed in genetic algorithms—a random exchange applied to a given element in solution (e.g., a variable set to 1 may be randomly changed to a value of 0).

Example 12.5: The Application of Genetic Algorithms to Clustering

It is probably easiest to explain the basics of the genetic algorithm approach by means of a small numerical example. Because several of our heuristic programs in this chapter have been applied to clustering, let us use a small problem in which various objects are to be clustered. We list the data associated with our example in Table 12.12. Note that we wish to find clusters for five objects using two attributes (features, scores).

TABLE 12.12
CLUSTER
EXAMPLE

Object	Scores	
	X	Y
1	1	1
2	2	1
3	3	4
4	3	3
5	4	3

Should the reader care to graph this data, it should be immediately apparent that it would appear that two clusters exist in terms of the physical position of the points: one for the first two objects and the other for the last three. Thus, to make things simple, we shall employ the notion of physical closeness as our measure of cluster effectiveness.

Recall that in cluster analysis, the number of clusters is not predetermined. However, we generally impose some upper limit on that number. For this example, we shall assume that we will not employ more than two clusters. Further, the measure of physical closeness will be developed by means of computing the sum of the squares of the distances of the objects in each cluster from the centroid of that cluster.

Obviously, there can only be one solution if but one cluster is employed. That is, all objects are assigned to that cluster. If we do this, the sum of the squares (SS) is simply 12.4 units. Obviously, no search routine is needed for this phase.

Next, we shall attempt to determine the best assignment of objects to two clusters and here we shall employ the genetic algorithm. The first step in the process is to find some way in which to code the solution as a string (in the literature, such a string is typically composed of but zeros and ones, but this is not the only type of strings permitted). Let us use strings of zeros and ones, where a zero in position "p" of the string denotes that the object in that position will be assigned to cluster A. If the element has a one in this position, then the object is assigned to cluster B. For example:

$$0\ 1\ 0\ 0\ 1$$

is used to represent the assignment of objects 1, 3, and 4 to cluster A and objects 2 and 5 to cluster B.

Once the coding scheme has been determined, we then randomly generate an initial "population," that is, a number of candidate solutions. Let us assume that this initial population is

$$0\ 1\ 0\ 0\ 1$$

$$0\ 0\ 1\ 0\ 0$$

$$1\ 0\ 1\ 1\ 0$$

$$0\ 0\ 1\ 0\ 1$$

We then compute the performance measure (i.e., sum of squares of distances) for each candidate.

In order to more clearly tabulate the process, we shall employ sets of tables. The first table, Table 12.13, represents the computations associated with the initial population. The first step is to find the sum of the squares for each candidate solution

TABLE 12.13 INITIAL POPULATION

String number	Initial population	Equivalent clusters	SS	$p/\Sigma p$
1	0 1 0 0 1	(1, 3, 4) (2, 5)	11.33	0.303
2	0 0 1 0 0	(1, 2, 4, 5) (3)	9	0.241
3	1 0 1 1 0	(2, 5) (1, 3, 4)	11.33	0.303
4	0 0 1 0 1	(1, 2, 4) (3, 5)	5.67	0.152
			$\Sigma p = 37.33$	

(and equivalent set of clusters). This value is shown in the fourth column of the table, under SS. All these values are then summed, as indicated in the last row of the table. Finally, the last row of the table is simply the individual sum of squares divided by the sum of these sums. We use "p" to indicate that it is the performance measure being evaluated.

As may be noted in the previous table, the strongest candidate is string 4, with string 2 in second place. At this point, we develop a new population—the reproduction phase of the method. However, we do so by giving more weight to the stronger candidates and less to the weaker. Thus, those candidates with the smallest values in the last column of the table should have a *higher* probability of being reproduced in the next generation. We may assign a portion of the numbers from 00 to 99 to each candidate and then, using a random-number table or generator, determine which members appear (and at what frequency) in the next generation. Table 12.14 depicts the result of this phase wherein the second column of the table depicts the "mating pool" that shall be used in the reproduction process. It may be noted that string (0 0 1 0 1), the strongest string from Table 12.13, appears twice in the mating pool and string (0 1 0 0 1) has disappeared (or "died out").

TABLE 12.14   REPRODUCTION—SECOND GENERATION

String number	Mating pool	Mate	Crossover	New generation	Equivalent clusters	SS	$p/\Sigma p$
1	0 0 | 1 0 1	4	2	0 0 1 1 0	(1, 2, 5) (3, 4)	7.83	0.228
2	0 0 1 0 | 1	3	4	0 0 1 0 0	(1, 2, 4, 5) (3)	9	0.262
3	0 0 1 0 | 0	2	4	0 0 1 0 1	(1, 2, 4) (3, 5)	5.67	0.165
4	1 0 | 1 1 0	1	2	1 0 1 0 1	(2, 4) (1, 3, 5)	11.833	0.345
						$\Sigma p = 34.333$	

The "mating process" is conducted randomly. That is, strings are paired off on a strictly random basis. Thus, we note from the third column of Table 12.14 that string 1 is mated with string 4 and string 2 is mated with string 3. We must then employ the crossover heuristic to generate the results of such matings. This process is conducted by again randomly selecting "crossover" points for the strings that have been selected for mating. We have noted such crossover points by means of the insertion of a vertical line (|) in the strings within the mating pool. Notice that this crossover point is 2 for strings 1 and 4 and the crossover point is 4 for strings 2 and 3.

Crossover (i.e., for this particular problem—the process must be adapted to the specific type of problem under consideration) simply involves an exchange, or swap, for all elements to the right of the crossover point. For example, the crossover point designated for strings 1 and 4 is 2—which indicates that all elements to the right of the second element in each string are to be swapped. This swap results in the two new strings 1 and 4, as noted in the fifth column (new generation) of the table. A similar process is used to swap the elements of the mating pair of strings 2 and 3 (where the crossover point for these swaps is set after the fourth element of the string).

Before proceeding further, it should be noted that the crossover heuristic is usually only applied to produce a *portion* (usually around 60%) of the members of the next generation. Consequently, it is typical to carry one or more "old" strings into the next generation. However, in this particular illustration, we have chosen to use crossover in the reproduction of all next generation strings.

Further, we have also failed to note one other step used in the reproduction process, that of mutation. Mutation is performed randomly on each element of each string. In the set of four strings, there are 20 elements—each with a small probability of mutation (e.g., perhaps 0.001). Thus, after the swapping process, we should determine if any element is to be mutated. In our case, none were, so the strings under the new generation column of Table 12.14 remain as shown.

Next, we simply compute the performance values (i.e., sum of squares) for each new string and proceed to the next generation. This is shown in the last two columns of Table 12.14. At this point no better candidate solution has been generated, but the total of the sum of the squares has been reduced from 37.33 to 34.333. Thus, we appear to have generated a more "fit" population overall. Using the same reproduction heuristic as before, we generate the new mating pool, mates, and crossover points. The result of all this is shown in Table 12.15. It might be noted that by sheer chance, the selection of mates turns out to be identical to that of the previous table (the crossover points are different, however).

TABLE 12.15   REPRODUCTION—THIRD GENERATION

String number	Mating pool	Mate	Crossover	New generation	Equivalent clusters	SS	$p/\Sigma p$
1	0 0 1 0	1 4	4	0 0 1 0 0	(1, 2, 4, 5) (3)	9	
2	0 0 1	0 1 3	3	0 0 1 0 0	(1, 2, 4, 5) (3)	9	
3	0 0 1	0 0 2	3	0 0 1 0 1	(1, 2, 4) (3, 5)	5.67	
4	0 0 1 1	0 1	4	0 0 1 1 1	(1, 2) (3, 4, 5)	1.833	

$$\Sigma p = 25.503$$

By comparing the mating pool of this most recent table with the previous table, it may be observed that not much has changed—a result that often occurs in problems of such a small size. However, we did manage to generate a much better cluster (i.e., string 4 of the new generation column with a SS of 1.833) and, once again, the total of the sum of squares for all four strings has been reduced. We shall terminate the process at this point; however, in actual practice, the procedure would continue until a specific termination rule was invoked (e.g., no improvement after a certain number of generations). *However, realize that whatever rule that we use to terminate the process, there is no assurance that we have reached an optimal solution—or, in fact, that some other heuristic programming procedure could not have solved the particular problem for a better solution, and more efficiently.* And this last point has seemed to completely escape the notice of some of the users and/or advocates of genetic algorithms.

Some Observations

It might appear to some readers that we have been unduly harsh with regard to some of our statements with regard to genetic algorithms. We want to stress that it is not our intention to unfairly criticize genetic algorithms as we are, in fact, quite enthused about the concept and have used it, with some success, for the solution of at least two very difficult problems (one of scheduling and one of deployment). What concerns us, however, is the tremendous amount of hype that has recently been generated with regard to this particular approach—and of certain unsubstantiated claims that have been made with regard to its performance. We are convinced that the method is a very good, exceptionally interesting, but strictly heuristic approach; and this last fact must be squarely faced by those who would choose to select the concept for real-world problem solution. More specifically, those who might consider genetic algorithms should consider the following facts:

- the method is a strictly heuristic approach that, like all heuristic methods, cannot guarantee an optimal—or, in some cases, even a very good solution
- although certain very appealing terminology is used (e.g., reproduction, mutation, genes, and chromosomes), the heuristic methods and elements so named only have a certain—and very remote—analogy to the mechanics of actual genetics
- there is no evidence that genetic algorithms are significantly better, or worse, than alternative heuristic programs; quite simply, the sum total of all comparisons made on heuristic methods thus far is insufficient to draw any particularly strong conclusions

To elaborate on the last point: although numerous heuristic programming methods have been developed, proposed, and in some cases implemented, relatively little has been accomplished with regard to a fair, comprehensive, objective, and defensible comparison of these approaches. In our limited experience with genetic algorithms, we have noted three aspects of particular interest.

First, the approach would appear to be more robust than most other heuristic methods, which, for the most part, have been specifically tailored to a very particular type of problem. Although this is certainly an attractive attribute, it does have its drawbacks. Specifically, when we have compared the performance of genetic algorithms with alternative heuristic methods, *on those types of problems for which the alternative methods were specifically developed*, it has been observed that in most cases, the genetic algorithms did not perform as well (i.e., in terms of solution accuracy and computational efficiency). However, once again, we emphatically stress that such observations are preliminary and that far more testing should be done before drawing any final conclusions.

Second, the performance of genetic algorithms on the problems to which we applied the methodology appeared to be *very* sensitive to various choices that were made (and must be made in any application of the approach). Specifically, one must rather arbitrarily select such things as the mutation rate, crossover rate, number of strings in the initial population, and the specific manner in which the strings are coded. Under some combinations of such choices, we observed excellent performance, whereas, under others, we encountered mediocre to very poor results.

Third, and finally, we noted that genetic algorithms seemed to perform better—on the limited number of cases that we investigated—on randomly generated problems than on real-life data. And this is a somewhat disturbing result that certainly warrants further investigation.

SIMULATED ANNEALING*

Simulated annealing shares certain features in common with both tabu search and genetic algorithms. As in the case of these alternative methods, one is attempting to avoid cycling. And, of course, all three approaches are strictly heuristic methods.

Simulated annealing uses ideas from statistical mechanics as the basis for an approach to large integer/discrete optimization problems. As with genetic algorithms, implementations of simulated annealing cannot guarantee attainment of an optimal solution, although they may produce very good solutions in reasonable times. Also, like genetic algorithms, simulated annealing is a general procedure that can be used across a wide variety of combinatorial optimization problems rather than a procedure tailored to a specific problem. Finally, simulated annealing is relatively easy to program and use. The combination of these factors—good solutions to hard problems, general applicability, and ease of use—have made simulated annealing a popular choice among practitioners in many fields.

As the name implies, simulated annealing draws an analogy between the physical annealing of solids and the solution of large integer or combinatorial optimization problems. The process of annealing begins with heating a solid until it becomes liquid. The temperature of the material is then decreased *slowly* until the lowest energy state (ground state) of the solid is obtained. If the liquid is cooled too rapidly or if the initial temperature is too low, the solid can end up in a higher energy state (metastable state) than the ground state. Solids in a metastable state are susceptible to fracture.

In 1953 physicists began to simulate this annealing process for solids on digital computers (Metropolis et al., 1953). Almost 30 years later, another group of physicists working on discrete optimization problems for VLSI layout began to explore the analogy between these discrete optimization problems and the simula-

---

* The material on simulated annealing has been contributed by Dr. Donald E. Brown, Department of Systems Engineering, University of Virginia.

tion of annealing in solids. They observed the following linkages:

Simulated annealing $\Rightarrow$ Discrete optimization procedure
Ground state $\Rightarrow$ Global optimum
Metastable states $\Rightarrow$ Local optima
Energy $\Rightarrow$ Cost

They exploited these linkages to develop the simulated annealing approach to discrete optimization (Kirkpatrick, Gelatt, and Vecchi, 1983). Since this introduction, the number of applications of simulated annealing has exploded and includes image processing (Geman and Geman, 1984), computer-aided design (Fleisher, Tavel, and Martin, 1985), and clustering (Brown and Huntley, 1992) to name a few.

A key characteristic of the simulation of annealing developed by Metropolis (Metropolis et al., 1953) is the occasional movement of the system of particles in the material to higher energy states. The number of these moves away from the ground state is a function of both the temperature and the increase in energy. At very high temperatures, the number of moves to higher-energy states almost equals the number of moves toward the ground state. At low temperatures, the movement of the particles tends to lower energy states. At a fixed temperature, the system is more likely to move to a new state with a slightly higher energy level than to one with a large increase in energy. A slow cooling process both keeps the material at the same temperature for a long period of time and also decreases the temperature slowly when a change in temperature occurs. The slow cooling in the annealing process allows the solid to escape metastable states through the occasional jumps away from low-energy states. If the liquid were cooled immediately, the particles would arrange themselves in the nearest low-energy state, which would likely be a metastable state.

Simulated annealing for discrete optimization also exploits parameter- (temperature-) controlled acceptance of nonimproving moves. Thus, the search in simulated annealing initially allows for moves that *increase* the cost for a cost-minimization problem (obviously, a maximization problem proceeds in the other direction). As the search proceeds (and the temperature parameter becomes smaller), the simulated annealing procedure accepts fewer and fewer moves that increase cost. The motivation for this strategy is identical with the annealing process—we want to escape local optima. However, one should not become too enamored with this analogy because in one case we are modeling a physical process with known characteristics, while in the other case we are attempting to optimize a discrete function about which we know very little. Without further justification, there is no reason to believe that this strategy should be successful for general discrete optimization problems.

There are two justifications provided for the use of simulated annealing in discrete optimization, one is theoretical and the other is empirical. A complete description of the theoretical argument is in van Laarhoven and Aarts (1987) and Aarts and Korst (1989). Essentially, this justification shows that simulated annealing will converge to the global minimum of a discrete optimization problem if it proceeds through a *suitable* sequence of temperatures. Unfortunately, from a practical standpoint, we do not know how to find this *suitable* sequence of temperatures for any problem instance. Most practitioners have used this theoretical result to justify a slow cooling process.

Empirical justifications have compared simulated annealing with random restart. In random restart, we begin the search for a global optimum (or at least a very good solution) from a randomly chosen starting point. We then proceed greedily accepting only improving moves until we no longer have any such moves left. Clearly, this approach is equivalent to instantaneous cooling in annealing and will likely lead to a local optimum. However, if we restart out local searches from many randomly chosen points won't we accomplish the same thing as simulated annealing? Surprisingly, the answer is no (Lundy and Mees, 1986). Tests of this procedure have shown the superiority of simulated annealing to random restart with equivalent numbers of objective function evaluations (Johnson, Aragon, and McGeoch, 1989).

The simulated annealing procedure contains the following basic steps:

1. Select initial parameters and the current solution.
2. Obtain a neighbor to the current solution.
3. Evaluate this neighboring solution against the current solution.
   (a) If it is better (e.g., lower cost), then make it the current solution.
   (b) If it is not better, then make it the current solution according to some probability; otherwise, keep the current solution.
4. Revise parameters and return to step 2.

There are two parameters that must be set in simulated annealing, the initial temperature and the chain length. The temperature controls the probability that a nonimproving move will be accepted in step 3(a). The chain length specifies the number of iterations that will be run at a specified temperature (acceptance probability). Most researchers use a third parameter to control the decrease in temperatures, called the cooling schedule. This parameter specifies the fraction of the current temperature that will be used as the next temperature once we have completed a number of iterations equal to the chain length.

The second step of the algorithm requires the selection of a neighbor to the current solution. Unlike genetic algorithms, which perform this selection process using crossover, simulated annealing requires a problem-specific selection procedure. In most cases, this is easily accomplished by simply permuting one of the decision variables.

Example 12.6: The Application of Simulated Annealing to Clustering

We can illustrate the use of simulated annealing using the clustering problem from Example 12.5. We again use the binary-string representation, where a 0 in position i of the string indicates that object i is assigned to cluster A and a 1 in that position indicates assignment to cluster B. A convenient neighborhood structure in this representation consists of all strings that are but one bit different from the target string. For example, the neighbors of the string, 0 0 0 0 0, are 1 0 0 0 0, 0 1 0 0 0, 0 0 1 0 0, 0 0 0 1 0, and 0 0 0 0 1.

In this problem, we are again trying to minimize the sum of squares (SS) for each candidate solution. We compute this by computing the squared distance from each point in a cluster to the cluster centroid. The centroid of a cluster is given by the vector of mean values for each of the two dimensions. Thus, the string 0 1 0 0 1 (denoting objects 2 and 5 in cluster B and the rest in cluster A) has a sum of squares of 11.33.

Table 12.16 shows the progress of simulated annealing with an initial string of 1 1 0 0 1, a chain length of 3, and a high temperature. The last column in the table shows the difference between the sum-of-squares criterion for the selected neighbor (SS_N) and the current solution (SS_C). Because this is a minimization problem (we want clusters with small sum of squares), improving moves have negative (decreasing) values in the last column and nonimproving moves have positive values.

TABLE 12.16 INITIAL ITERATIONS AT HIGH TEMPERATURE

Iteration	Current solution	Neighbor solution	SS_N-SS_C
1	1 1 0 0 1	0 1 0 0 1	3.5
2	0 1 0 0 1	0 1 0 0 0	-1.81
3	0 1 0 0 0	0 1 0 1 0	2.31

Because the temperature is high, we have an increased probability of accepting a nonimproving move. For this example, the temperature is high enough that 50% of the time we will accept a move that increases the sum of squares by two or more. In the first iteration, the neighbor solution shows an increase in the sum of squares of 3.5. We draw a random number between 0 and 1 and accept this nonimproving move only if the random number is less than 0.5. On this iteration, we draw a 0.1, hence, we accept the nonimproving move. On the next iteration, the selected neighbor shows improvement over our current solutions, so we accept it. On the third and final iteration at this temperature, we select a neighbor that again is nonimproving (an increase of 2.31 in our sum-of-squares criterion). We draw a random number, 0.3 in this case, and accept the nonimproving move.

Table 12.17 shows a continuation of this process. In this table, the temperature has dropped and so has the likelihood of accepting a nonimproving move. We now accept only 20% of the nonimproving moves with changes of two or more.

Iterations 4 and 5 are both improving moves and are accepted. In iteration 6, a nonimproving move is contemplated, but when we draw a random number of 0.6, we reject it. Simulated annealing terminates when no changes have occurred within some specified time period or, more commonly, at the completion of all iterations at a specified final temperature.

TABLE 12.17 SUBSEQUENT ITERATIONS AT LOWER TEMPERATURE

Iteration	Current solution	Neighbor solution	$SS_N - SS_C$
4	0 1 0 1 0	1 1 0 1 0	−6.16
5	1 1 0 0 1	1 1 0 0 0	−1.17
6	1 1 0 0 0	1 1 1 0 0	4.0

From Table 12.17, we should note that the best solution thus identified is that consisting of the string 1 1 0 0 0—as determined in iteration 5. This is in fact equivalent to the same solution obtained by genetic algorithms in Table 12.5. That is, both methods found the (optimal) solution of assigning the first two objects to one cluster and the last three to another cluster. In actual practice, of course, different solutions may be derived by the two different approaches—and neither can guarantee optimality.

Some Observations

From this discussion, we may conclude that simulated annealing has applicability to a broad range of problems, obtains good (although not necessarily optimal) solutions, and is easily implemented. When these advantages are coupled with the well-developed (but not particularly practical) theoretical foundation and objective empirical comparisons, it would appear that simulated annealing represents a promising approach for large-scale discrete optimization.

However, most practitioners who have employed simulated annealing would undoubtedly stop far short of overwhelming praise. There are in fact good reasons for a tempered view of simulated annealing. The first concerns run time. We began this section by noting that simulated annealing produces good solutions in reasonable time. Reasonableness, like beauty, is in the eye of the beholder. Simulated annealing can be notoriously slow and this slowness is even encouraged as we noted by theoretical convergence results. If a problem requires extensive evaluation time, the run time for simulated annealing can become intolerable. For classical problems from the literature, like graph partitioning or the traveling salesman problem, evaluations are quick and users can tolerate slow cooling schedules. However, for practical problems, where evaluations are more involved, the temptation is to speed cooling and thus risk attaining a solution of poorer quality rather than endure a lengthy cooling schedule.

A second problem with simulated annealing and one that it shares with other heuristic programming techniques concerns the need to set parameter values. Although simulated annealing has fewer parameters than genetic algorithms or tabu search, nonetheless, a user must choose an initial temperature, chain length, and cooling schedule. Chosen poorly, simulated annealing will not work well. However, some analysts suggest that choosing parameter values in simulated annealing is far easier than in many heuristic programming methods (in particular, when compared with genetic algorithms). Further, it has been noted that when people have difficulty with simulated annealing, they generally do not have problems with parameter settings (rather they usually have difficulties getting their evaluation to run fast enough or with devising an appropriate neighborhood structure, as discussed in what follows). Nonetheless, the need for the user to correctly select parameter settings is one of the reasons that using simulated annealing or any of the other heuristic programming techniques is something of an art.

Finally, simulated annealing requires the user to define an appropriate neighborhood structure to generate feasible solutions. As we noted, in classical problems from the literature, this choice is typically simple. However, for many practical problems, the neighborhood structure might not be quite so obvious. The proper selection of a neighborhood structure can influence significantly the performance of simulated annealing and can present a major challenge to the new user.

Despite these difficulties, simulated annealing would appear to remain a good choice for a surprisingly broad range of problems. Our experience suggests that simulated annealing is an excellent choice for large, discrete optimization problems for which an algorithm to generate good answers must be quickly coded.

TABU SEARCH, EXPERT SYSTEMS, AND NEURAL NETWORKS

In this section, we shall provide an extremely brief overview of three other heuristic methods that have received, as of late, considerable interest. Those interested in a more detailed presentation are directed to the references cited.

Tabu Search

We perceive tabu search as a metaheuristic, that is, a heuristic procedure that may be used to guide and possibly improve the performance of many other heuristic methods. Tabu search, as in the case of genetic algorithms, has been associated with artificial intelligence primarily on the basis of certain terminology (i.e., short-term and long-term memory). However, we perceive it as simply another example of heuristic programming. We will only provide a very general overview of the approach and those readers wishing to learn more about the method are directed to the references (Glover, 1986; Glover et al., forthcoming).

As was mentioned, tabu search is essentially a metaheuristic—a heuristic used to guide other heuristic programming methods in the conduct of their search. By means of the use of tabu search, one may be able to escape from local optimal solutions and seek improved solutions.

In general, most heuristics are "hill-climbing" procedures. They begin with a feasible solution and then conduct a local search in the neighborhood of this solution to find a local optimum. The problem with such an approach is that when a local optimum is reached, movement in any direction will cause the value of the objective function to worsen. In most cases, the typical search process stops here. One option to obtain a better solution is, of course, to start from a different starting solution and generate another local optimum. Tabu search allows the user to get away from such local optima to continue with the search.

Tabu search essentially works on the strength of two concepts: short-term memory and long-term memory. Short-term memory keeps track of a few past moves (the actual number is defined by the length of the tabu list) to prevent search in these areas, thus attempting to avoid cycling. A tabu list is maintained, and moves that are members of this list are not allowed. The tabu list is a cyclic list, where the latest move is added and the earliest move in the list deleted. Also, the best move, among all available moves, is chosen. It should be understood here that the best possible move need not necessarily be an improving one. This enables the system to move out of local optima.

The short-term memory part of the algorithm is run for a prescribed number of iterations. At the end of these iterations, the user has one of the following choices:

1. Start from the beginning with different values for the tabu size and maximum number of iterations.
2. Invoke long-term memory and continue the procedure.
3. Stop with the best solution generated thus far.

The long-term memory part of the tabu search may be viewed as "learning" and "unlearning" in a system. The philosophy is to use the past searches to bias the new search, so that the search space is reduced. Learning can work either way; it can either block out those areas of the state space that have been visited before or concentrate the search in the areas that have been visited. Rationale for both philosophies can be given.

Expert Systems

Although expert systems, like genetic algorithms, tabu search, and simulated annealing, are often considered to be tools of artificial intelligence, they, too, simply comprise another heuristic procedure (Ignizio, 1991). The most meaningful difference between other heuristic programming methods and expert systems is that knowledge (i.e., heuristic rules) is separated from the procedures (i.e., the steps used to process the knowledge so as to develop new facts). Unfortunately, this fact is all too often lost in the confusion of new terminology (e.g., inference engines, production rules, frames, and objects) and further obscured by the use of so-called AI programming languages (e.g., LISP and PROLOG).

Since heuristic programming is more or less a philosophy, rather than a specific tool for a specific problem, expert systems also comprise a philosophy of problem solving. At this point in time, expert systems have been much more widely (and successfully) applied to problems of association (e.g., prediction, forecasting, diagnosis, classification, and pattern recognition) than to problems of construction (e.g., scheduling, planning, configuration, and layout). Although there are certainly some exceptions (e.g., the XCON expert system for config-uring VAX computers), this is generally the case—and there is, as will be dis-cussed, a reason for this.

Problems of association are far easier for the typical expert system than are problems of construction because the vast majority of expert systems employ production rules—that is, the transformation of heuristic rules into if–then state-ments. And production rules are themselves rules of association, that is:

- *if* interest rates rise, *then* bond prices will fall
- *if* it is cloudy, *then* rain is likely

Putting such rules together permits the solution of more complex problems of association. Further, to solve a problem (or, in expert systems terminology, to conduct a consultation session), all that is typically needed is the substitution of the facts concerning the particular problem under consideration.

On the other hand, problems of construction involve—as mentioned several times before—feedback and comparison, operations that are more easily handled by procedures. Procedures involve such things as clearing old facts, crunching numbers, and jumping to a new step, and whereas they may themselves some-times be represented as if–then statements, they are *not* the same as a production rule based on some human expert's judgment and experience. Further, in the design of an expert system, heuristics and procedures *ideally* should be sepa-rated. Heuristics are placed in the knowledge base, whereas procedures typically appear in the inference engine (knowledge processor). Further, most expert sys-tems software has a fixed inference package. Thus, to incorporate any procedures associated with the procedure either involves placing them in the knowledge base (which serves to defeat one of the main advantages of expert systems) or to develop external programs that invoke the procedures (which tend to complicate the process). Frankly, for many problems of construction, it is probably easier to simply develop a heuristic programming method such as those indicated previ-ously rather than attempting to force the problem into a particular expert system software package.

Neural Networks

In Chapter 8, we introduced the use of neural networks for solution of problems of association (specifically, prediction and pattern classification). Such networks have also been proposed for the solution of problems of construction—including integer programming. It would appear that the basic approach employed involves the construction of a particular network architecture that may be used to represent, in some fashion, the particular problem under investigation. Typically, we also construct an "energy function," which would likely appear to those with a background in mathematical programming, to be a type of objective or penalty function. An input is then applied to the network and, by means of feedback and comparison, network stability is, we hope, reached. If one is fortunate, the solution represented by the stabilized network corresponds to the global optimal solution for the problem being modeled. However, for most problems, the approach is heuristic and one can hope only for an acceptable solution. Despite the heuristic nature of the approach, it would appear that neural networks hold considerable promise in problem solving. The reader seeking further details is directed, in particular, to the special issue of *Computers and Operations Research* dealing with neural networks (Ignizio and Ignizio-Burke, 1992).

ASSESSMENT OF HEURISTIC PROGRAMMING METHODS

As may be noted from the material presented herein, heuristic programming methods—even those intended for extremely complex problems of massive sizes—are neither complex nor particularly sophisticated. This observation has led the more naive practitioner to propose and employ heuristic methods with little regard to either the logic underlying the procedure or of the assessment of the process.

Prior to presenting our opinions as to how one might best go about assessing the performance of any heuristic procedure, let us list some of the things one must *not* do.

- Do not readily accept any heuristic programming procedures whose logic is suspect. That is, try to assess the plausibility of both the individual heuristics employed and their combination. If something appears irrational, or naive, it is unlikely that the resultant method will be of any value. As just one example of a naive heuristic, consider the case in which the number of bugs splattered over your car windshield is observed to increase when the stock market rises and to decrease when it falls. Although one might note a surprisingly good level of correspondence between these two, very unrelated events, it is most likely due only to chance—and the misfortune of a number of innocent bugs. On the other hand, one might not be too quick to discard such observations—particularly with regard to the movement of the stock market, an event that has defied all rational explanation.

- Never limit your assessment of a heuristic procedure to just those problems of sizes and complexity for which exact solution methods are possible (i.e., so that one might compare the heuristic program's solution with a known optimal solution). Heuristic methods are only appropriate for problems of size and complexity *beyond* that of existing exact methods. *As such, the performance that one should be interested in is that achieved on problems beyond the capability of algorithms.* And it might be noted that this is a rule that is violated over and over in most evaluations and/or justifications of heuristic methods.

- Do not rely on evaluations that are performed on (only) randomly generated problems. Remember that a heuristic procedure is a pragmatic approach intended for real-world problems of size and complexity beyond the capabilities of exact methods. Further, most observations (and virtually all that we have personally experienced) indicate that well-designed heuristic methods perform best on real-world examples, and not nearly so well on randomly generated artifacts.

- Do not attempt to assess the performance of a heuristic method without *first* establishing aspirations as to its level of achievement for whatever performance measures are deemed important.

It is our opinion that the characteristics associated with "good" heuristic programming methods are as follows:

- Simplicity of approach.
- Simplicity of computational procedures (i.e., operations employed).
- Low-order polynomial growth in computational time required as a function of problem size.
- The achievement of results, by means of a systematic and rational evaluation, that are acceptable and, it is hoped, close to those aspirations held prior to the evaluation.

Let us now reflect on just how one may evaluate the performance of a heuristic method on problems *beyond the capability of exact methods.* In essence, there are three typical ways to do this.

1. Determine the computational complexity of the heuristic program.
2. Compare its performance against those achieved in the past by other means, particularly those results that are considered to represent "good" solutions.
3. Compare its performance by means of solving massive problems with known solutions.

We shall only briefly discuss the first concept. Here, we seek to determine the computational performance of the heuristic program in terms of the solution times (or number of operations) required as a function of problem size. On "hard" combinatorial problems, the performance of exact methods is typically some exponential function of problem size. Thus, if it is possible to prove that a heuristic procedure has superior performance—typically low-order polynomial growth as a function of problem size—then it becomes easier to justify the use of the heuristic. *Of course, low-order polynomial growth is, by itself, unimportant if the results achieved by the heuristic method are considered poor or unacceptable.*

In some cases, it may be possible to prove that the performance of a heuristic, for the worst case, is less than some low-order polynomial. However, in many instances, this cannot be achieved other than by means of an empirical study. Further, it may well be that instances of the worst-case performance of a method are extremely rare, or limited to randomly generated problems, and thus not of prime importance. Thus, let us proceed to a discussion of the remaining two approaches to assessment.

In most cases, it will be possible to compare the performance of a heuristic method with those derived by other means, such as

- randomly generated solutions
- solutions developed by human experts
- solutions developed by alternative heuristic methods

In those instances in which heuristic programs are assessed, the use of randomly generated solutions is often noted. However, this approach has a number of drawbacks. First, it may be difficult (e.g., in the case of complex, resource-constrained scheduling problems) to generate *feasible* random solutions. Second, just because we generate what appears to be a large number of random solutions, it is by no means an assurance that we have examined even some miniscule fraction of all possible solutions. Still, when tempered by good judgment, such an approach is at least worth considering.

The second approach is that of comparing the solutions achieved by the heuristic program with those derived by human experts. From a strictly pragmatic point of view, if we can generate solutions that are better than those now being derived (i.e., by other means), then we have certainly accomplished something positive. However, it is unlikely that much confidence may be obtained in a given method through such means, as it is usually difficult to impossible to make a reasonably large number of such comparisons. Still, the approach has merit and, no matter what other approaches are taken, should be included in the assessment process if possible.

To demonstrate the use of the approach involving comparisons with human experts, recall the first heuristic method described in this chapter: the add/drop procedure for cluster/site location. As was discussed then, the largest problem solved by the procedure involved the siting of up to three (i.e., $K \leq 3$) sites from among 306,252 candidate sites so as to provide coverage of 25,521 sectors of airspace. The solution derived by the heuristic program for this problem was compared with that derived by human experts—military officers trained in the deployment of air defense missile systems. In addition, random solutions were generated. The results achieved in this comparison are listed in Table 12.18. As may be seen, the coverage provided by the heuristic method was far beyond the best solutions of either the human experts or, in particular, the randomly generated solutions (Ignizio, 1968, 1971).

TABLE 12.18 COMPARISON OF RESULTS
FOR ADD/DROP HEURISTIC

Method	Airspace covered (%)
Add/drop heuristic	75.9
Human experts	38.8 to 70.8
Random solutions	25.1 to 39.1

The third approach is to simply compare the results of your heuristic programming method against those that have been developed to solve the same type of problem. Such a comparison, when possible, should permit a reasonably effective means to evaluate the performance of your method against alternative methods using the same computational resources.

However, all too often, such comparisons are made against the results of alternative methods as reported in the literature. And this may lead to several problems. First, the alternative method in the literature may not represent the most recent version of the approach. Second, any comparisons made are limited to application to the same problems as cited in the literature. And, third, computational comparisons, such as speed and storage requirements, can only be estimated—and this is true even when the same types of computers are employed.

Finally, let us briefly describe yet another approach to the assessment of heuristic programs. As mentioned earlier, an approach suggested is that of generating massive-size problems with known solutions. As an indication of how this may be accomplished, let us once again focus on the add/drop heuristic for clustering/site location.

Although the add/drop heuristic for clustering/site location is intended primarily for those types of problems, we may note that it is possible to transform a type of assignment problem—known as the matching problem—into a special type of clustering/site-location problem. Specifically, it is a straightforward process to transform a relatively small matching problem with a known solution into a much larger total cover problem with a known solution. We shall not describe this method here as it is readily available in the literature (Ignizio, 1971; Ignizio and

Case, 1972). By using this procedure, total cover problems were generated (with known solutions) of sizes from 16 rows (customers) and 34 columns (sites) up to 40 rows and 100 columns. With today's computational resources, far larger test problems could be generated if so desired.

SUMMARY AND CONCLUSIONS

There are a host of heuristic methods that have been developed for application to a wide variety of problem types. In many cases, such methods have been designed specifically for a particular type of problem (e.g., the add/drop heuristic for clustering/site location); in others, a more robust approach has been developed (e.g., genetic algorithms, tabu search, simulated annealing). However, none of these methods should be considered unless the size and complexity of the particular problem being faced precludes the use of exact methods.

We know of no evidence that would serve to substantiate certain recent claims as to the superiority of one heuristic programming method versus another—and this includes the so-called AI techniques such as genetic algorithms, tabu search, simulated annealing, expert systems, or neural networks. All such methods are heuristic and thus provide no guarantee as to solution optimality—and there has yet to be performed any thorough, objective comparison of these approaches.

EXERCISES

12.1. Apply the add/drop heuristic to the problem depicted in Table 12.19. Note that a 1 in the table represents complete coverage and a 0 indicates no coverage. We seek to find a set of clusters that will maximize the coverage provided to each customer, using no more than (a) two sites and (b) three sites. Break all ties by selecting the lowest-numbered site. What is the best single site? Best two sites? Best three sites?

TABLE 12.19

Customer	Alternatives			
	Site 1	Site 2	Site 3	Site 4
1	1	0	1	0
2	1	1	1	0
3	0	0	0	1
4	0	0	1	0
5	1	1	0	1

12.2. Apply the exchange heuristic to the problem of Example 12.2, where a starting solution of $\mathbf{x} = (0, 1, 0, 0, 1)$ is used.

12.3. Apply a permutation search to the problem of Example 12.3 if the initial permutation is [3, 1, 4, 2].

12.4. Describe how a permutation search might be applied to the site-location problem of Example 12.1.

12.5. Develop the conflict matrix for the eight-queen problem. Solve the model by the add (column-sum) heuristic.

12.6. Apply the add (column-sum) heuristic to the conflict matrix of Figure 12.3.

12.7. Use the genetic algorithm to solve the problem in Exercise 12.2. Compare results.

12.8. Assume that you wish to solve, via the genetic algorithm, the site-location problem of Example 12.1. Describe the coding scheme that you might use.

12.9. Assume that you wish to solve, via the genetic algorithm, a clustering problem in which up to three (3) clusters are permitted. How would you go about coding the candidate solution set (i.e., population)?

12.10. Consider the traveling salesperson problem where the solution representation is the order in which each city is visited [e.g., (3, 5, 2, 4, 1) indicates a tour from city 3 to 5 to 2 to 4 to 1 and then back to city 3]. If we use genetic algorithms to solve this problem, how might crossover be conducted so as to assure that any crossover results in a *feasible* (i.e., one in which all cities are visited and no city is visited—other than the first—more than once) tour?

12.11. Read the papers on DMES (Cochard and Yost, 1985) and AALPS (Anderson and Ortiz, 1987) and attempt to explain why the first method is termed a *heuristic program* and the second is called a *knowledge-based system* (i.e., expert system). Which of these methods, *according to these papers*, was actually used for aircraft loading for the Grenada invasion? These two papers have generated considerable controversy. Can you list some of the possible reasons for this?

12.12. Recently, the add/drop heuristic for clustering/site location—as described in this chapter—was coded via the LISP programming language and called an expert system for site location. Can you explain why the FORTRAN version is called a heuristic program and the LISP version is alleged to be an expert system?

CHAPTER **13**

MULTIOBJECTIVE OPTIMIZATION

CHAPTER OVERVIEW

In the previous 12 chapters, our focus has been restricted, for the most part, to the "traditional" (i.e., single-objective) linear programming model. And, in fact, the majority of texts on linear programming, or on the more general area of mathematical programming, limit their coverage to such traditional, and possibly more comfortable, concepts. However, in this chapter, and in those to follow, we broaden this perspective so as to consider (linear) mathematical programming problems that involve multiple, conflicting objectives subject to both "hard" and "soft" constraints. In particular, it is our purpose to address the means through which one may effectively and efficiently deal with mathematical programming models that are, quite frankly, considerably more representative of the inherent characteristics of typical, real-world problems.

Let us pause to reflect on the last statement. To some, the very notion of *multiple*-objective models is controversial.* They feel that single-objective mathematical programming has an indisputable record of accomplishment—and has most certainly found wide, and extremely successful, application to an enormous number of real-world problems. And, of course, they are absolutely right. As such, one might ask, why even bother with the notion of models involving multi-

---

* Much, if not most of this controversy is based upon certain myths and misconceptions—and these shall be addressed toward the conclusion of this chapter.

ple objectives—and why, in particular, address a topic for which there are so many conflicting proposals—and for which there is so little agreement?

Our purpose, in Part 3, is not to diminish either the importance or usefulness of the traditional approach. Further, we recognize that traditional methods may certainly be applied to multiobjective models by means of judicious and thoughtful modeling, coupled with the prudent use of sensitivity and postoptimality analysis. However, as we shall see, multiobjective approaches often serve to uncover certain aspects of the problem that are inadvertently masked by the traditional approach. As such, we may be able to obtain, through the use of multiobjective methods, both a better appreciation of the actual problem as well as an improved understanding of the resultant solution. Specifically, we believe that, in many cases, the multiobjective approach leads to *increased understanding and insight*—and, for this reason alone, the study of the multiobjective concept is worthwhile.

Consequently, we shall address and illustrate, in this chapter, the various *philosophies* that have been proposed for dealing with multiobjective models— and list their advantages and disadvantages.* We then shall briefly note some of the more prevalent myths and misconceptions that surround multiobjective optimization—and seek to dispel these notions. Finally, we shall very briefly introduce the multiplex model (Ignizio, 1985b). It is this model that we shall then employ, in the remaining chapters, to model, solve, and analyze a (modest) variety of both single-objective and multiobjective models. First, however, let us compare the fundamental differences that exist between traditional (single-objective) and unconventional (multiobjective) concepts.

THE FAMILIAR VERSUS THE NONCONVENTIONAL

The single-objective approach has become so ingrained, and so widely accepted, that it may seem hard to believe that it has only seen widespread use since 1947. Further, it is easy to forget (or, for those who never knew, to overlook) the fact that in 1947 the very notion of even a single-objective function was considered quite revolutionary. Specifically, until the development of LP, the typical mathematical model consisted of either a system of equations or a system of inequalities—and, for the most part, one's attention was directed toward the determination of just a feasible solution (i.e., one that satisfied the system of constraints as opposed to one that both satisfied the constraint set *and* optimized a single measure of performance). As such, in 1947, the concept of the inclusion of an objec-

---

* It is *absolutely essential* that the reader appreciate that *each and every* approach to multiobjective optimization (or to the closely related topic of multicriteria decision making, or MCDM) is itself a purely subjective method, based upon certain philosophies (with which the reader may or may not agree). Further, there simply is no "one right way" to approach a problem involving multiple, conflicting objectives.

tive function was considered just as radical as some now view the inclusion of multiple-objective functions.

However, although the consideration of multiple objectives may seem a novel concept, consider virtually any nontrivial, real-world problem. Real problems invariably involve multiple objectives. Or, stated another way, we evaluate the performance of most any real-world system by means of an *array* of performance measures. For example, the success of an airplane is determined by such things as its

- cost (to be minimized)
- payload (to be maximized)
- speed (to be maximized)
- maximum range (to be maximized)
- weight (to be minimized)
- survivability (to be maximized)
- etc.

And, in the design of an aircraft, we may actually hope to optimize *each and every one* of these parameters.

Restricting our attention, however, to a single objective (i.e., a single measure of effectiveness) does serve to (at least apparently) make things much more simple. Consider once again the example of an airplane, specifically a military fighter plane. Now, if we have one design that costs $30M per plane, and another costs $35M, then the first design is obviously preferred—*with respect to cost alone*. However, what if the $30M plane has a maximum effective range of 400 nautical miles and the $35M plane has a range of 500 (and both satisfy the *minimum* range requirements)? By introducing just one more measure of effectiveness, the decision as to the "best" plane has been changed from one that is quite simple, if not apparently trivial, to one that is surprisingly difficult. That is, if we only consider cost, the smaller the cost, the better. However, as soon as we add range, *there is no easy answer*.

In essence, by moving from a single objective to multiple objectives, we move from the evaluation of a *scalar* (i.e., the performance of a single measure of effectiveness) to that of a *vector* (i.e., the performance of an *array* of measures of effectiveness). And, in doing so, we move from decisions in which there is but one "correct" answer to those in which the "correct" answer *is a matter of philosophy*. This last factor alone may serve to indicate just why some analysts are so nonplussed by the multiobjective concept.

Now, let us add yet a further complication. In the traditional LP model, each and every constraint is considered to be absolutely rigid. That is, a solution that does not satisfy each and every constraint is termed *infeasible*. However, in real-world problems, the notion of strictly rigid constraints does not necessarily hold—at least not for every constraint function. For example, we may have a raw material resource "constraint" such as

$$3x_1 + 7x_2 + 9x_3 \leq 100$$

This would imply, for example, that the program $\mathbf{x} = (20, 3, 2)$ is (mathematically) feasible and $\mathbf{x} = (20, 3, 3)$ is (mathematically) infeasible. However, in real-world problems, we just may be able to tolerate a certain level of "violation" of a constraint. That is, although we might have only 100 units of a raw material readily available (i.e., in the last constraint), if we really had to, we just might be able to obtain eight or nine more units (which would then render the second program feasible). Thus, the last constraint may not be nearly so rigid as is implied by its mathematical appearance. Such flexible constraints are termed "soft" constraints (or soft goals)—and are *frequently* encountered when we deal with actual problems. Thus, a soft constraint is one that we would like to satisfy, but for which we would be able to accept some degree of "violation." On the other hand, a hard constraint (or hard goal) is one for which *any* degree of violation would be absolutely intolerable. However, from a traditional LP point of view, such notions as multiple objectives and soft constraints only serve to complicate the situation (or, and perhaps even worse, to implicitly question the underlying assumptions of the traditional model).

Multiple, conflicting objectives and hard and soft constraints are, we believe, characteristic of most real problems. However, far and away, the bulk of research in LP (or mathematical programming in general) has been concentrated upon the *enhancement of algorithms*—at, quite possibly, the expense of the *enhancement of models*. In fact, in many quarters the (traditional) LP model is so sacrosanct that one dares not even question its format. And this may explain just why there are such ardent feelings and emotions whenever one even raises the notion of the multiobjective approach.

Despite such opinions, perceptions, and passions, we intend to address the multiobjective approach in this and forthcoming chapters—and to indicate just how a single, unified concept may be used to approach a variety of multiobjective models (as well encompass the traditional model). Further, we shall show that this can be achieved without any appreciable sacrifice of computational capability or efficiency.* To accomplish this, let us first examine a simple example consisting of but two objectives.

AN ILLUSTRATIVE EXAMPLE

In order to introduce some of the more popular of the multiobjective concepts, we shall employ a simple problem involving but two objectives and two variables. Specifically, consider the problem in which a firm produces just two products.

---

* More specifically, one may solve problems that are just as large as those that may be dealt with through conventional means—without any significant degradation of computational efficiency (i.e., speed and accuracy).

Further, management has determined that it wishes to find a production scheme that will

- maximize total profits
- maximize the expected number of market shares "captured"
- satisfy processing limitations (i.e., raw material availability)
- avoid saturation of the market (i.e., to be able to sell all items produced)

This problem is immediately unusual (i.e., with respect to problems encountered in textbooks dealing solely with traditional models) in that *two* objectives have been cited. Let us further assume that x_1 = number of units of product 1 produced per time period and x_2 = number of units of product 2 produced per time period. Further, each unit of product 1 provides a profit of 3 monetary units and each unit of product 2 returns 1 monetary unit. Also, for every unit of product 1 sold, it is estimated that the firm will gain 2 units of market share, and for every unit of product 2, it will gain 3 units of market share. Product 1 requires 2 units of raw material per unit produced and product 2 requires 1 unit per unit produced—and only 50 units of raw material are available each time period. Finally, market surveys indicate that no more than 20 units of product 1 and no more than 30 units of product 2 should be produced each time period. In fact, if we produce any units above these limits, not only will we not be able to sell them, we shall be forced to destroy them (at a prohibitive cost to the firm). As such, we may write the model representing this problem as follows:

Find x_1 and x_2 so as to

maximize $z_1 = 3x_1 + x_2$ (i.e., total profit per time period)   (13.1)

maximize $z_2 = 2x_1 + 3x_2$ (i.e., market shares captured per time period)   (13.2)

satisfy:

$$2x_1 + x_2 \leq 50 \text{ (raw material limitations)} \tag{13.3}$$

$$x_1 \leq 20 \text{ (market saturation level, product 1)} \tag{13.4}$$

$$x_2 \leq 30 \text{ (market saturation level, product 2)} \tag{13.5}$$

$$\text{and } x_1, x_2 \text{ are nonnegative} \tag{13.6}$$

Figure 13.1 depicts this problem graphically. Note that the cross-hatched area is the region that satisfies both the raw material and market-level constraints and the dashed lines are, in essence, representative "iso-profit" and "iso-market-share" lines for the two objectives. Point A represents the solution that would optimize the second objective (market shares) and point B is the solution that optimizes the first objective (profit). It should be clear to the reader that there is no way in which we may optimize both objectives simultaneously. Let us examine *some* of the ways in which this problem might be modeled.

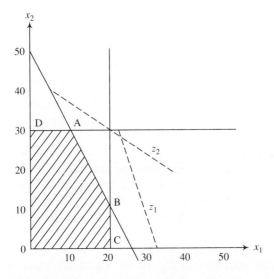

Figure 13.1 Graph of illustrative example.

Conversion to a Linear Program via Objective Function Transformation (or Deletion)

The traditionalist would most likely decide that, regardless of what management may have stated, a single objective model is going to be employed.* Thus, one way to *force* the problem into the single objective format is to select one of the objectives (i.e., either profit or market shares), use it as *the* single objective, and then either ignore the other objective *or* treat it as a (rigid) constraint. Because most analysts are trained (by means of textbook examples) to think of profit (or cost) as being "the" objective in LP models, let us pick the profit objective as our single objective and treat the market-share objective as a constraint. We may accomplish this latter transformation by assuming some desired, or acceptable, level of market shares and using that value as the right-hand side of the new constraint. For example, let us assume that we use a value of 100 market-share units for the right-hand side of the market-share "constraint." The LP representation of our problem becomes the following.

Find x_1 and x_2 so as to

maximize $3x_1 + x_2$ (total profit per time period)   (13.7)

subject to:

$2x_1 + 3x_2 \geq 100$ (market shares desired)   (13.8)

$2x_1 + x_2 \leq 50$ (raw material limitations)   (13.9)

---

* It is in fact argued that too many mathematical programming analysts simply do not listen to the decision maker. Instead, they hear only what they want to hear—and consider only those models (and methods) with which they are comfortable.

$$x_1 \leq 20 \qquad \text{(market saturation level, product 1)} \qquad (13.10)$$

$$x_2 \leq 30 \qquad \text{(market saturation level, product 2)} \qquad (13.11)$$

$$\text{and } x_1 \text{ and } x_2 \geq 0 \qquad \text{(nonnegativity conditions)} \qquad (13.12)$$

Notice carefully that the maximization objective of (13.2) has been converted into the type II constraint ($\geq$) of (13.8). Had we transformed a minimization objective, it should be obvious that we would have employed a type I constraint ($\leq$).

The obvious advantage of using the traditional LP model is that one may immediately employ any existing LP algorithm or software to solve the *model*. Unfortunately, it is all too often forgotten that the solution so obtained is only appropriate for the transformed model and not necessarily for the original model (or, in particular, for the actual problem). Three other disadvantages associated with the use of the LP model in the representation of multiobjective problems follow.

- Unless we are careful (and/or lucky), conversion of objectives into (rigid) constraints may lead to models that are mathematically infeasible (i.e., had we used a value of, say, 120 instead of 100 for the right-hand side of the market-share constraint, the resultant model would have been mathematically infeasible).

- The converted objective, or objectives, as well as any *soft* constraints, are treated as *hard* constraints by the LP algorithm. Thus, even though we may well get along with less than 100 market shares per period, no such solution can be generated by the associated algorithm (i.e., without the employment of postoptimality analysis).

- There is a great deal of subjectivity involved in the selection of the single objective to be employed in the transformed LP model—and results may differ considerably depending on the choice made.

Finally, notice that should we solve this LP model listed, our solution will lie at $\mathbf{x} = (12.5, 25)$. This solution is depicted as point $\mathbf{x}^*$ in Figure 13.2. Had we simply deleted the second objective (function 13.2) and solved the resultant LP, the solution arrived at would have been at point B of Figure 13.1.

Conversion to a Linear Program via Utility Theory: A Method of Aggregation

Theoretically (and only theoretically), it should be possible to combine any number of objectives into an equivalent, single objective—if we can determine a common measure of effectiveness (i.e., a so-called "proxy") by means of which each of the objectives may be expressed. The basis of such an approach is the *aggregation* of multiple objectives into a single and, it is hoped, equivalent function.

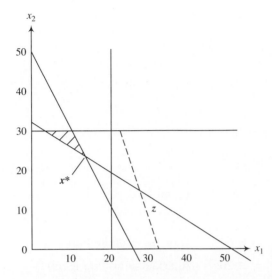

Figure 13.2 LP conversion solution.

In our illustrative example, one objective is measured in dollars (of profit) and the other is measured in market shares gained (e.g., some measure of the "loyalty" of the customers for a given product or brand—resulting in a greater likelihood of repeat sales of that product or brand in the future). Now, if one could convert, say, *market shares captured* into *dollars of profit* (or, alternatively, *profit in dollars* into *units of market shares*), we could then combine both objectives into one that is measured in common units. And, in fact, if such a combination appears reasonable and can be accomplished, it should certainly be utilized in any multiobjective model.

However, what if it is not so easy to find a common measure? For example, it is well known that the wider a highway, the safer the highway. Thus, in the design of a new highway system, we might have one objective that is concerned with the minimization of costs (measured in dollars—and increasing as the width of the highway increases) and another may focus on the minimization of accidents (measured in pain and injury, and the loss of life—factors that are decreased as highway width is increased). Now, if we are to combine these two objectives, we need to find a way to convert the loss of a human life into dollars—or, alternatively, find a proxy measure by means of which both objectives may be expressed. Despite the fact that numbers are often employed to represent the value of a human life, it should be obvious that any values used are highly subjective (and virtually impossible to defend) and that this is no easy problem.

Despite such obstacles, there are those who advocate the use of utility theory to combine multiple objectives. They do so by using a single measure, "utiles," to represent the values of each objective. Entire books have been written on this single subject (see, e.g., Keeney and Raiffa, 1976), so we can hardly hope to do it justice here. Nor is it our intention to spend a great deal of time on methods of aggregation. However, in brief, one attempts to elicit, from the "decision maker," his or her "utility function" by means of a series of carefully considered questions. Such a function may be of various forms, and is subject to several restrictive assumptions. However, if we can assume that we have some-how determined this function, *and* that the utilities are additive, one can change each objective into one measured in utiles, and then add them together.

In our example, let us assume that we are able to find these utility functions, and that the two objectives may be combined by weighting the first by 0.6 and the second by 0.4. Adding the weighted objectives, we are able to form the following LP:

Find x_1 and x_2 so as to

$$\text{maximize } 2.6x_1 + 1.8x_2 \qquad \text{(combined objectives, in utiles)} \qquad (13.13)$$

subject to:

$$2x_1 + x_2 \le 50 \qquad \text{(raw material limitations)} \qquad (13.14)$$

$$x_1 \le 20 \qquad \text{(market saturation level)} \qquad (13.15)$$

$$x_2 \le 30 \qquad \text{(market saturation level)} \qquad (13.16)$$

$$\text{and } x_1 \text{ and } x_2 \ge 0 \qquad \text{(nonnegativity conditions)} \qquad (13.17)$$

Solving this LP, we find that the "optimal" solution is at $\mathbf{x} = (10, 30)$. This is represented as point A on Figure 13.1.

The one obvious advantage of the utility theory approach is that it permits one to use conventional, single-objective methods to solve the converted model. The disadvantages include the following.

- Considerable time and care is required to determine the utility functions necessary for conversion.
- The various assumptions employed in, and underlying, utility theory simply may not hold for the specific situation addressed.
- One implicitly assumes that all objectives are commensurable (i.e., they may be expressed in a common unit), an assumption difficult to justify for certain cases (e.g., as in combining an objective measured in the loss of human life with that of one measured in terms of dollars of profit).
- The utility function derived for an *individual* may not be the utility function of the *firm*, as a whole (i.e., the decisions one makes as an individual can— and usually will—differ significantly from those made as part of a team, or as a representative of a firm).
- The utility function derived for an individual, from a set of hypothetical cases and questions, is unlikely to be the same utility function that the individual uses in actual decision making (further, it is suspected that an individual's utility function changes over time).

- The resultant single objective is an aggregation of all those objectives (with each expressed in terms of a variety of measures of effectiveness) into a single function that serves to hide, or distort, the original information.

Further, we should note that methods of aggregation are most generally extremely sensitive to the weights ultimately derived. For example, studies have shown that even a slight variation of the weights selected (e.g., a change of a few percent or less in value) may result in a completely different answer (i.e., a different extreme point of the convex set).

Finally, the reader may note that of the three LP models so far developed (i.e., in this and the previous section), each one has produced an entirely different solution.

Conversion to a Goal Program (GP)

When one employs utility theory, the bulk of one's efforts is typically dedicated toward obtaining an adequate and rational *representation of the decision maker's* (theoretical) *preference function*. However, when one uses goal programming, the effort shifts toward that of obtaining a better *representation of the actual problem*, through the development of the goal-programming model. Whichever approach is deemed "best" is strictly a function of one's *personal* perspective.

There are actually a number of types of goal programs, each espousing a somewhat different philosophy (i.e., with respect to how to measure the "goodness" of a solution to a problem involving multiple, conflicting goals). Three of the most popular (as well as the most practical) forms of GP are Archimedean GP (i.e., weighted GP), non-Archimedean GP (i.e., lexicographic GP, or preemptive GP), and Chebyshev GP (or minimax GP, or fuzzy GP) (Charnes and Cooper, 1961; Ignizio, 1983, 1985a). In this section, we shall focus on the first two approaches.

To form a goal-programming model, the very first thing that must be done is to convert *all* objectives into goals. Actually, we have already demonstrated just how this might be accomplished when we converted our illustrative example into an LP by means of changing the second objective into a "constraint." Specifically, we adhere to the following guidelines:

- A maximizing objective is converted into a type II ($\geq$) inequality, by means of the establishment and inclusion of a right-hand side, or aspiration-level value.
- A minimizing objective is converted into a type I ($\leq$) inequality, by means of the establishment and inclusion of a right-hand side, or aspiration-level value.

Thus, to convert the illustrative example of (13.1–13.6) into a goal program, we must first establish aspiration levels for each objective. These aspiration levels are values that we might hope to achieve in the final solution—or that represent "acceptable" levels for profit and market shares. For example, if our present (nonoptimized) production policy is returning 40 units of profit and 50 units of market share, we might realistically hope to increase each of these amounts by 10%—or perhaps even 20%. However, one should try to avoid unrealistically high values in the case of maximization, or unrealistically low values for minimizing objectives.*

The use of aspiration levels to transform objectives (which are to be optimized) into goals (which are to be achieved) is known as the concept of "satisficing." Satisficing, in turn, is a pragmatic approach based upon the manner in which most organizations, and most individuals, approach real-world decision making (Simon, 1957; March and Simon, 1958). That is, rather than attempting to achieve solution optimality (which is actually only meaningful for static, deterministic, error-free, single-objective problems), we hope to find a solution that comes "as close as possible" to satisfying our goals.

For sake of discussion, let us assume that we establish an aspiration level of 50 units for profit and 80 units for market shares. Consequently, our two objectives are changed into the following goals:

$$3x_1 + x_2 \geq 50 \qquad \text{(profit goal)}$$

$$2x_1 + 3x_2 \geq 80 \qquad \text{(market-shares goal)}$$

And these two functions may be combined with the other functions (i.e., the original constraint set) to form a model consisting solely of goals.

Find x_1 and x_2 so as to satisfy

$$3x_1 + x_2 \geq 50 \qquad \text{(profit goal)}$$

$$2x_1 + 3x_2 \geq 80 \qquad \text{(market-shares goal)}$$

$$2x_1 + x_2 \leq 50 \qquad \text{(raw material limitations)}$$

$$x_1 \leq 20 \qquad \text{(market-level goal, product 1)}$$

$$x_2 \leq 30 \qquad \text{(market-level goal, product 2)}$$

However, we are not yet finished. Specifically, in a typical model, some of the goals will be hard (i.e., they absolutely *must* be satisfied) and some will be soft (i.e., some deviation is tolerable). Thus, we need a means to indicate the deviations from the right-hand sides of each goal—*whether hard or soft*. To accom-

---

* It is important that one not become too "hung up" on the precise values of the aspiration levels employed. Although it would be best if they were realistic (i.e., not too high or too low), we may always investigate increases or decreases in such values through postoptimality analysis. One way to determine aspiration values is to solve the problem using just one objective at a time. In this manner, you should be able to determine the best possible levels (i.e., the "utopian" levels) of each objective— and then estimate your aspiration levels using this knowledge. However, in most cases, a reasonable, subjective estimate will provide an appropriate starting point in the analysis.

plish this, we shall add negative deviations (η's) *and* subtract positive deviations (ρ's) from the left-hand sides of each goal (and constraint). This results in the following model.

Find x_1 and x_2 so as to minimize all unwanted deviations, wherein

$$2x_1 + x_2 + \underline{\eta_1} - \underline{\rho_1} = 50 \qquad \text{(raw material limitations)} \qquad (13.18)$$

$$x_1 + \underline{\eta_2} - \underline{\rho_2} = 20 \qquad \text{(market-level goal, product 1)} \qquad (13.19)$$

$$x_2 + \underline{\eta_3} - \underline{\rho_3} = 30 \qquad \text{(market-level goal, product 2)} \qquad (13.20)$$

$$3x_1 + x_2 + \underline{\underline{\eta_4}} - \rho_4 = 50 \qquad \text{(profit goal)} \qquad (13.21)$$

$$2x_1 + 3x_2 + \underline{\underline{\eta_5}} - \rho_5 = 80 \qquad \text{(market-shares goal)} \qquad (13.22)$$

$$\text{and all } x_j, \eta_i, \text{ and } \rho_i \text{ are nonnegative*} \qquad (13.23)$$

Notice carefully that all *unwanted* deviations have double underlines. This serves to provide a quick visual clue as to just which deviation variables should be minimized—and it is a modeling practice that is strongly recommended for the novice.

Now, although the model of (13.18–13.23) is expressed solely in terms of goals (where some are hard and some are soft), we next need a function by means of which the achievement of the minimization of the *unwanted* goal deviations may be measured. This function, in fact, is termed the goal-programming *achievement function.*** Further, we need a philosophy upon which to develop such a function. We shall now address two approaches to this problem, the first of which is denoted as Archimedean GP.

Archimedean goal programming. In Archimedean, or weighted, GP, we shall form an achievement function consisting of precisely two terms. The first term represents the sum of all *unwanted* deviations for those goals that are hard (i.e., the rigid constraints). The second is composed of the weighted sum of all *unwanted* deviations for those goals that are soft. Thus, the achievement function for the general Archimedean GP model is given as

$$\text{lexmin } \mathbf{u} = \{(\boldsymbol{\mu}^{(1)}\boldsymbol{\eta}^{(1)} + \boldsymbol{\omega}^{(1)}\boldsymbol{\rho}^{(1)}), (\boldsymbol{\mu}^{(2)}\boldsymbol{\eta}^{(2)} + \boldsymbol{\omega}^{(2)}\boldsymbol{\rho}^{(2)})\} \qquad (13.24)$$

where

$$\text{lexmin} = \textit{lexicographic} \text{ minimum (of an ordered vector)}$$

$$\mathbf{u} = \text{achievement vector (or, achievement function)}$$

$$\boldsymbol{\eta}^{(k)} = \text{vector of negative deviations, at priority level } k$$

---

* Even further, it should be noted that $\eta_i\rho_i = 0$ for all *i*. However, this requirement will automatically be taken care of when employing the associated algorithm.

** Actually, as we shall see, the function shall be used to measure the *non*achievement of the goals. For this reason, we shall employ the letter "**u**" to represent the "*un*achievement" function.

$\boldsymbol{\rho}^{(k)}$ = vector of positive deviations, at priority level k

$\boldsymbol{\mu}^{(k)}$ = vector of weights for all negative deviations at priority k

$\boldsymbol{\omega}^{(k)}$ = vector of weights for all positive deviations at priority k

The lexicographic minimum may be defined as follows. Given an ordered array, say, $\mathbf{u}$, of nonnegative elements, the solution given by $\mathbf{u}^{(r)}$ is preferred to $\mathbf{u}^{(s)}$ if $u_k^{(r)} < u_k^{(s)}$ and all higher-ordered terms (i.e., $u_1, u_2, \ldots, u_{k-1}$) are equal. If no other solution is preferred to $\mathbf{u}$, then $\mathbf{u}$ is the lexicographic minimum. Thus, for example, if we have two arrays: (0, 17, 500, 477) and (0, 18, 2, 7), then the first is (lexicographically) preferred to the second.

Further, it should be noted that

- all weights are nonnegative
- the only nonzero-valued weights are those associated with *unwanted* deviations (i.e., those that are to be minimized)
- the weights for those unwanted deviations at priority level one are (normally) all set to unity (i.e., because no deviations whatsoever are permitted for rigid goals, there is usually no need to differentiate their "importance")
- the weights for those unwanted deviations at the second priority level are, for the most part, established by means of a purely subjective process (and any increases or decreases in these weights may be examined during postoptimality analysis)

Simply for purpose of discussion, let us assume that the firm in question feels that it is twice as important to achieve the market shares goal than it is to achieve the profit goal (i.e., it is more interested in long-term survival than in immediate profit). Consequently, the resultant Archimedean LGP model for the illustrative example becomes the following.

Find x_1 and x_2 so as to

$$\text{lexmin } \mathbf{u} = \{(\rho_1 + \rho_2 + \rho_3), (\eta_4 + 2\eta_5)\} \qquad (13.25)$$

satisfy:

$$2x_1 + x_2 + \eta_1 - \rho_1 = 50 \qquad \text{(raw material limitations goal)} \qquad (13.26)$$

$$x_1 + \eta_2 - \rho_2 = 20 \qquad \text{(market-level goal, product 1)} \qquad (13.27)$$

$$x_2 + \eta_3 - \rho_3 = 30 \qquad \text{(market-level goal, product 2)} \qquad (13.28)$$

$$3x_1 + x_2 + \eta_4 - \rho_4 = 50 \qquad \text{(profit goal)} \qquad (13.29)$$

$$2x_1 + 3x_2 + \eta_5 - \rho_5 = 80 \qquad \text{(market-shares goal)} \qquad (13.30)$$

$$\text{and all } x_j, \eta_i, \text{ and } \rho_i \text{ are nonnegative} \qquad (13.31)$$

When this model is solved, the resultant solution is found not necessarily at a point (as is typical in the case of LP), but within a *region*. Specifically, the solution to the Archimedean LGP is the entire cross-hatched *region* shown in Figure 13.3. *That is, any point within this region, or on its boundaries, serves to satisfy all soft and hard goals.* This region is defined by the extreme points denoted as A, E, F, and G.

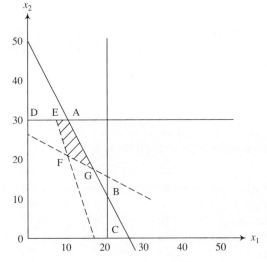

Figure 13.3 Goal-programming solutions.

The advantages of Archimedean LGP include the following.

- Each goal (and objective) is individually represented in the model (i.e., aggregation is avoided) and thus one deals with an array of performance measures rather than a single, proxy, measure.
- We may deal directly with both hard and soft goals.
- The decision maker is forced to estimate aspiration levels for his or her objectives, and this serves to force one to give additional consideration to the model developed (this could also be considered by some to be a disadvantage).*
- The approach may be extended to encompass a variety of other very important, and very practical decision models, including curve fitting and prediction, pattern recognition or classification, and cluster analysis (recall the material covered in Chapter 8).

*Some years ago, one of the authors received a letter in which the writer commented on the use of goal programming. Specifically, he cited his primary complaint about GP. He stated: "When I use goal programming, I have to think about the problem. However, when I use LP, I don't need to think. Thus, I believe that I will continue to use LP." Unfortunately, he completely failed to realize that he had identified one of the main benefits of the GP approach.

- Solution to the Archimedean LGP model, as well as a complete sensitivity analysis, is possible by means of any conventional LP algorithm or software.

And the disadvantages include the following.

- More time, and thought, is required in the construction of the model (as mentioned, this could—and, quite frankly, should—be considered an advantage).
- More decision-maker involvement is required (i.e., in the establishment of aspiration levels and weightings—although, again, this could also be considered an advantage).
- The subjectivity regarding the weights given to priority level two goal deviations may be of concern.
- It may be difficult, if not impossible, to develop the weights necessary to add the goal deviations at priority level two (a problem addressed in non-Archimedean GP, as discussed in what follows).

Non-Archimedean goal programming. In non-Archimedean GP (also called lexicographic or preemptive GP*), we also form an achievement function. However, the number of terms in this achievement function will always be *three or more*. As before, the first term represents the sum of all unwanted deviations for those goals that are hard (i.e., the rigid constraints). The second is composed of the weighted sum of all unwanted deviations for those goals at priority level two. The third is composed of the weighted sum of all unwanted deviations at priority level three, and so on. The general form of the achievement function for a non-Archimedean GP is given as

$$\text{lexmin } \mathbf{u} = \{(\boldsymbol{\mu}^{(1)}\boldsymbol{\eta}^{(1)} + \boldsymbol{\omega}^{(1)}\boldsymbol{\rho}^{(1)}), \ldots, (\boldsymbol{\mu}^{(K)}\boldsymbol{\eta}^{(K)} + \boldsymbol{\omega}^{(K)}\boldsymbol{\rho}^{(K)})\} \qquad (13.32)$$

wherein the total number of priority levels is K (i.e., $k = 1, 2, \ldots, K$). As in the case of Archimedean GP, the nonzero weights at priority level one are (normally) set to unity.

For sake of discussion, let us assume that our firm's decision makers agree that market shares are more important than profit, but that they are unable to agree on just how much more. Thus, rather than weighting the associated goal deviations, they are only able to rank order these deviations. In such a case, we *might* resort to non-Archimedean GP and form the following model.

Find x_1 and x_2 so as to

$$\text{lexmin } \mathbf{u} = \{(\rho_1 + \rho_2 + \rho_3), (\eta_5), (\eta_4)\} \qquad (13.33)$$

---

* More properly, one should call this form of GP as *non*preemptive GP. However, the term preemptive GP is so widely employed, and has become such a tradition, that we, too, will use it to describe the non-Archimedean approach.

and satisfy:

$$2x_1 + x_2 + \eta_1 - \underline{\rho_1} = 50 \qquad \text{(raw material limitations goal)} \qquad (13.34)$$

$$x_1 + \eta_2 - \underline{\rho_2} = 20 \qquad \text{(market-level goal, product 1)} \qquad (13.35)$$

$$x_2 + \eta_3 - \underline{\rho_3} = 30 \qquad \text{(market-level goal, product 2)} \qquad (13.36)$$

$$3x_1 + x_2 + \underline{\eta_4} - \rho_4 = 50 \qquad \text{(profit goal)} \qquad (13.37)$$

$$2x_1 + 3x_2 + \underline{\eta_5} - \rho_5 = 80 \qquad \text{(market-shares goal)} \qquad (13.38)$$

and all x_j, η_i, and ρ_i are nonnegative $\qquad (13.39)$

When this model is solved, we will arrive at precisely the same solution space (i.e., the cross-hatched area of Figure 13.3) as we developed for the Archimedean LGP model.* Thus, in this particular instance, the additional priority level made no difference. In general, this will not be the case.**

When first faced with a problem involving the lexicographic minimum (or preemptive priorities), some individuals exhibit a certain degree of discomfort. The notion seems alien—and certainly is when compared to the manner in which traditional methods *appear* to approach mathematical programming problems. However, it should be noted that the lexicographic notion is, actually, an implicit part of traditional methods—as well as a common approach to real-world decision making.

To illustrate, consider first a traditional, single-objective LP model. Although we may like to think of such a model as having but a single priority, this is not the case. Single-objective problems actually have two (preemptive) priorities. First, one must satisfy the set of rigid constraints. Then, *and only then*, one may address the optimization of the single objective.

Further, as mentioned, the lexicographic notion has been demonstrated to be a typical method of dealing with real-world, multiobjective (or multicriteria) decisions. As just one example, consider the shopper in a grocery store—and let us assume that he or she wishes to purchase a can of green beans. A typical "solution process" is to first ignore all cans with unfamiliar, or (supposedly) inferior brands. Thus, our *first priority* is, implicitly, to restrict our selection to just a subset of brands of green beans. Next, the shopper may evaluate the purchase based upon sodium content. That is, the less salt, the better. Thus, our *second priority* is to minimize salt content. This may result in just a few brands from which to choose. Finally, the can is selected, from the remaining set of brands, that is the cheapest. That is, our *third priority* is cost. And this is hardly an unusual example. Studies have shown that the lexicographic process is used in

---

* The solution to either an Archimedean or non-Archimedean LGP will either exist as a region or as a point. Further, as soon as one has converged (in the associated algorithm for Archimedean or non-Archimedean LGP) to a point, the solution process may be terminated.

** For example, had we used the achievement vector $\{(\rho_1 + \rho_2 + \rho_3), (\eta_4 + \eta_5), (\eta_3)\}$, the resultant solution would be the line segment connecting points A and E.

a host of real-world decision making, including the purchase of homes and automobiles, the hiring of employees, the choice of schools, and so forth. In essence, the approach serves as a means through which one can sequentially filter out alternatives, until one is left with a reasonable number from which to choose.

The advantages of non-Archimedean LGP include the following.

- Each goal (and objective) is individually represented in the model (i.e., aggregation is avoided) and thus one always deals with an array of performance measures rather than a single, proxy, measure.
- We may deal directly with both hard and soft goals.
- The decision maker is forced to estimate aspiration levels for his or her objectives, and this serves to force one to give additional consideration to the model developed (this could also be considered, by some, to be a disadvantage).
- Problems involved with determining numerical weightings for goal deviations (as in the case of Archimedean GP) may possibly be avoided by, instead, ranking these deviations.
- The approach may be extended to encompass a variety of other very important, and very practical, decision models, including curve fitting and prediction, pattern recognition or classification, and cluster analysis (again, recall the material in Chapter 8).
- Solution to the non-Archimedean LGP model, as well as a complete sensitivity analysis, is possible by means of straightforward extensions to any conventional LP algorithm or software.

And the disadvantages include the following.

- More time, and thought, is required in the construction of the model (this could—and indeed should—be considered an advantage).
- More decision-maker involvement is required (i.e., in the establishment of aspiration levels, weightings, and rankings—although, again, this could also be considered an advantage).
- The subjectivity regarding the weights given to the deviation variables *within* a lower priority (i.e., within levels 2 through K) may be of concern.
- The ranking of goals, coupled with the notion of the lexicographic minimum, means that any goals at priority level k are *preemptively* preferred to those at priority level $k + 1$, or lower (i.e., no matter what numerical weighting is given to the goals at level $k + 1$, those at level k are considered more important).

Finally, while not demonstrated in our example, it should be appreciated that any number of goals (i.e., as represented by their associated deviation variables) may appear in any priority level of the non-Archimedean model. However, except for priority level one (which is reserved for rigid constraints), goals combined within a given priority level must be commensurable (i.e., it must be possible to find a weighting so that all the deviation variables within a given level are additive).

Conversion to a Chebyshev Goal Program

We may think of Archimedean GP as a method whose underlying philosophy is to find a solution that *minimizes the weighted sum* of all *unwanted* goal deviations. Non-Archimedean GP, on the other hand, involves a philosophy that seeks the solution that *minimizes*, in a strictly lexicographic sense, *an ordered vector* of all *unwanted* goal deviations. Chebyshev GP involves yet another notion of the "best" solution to a multiobjective problem. Specifically, the underlying philosophy of Chebyshev GP is to find that solution that serves to minimize the *single* worst *unwanted* deviation from any (soft) goal. This particular notion also provides the basis of what is called minimax GP and fuzzy programming, or fuzzy GP. Here, we shall simply address the most basic form of the Chebyshev GP model—and then note how it relates to the notion of fuzzy programming. To accomplish this, we shall again utilize our numerical example.

As with any GP approach, the first step is to convert the problem into one containing nothing but goals. However, instead of employing a subjective approach to establish the aspiration levels for the objectives, let us first find the "best" and "worst" values of these objectives. The example has been rewritten in what follows, where, for purpose of discussion, we will assume that the last four goals (13.42–13.45) are hard (i.e., rigid constraints).

Find x_1 and x_2 so as to

$$\text{maximize } z_1 = 3x_1 + x_2 \qquad \text{(profit objective)} \qquad (13.40)$$

$$\text{maximize } z_2 = 2x_1 + 3x_2 \qquad \text{(market-shares objective)} \qquad (13.41)$$

subject to:

$$2x_1 + x_2 \leq 50 \qquad \text{(raw material limitations goal)} \qquad (13.42)$$

$$x_1 \leq 20 \qquad \text{(market-level goal, product 1)} \qquad (13.43)$$

$$x_2 \leq 30 \qquad \text{(market-level goal, product 2)} \qquad (13.44)$$

$$\text{and all } x_j \text{ are nonnegative} \qquad (13.45)$$

Next, we shall solve this model as a conventional LP, *using but one objective at a time*. Once we have solved such a problem, we have determined the best possible value of the objective being considered. We may also substitute the resulting program into any other objectives (where there so happens to be only one other objective in this case) so as to determine their values when this particular objective is optimized. Thus, in our example, we shall solve one LP using objective (13.40) and constraints (13.42–13.45) and another using objective (13.41) and constraints (13.42–13.45). The results are summarized in Table 13.1. Notice that when the model is solved using just the first objective (z_1), the resulting program is $\mathbf{x} = (20, 10)$ and the associated value for z_1 is 70 and z_2 (for this program) is also 70. (Thus, this solution occurs at point B of Figure 13.1.) When the model is solved using just the second objective (z_2), the resulting program is $\mathbf{x} = (10, 30)$ and the associated values of z_1 and z_2 are 60 and 110, respectively. (And this coincides with point A of Figure 13.1.)

TABLE 13.1 SUMMARY OF RESULTS

	Optimization with respect to z_1	Optimization with respect to z_2
Value of z_1	**70**	60
Value of z_2	70	**110**

We have now determined that the value of the first objective can never be larger than 70 units and that of the second can never be greater than 110. (We have also determined, in fact, the "worst" values of the objectives.*) Thus, let us use these values as our "aspiration levels."

We may now write the associated Chebyshev GP model as

$$\text{minimize } \delta \tag{13.46}$$

subject to:

$$2x_1 + x_2 \leq 50 \qquad \text{(raw material limitations)} \tag{13.47}$$

$$x_1 \leq 20 \qquad \text{(market-level goal, product 1)} \tag{13.48}$$

$$x_2 \leq 30 \qquad \text{(market-level goal, product 2)} \tag{13.49}$$

$$3x_1 + x_2 + \delta \geq 70 \qquad \text{(profit goal)} \tag{13.50}$$

$$2x_1 + 3x_2 + \delta \geq 110 \qquad \text{(market-shares goal)} \tag{13.51}$$

$$\text{and } \delta \text{ and all } x_j \text{ are nonnegative} \tag{13.52}$$

And notice that if we find the program wherein δ is minimized, we have minimized the worst deviation from any single-goal aspiration level.

The solution to the Chebyshev GP model just given is found at the point designated *Ch* in Figure 13.4. Notice that the final solution to a Chebyshev model will be on the boundary of the original "constraint" set, but does not necessarily

---

* Strictly speaking, the worst values for both objectives are zero, and occur at the origin (i.e., $\mathbf{x} = 0, 0$). Here, we simply mean that this is the worst value of those determined through the solution of the various LPs.

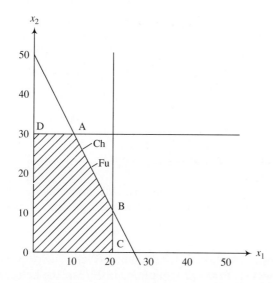

Figure 13.4 Chebyshev and fuzzy programming solutions.

have to be an extreme point for the original problem. This point, in fact, is at $\mathbf{x} = (12, 26)$, with a δ value of 8.

As with any other method, the Chebyshev GP approach has its own set of advantages and disadvantages. The advantages of the approach include the following.

- Each goal (and objective) is individually represented in the model (i.e., aggregation is avoided).
- The decision maker is not required to estimate aspiration levels for his or her objectives, as this is achieved by means of solving a series of LP models.
- No weighting nor ranking procedure need be employed.
- Only a single new variable (δ) is required.
- Solution to the Chebyshev LGP model is possible by means of any conventional LP algorithm or software.

And the disadvantages include the following.

- One must solve as many LP models as there are objective functions (although this is rarely a significant problem).
- Just one goal can serve to dictate the final solution.

To better appreciate the last comment, consider a hypothetical problem involving five objectives (which, in the Chebyshev GP model, have been transformed into goals). Further, consider the two solutions to this problem shown in Table 13.2. Notice carefully that the second solution serves to minimize the worst deviation (99.5 units, compared to 100.0 for the first solution), but yet it provides a

TABLE 13.2

Solution one		Solution two	
Objective	Value of δ	Objective	Value of δ
z_1	0.0	z_1	99.5
z_2	0.1	z_2	89.9
z_3	0.2	z_3	98.9
z_4	0.1	z_4	99.2
z_5	100.0	z_5	87.8

result that is relatively poor for *all* objectives. On the other hand, solution one does exceptionally well for all but the fifth objective. Despite this, it is indeed the solution with the single worst goal deviation.

Fuzzy programming.   As noted, Chebyshev GP and fuzzy programming are (closely) related. However, fuzzy programming is, by far, the better known of the two concepts and has, in fact, established a wide following in the multiobjective optimization and MCDM (multicriteria decision making) communities— wherein numerous real-world problems have been approached and successfully solved by the methodology. Oddly enough, however, many fuzzy programming advocates seem unaware of the fact that the method may be regarded as simply another form of goal programming. As such, let us very briefly describe fuzzy programming—and then show that it is nothing more than a *very slight* modification of Chebyshev GP. Using Zimmermann's (1978, 1985) approach to fuzzy programming, and assuming that all objectives are of the maximizing type, we may represent the general fuzzy linear programming model as

$$\text{minimize } \delta \tag{13.53}$$

subject to:

$$\delta \geq (U_k - z_k)/d_k, \quad \text{for all } k \text{ objectives} \tag{13.54}$$

$$\mathbf{Ax} \ (\leq, \geq, \text{ or } =)\mathbf{b}, \quad \text{for all } m \text{ constraints} \tag{13.55}$$

$$\text{and } \delta \text{ and all } x_j \text{ are nonnegative} \tag{13.56}$$

where

δ = dummy variable representing the worst deviation level

z_k = function representing the kth objective

U_k = maximum value that z_k can take on (by means of solving all LPs)

L_k = minimum value that z_k can take on (by means of solving all LPs)

$d_k = U_k - L_k$

and where the right-hand side of (13.54) is termed the *fuzzy membership function*.

The purpose of the fuzzy programming approach is to find the solution that serves to minimize the largest fuzzy membership function. However, the observant reader may have already noted that the fuzzy programming model is identical to the Chebyshev model *except for the weight given to* δ. That is, we may rewrite (13.54) as simply

$$z_k + d_k \delta \geq U_k, \qquad \text{for all } k$$

Using fuzzy programming to represent our illustrative example, we may form the following model:

$$\text{minimize } \delta \qquad\qquad (13.57)$$

$$\text{subject to:}$$

$$\delta \geq [70 - (3x_1 + x_2)]/(70 - 60) \qquad \text{(from the profit goal)} \qquad (13.58)$$

$$\delta \geq [110 - (2x_1 + 3x_2)]/(110 - 70) \qquad \text{(from the market-shares goal)} \qquad (13.59)$$

$$2x_1 + x_2 \leq 50 \qquad \text{(raw material limitations)} \qquad (13.60)$$

$$x_1 \leq 20 \qquad \text{(market-level goal, product 1)} \qquad (13.61)$$

$$x_2 \leq 30 \qquad \text{(market-level goal, product 2)} \qquad (13.62)$$

$$\text{and } \delta \text{ and all } x_j \text{ are nonnegative} \qquad (13.63)$$

And, once again, this is identical to the Chebyshev GP model originally developed except for the weights given to δ. That is, we may rewrite (13.58) and (13.59) as

$$(3x_1 + x_2) + (70 - 60)\delta \geq 70$$

$$(2x_1 + 3x_2) + (110 - 70)\delta \geq 110$$

Comparing these two functions to those of (13.50) and (13.51) respectively, we note the sole difference is the weight given to δ in each function (i.e., a value of 10 in the first and 40 in the second).

Should we solve this particular model, the solution will be found at $\mathbf{x} = (15, 20)$, with a δ value of 5. This solution is depicted as point Fu in Figure 13.4. Again, this solution is on the boundary of the original constraint set but is not an extreme point of that set.

The Generating Method: The (Not So) Perfect Approach

As you may have noted, every approach described thus far has required the employment of various assumptions—some of which may have seemed a bit difficult to satisfy. In particular, in certain cases, we have had to assume that we could (1) find weights so as to combine objectives, or (2) find aspiration levels so as to transform objectives into goals, or (3) find weights to combine goal deviations, or (4) determine the ranking of the various goals.

Those who would advocate the use of the generating method (also known as the vectormin, or vectormax approach) oftentimes claim that this particular method is "assumption-free." More specifically (and more correctly), the generating method avoids the use of either weights or ranks. It does so by generating, at least in theory, all of the "nondominated" (or "efficient") solutions to a given model—and then leaving it up to the decision maker to choose from among these results (Gal, 1977; Hwang and Masud, 1979).

Before exploring this approach further, let us explain the notion of nondominated solutions. If we have a problem with multiple objectives, and if these are all maximizing objectives, then a nondominated solution is one for which there is no other solution having equal or greater values of each and every objective (in the case of minimizing objectives, it would simply be a solution for which there is no other solution having equal or lesser values of each and every objective). For example, consider a problem with three maximizing objectives (z_1, z_2, and z_3) and the four solutions listed in Table 13.3. Notice carefully that solution 1 is dominated by solution 4. All remaining solutions are, at least for the solution set given, nondominated.

TABLE 13.3

	Solution 1	Solution 2	Solution 3	Solution 4
z_1	100	150	90	145
z_2	20	60	55	50
z_3	110	90	95	110

Consider for a moment the features of a dominated solution. A dominated solution is one for which there is some other solution that provides results that are just as good, or better, for each and every objective. The advocates of the generating method would state that no *rational* decision maker would ever (knowingly) select a dominated solution. And, on the face of it, this would seem to be a reasonable conclusion. However, some advocates of the generating method (or of other methods that guarantee nondominated solutions) have gone even further. Specifically, they claim that neither Archimedean or non-Archimedean GP should be even considered for multiobjective optimization because the use of such methods cannot guarantee a nondominated solution. And it is indeed true that these two approaches indeed can lead to solutions that are dominated (e.g., any point strictly *within* the cross-hatched region of Figure 13.3 is, in fact, dominated).

Although such an argument may, on the surface, appear to have some validity, it is founded on assumptions that are erroneous and that ignore real-world practicality. First of all, GP is based upon the philosophy of *satisficing*, and *not on that of optimization*. Consequently, it is unreasonable to criticize a method for failing to accomplish something for which it was never intended. Second, in even the smallest problem, the number of nondominated solutions may be infinite. This is because all points on the line between two nondominated extreme points are themselves nondominated. As such, it is impossible to list all of the nondominated solutions to any but the most trivial problem.

To counter this difficulty, those who employ the generating method usually restrict the nondominated solutions developed to those that are extreme points. This guarantees that the number of nondominated solutions will be finite. Of course, and as should be obvious, this is still not of much help if one faces a typical, real-world problem. Specifically, the total number of extreme points for a problem with n variables and m constraints can approach $\{[(n + m)!]/[m!n!]\}$, where n is the number of variables and m the number of rigid constraints. Thus, for just a very small problem, say, one with just 20 variables and 10 constraints, the total number of extreme points can approach 184,756. Although not all extreme points will be nondominated (or even feasible), this relationship should still indicate the utter futility of attempting to derive all, or even a significant portion, of the nondominated extreme points for any problem of realistic size.

As a result of the immense number of nondominated solutions in actual-size problems, the only successful generating methods have been those that generate but a small portion of the nondominated solutions. Consequently, although such methods do indeed avoid weighting, ranking, and dominated solutions, they can hardly be considered anything other than heuristic approaches—as are *all* the methods for multiobjective optimization.

Let us now address the matter of just how to solve our example problem by means of the generating approach—in its most simple, and most idealistic form. The model to be employed is as follows.

Find x_1 and x_2 so as to

$$\text{maximize } z_1 = 3x_1 + x_2 \qquad \text{(profit objective)} \qquad (13.64)$$

$$\text{maximize } z_2 = 2x_1 + 3x_2 \qquad \text{(market-shares objective)} \qquad (13.65)$$

subject to:

$$2x_1 + x_2 \leq 50 \qquad \text{(raw material limitations)} \qquad (13.66)$$

$$x_1 \leq 20 \qquad \text{(market-level goal, product 1)} \qquad (13.67)$$

$$x_2 \leq 30 \qquad \text{(market-level goal, product 2)} \qquad (13.68)$$

$$\text{and all } x_j \text{ are nonnegative} \qquad (13.69)$$

To solve, we first need to develop an initial starting solution—for example, the solution of this model subject to just one of the objectives (i.e., ignore the other objective and solve the resultant LP). Next, using postoptimality analysis, we move to any adjacent extreme points that are nondominated. That is, we conduct a series of simplex pivots, moving from one basis to another *if such a move will lead to a nondominated extreme point*. In theory, we then continue this procedure until we have generated all the nondominated extreme points.

If we were to conduct such a search, the resultant solutions would be found at points A and B of Figure 13.5 (and, in fact, for the entire line segment connecting points A and B). Our next step is to provide these nondominated solutions to the decision maker and ask him or her to select that one that is considered most preferable.

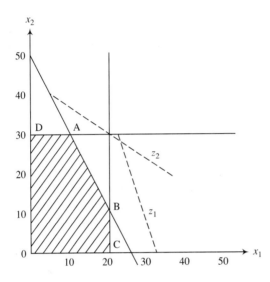

Figure 13.5   Generating method solutions: points A and B.

In practice, the solution process used in generating methods is somewhat more (to considerably more) complicated than that implied before. Further, as mentioned earlier, all *practical* generating methods limit their search to but a subset of nondominated solutions. The reader is thus advised to examine the references (see, e.g., Cohon, 1978; Hwang and Masud, 1979) for specific details.

In summary, the main advantage of generating method approaches is that one avoids the use of weights and ranks. The disadvantage is that, for other than trivial problems, one must resort to fairly complex and time-consuming means for generating but a small subset of nondominated extreme points. As such, one cannot guarantee that the solution ultimately selected by the decision maker is the most preferred of all the possible nondominated solutions. Even further, the entire foundation of the argument for the generation of nondominated solutions is based, implicitly, upon error-free, deterministic, and linear mathematical models—models rarely if ever encountered in actual practice.

To clarify the last statement, consider, for example, the model provided in (13.64)–(13.69). Now, if there is *any* error with regard to the coefficients of that model—and there will be in any real-life problem—then it is quite possible that the solutions generated will actually be *dominated*. For example, if the constraint functions of (13.66)–(13.68) are actually

$$2x_1 + x_2 \leq 55 \qquad \text{(raw material limitations)}$$

$$x_1 \leq 22 \qquad \text{(market-level goal, product 1)}$$

$$x_2 \leq 33 \qquad \text{(market-level goal, product 2)}$$

then the nondominated solutions of the original model are *dominated* for the correct model. This may be seen in Figure 13.6, where we may note that points A and B are now in the *interior* of the convex solution space formed by the "correct" values of the constraint set.

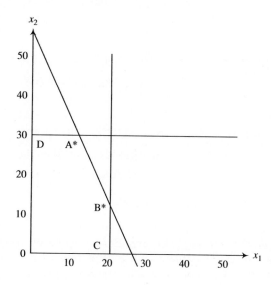

Figure 13.6 "Dominated" solutions in the presence of measurement error.

The Lexicographic Vectormax (Vectormin) Approach

The lexicographic vectormax (or, lexicographic vectormin, depending on the form of the objectives employed) is yet another concept that has been proposed for the solution of multiobjective models. It is also a method that has, unfortunately, been confused with that of goal programming (specifically, non-Archimedean GP) even though the method is, most definitely, *not* a goal-programming approach.

The thrust of the lexicographic vectormax/min approach is to first construct a multiobjective model consisting of but objectives and rigid constraints. Next, all the objectives are converted to the same form (i.e., either all maximizing or all minimizing), and, finally, every objective is ranked according to preference. In particular, a preemptive ranking procedure is used. The result is the lexicographic vectormax, or lexicographic vectormin, model.

To solve such a model, we first employ only the top-ranked objective and solve the equivalent LP. We then move to the next highest-ranked objective and use it, and the original constraint set, to form a new LP. If a solution may be found to this new LP, and this solution does not degrade the values previously found for higher-ranked objectives, we continue the process. Otherwise (i.e., if the solution found to any subsequent LP would degrade the solution previously found for all higher-ranked objectives), we terminate the solution process.

It should be noted that the solution procedure can only continue if there are alternate optimal solutions for the problem being addressed. As such, the lexicographic vectormax approach often amounts to nothing more than solving the model formed by using the top-ranked objective—and then stopping at that point.

To demonstrate, let us use the following model and assume—for the sake of discussion—that the profit objective is the highest-ranked objective.

Find x_1 and x_2 so as to

$$\text{maximize } z_1 = 3x_1 + x_2 \qquad \text{(profit objective)} \qquad (13.70)$$

$$\text{maximize } z_2 = 2x_1 + 3x_2 \qquad \text{(market shares objective)} \qquad (13.71)$$

subject to:

$$2x_1 + x_2 \leq 50 \qquad \text{(raw material limitations)} \qquad (13.72)$$

$$x_1 \leq 20 \qquad \text{(market level, product 1)} \qquad (13.73)$$

$$x_2 \leq 30 \qquad \text{(market level, product 2)} \qquad (13.74)$$

$$\text{and all } x_j \text{ are nonnegative} \qquad (13.75)$$

We would thus solve the LP consisting of the highest-ranked objective (i.e., function 13.70) and the constraint set. This results in a solution at point B in Figure 13.1. We next use the second objective (i.e., function 13.71) and the constraint set to form the next LP. However, the solution to this LP would result in a program that would degrade the solution (i.e., the value of the top-ranked objective) previously obtained for the first LP. Thus, the resultant solution for this problem is simply point B of Figure 13.1. In essence, we have deleted the second objective from consideration.

As noted, the lexicographic vectormax approach should not be confused with that of goal programming (specifically, non-Archimedean, or lexicographic GP). First, a goal-programming model consists solely of goals—whereas the lexicographic vectormax approach contains both objectives and hard goals (rigid constraints). Second, the solutions derived for the (linear) lexicographic vectormax approach will always lie on the boundaries of the convex set formed by the constraints. In contrast, the solution set for a GP model can be a region, consisting of points in the interior of the convex space.

Frankly, while the lexicographic vectormax approach has a (small) following, there is little to recommend about the concept. Its sole advantage is that it avoids the use of weights or aspiration levels. Its main disadvantage is that, for most problems, it is equivalent to nothing more than picking a single-objective function and then solving the resultant LP model. And this is precisely what happened when we employed it to solve our example problem.

Interactive Methods

The final philosophy to be mentioned is that known as interactive programming. In essence, interactive methods employ the decision maker in the solution process. Of course, any rational (and successful) analyst deals with the decision maker throughout the modeling and solution process—*no matter what approach is used*. However, interactive methods are based upon *extensive* employment of the decision maker—particularly so throughout the solution process.* Continual feedback from the decision maker is, in fact, employed as a means to direct the search process.

Interactive methods take on a variety of forms and are described in the literature (Cohon, 1978; Hwang and Masud, 1979). Although such methods are founded upon certain admirable intentions, they often tend to be somewhat restricted in their practical use. Specifically, to be successful, one must be dealing with a very special (and, from our experience, very unusual) decision maker—one that is willing to commit a great deal of time and effort to the (technical aspects of the) solution process. At this point in time, the majority of these approaches are also limited to problems of relatively small sizes. As such, the reader who wishes to pursue this topic in more detail is directed to the references.

MYTHS AND MISCONCEPTIONS

Having introduced various forms of multiobjective models, let us pause for a moment to address some of the more prevalent, and more erroneous, myths and misconceptions that surround such concepts. Some of these notions are the result of the distrust, on the part of the traditionalist, of multiobjective methods in general. However, for the most part, they come about as a consequence of various erroneous and/or misleading comments made about one multiobjective approach by certain advocates of another.

The following are just a few of these myths and misconceptions.

- **Multiobjective methods have seen little if any real-world application.** This myth is likely the result of the fact that some of the better known journals (e.g., *Operations Research*, *Management Science*), at least in the United

---

* This differs from the manner in which other methods, in particular goal programming, interface with the decision maker. Specifically, when employing goal programming, the primary interfacing occurs during model development (wherein the input of the decision maker is deemed vital to the construction of the model) and after problem solution (where the decision maker's reaction to the solution, or alternative solutions, is sought).

States, rarely publish articles dealing with multiobjective approaches. One may thus be led to believe that such methods find little application. However, the exclusion of such articles in certain journals simply reflects the attitudes and perceptions of the editorial boards of these publications—which are not necessarily consistent with those of either the practitioner or decision maker. Other journals, and particularly those published in other countries (e.g., *European Journal of Operational Research, Journal of the Operational Research Society, International Journal of Computers and Operations Research*), do publish articles on multiobjective methods and their application—and should serve as the primary references for the reader seeking more information on the topic. Finally, multiobjective methods—in particular the various goal-programming techniques—have found wide application to a host of real-world problems.

- **Multiobjective methods cannot solve anything other than very small problems.** Oddly enough, this misconception is prevalent both within and outside the multiobjective optimization and MCDM communities. In fact, even in the most recent literature, there are those in the multiobjective and MCDM sector who deplore the "lack of computational capabilities" for virtually all of the related methods. As just one example, in a recent joint ORSA/TIMS (Operations Research Society of America/The Institute of Management Sciences) meeting, one speaker was heard to say that what is most needed to gain acceptance for goal programming is software that can solve other than toy problems—and this speaker is an individual whose education and professional experience has been centered about GP. However, the fact of the matter is that one can solve GP models of sizes virtually identical to those solved by commercial, single-objective software—and this has been the case since, at least, the late 1960s (Ignizio, 1967). This is because, quite simply, one can employ any commercial, single-objective software package (e.g., LP software) to solve GP models—either directly (i.e., in the case of Archimedean or Chebyshev LGP) or (in the case of non-Archimedean LGP) by means of obvious and simple refinements (which shall be further explored in Chapter 16). Further, such approaches can be extended (i.e., by means of the multiplex approach) to a variety of other forms of multiobjective models. Consequently, this is yet another myth that simply has no basis in fact.

- **A rational decision maker should not accept a dominated solution.** This notion, if accepted, implies that any multiobjective method that cannot guarantee a nondominated solution is of no use. It has been used, for the most part, as a criticism of goal programming (although, in actuality, it applies only to the most elementary forms of Archimedean and non-Archimedean GP—and has never been a valid criticism of either Chebyshev or fuzzy programming). However, the criticism could just as well be applied to *any* multiobjective approach as, unless the model being employed is error-free and deterministic, one cannot guarantee the generation of nondominated solutions (recall our earlier discussion with regard to generating methods). This argument is, quite frankly, simply a *red herring*. The fact is that decision makers, including "rational" decision makers, do indeed accept dominated solutions. Real-world decision makers, dealing with real-world problems, almost always seek either improved solutions or else solutions that come "close" to satisfying their aspirations. As Herbert Simon has noted [Simon, 1957; March and Simon, 1958], organizations, *of necessity*, "satisfice" rather than optimize. Even further, most real-world problems are large, complex, dynamic, and invariably include considerable errors in measurement. As such, the individual who will settle for nothing less than a nondominated solution is, we believe, hopelessly idealistic.

- **Ranking and/or weighting methods (of GP) are too subjective, and require too much time of the decision maker.** This argument has, on the surface, a certain basis in fact. Indeed, it does take time to rank and/or weight goals and the associated goal-deviation variables. However, the misconception lies in the notion of taking "too much" time. Although modeling approaches that avoid ranking and weighting certainly require less time, it may well be that this apparent savings is offset by a lack of model credibility. Further, although no one claims that such ranks and weights are perfect, it should also be realized that (1) variations in rankings and weightings can be explored as a part of postoptimality analysis, and (2) there are errors and subjectivity in *any* quantitative model used to represent *any* real-world problem. Finally, if methods that involve ranking and/or weighting require "too much" time of the decision maker, then why is it that these methods (i.e., GP methods and utility theory) are the ones with the longest and most extensive record of actual application?

THE MULTIPLEX APPROACH: A PREVIEW

As may be noted from the previous discussion, there is a wide variety of methods proposed for dealing with multiobjective problems—and we have not even touched upon all of those that have been proposed. Further, there is little agreement as to which is the "best approach." And that situation is unlikely to change in the foreseeable future. However, it is generally agreed—even by its critics—that goal programming is the "workhorse" of the multiobjective optimization methods. This is due to a number of factors, including those listed in what follows.

- The GP method, with its fundamental concept of "satisficing" (i.e., the employment of aspiration levels), is consistent with the approaches used in most real-world decision making.

- The method has a relatively long history (i.e., since the mid-1950s) and an extensive record of actual, successful, application (e.g., it was employed, in the early 1960s, to design the launch vehicle antennas for the Saturn/Apollo moon landing mission (Ignizio, 1963)).
- The modeling process avoids objective function aggregation.
- The modeling procedure is relatively straightforward, and often lends itself to increased understanding of, and insight into, the problem under consideration.
- The solution process either utilizes conventional single-objective algorithms directly (as in the case of Archimedean or Chebyshev GP), or involves relatively minor revisions to such methods (as in the case of non-Archimedean GP).

Yet another significant feature of goal programming lies in that a fair portion of alternative approaches to multiobjective optimization may be encompassed by a modeling and solution process that is, in essence, but a slight extension of that employed in non-Archimedean GP. This particular approach, denoted as *multiplex*, was developed by Ignizio in the 1970s and early 1980s (Ignizio, 1985b), and is the procedure (in, of course, an updated form) that will be addressed in the material to follow.

The use of the multiplex concept will provide us with a single, unified process for the modeling, solution, and analysis of many of the multiobjective approaches (including all of the GP concepts) as well as that of single-objective optimization. As such, we need learn only a *single* model and employ but a *single* algorithm. Further, we may employ the multiplex concept using minor extensions to virtually any single-objective algorithm or software program. This means that multiobjective models of sizes identical to that of single-objective models may be approached and solved—with equivalent computational efficiencies.

To accomplish this, it is necessary that a single model (the multiplex model) be used to represent the various optimization concepts. The general form of the multiplex model is

$$\text{lexmin } \mathbf{u} = \{\mathbf{c}^{(1)}\mathbf{v}, \ \mathbf{c}^{(2)}\mathbf{v}, \ \ldots, \ \mathbf{c}^{(K)}\mathbf{v}\} \tag{13.76}$$

subject to:

$$\mathbf{A}\mathbf{v} = \mathbf{b} \tag{13.77}$$

$$\mathbf{v} \geq \mathbf{0} \tag{13.78}$$

where

lexmin = lexicographic minimum (of an ordered vector)

    $\mathbf{u}$ = achievement vector (or achievement function)

    $\mathbf{c}^{(k)}$ = vector of coefficients, or weights, of $\mathbf{v}$ in the kth term of the achievement vector

$\mathbf{v}$ = vector composed of all structural *and* deviation variables

$\mathbf{A}$ = technological coefficients matrix of the goal set

$\mathbf{b}$ = right-hand side vector of the goal set

We shall, in the next chapter, show how a single-objective model, as well as many of the multiobjective models, may be (easily) transformed into the formulation indicated before.

OVERVIEW OF MATERIAL TO FOLLOW

In the next chapter, we shall demonstrate just how to employ the multiplex model, as listed before, so as to encompass conventional LP models, Archimedean LGP models, Chebyshev (and thus fuzzy) LGP models, non-Archimedean LGP models, and lexicographic vectormax (or lexicographic vectormin) models. Then, in Chapter 15, we shall address the topic of the multiplex algorithm—including both the multiphase and sequential approach—and computational considerations. In Chapter 16, we shall discuss postoptimality analysis using multiplex, as based upon the multidimensional dual (a dual formulation for the multiplex primal). Finally, in Chapter 17, the notions of integer and nonlinear multiobjective models shall be briefly discussed, and methods for their solution briefly outlined.

SUMMARY

Unlike single-objective optimization, there is no one correct philosophy or a single best approach to multiobjective optimization. In fact, some of the most successful multiobjective approaches do not actually even seek optimality per se. Instead, *acceptable* (i.e., satisficing) solutions are sought. Unfortunately, this wide variety of philosophies, concepts, models, algorithms, and terminology has served to cloud and confuse the many shared ideas of both single and multiobjective methods. In this chapter, we have described some of the more well-known approaches to multiobjective optimization. We shall now address a single, unified approach (i.e., multiplex) that will permit us to employ the bulk of such methods in actual problem solving.

EXERCISES

We shall use the term "baseline model" to describe a mathematical model that has been formulated *without regard to solution philosophy*. The purpose of a baseline model is to attempt to represent, as closely as possible, the real-world situation. One example of a baseline model is the formulation provided by relationships (13.1) through (13.6). In Exercises 13.1 through 13.6, simply develop the associated baseline model (i.e., do not yet force the model into any particular type—such as a conventional LP, goal program, and so on).

13.1. An investor has decided to invest a total of $50,000 among three different investment opportunities: savings certificates, municipal bonds, and common stocks. The estimated annual, after-tax return on each investment is 7%, 9%, and 14% respectively. The investor does not intend to reinvest these annual returns; instead he wishes to use these returns to finance a yearly vacation. When pressed for more details, our investor states that he hopes
   • to obtain a yearly return of about $5000
   • to invest a minimum of $10,000 in municipal bonds
   • to invest an amount in stocks that does not exceed more than the combined total of his investments in bonds and savings certificates
   • to invest between $5,000 and $15,000 in savings certificates
   Construct the "baseline model" for this problem. In doing so, notice carefully that the investor has not made any statements whatsoever as to a desire to maximize or minimize any measure of performance.

13.2. A firm has been asked to produce a number of packing containers. These containers are rectangular and made from two types of material: particle board and reinforced cardboard. Particle board weighs 0.5 pound per square foot and cardboard weighs 0.4 pound per square foot. The top and bottom of the container must be made from cardboard and the sides must be particle board. The length of the container must be less than 4 feet, and the girth is restricted to 10 feet. The customer wishes to maximize the volume of the container and minimize its weight. Form the baseline model—and note that the resulting model will not be linear.

13.3. A satellite is to be placed into low earth orbit, and it has been discovered that there are enough unused resources so as to permit the inclusion of a number of additional experimental packages. Five experiments, with specifications as cited in Table 13.4, have been proposed for inclusion into the satellite. A panel of experts has rated these experiments according to two measures of performance: scientific value and political value—and it is desired to find a combination of experiments that will maximize both measures. The estimated scientific and political value for each experiment is also listed in the table, where such values lie between 0 and 10 (the higher the value, the better). The limits on the combination of experiments flown are 100 pounds, 20 cubic feet, and 1000 watts. Form the baseline model. (*Hint:* Use zero–one variables to represent the structural variables.)

TABLE 13.4

Experiment	Weight	Volume	Power	Scientific value	Political value
1	20	1.5	100	5	8
2	25	4	200	3	7
3	40	6	150	8	5
4	30	6	300	2	9
5	12	7	500	9	4

13.4. Two recent graduates have decided to enter the field of personal computers. They have decided to manufacture (actually, to simply assemble—from parts ordered from overseas suppliers) two types of computers: Autoaccount and Scicomp. It is expected that they can, in the foreseeable future, sell as many computers as they could possibly produce. However, they wish to size the production rate so as utilize no more than a small group of production workers. Specifically, they want to keep assembly hours to 150 per week and test hours to 70 hours per week. Autoaccount requires 4 hours of assembly and 3 of testing and the Scicomp consumes 6 hours and 2 hours, respectively. Profit for the Autoaccount is estimated at $300 per unit and that of the Scicomp is $450 per unit. The partners want to maximize profit *and* maximize the total number of units produced each week (i.e., so as to achieve market visibility). Under no circumstances will they accept an increase in assembly hours beyond 150, or testing hours beyond 70, as this would require hiring additional workers or the use of overtime hours for workers who are already loaded to capacity. Develop the baseline model.

13.5. An automotive firm produces three types of cars: a large luxury car, a medium-sized car, and a compact car. The gasoline mileage figures, predicted sales, and profit figures for each type of car are given in Table 13.5. Government regulations state that the average gasoline mileage for the company's entire line of cars (i.e., the average mileage of the total number of cars produced in a year) should equal or exceed 30 miles per gallon (mpg). For every mpg below 30, the company must pay a penalty of $200 per car produced. The firm wishes to maximize its profits, but it also wants to minimize the mpg rate below 27 mpg because, not only would it pay a penalty, it will receive bad publicity. Develop the baseline model for this problem. Finally, the firm does not wish to produce more cars of any type than will be sold according to the market estimates. (*Hint:* Although there are various ways to model this problem, the best results will be obtained through the use of the specification of some "auxiliary" variables—variables that serve to specify the amount over or the amount under the fleet's average mpg rate.)

TABLE 13.5

Car	Miles per gallon (mpg)	Profit/car	Market
Large	18	$600	600,000
Medium	29	$400	800,000
Compact	36	$300	700,000

13.6. The preliminary layout of a communication network has been determined and is illustrated in the accompanying network diagram. The links of this network are depicted as branches, or arcs between node pairs, and the cost of transmitting a single message unit across a link is indicated in parentheses beside each link. The nodes of the network (with the exception of nodes 1 and 5, which are simply transmission or reception terminals) represent repeater stations. These stations receive, amplify, error check, and then transmit any messages incident to the node. The cost of transmission through a repeater station is a linear function of the maximum num-

TABLE 13.6

Link	Maximum capacity (messages/second)	Transmission cost per message	Repeater station, node number	Cost per messages per second
1–2	300,000	$3	2	$2.5
1–3	400,000	6	3	2
2–4	400,000	4	4	3
3–4	100,000	3		
3–5	300,000	7		
4–5	300,000	3		

ber of message units per second to be transmitted through that station. Further, there are limits to the message units per second that may be transmitted through a link—and this is listed as the maximum capacity in Table 13.6. The system should be designed and evaluated as based upon the maximum (i.e., worst-case) message load between terminals (i.e., between nodes 1 and 5). The maximum load is expected to be 500,000 messages per second. Formulate the baseline model for this problem.

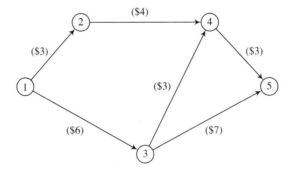

13.7. Develop the baseline model for Exercise 13.6 if, in addition to minimizing total cost, you wish to do the following:
   (a) Balance the communication loading across all transmission links.
   (b) Minimize the variation in repeater station sizing.
   (c) Minimize the number of messages sent through link 3–4.

CHAPTER 14

MULTIOBJECTIVE MODELS

CHAPTER OVERVIEW

In the previous chapter, we introduced certain notions with regard to the fundamental philosophies employed in multiobjective optimization—primarily by means of illustrative examples. However, that discussion was intended simply to provide a general overview of the topic. In this chapter, we address a far more systematic approach to the modeling of multiobjective problems: an approach that involves the use of a methodology that has been denoted as *multiplex*. Here, we shall illustrate the manner in which multiplex may be used as a unified approach to the modeling of a certain portion of mathematical programming problems, involving either a single or multiple objectives.* However, before proceeding to the multiplex approach, let us first discuss the fundamental building blocks, and the associated terminology, necessary to better appreciate both multiobjective optimization and its solution methods.

TERMINOLOGY

In the past, each type of multiobjective model (e.g., utility theory model, goal-programming model, generating-method model, lexicographic vectormin model, and fuzzy programming model) has been dealt with by means of its own individual

---

* Although we shall focus, for the most part, on linear models, it should be noted that the multiplex approach can be, and has been, extended to both discrete and nonlinear models.

approach, and described by its own specific terminology. As such, it has been the real or (mostly) imagined differences between each approach that have most often been emphasized, rather than the many similarities that actually exist. In order to better appreciate these similarities, as well as achieve an understanding of the modeling and solution process that will be employed, we shall first discuss the basic terminology to be used throughout the remainder of the text. Although some of this terminology has already been presented in the sections dedicated to single-objective models, one should carefully reflect on the different interpretations (some of which can be quite subtle) that may exist when dealing with multiple objectives. A listing of the terminology to be defined follows:

- variables
- objectives
- constraints
- aspiration levels
- goals
- ranking
- lexicographic minimum
- feasibility
- implementability
- optimization
- unbounded models

VARIABLES   Theoretically, variables are of two types: (1) variables that can be (directly) controlled—also known as *decision variables*, *control variables*, or *structural variables*—and (2) variables over which we have no control, designated as *uncontrolled* or *independent variables*. In practice, of course, the distinction between these two classes is not always so clear-cut. Specifically, we may encounter variables that fall into some "gray area." For example, the temperature inside one's house is a function of uncontrolled variables such as the outside temperature and the shading of the house (i.e., by trees or other buildings). Variables that also serve to determine the inside temperature, and over which we do have control, include the setting of the thermostat, the amount of home insulation, and the number and types of window blinds and drapes installed. However, the inside temperature is also a function of the opening and closing of the doors leading to the outside. And, if one has children, this is a variable (depending on the degree of parental control) that is not completely controllable—nor is it (we hope!) totally unmanageable.

　　In addition to variables that can be (directly) controlled and those that cannot, we must consider a third class of variable: *logical variables*. Such variables are *indirectly* controlled (i.e., their values are determined by the values ultimately given to the structural variables). In single-objective optimization, the logical variables are those designated as *slack*, *surplus*, and *artificial* variables—and these variables serve, among other things, to permit the speedy development of a basic feasible solution. In the multiplex method, we also employ logical variables, but these are denoted as *deviation variables*. Specifically, we have negative and positive deviations—that are analogous to the slack and surplus variables, respectively, used in single-objective models.

OBJECTIVES An objective is a function that is expressed in terms of the structural variables of a problem, in conjunction with certain coefficients (e.g., the cost or profit coefficients). Further, it is a function for which we seek an *optimal* value, that is, a function that is to be maximized or minimized. There is no difference (other than in terms of number) between the objectives of traditional models and those of multiobjective models.

CONSTRAINTS A constraint is expressed as either an equality or inequality. The left-hand side of a constraint is a function of the structural variables; the right-hand side is a constant. In our approach to multiobjective optimization, the term "constraint" is strictly reserved for those equalities or inequalities that absolutely *must* be satisfied. Its alternative designation is that of a "hard goal."

ASPIRATION LEVELS An aspiration level is employed in order to convert an objective into a goal. It represents a target level for the given objective—a level that is desired and/or acceptable. When one employs aspiration levels, he or she is implicitly using the notion of *satisficing*, as advocated by Simon (Simon, 1957; March and Simon, 1958).

GOALS In casual conversation, individuals often use the notions of objectives and goals interchangeably. However, we must be much more precise. A goal is a function that we wish to achieve. An objective is a function for which we seek the optimal solution. The value to be achieved by a goal is represented by a constant on the right-hand side of the goal (i.e., the aspiration level). The left-hand side of the goal is a function of the structural variables. It is vital to realize that a goal may be either hard or soft. A hard goal is one that *must* be achieved (i.e., a rigid constraint); a soft goal is simply one that we only *hope* to achieve.

Mathematically, goals and constraints appear identical. That is, both are represented by either equalities or inequalities. However, always remember that a constraint is a hard goal—or, alternatively, that a constraint is simply a subset of the notion of a goal.

RANKING A ranking is simply an ordering of objects. Commonly, we rank objects from "first to last" or from "top to bottom." Thus, we generally consider a goal at rank 2 to be "more important" than one of rank 3. Using the multiplex method, we shall always consider hard goals (i.e., rigid constraints) to be top ranked (or of "first priority"). The term *preemptive ranking* has been employed in multiobjective optimization to describe that situation in which a ranking is unchanged despite any weights given to the individual elements that have been ranked. As was mentioned, this is actually a misuse of the term—as the notion actually being described is one of *non*preemptive ranking. However, because this particular terminology has become so ingrained, and is so widely employed, we shall continue to use the term "preemptive ranking" to describe a ranking wherein a lower-ranked object can never preempt one at a higher rank.

LEXICOGRAPHIC MINIMUM We have already discussed and defined the notion of the lexicographic minimum in Chapter 13. Here, let us simply emphasize the fact that the lexicographic minimum

- deals with an ordered vector
- employs a lexicographic, or "preemptive," ranking
- serves to measure the achievement of a solution to a multiobjective model

With respect to the last item, recall that an ordered vector such as (0, 17.5, 99999, 9120) is preferred to the vector (0, 18, 2, 3). That is, we examine the first term of each vector and note that both values are equal. In such an instance, we move to the second term. Here, the first vector has a lower value (of the second term) than the second—and thus the first vector is lexicographically minimal (and there is no need to evaluate any lower-ranked terms). If this notion still seems unusual, consider the two vectors

$$\mathbf{u}^r = (0.000001, 5)$$

$$\mathbf{u}^s = (0, 99999)$$

Further, let us assume that we have a conventional LP problem that is to be minimized, and that there exist two solutions, r and s, that are to be compared. For solution r, the value of the objective function is 5 units—however, the constraint set is *violated* (by a total of 0.000001 units). For solution s, the value of the objective is much worse, 99999 units, but the constraint set is satisfied (i.e., the violation is 0 units). These results have been, in fact, summarized in the preceding two vectors. For such a case, solution s would indeed be preferred to solution r—and this is precisely the interpretation of the lexicographic minimum. Thus, whether the reader has realized it or not, he or she has been (implicitly) using the lexicographic minimum throughout in the study of traditional mathematical programming.

FEASIBILITY In single-objective optimization, a solution (i.e., program) is feasible as long as no constraint is violated. However, in the multiplex model, we employ goals—and not all goals will necessarily be achieved in any nontrivial problem. As such, in multiplex, a feasible solution is defined as simply one in which all variables (i.e., structural and deviation) are non-negative.

IMPLEMENTABILITY An implementable solution is one that can be used in actual practice. On the other hand, an unimplementable solution is one that absolutely cannot be used, or implemented. If the first term of the achievement vector of the multiplex model is zero (i.e., all hard goals satisfied), then the solution is considered implementable; otherwise, it is unimplementable.

OPTIMIZATION In single-objective optimization, we need only address a scalar: the value of the objective function. In multiobjective optimization, and by using the multiplex concept, an "optimal" solution is one that is lexicographically minimal—and we deal with an ordered vector. However, there exist other forms of multiobjective optimization, specifically the generating method, in which an optimal solution is defined as the particular nondominated solution that is preferred by the decision maker.

UNBOUNDED MODELS If an objective function can be increased (in the case of maximization) or decreased (in the case of minimization) without bound, and the constraints (i.e., hard goals) satisfied, the associated model is termed unbounded. However, another type of unboundedness exists with regard to models in which one or more structural variables may be increased without bound, but wherein the value of the objective function itself is bounded.* If the multiobjective model is anything other than a goal-programming model, it may be unbounded. However, if the model addressed is indeed a goal-programming model (of either the Archimedean or non-Archimedean class), *an unbounded achievement function is impossible*—although unbounded *programs* can be generated.** In practice, the existence of an unbounded model indicates that one has either formed the model improperly or has overlooked certain hard goals.

MODELING BASICS

With some appreciation of the terminology to be used, let us next address the fundamental tasks that may be necessary in the development of a model for a multiobjective problem, and in particular for implementation via the multiplex

---

* As an example, consider the LP in which we seek to minimize $x_1 - 2x_2$ subject to: $-2x_1 + x_2 \le 8$ and $-0.5x_1 + x_2 \le 12$, with all x_j nonnegative. Notice that z is finite (a minimum value of -24), whereas the values for the structural variables are unbounded.

** Min and Storbeck (1991) allege that unboundedness of the achievement vector *can* occur in goal-programming models. To accomplish this, however, one would have to *maximize* the unwanted deviations! And, if one were so foolish as to construct such a nonsensical model, we do agree that the achievement vector could then indeed contain unbounded terms. However, if one properly models a GP problem, unboundedness of the achievement vector is, we repeat, an impossibility.

approach. Specifically, we shall describe just how one may

- weight (and thus aggregate) objectives (as in utility theory)
- establish aspiration levels for objectives
- convert goals to the standard form (i.e., goal processing)
- weight goal deviations (as in goal programming)
- rank goals
- scale the model and/or normalize goals
- develop the achievement vector

The Weighting of Objectives

As discussed in Chapter 13, methods of aggregation seek weightings for each and every objective, so as to combine them into a single objective—and single measure of performance (i.e., a proxy such as "utiles"). This allows one to then approach the resultant model by means of conventional, single-objective methods. Such a philosophy is the basis for utility theory approaches, and we shall but briefly indicate the procedure employed for that process in this section.

If we let f_r represent the rth objective function (with $r = 1, \ldots , R$), then our purpose is to develop

$$U(\mathbf{f}) = U(f_1, f_2, \ldots , f_r, \ldots , f_R)$$

where $U(\mathbf{f})$ is the utility function for the R objectives. However, for this to be accomplished, the specific form of the utility function must be known. And the determination of a form that supposedly represents the utility function of the actual decision maker is no trivial task (see, e.g., the discussions in [Keeney and Raiffa, 1976; Cohon, 1978; Hwang and Masud, 1979]).

A common assumption employed in the use of weighting by means of utility functions is that the utility function is additively separable. For this case:

$$U(\mathbf{f}) = \Sigma \, U_r(f_r)$$

or
$$U(\mathbf{f}) = \Sigma \, w_r \cdot f_r$$

where w_r is the weight given to the rth objective. And a routine approach, based upon this form, is to use weights (w_r) to indicate the decision maker's preference for each objective. Although the use of this form makes things considerably simpler, there is no assurance that these (and other) assumptions upon which it is based actually hold.

We shall not pursue this particular approach any further, but we shall describe an alternative method of determining objective function weights—or goal-deviation variable weights—in a later section. There, a description of Saaty's

(1980) AHP method for the determination of such weightings shall be presented.

The Establishment of Aspiration Levels

In many instances (and always, in the case of any type of a goal-programming model), it will be necessary to convert an objective function into a goal. To accomplish this, we must establish aspiration levels to serve as the right-hand side of the resultant goal. As mentioned, such a conversion is based, either explicitly or implicitly, on the notion of satisficing. That is, we reject the utopian notion of optimality and seek, instead, to achieve an acceptable, or desirable, level for a given attribute. Thus, if one has a maximizing objective, we convert it to a type II ($\geq$) inequality. If one is dealing with a minimizing objective, we transform it into a type I ($\leq$) inequality.

The major consideration in such a transformation is that of providing a "good" right-hand side, or aspiration-level, value. However, some critics of the satisficing philosophy (i.e., goal programming) seem unduly concerned with the establishment of "precise" values—while remaining oblivious to this same lack of preciseness in optimization models. In practice (i.e., extensive experience in real applications since the mid-1950s), it has been shown that one need only develop a *reasonable* estimate of the desired aspiration level. If the problem being modeled is one that represents an existing system (e.g., the enhancement of the scheduling of operations in a factory), then we normally set an aspiration level at some percentage increase (or decrease) above (or below) the existing level of operation. For example, if we are attempting to decrease production costs, and these costs are presently $1 million per year, we might set the aspiration level to $800,000—a decrease of 20%. Because most organizations only operate at but a fraction of their theoretically optimal efficiency, such a decrease may be realistic. And, of course, once we have developed a solution, we may employ postoptimality analysis to investigate the feasibility of achieving even further reductions in cost.

Generally, it is inefficient to set one's aspiration levels "too high" in the conversion of maximizing objectives or "too low" when converting minimization objectives. For example, in the production scheduling problem cited earlier, one might be tempted to set the aspiration level for production costs to $0. However, in doing so, we face two problems:

- We have clearly not given sufficient thought and attention (if any) to the problem being modeled (i.e., as would be necessary to develop a reasonable target level). Thus, we have failed to gain any real insight into the actual problem.
- The resultant (unrealistic) goal can have a negative impact on other (i.e., lower ranked) goals. For example, if the cost goal is ranked or weighted

higher than, say, the customer-satisfaction goal (e.g., satisfy the customer's product demand), then we may focus on an unreasonable level of cost reduction at the expense of customer orders. Again, this can be adjusted in post-optimality analysis, but it is always better to begin such an analysis with a more realistic and more practical solution.

Aspiration levels should be set for objectives in conjunction with the decision maker. That is, his or her advice and input must be considered. Usually, the decision maker is asked questions such as

- What is the minimal level of profit that you desire?
- What is your target for cost reduction?
- How many defective products can you tolerate, per period, in your production process?
- What level of throughput can you achieve with the present system—and what amount of increase would you deem realistic?

And the decision maker should be told that any responses will be considered as simply estimates (and, if he or she wishes, kept strictly confidential). In particular, one should avoid putting the decision maker "on the spot" and in a position that may lead to his or her embarrassment.

One way in which to gauge the "reasonableness" of such estimates is to decompose the multiobjective model into a series of single-objective models (e.g., a series of linear programs in the case of a linear multiobjective model) and then determine the best possible values for each objective. If the decision maker has set an aspiration level that far exceeds the best possible level, the estimate obviously needs to be reconsidered.

Alternatively, one could simply provide the decision maker with the set of best possible values *before* soliciting his or her estimates (and there are some analysts who advocate such an approach). However, when this is done in practice, it has been noted that the typical decision maker will simply use the set of best possible values rather than attempt to voice any independent estimates.

The Processing of Goals

Unprocessed goals (either hard or soft) are represented as simply equations or inequalities, expressed solely in terms of the structural variables and a (positive-valued) constant (i.e., the right-hand side). Before we employ such goals in a model that is to be solved, we must first process them. Table 14.1 lists the guidelines that are to be followed in such processing. Further, the unwanted deviation variables, for each type of goal, are cited. And it is these unwanted deviation variables that will ultimately appear in the achievement vector. Notice that the *unwanted* deviation variables are flagged by double underlines. Again, it is to be

TABLE 14.1   GOAL PROCESSING

Unprocessed goal form	Processed form	Unwanted deviation variables (those to be minimized)
$f_i(\mathbf{x}) \leq b_i$	$f_i(\mathbf{x}) + \underline{\eta}_i - \rho_i = b_i$	ρ_i
$f_i(\mathbf{x}) \geq b_i$	$f_i(\mathbf{x}) + \eta_i - \underline{\rho}_i = b_i$	η_i
$f_i(\mathbf{x}) = b_i$	$f_i(\mathbf{x}) + \eta_i - \rho_i = b_i$	$\eta_i + \rho_i$

emphasized that the guidelines listed in Table 14.1 are to be followed whether the goal to be processed is hard or soft.

Let us next introduce a new vector, $\mathbf{v}$, where

$$\mathbf{v} = \begin{bmatrix} \mathbf{x} \\ \boldsymbol{\eta} \\ \boldsymbol{\rho} \end{bmatrix}$$

That is, $\mathbf{v}$ is composed of the structural and deviation variables, in the order shown. Thus, we may represent a set of goals that have been completely processed as

$$\mathbf{Av} = \mathbf{b}$$

where
$$\mathbf{A} = [\mathbf{A'} | \mathbf{I} | -\mathbf{I}]$$

Note carefully that $\mathbf{A'}$ is the set of technological coefficients for the original goal set (and associated with the structural variables of that goal set), $\mathbf{I}$ is the identity matrix associated with the set of negative deviation variables, and $-\mathbf{I}$ is the negative of the identity matrix, as associated with the set of positive deviation variables.

To demonstrate, consider a numerical example. Here, let us assume that we have the following three goals:

$$2x_1 + 3x_2 + x_3 \geq 100 \tag{14.1}$$

$$2x_1 + x_2 + x_3 \leq 80 \tag{14.2}$$

$$x_1 + x_2 = 50 \tag{14.3}$$

These may be converted, using the guidelines of Table 14.1, into the processed goals:

$$2x_1 + 3x_2 + x_3 + \underline{\eta}_1 - \rho_1 = 100 \tag{14.4}$$

$$2x_1 + x_2 + x_3 + \eta_2 - \underline{\rho}_2 = 80 \tag{14.5}$$

$$x_1 + x_2 + \underline{\eta}_3 - \underline{\rho}_3 = 50 \tag{14.6}$$

Or, by using matrix notation, as simply

$$\mathbf{Av} = \mathbf{b}$$

where

$$\mathbf{A} = \begin{bmatrix} 2 & 3 & | & 1 & 0 & 0 & | & -1 & 0 & 0 \\ 2 & 1 & | & 0 & 1 & 0 & | & 0 & -1 & 0 \\ 1 & 1 & | & 0 & 0 & 1 & | & 0 & 0 & -1 \end{bmatrix} \qquad \mathbf{b} = \begin{bmatrix} 100 \\ 80 \\ 50 \end{bmatrix} \qquad \mathbf{v} = \begin{bmatrix} x_1 \\ x_2 \\ x_3 \\ ---- \\ \eta_1 \\ \eta_2 \\ \eta_3 \\ ---- \\ \rho_1 \\ \rho_2 \\ \rho_3 \end{bmatrix}$$

The Weighting of Unwanted Goal Deviations

Goal-programming approaches avoid the aggregation of objectives as each goal is always maintained, individually, in the GP formulation. However, one must still determine the weights to be assigned to the *unwanted* goal deviations that appear in the associated achievement vector. As discussed, the unwanted deviations for hard goals (i.e., priority level one) are normally just equally weighted—as any deviation whatsoever cannot be tolerated. However, in the remaining priority levels, the weights should reflect an estimate of the degree of importance given, by the decision maker, to the satisfaction of the associated goal. For the most part, such *intrapriority* weights have been established on a purely subjective basis, and the results—as evidenced by decades of successful application—have generally been favorable. Further, as mentioned, we may always (and, in fact, should always) analyze any changes in these weights during postoptimality analysis.

However, there are more systematic means through which such weights may be developed. Of course, it still must be kept in mind that, no matter how systematic, or clever, or complex, all such methods are still heuristics and cannot guarantee the validity—or preciseness—of the resultant weights that are developed. For the sake of illustration, we shall present just one of these approaches: the AHP (Analytic Hierarchy Process) method as proposed by Saaty (1980). Although our focus in this section shall be on the use of the AHP approach to establish the weights on the unwanted goal deviation variables *within* a given priority level (i.e., priority levels 2 through K), it should be noted that precisely the same approach may be applied to the weighting of objective functions (i.e., as an adjunct to the utility theory method).

We may use the AHP method to systematically (but still heuristically) estimate goal-deviation weights by means of a relatively straightforward process. Detailed descriptions, including numerous examples, are provided in the references (Saaty, 1980; Gass, 1986). Here, we shall simply summarize the basic elements of the process.

We begin by having the decision maker compare the m goals in a pairwise fashion. By letting G_i represent the ith goal and w_i the weight assigned to that goal (i.e., to its deviation variable, or variables), the pairwise comparisons may be represented in matrix form as

$$
\mathbf{A} =
\begin{array}{c|cccc}
 & G_1 & G_2 & \cdots & G_m \\
\hline
G_1 & w_1/w_1 & w_1/w_2 & \cdots & w_1/w_m \\
G_2 & w_2/w_1 & w_2/w_2 & \cdots & w_2/w_m \\
\vdots & \vdots & \vdots & \ddots & \vdots \\
G_m & w_m/w_1 & w_m/w_2 & \cdots & w_m/w_m
\end{array}
$$

By using matrix $\mathbf{A}$, the AHP approach seeks the solution to $\mathbf{Av} = \lambda\mathbf{v}$. This is, of course, the well-known eigenvalue problem. A solution, $\mathbf{v} = (v_1, v_2, \ldots, v_m)$ for a given eigenvector, λ, is termed an eigenvector. It may be shown, subject to certain minor assumptions, that for λ equal to the largest eigenvalue, one may interpret the normalized v_i coefficients as weights that represent the importance of one goal with respect to another. Thus, to find weights for our goals, one need only conduct a pairwise comparison of all goals, solve for the eigenvalues, select the largest eigenvalue, and then solve for the associated eigenvector. Then, by normalizing this vector, the goal weights are determined.

To facilitate the comparison process, Saaty advocates the use of a scale from 1 to 9 (and, as required by the process, the reciprocals) to reflect the decision maker's opinion as to the importance of one goal with respect to another. Specifically, the following scale is employed:

$1 \Rightarrow$ equally important

$3 \Rightarrow$ moderately more important

$5 \Rightarrow$ strongly more important

$7 \Rightarrow$ very strongly more important

$9 \Rightarrow$ extremely more important

where the values 2, 4, 6, and 8 are used to represent a compromise in judgment (e.g., a value of 2 would be the compromise between 1 and 3). Thus, the question to be asked, in reference to each row of this table, is: When the goal on the left of the table is compared to the goal on the top of the table, how much more important is it?

To demonstrate the development of such a table, let us consider three goals:

- G_1: in terms of profit
- G_2: in terms of market share
- G_3: in terms of company prestige

We may then compare these, a pair at a time, and enter the results in the pairwise comparison matrix. Let us assume that, for our example:

- G_1 is very strongly more important than G_2 (i.e., a value of 7)
- G_1 is extremely more important than G_3 (i.e., a value of 9)
- G_2 is strongly more important than G_3 (i.e., a vaue of 5)

The resultant matrix is then given as

$$
\mathbf{A} = \begin{array}{c|ccc} & G_1 & G_2 & G_3 \\ \hline G_1 & 1 & 7 & 9 \\ G_2 & \frac{1}{7} & 1 & 5 \\ G_3 & \frac{1}{9} & \frac{1}{5} & 1 \end{array}
$$

Solving the associated eigenvalue problem, we may determine that the largest eigenvalue is 3.21.* Thus, the normalized eigenvector is given as: (.77, .17, .05).

Using these results, we might weight each goal as follows: G_1 by 0.77, G_2 by 0.17, and G_3 by 0.05. And some analysts have, in fact, used such an approach. However, others note that the eigenvectors, being normalized, are "too close" to one another. Gass proposes that a translation be used; this translation is summarized in Table 14.2.

TABLE 14.2 EIGENVECTOR TRANSLATION TABLE

	Eigenvector scale		Multiplier		GP weights
(Low)	0.00–0.10	×	10	=	0–1
	0.11–0.20	×	20	=	2–4
	0.21–0.30	×	40	=	8–12
	0.31–0.40	×	60	=	19–24
	0.41–0.50	×	100	=	41–50
	0.51–0.60	×	150	=	76–90
	0.61–0.70	×	200	=	122–140
	0.71–0.80	×	400	=	284–320
	0.81–0.90	×	600	=	486–630
(High)	0.91–1.00	×	1000	=	910–1000

If we use this approach, the associated weights for our goals would be given as (308, 3.4, 0.5). However, we once again warn the reader that although such

---

* Strictly speaking, this value should be equal to the rank of the matrix (the number of rows, or *m*). If not, the matrix is considered inconsistent. In practice, as long as the eigenvalue is not too much larger than *m*, the process seems appropriate.

approaches as the AHP method, or the AHP method in conjunction with the translation table of Gass, are systematic and analytical, there is no real proof that the results obtained are any better—or any worse—than those arrived at by a simple, subjective estimate. Regardless of the approach used to develop such weights, one must always perform a final postoptimality analysis to investigate the apparent rationality of the results.

The Ranking of Goals and/or Objectives

The non-Archimedean GP approach requires that one rank order the goal set, whereas the lexicographic vectormin procedure requires one to rank order the objectives. Further, once these functions are ranked, it is assumed that the ranking is lexicographic (i.e., preemptive). If we are dealing with a single decision maker, we may accomplish such a ranking, for either goals or objectives, by means of *pairwise comparisons*.

The paired comparison method for ranking simply requires that the decision maker compare two functions (i.e., goals or objectives, based upon the model to be employed) at a time—and to indicate which particular function is preferred. This is repeated until all possible pairs of functions have been considered. As such, the total number of comparisons necessary is given by $m!/[2!(m - 2)!]$, where m is the total number of functions.

As an example, consider the case in which four goals are to be ranked. This will require six comparisons. Let us assume that the result is

$$G_1 > G_2 \qquad G_4 < G_1 \qquad G_2 > G_4$$
$$G_1 > G_3 \qquad G_2 < G_3 \qquad G_3 > G_4$$

where $G_1 > G_2$ is read as "goal one is preferred to goal two." Rearranging these expressions so that all preference signs point to the right, we have

$$G_1 > G_2 \qquad G_1 > G_4 \qquad G_2 > G_4$$
$$G_1 > G_3 \qquad G_3 > G_2 \qquad G_3 > G_4$$

Now, the most preferred goal will (or, at least, should) be preferred to exactly $m -$ 1 goals. Thus, the most preferred goal should appear to the left of our expression three times. This is true for G_1. The next most preferred goal must preferred to m $- 2$ (or, for the example, two goals) goals, and this is G_3. The final rankings are thus

Rank	Goal
1	G_1
2	G_3
3	G_2
4	G_4

The paired comparison approach works if the decision maker is consistent in his or her rankings. If an inconsistency exists (e.g., $G_3 > G_2$, $G_2 > G_4$, and $G_4 > G_3$), the problem should be reviewed by the analyst and decision maker—so as to attempt to resolve the situation.

When a committee, rather than an individual, is used to rank functions, one may use a technique to form *composite* rankings. A number of such approaches exist, including the Kendall array method, Thurstone's procedure, and others. The reader may consult the references for the details of such approaches [Siegel, 1956; Churchman, Ackoff, and Arnoff, 1957; Hall, 1962; Ackoff and Sasieni, 1968].

There is yet one other point that needs to be emphasized. If we are employing non-Archimedean GP, there may be a temptation to put each individual goal (i.e., its associated unwanted goal deviation variable) into its own individual priority level. This is usually not necessary—nor appropriate. Further, there are two problems with such an approach. First, we can wind up with a large number of priority levels—with a very small likelihood that any goal at the lower levels will have any influence whatsoever on the resultant solution. As such, it is generally recommended that one not employ more than about five priority levels. The second problem is that, if we place each goal into a separate priority, we may be overlooking goals that could be combined by means of numerical weights—and we should always attempt to combine goals within a level whenever possible—and, of course, reasonable.

Scaling and Normalization of Goals

Regardless as to whether we employ traditional, single-objective models or multiobjective forms, we need to consider problem scaling. However, much has been made of the alleged potential problems involved in GP models (i.e., even though similar problems exist for all alternative approaches). For example, consider the following two (soft) goals:*

$3x_1 + x_2 \geq 5,000,000$ (desired profit per time period, in dollars)

$x_1 + x_2 \leq 10$ (amount of toxic discharges, per period, in tons)

Further, let us assume that the firm in question has stated that the profit goal is twice as important as the toxic discharge goal. Thus, $2 of profit might appear to be equal to 1 ton of toxic waste discharged—but this certainly doesn't seem to be reasonable. However, even if we employed such a naive weighting scheme, this problem would be readily identified during postoptimality analysis—and would likely result in a new, and more appropriate, weighting scheme.

---

* Because the disparity between the largest (5,000,000) and smallest (1) numbers in this model is so great, any experienced analyst would modify the formulation so to alleviate this problem. For example, we could simply measure profit in *millions* of dollars, rather than just dollars. This results in a scaled goal for profit that is $3x_1 + x_2 \geq 5$.

If we wish to try to catch the problem before dealing with the postoptimality phase, we might try to use a more systematic approach to the development of weights—and one such method has already been described in a preceeding section (i.e., the AHP procedure). Others have advocated a "normalization" procedure. Unfortunately, most of these latter approaches may do more harm than good. For example, one particularly naive proposal has been to simply divide each goal by its right-hand side. For the two goals given previously, this would result in

$$(6\text{E-}7)x_1 + (2\text{E-}7)x_2 \geq 1 \qquad \text{("normalized" profit goal)}$$

$$0.1x_1 + 0.1x_2 \leq 1 \qquad \text{("normalized" toxic discharge goal)}$$

As should be immediately noted, this approach is not only exceptionally naive, it also accomplishes absolutely nothing, other than making the model more difficult to read.

Another normalization approach, and one that more accurately represents a true normalization process, is to apply the Euclidean norm associated with the coefficients of the goal, that is,

$$N_i = \|\mathbf{c}^i\| = (\mathbf{c}^i \cdot \mathbf{c}^i)^{1/2}$$

where N_i is the Euclidean norm, and $\mathbf{c}^i$ is the vector of coefficients for the structural variables of the ith goal. The use of this norm may be demonstrated by means of an example and argument originally presented by Gass (1987). Consider the goal

$$3x_1 + 2x_2 + 4x_3 + \eta - \rho = 6 \qquad (14.7)$$

By using the Euclidean norm, this goal may be written as

$$\delta = \eta' - \rho' = 6/\sqrt{29} - (3/\sqrt{29}x_1 + 2/\sqrt{29}x_2 + 4/\sqrt{29}x_3 \qquad (14.8)$$

where the unrestricted deviation variable, δ, measures the true, dimension-free distance of any point from the defining hyperplane. The normed goal may then be written in standard form as

$$3/\sqrt{29}x_1 + 2/\sqrt{29}x_2 + 4/\sqrt{29}x_3 + \eta' - \rho' = 6/\sqrt{29} \qquad (14.9)$$

It is then suggested that one determine the achievement vector weights for the unwanted goal-deviation variables.

However, as Gass demonstrates, such a process is equivalent to simply using the original, unnormed goal (14.7) and then multiplying its achievement-vector weightings by the reciprocal of the goal norm. Further, he notes that such norms tend to be small, more or less equal, and of the same magnitude. *As such, their influence on the model is minimal—and buried in the overall uncertainty associated with the selection of the weights themselves.* As Gass states: "there is no need to adjust this constraint [goal] in that any norming is equivalent to adjusting the [uncertain] objective function [achievement vector] weights of the normed expression." Thus, although the norming process may give the appearance of improving the model, it accomplishes little of significance. Consequently, although we certainly advocate the scaling of any model, we believe that any efforts devoted to a norming process are exertions that are wasted.

The Development of the Achievement Vector

Rather than employing an objective function to measure the goodness of a solution, as in single-objective optimization, we employ an achievement function that serves to indicate just how well a given program is able to achieve the goals and objectives of the respective model under consideration. And it is vital to realize that this function is an ordered *vector*, rather than a scalar. As noted in the previous chapter, the general form of the achievement vector is

$$\text{lexmin } \mathbf{u} = \{\mathbf{c}^{(1)}\mathbf{v}, \ldots, \mathbf{c}^{(K)}\mathbf{v}\} \tag{14.10}$$

where

lexmin = lexicographic minimum (of an ordered vector)

K = total number of priority levels ($k = 1, 2, \ldots, K$)

m = total number of goals in the associated model

$\mathbf{u}$ = achievement vector (or achievement function); a ($K \times 1$) column vector

$\mathbf{c}^{(k)}$ = vector of coefficients, or weights, of $\mathbf{v}$ in the kth term of the achievement vector; a $[1 \times (n + 2m)]$ row vector

$\mathbf{v}$ = vector of all structural *and* deviation variables; an $[(n + 2m) \times 1]$ column vector

As we shall see, each particular class of model has its own specific achievement-vector characteristics. This is summarized in Table 14.3. These differences may be better appreciated when we demonstrate the construction of the complete multiplex model.

MULTIPLEX FORMULATIONS (IGNIZIO, 1985b)

The formulation of the multiplex model for each problem class is best illustrated by means of examples, and we shall provide illustrations in this section. However, first recall that the general form of the multiplex model is given as

$$\text{lexmin } \mathbf{u} = \{\mathbf{c}^{(1)}\mathbf{v}, \mathbf{c}^{(2)}\mathbf{v}, \ldots, \mathbf{c}^{(K)}\mathbf{v}\} \tag{14.11}$$

subject to:

$$\mathbf{Av} = \mathbf{b} \tag{14.12}$$

$$\mathbf{v} \geq \mathbf{0} \tag{14.13}$$

where all terms have been previously defined.

TABLE 14.3 ACHIEVEMENT VECTORS

Type of model	Achievement-vector characteristics
Linear program (as formed by whatever method desired)	Two priority levels, wherein the first is composed of the unwanted goal-deviation variables for the constraint set, and the second is the original objective, in minimization form
Archimedean LGP	Two priority levels, wherein the first is composed of the unwanted goal-deviation variables for all hard goals, and the second contains the unwanted (and weighted) goal-deviation variables for all soft goals
Non-Archimedean LGP	K priority levels, where $K > 2$. The first level is composed of the unwanted goal-deviation variables for all hard goals, and each of the remaining terms contains the unwanted (and weighted) goal-deviation variables associated with those goals at the priority level under consideration
Chebyshev LGP, or fuzzy LGP	Two priority levels (for the most elementary forms of these approaches), wherein the first is composed of the unwanted goal-deviation variables for all hard *and* soft goals, and the second contains the single-deviation variable (δ) representing the worst-case deviation
Lexicographic vectormin	K priority levels, where $K = p + 1$, and p = number of (minimizing) objectives. The first level is composed of the unwanted goal-deviation variables for all constraints, and each of the remaining terms contains the original objectives, listed in order of their respective rank

We shall now demonstrate how one may employ this form to represent: (1) linear programs, (2) Archimedean goal programs, (3) non-Archimedean goal programs, (4) Chebyshev—or fuzzy—goal programs, and (5) the lexicographic vectormin problem. We begin with the representation of a conventional linear program.

Multiplex Model: Linear Programs

Consider a linear program as given in the following general form:

$$\text{minimize } z = \mathbf{dx} \tag{14.14}$$

subject to:

$$\mathbf{A'x} \ (\leq, \geq, \text{ or } =) \ \mathbf{b} \tag{14.15}$$

$$\mathbf{x} \geq \mathbf{0} \tag{14.16}$$

Note carefully that the primed notation (i.e., $\mathbf{A'}$) is used simply to indicate the original form of this matrix (i.e., the technological matrix prior to conversion to the multiplex format). Further, $\mathbf{d}$ represents the coefficients of the objective function "cost" vector, $\mathbf{b}$ is the right-hand-side vector, and $\mathbf{x}$ is the vector of structural variables.

The equivalent multiplex representation is then

$$\text{lexmin } \mathbf{u} = \{\mathbf{c}^{(1)}\mathbf{v}, \ \mathbf{c}^{(2)}\mathbf{v}\} \tag{14.17}$$

subject to:

$$\mathbf{Av} = \mathbf{b} \tag{14.18}$$

$$\mathbf{v} \geq \mathbf{0} \tag{14.19}$$

where

$$\mathbf{A} = [\mathbf{A}'\,|\,\mathbf{I}\,|\,-\mathbf{I}] \qquad \mathbf{c}^{(1)} = [\mathbf{0}\,|\,\boldsymbol{\mu}^{(1)}\,|\,\boldsymbol{\omega}^{(1)}] \qquad \mathbf{c}^{(2)} = [\mathbf{d}\,|\,\mathbf{0}\,|\,\mathbf{0}]$$

$$\mathbf{v} = \begin{bmatrix} \mathbf{x} \\ ---- \\ \boldsymbol{\eta} \\ ---- \\ \boldsymbol{\rho} \end{bmatrix}$$

and

$\boldsymbol{\eta}^{(k)}$ = vector of negative deviations at priority level k

$\boldsymbol{\rho}^{(k)}$ = vector of positive deviations at priority level k

$\boldsymbol{\mu}^{(k)}$ = vector of weights for all negative deviations at priority k

$\boldsymbol{\omega}^{(k)}$ = vector of weights for all positive deviations at priority k

As may be noted, our first priority is to satisfy the constraint set (via the minimization of the first term in the achievement vector) and our second is to minimize the single-objective function (if the objective is to be maximized, it must be converted to a minimization form). To illustrate this process, let us use the linear programming problem of Chapter 13, as represented by functions (13.7) through (13.12), and repeated as follows:

Find x_1 and x_2 so as to

$$\text{maximize } 3x_1 + x_2 \qquad \text{(total profit per time period)} \tag{14.20}$$

subject to:

$$2x_1 + 3x_2 \geq 100 \qquad \text{(market shares desired)} \tag{14.21}$$

$$2x_1 + x_2 \leq 50 \qquad \text{(raw material limitations)} \tag{14.22}$$

$$x_1 \leq 20 \qquad \text{(market saturation level, product 1)} \tag{14.23}$$

$$x_2 \leq 30 \qquad \text{(market saturation level, product 2)} \tag{14.24}$$

$$\text{and } x_1 \text{ and } x_2 \geq 0 \qquad \text{(nonnegativity conditions)} \tag{14.25}$$

This problem, in multiplex form, is then given as

$$\text{lexmin } \mathbf{u} = \{(\eta_1 + \rho_2 + \rho_3 + \rho_4),\ (-3x_1 - x_2)\} \tag{14.26}$$

subject to:

$$2x_1 + 3x_2 + \underline{\eta}_1 - \rho_1 = 100 \tag{14.27}$$

$$2x_1 + x_2 + \eta_2 - \underline{\rho}_2 = 50 \tag{14.28}$$

$$x_1 + \eta_3 - \underline{\rho}_3 = 20 \tag{14.29}$$

$$x_2 + \eta_4 - \underline{\rho}_4 = 30 \tag{14.30}$$

$$\mathbf{v} \geq \mathbf{0} \tag{14.31}$$

where

$$\mathbf{A} = \begin{bmatrix} 2 & 3 & 1 & 0 & 0 & 0 & -1 & 0 & 0 & 0 \\ 2 & 1 & 0 & 1 & 0 & 0 & 0 & -1 & 0 & 0 \\ 1 & 0 & 0 & 0 & 1 & 0 & 0 & 0 & -1 & 0 \\ 1 & 0 & 0 & 0 & 0 & 1 & 0 & 0 & 0 & -1 \end{bmatrix} \quad \mathbf{b} = \begin{bmatrix} 100 \\ 50 \\ 20 \\ 30 \end{bmatrix} \quad \mathbf{v} = \begin{bmatrix} x_1 \\ x_2 \\ \text{----} \\ \eta_1 \\ \eta_2 \\ \eta_3 \\ \eta_4 \\ \text{----} \\ \rho_1 \\ \rho_2 \\ \rho_3 \\ \rho_4 \end{bmatrix}$$

$$\mathbf{c}^{(1)} = [0 \quad 0 \,|\, 1 \quad 0 \quad 0 \quad 0 \,|\, 0 \quad 1 \quad 1 \quad 1]$$

$$\mathbf{c}^{(2)} = [-3 \quad -1 \,|\, 0 \quad 0 \quad 0 \quad 0 \,|\, 0 \quad 0 \quad 0 \quad 0]$$

Multiplex Model: Archimedean Linear Goal Programs

The general form for the Archimedean LGP is given as

$$\text{lexmin } \mathbf{u} = \{(\boldsymbol{\mu}^{(1)}\boldsymbol{\eta}^{(1)} + \boldsymbol{\omega}^{(1)}\boldsymbol{\rho}^{(1)}),\ (\boldsymbol{\mu}^{(2)}\boldsymbol{\eta}^{(2)} + \boldsymbol{\omega}^{(2)}\boldsymbol{\rho}^{(2)})\} \tag{14.32}$$

subject to:

$$\mathbf{A}'\mathbf{x} + \boldsymbol{\eta}^{(1)} - \boldsymbol{\rho}^{(1)} = \mathbf{r}^{(1)} \tag{14.33}$$

$$\mathbf{D}\mathbf{x} + \boldsymbol{\eta}^{(2)} - \boldsymbol{\rho}^{(2)} = \mathbf{r}^{(2)} \tag{14.34}$$

$$\mathbf{x},\ \eta,\ \rho \geq \mathbf{0} \tag{14.35}$$

where

$\boldsymbol{\eta}^{(k)}$ = vector of negative deviations at priority level k

$\boldsymbol{\rho}^{(k)}$ = vector of positive deviations at priority level k

$\boldsymbol{\mu}^{(k)}$ = vector of weights for all negative deviations at priority k

$\boldsymbol{\omega}^{(k)}$ = vector of weights for all positive deviations at priority k

$\mathbf{x}$ = vector of the structural variables

$\mathbf{A}'$ = matrix of technological coefficients for all goals at priority 1

$\mathbf{D}$ = matrix of technological coefficients for all goals at priority 2

$\mathbf{r}^{(k)}$ = right-hand-side vector for the goals at priority level k

Actually, the Archimedean LGP model is, in essence, already in multiplex form. This is made more obvious when we utilize the more general multiplex form:

$$\text{lexmin } \mathbf{u} = \{\mathbf{c}^{(1)}\mathbf{v}, \, \mathbf{c}^{(2)}\mathbf{v}\} \tag{14.36}$$

subject to:

$$\mathbf{Av} = \mathbf{b} \tag{14.37}$$

$$\mathbf{v} \geq \mathbf{0} \tag{14.38}$$

where

$$\mathbf{A} = \begin{bmatrix} \mathbf{A}' & \mathbf{I} & \mathbf{0} & -\mathbf{I} & \mathbf{0} \\ \mathbf{D} & \mathbf{0} & \mathbf{I} & \mathbf{0} & -\mathbf{I} \end{bmatrix} \qquad \mathbf{b} = \begin{bmatrix} \mathbf{r}^{(1)} \\ \mathbf{r}^{(2)} \end{bmatrix} \qquad \mathbf{v} = \begin{bmatrix} \mathbf{x} \\ \hline \boldsymbol{\eta} \\ \hline \boldsymbol{\rho} \end{bmatrix}$$

$$\mathbf{c}^{(1)} = [\mathbf{0} \mid \boldsymbol{\mu}^{(1)} \mid \boldsymbol{\omega}^{(1)}] \qquad \mathbf{c}^{(2)} = [\mathbf{0} \mid \boldsymbol{\mu}^{(2)} \mid \boldsymbol{\omega}^{(2)}]$$

We may use the model of Chapter 13, (13.25–13.31), to demonstrate the modeling process. This model is repeated in what follows.

Find x_1 and x_2 so as to

$$\text{lexmin } \mathbf{u} = \{(\rho_1 + \rho_2 + \rho_3), \, (\eta_4 + 2\eta_5)\} \tag{14.39}$$

satisfy:

$$2x_1 + x_2 + \eta_1 - \rho_1 = 50 \qquad \text{(raw material limitations goal)} \tag{14.40}$$

$$x_1 + \eta_2 - \rho_2 = 20 \qquad \text{(market-level goal, product 1)} \tag{14.41}$$

$$x_2 + \eta_3 - \rho_3 = 30 \qquad \text{(market-level goal, product 2)} \tag{14.42}$$

$$3x_1 + x_2 + \eta_4 - \rho_4 = 50 \qquad \text{(profit goal)} \tag{14.43}$$

$$2x_1 + 3x_2 + \underline{\eta}_5 - \rho_5 = 80 \qquad \text{(market-shares goal)} \tag{14.44}$$

$$\text{and all } x_j, \eta_i, \text{ and } \rho_i \text{ are nonnegative} \tag{14.45}$$

Again, it may be noted that the normal form of the Archimedean LGP is already in multiplex form. The specific model components, in matrix notation, are as follows:

$$\mathbf{A} = \begin{bmatrix} 2 & 1 & 1 & 0 & 0 & 0 & 0 & -1 & 0 & 0 & 0 & 0 \\ 1 & 0 & 0 & 1 & 0 & 0 & 0 & 0 & -1 & 0 & 0 & 0 \\ 0 & 1 & 0 & 0 & 1 & 0 & 0 & 0 & 0 & -1 & 0 & 0 \\ \hline 3 & 1 & 0 & 0 & 0 & 1 & 0 & 0 & 0 & 0 & -1 & 0 \\ 2 & 3 & 0 & 0 & 0 & 0 & 1 & 0 & 0 & 0 & 0 & -1 \end{bmatrix}$$

$$\mathbf{c}^{(1)} = \begin{bmatrix} 0 & 0 & 0 & 0 & 0 & 0 & 0 & 1 & 1 & 1 & 0 & 0 \end{bmatrix}$$

$$\mathbf{c}^{(2)} = \begin{bmatrix} 0 & 0 & 0 & 0 & 0 & 1 & 2 & 0 & 0 & 0 & 0 & 0 \end{bmatrix}$$

$$\mathbf{v}^T = \begin{bmatrix} x_1 & x_2 & \eta_1 & \eta_2 & \eta_3 & \eta_4 & \eta_5 & \rho_1 & \rho_2 & \rho_3 & \rho_4 & \rho_5 \end{bmatrix}$$

$$\mathbf{b}^T = \begin{bmatrix} 50 & 20 & 30 & 50 & 80 \end{bmatrix}$$

Before proceeding further, let us note two things. First, the deviation variable for a particular goal may appear in more than one priority. Second, the order of the goals in the problem (i.e., their order of appearance—as in (14.40–14.45)) does not always have to be from highest to lowest—and in fact, it is often impossible to satisfy such an ordering. As such, the model characteristics listed should be considered as guidelines—and not be considered completely rigid. To clarify these remarks, consider the Archimedean LGP model that follows. Here, we wish to satisfy the hard goals of raw material limitations (100 units available) and processing hours (40 hours available). Further, we seek to achieve a profit of 50 units and also minimize the idle processing time—where the achievement of profit is evidently considered four times as important as the minimization of idle processing units. Note carefully that the second goal has one deviation variable (ρ_2) at priority level *1*, and one (η_2) at priority level *2*. We leave it to the reader to list the respective matrices and vectors of the multiplex form for this model.

Find x_1 and x_2 so as to

$$\text{lexmin } \mathbf{u} = \{(\rho_1 + \rho_2 + \rho_3), (4\eta_3 + \eta_2)\} \tag{14.46}$$

satisfy:

$$x_1 + x_2 + \underline{\eta}_1 - \underline{\rho}_1 = 100 \qquad \text{(raw material limitations goal)} \tag{14.47}$$

$$x_1 + 3x_2 + \underline{\eta}_2 - \underline{\rho}_2 = 40 \qquad \text{(processing hours goal)} \tag{14.48}$$

$$3x_1 + 5x_2 + \underline{\eta}_3 - \rho_3 = 50 \qquad \text{(profit goal)} \tag{14.49}$$

$$\text{and all } x_j, \eta_i, \text{ and } \rho_i \text{ are nonnegative} \tag{14.50}$$

Multiplex Model: Non-Archimedean Linear Goal Programs

The general form of the non-Archimedean LGP is given as

$$\text{lexmin } \mathbf{u} = \{(\boldsymbol{\mu}^{(1)}\boldsymbol{\eta}^{(1)} + \boldsymbol{\omega}^{(1)}\boldsymbol{\rho}^{(1)}), \ldots, (\boldsymbol{\mu}^{(K)}\boldsymbol{\eta}^{(K)} + \boldsymbol{\omega}^{(K)}\boldsymbol{\rho}^{(K)})\} \qquad (14.51)$$

subject to:

$$\mathbf{A}'\mathbf{x} + \boldsymbol{\eta}^{(1)} - \boldsymbol{\rho}^{(1)} = \mathbf{r}^{(1)} \qquad (14.52)$$

$$\mathbf{D}^{(2)}\mathbf{x} + \boldsymbol{\eta}^{(2)} - \boldsymbol{\rho}^{(2)} = \mathbf{r}^{(2)} \qquad (14.53)$$

$$\mathbf{D}^{(3)}\mathbf{x} + \boldsymbol{\eta}^{(3)} + \boldsymbol{\rho}^{(3)} = \mathbf{r}^{(3)} \qquad (14.54)$$

$$\vdots$$

$$\mathbf{D}^{(K)}\mathbf{x} + \boldsymbol{\eta}^{(K)} - \boldsymbol{\rho}^{(K)} = \mathbf{r}^{(K)} \qquad (14.55)$$

$$\mathbf{x}, \boldsymbol{\eta}, \boldsymbol{\rho} \geq \mathbf{0} \qquad (14.56)$$

where

$\mathbf{A}'$ = technological coefficient matrix for all hard goals

$\mathbf{D}^{(k)}$ = technological coefficient matrices for all soft goals at priority level k (i.e., the coefficients of the original objective set for those objectives at priority level k)

$\mathbf{r}^{(k)}$ = right-hand side for those goals at priority level k

and where all other terms have been previously defined.

The equivalent multiplex formulation may then be written as

$$\text{lexmin } \mathbf{u} = \{\mathbf{c}^{(1)}\mathbf{v}, \mathbf{c}^{(2)}\mathbf{v}, \ldots, \mathbf{c}^{(K)}\mathbf{v}\} \qquad (14.57)$$

subject to:

$$\mathbf{A}\mathbf{v} = \mathbf{b} \qquad (14.58)$$

$$\mathbf{v} \geq \mathbf{0} \qquad (14.59)$$

where

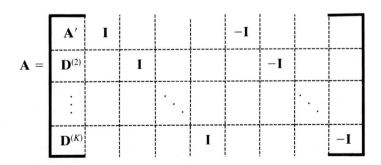

$$\mathbf{c}^{(1)} = \quad [\mathbf{0} \;\vdots\; \boldsymbol{\mu}^{(1)} \;\vdots\; \boldsymbol{\omega}^{(1)}]$$

$$\vdots$$

$$\mathbf{c}^{(k)} = \quad [\mathbf{0} \;\vdots\; \boldsymbol{\mu}^{(k)} \;\vdots\; \boldsymbol{\omega}^{(k)}]$$

$$\vdots$$

$$\mathbf{c}^{(K)} = \quad [\mathbf{0} \;\vdots\; \boldsymbol{\mu}^{(K)} \;\vdots\; \boldsymbol{\omega}^{(K)}]$$

$$\mathbf{b} = \begin{bmatrix} \mathbf{r}^{(1)} \\ \hline \vdots \\ \hline \mathbf{r}^{(K)} \end{bmatrix} \qquad \mathbf{v} = \begin{bmatrix} \mathbf{x} \\ \hline \boldsymbol{\eta} \\ \hline \boldsymbol{\rho} \end{bmatrix}$$

For the sake of clarity, we have not listed the zero submatrices in the foregoing representation of **A**.

To demonstrate, we shall employ the model from Chapter 13 as represented by functions (13.33–13.39), as repeated in what follows.

Find x_1 and x_2 so as to

$$\text{lexmin } \mathbf{u} = \{(\rho_1 + \rho_2 + \rho_3), (\eta_5), (\eta_4)\} \tag{14.60}$$

and satisfy:

$$2x_1 + x_2 + \eta_1 - \underline{\rho_1} = 50 \qquad \text{(raw material limitations goal)} \tag{14.61}$$

$$x_1 + \eta_2 - \underline{\rho_2} = 20 \qquad \text{(market-level goal, product 1)} \tag{14.62}$$

$$x_2 + \eta_3 - \underline{\rho_3} = 30 \qquad \text{(market-level goal, product 2)} \tag{14.63}$$

$$3x_1 + x_2 + \underline{\eta_4} - \rho_4 = 50 \qquad \text{(profit goal)} \tag{14.64}$$

$$2x_1 + 3x_2 + \underline{\eta_5} - \rho_5 = 80 \qquad \text{(market-shares goal)} \tag{14.65}$$

$$\text{and all } x_j, \eta_i, \text{ and } \rho_i \text{ are nonnegative} \tag{14.66}$$

Again, it may be noted that the normal form of the non-Archimedean LGP is already in multiplex form. The specific model components, in matrix notation, are as follows:

$$\mathbf{A} = \begin{bmatrix} 2 & 1 & 1 & 0 & 0 & 0 & 0 & -1 & 0 & 0 & 0 & 0 \\ 1 & 0 & 0 & 1 & 0 & 0 & 0 & 0 & -1 & 0 & 0 & 0 \\ 0 & 1 & 0 & 0 & 1 & 0 & 0 & 0 & 0 & -1 & 0 & 0 \\ \hline 3 & 1 & 0 & 0 & 0 & 1 & 0 & 0 & 0 & 0 & -1 & 0 \\ \hline 2 & 3 & 0 & 0 & 0 & 0 & 1 & 0 & 0 & 0 & 0 & -1 \end{bmatrix}$$

$$\mathbf{c}^{(1)} = [\,0 \quad 0 \quad \vdots \quad 0 \quad 0 \quad 0 \quad 0 \quad 0 \quad \vdots \quad 1 \quad 1 \quad 1 \quad 0 \quad 0\,]$$

$$\mathbf{c}^{(2)} = [\,0 \quad 0 \quad \vdots \quad 0 \quad 0 \quad 0 \quad 0 \quad 1 \quad \vdots \quad 0 \quad 0 \quad 0 \quad 0 \quad 0\,]$$

$$\mathbf{c}^{(3)} = [\,0 \quad 0 \quad \vdots \quad 0 \quad 0 \quad 0 \quad 1 \quad 0 \quad \vdots \quad 0 \quad 0 \quad 0 \quad 0 \quad 0\,]$$

$$\mathbf{v}^T = [\,x_1 \quad x_2 \quad \vdots \quad \eta_1 \quad \eta_2 \quad \eta_3 \quad \eta_4 \quad \eta_5 \quad \vdots \quad \rho_1 \quad \rho_2 \quad \rho_3 \quad \rho_4 \quad \rho_5\,]$$

$$\mathbf{b}^T = [\,50 \quad 20 \quad 30 \quad \vdots \quad 50 \quad \vdots \quad 80\,]$$

The observant reader may note that the elements of the **b** vector are not actually in priority order. That is, the first three elements are the right-hand sides for priority level 1. However, the fourth element is the right-hand side for the goal at the *third* priority level and the last element is the right-hand side for the *second*-level goal. If this bothers you, just interchange the fourth and fifth goals in the model. However, as noted earlier, it is not always necessary (or even possible) to maintain a completely rigid adherence to the format listed.

Multiplex Model: Chebyshev and Fuzzy Linear Goal Programs

The general form of the Chebyshev LGP model may be written as

$$\text{minimize } \delta \tag{14.67}$$

subject to:

$$\mathbf{A}'\mathbf{x} \ (\leq, \geq, \text{ or } =) \ \mathbf{r} \tag{14.68}$$

$$\mathbf{D}\mathbf{x} + \delta\mathbf{e} \geq \mathbf{r}' \tag{14.69}$$

$$\mathbf{x}, \delta \geq \mathbf{0} \tag{14.70}$$

where

    all original objectives are assumed to have been of the maximizing form

    $\mathbf{A}'\mathbf{x} \ (\leq, \geq, \text{ or } =) \ \mathbf{r}$ = original set of constraints (hard goals)

  $\mathbf{D}\mathbf{x}$ = original set of objective functions

    $\mathbf{r}'$ = right-hand sides, or aspiration levels, developed for the original set of objectives by means of solving a series of LP models

    δ = single worst deviation from any right-hand-side value for the soft goals

    $\mathbf{e}$ = column vector of 1's

The Chebyshev model, in multiplex form, is then

$$\text{lexmin } \mathbf{u} = \{\mathbf{c}^{(1)}\mathbf{v}, \ \mathbf{c}^{(2)}\mathbf{v}\} \tag{14.71}$$

subject to:

$$\mathbf{Av} = \mathbf{b} \tag{14.72}$$

$$\mathbf{v} \geq \mathbf{0} \tag{14.73}$$

where

$$\mathbf{A} = \begin{bmatrix} \mathbf{A}' & \mathbf{I} & \mathbf{0} & -\mathbf{I} & \mathbf{0} & \mathbf{0} \\ \hline \mathbf{D} & \mathbf{0} & \mathbf{I} & \mathbf{0} & -\mathbf{I} & \mathbf{e} \end{bmatrix}$$

$$\mathbf{c}^{(1)} = \begin{bmatrix} \mathbf{0} & \boldsymbol{\mu} & \boldsymbol{\omega} & \mathbf{0} \end{bmatrix}$$

$$\mathbf{c}^{(2)} = \begin{bmatrix} \mathbf{0} & \mathbf{0} & \mathbf{0} & \mathbf{1} \end{bmatrix}$$

$$\mathbf{v} = \begin{bmatrix} \mathbf{x} \\ \hline \boldsymbol{\eta} \\ \hline \boldsymbol{\rho} \end{bmatrix} \qquad \mathbf{b} = \begin{bmatrix} \mathbf{r} \\ \hline \mathbf{r}' \end{bmatrix}$$

To demonstrate, we shall employ the model of (13.33–13.39), as shown in what follows:

$$\text{minimize } \delta \tag{14.74}$$

subject to:

$$2x_1 + x_2 \leq 50 \qquad \text{(raw material limitations)} \tag{14.75}$$

$$x_1 \leq 20 \qquad \text{(market-level goal, product 1)} \tag{14.76}$$

$$x_2 \leq 30 \qquad \text{(market-level goal, product 2)} \tag{14.77}$$

$$3x_1 + x_2 + \delta \geq 70 \qquad \text{(profit goal)} \tag{14.78}$$

$$2x_1 + 3x_2 + \delta \geq 110 \qquad \text{(market-shares goal)} \tag{14.79}$$

$$\text{and } \delta \text{ and all } x_j \text{ are nonnegative} \tag{14.80}$$

Next, let us process the model via the addition of deviation variables. The resultant model follows:

$$\text{minimize } \delta \tag{14.81}$$

subject to:

$$2x_1 + x_2 + \eta_1 - \underline{\rho}_1 = 50 \tag{14.82}$$

$$x_1 + \eta_2 - \underline{\rho}_2 = 20 \tag{14.83}$$

$$x_2 + \eta_3 - \underline{\rho}_3 = 30 \tag{14.84}$$

$$3x_1 + x_2 + \underline{\eta_4} - \rho_4 + \delta = 70 \tag{14.85}$$

$$2x_1 + 3x_2 + \underline{\eta_5} - \rho_5 + \delta = 110 \tag{14.86}$$

$$\text{and } \delta \text{ and all } x_j, \eta_i, \text{ and } \rho_i \text{ are nonnegative} \tag{14.87}$$

When this model is placed into multiplex form, the representation is

$$\text{lexmin } \mathbf{u} = \{\mathbf{c}^{(1)}\mathbf{v}, \mathbf{c}^{(2)}\mathbf{v}\} \tag{14.88}$$

subject to:

$$\mathbf{Av} = \mathbf{b} \tag{14.89}$$

$$\mathbf{v} \geq \mathbf{0} \tag{14.90}$$

where

$$\mathbf{A} = \begin{bmatrix} 2 & 1 & 1 & 0 & 0 & 0 & 0 & -1 & 0 & 0 & 0 & 0 & 0 \\ 1 & 0 & 0 & 1 & 0 & 0 & 0 & 0 & -1 & 0 & 0 & 0 & 0 \\ 0 & 1 & 0 & 0 & 1 & 0 & 0 & 0 & 0 & -1 & 0 & 0 & 0 \\ \hline 3 & 1 & 0 & 0 & 0 & 1 & 0 & 0 & 0 & 0 & -1 & 0 & 1 \\ 2 & 3 & 0 & 0 & 0 & 0 & 1 & 0 & 0 & 0 & 0 & -1 & 1 \end{bmatrix}$$

$$\mathbf{c}^{(1)} = [\,0 \quad 0 \quad 0 \quad 0 \quad 0 \quad 1 \quad 1 \quad 1 \quad 1 \quad 1 \quad 0 \quad 0 \quad 0\,]$$

$$\mathbf{c}^{(2)} = [\,0 \quad 0 \quad 0 \quad 0 \quad 0 \quad 0 \quad 0 \quad 0 \quad 0 \quad 0 \quad 0 \quad 0 \quad 1\,]$$

$$\mathbf{v}^T = [\,x_1 \quad x_2 \quad \eta_1 \quad \eta_2 \quad \eta_3 \quad \eta_4 \quad \eta_5 \quad \rho_1 \quad \rho_2 \quad \rho_3 \quad \rho_4 \quad \rho_5 \quad \delta\,]$$

$$\mathbf{b}^T = [\,50 \quad 20 \quad 30 \quad 70 \quad 110\,]$$

Precisely the same procedure may be used to represent a fuzzy programming problem in multiplex format. The only difference will be that the deviation variable δ will be multiplied by a factor representing the difference between the best and worst objective function values as determined through the solution of the various LP models (refer to Chapter 13 for details).

Multiplex Model: Linear Lexicographic Vectormin Problems

Although it is possible to represent other models via the multiplex form, the final problem type that we shall describe herein is that of the lexicographic vectormin model. The general form of this model is

$$\text{minimize } \mathbf{d}^{(1)}\mathbf{x}$$

$$\text{minimize } \mathbf{d}^{(2)}\mathbf{x}$$

$$\vdots \tag{14.91}$$

$$\text{minimize } \mathbf{d}^{(p)}\mathbf{x}$$

subject to:

$$\mathbf{A}'\mathbf{x}\ (\leq,\ \geq,\ \text{or}\ =)\ \mathbf{b} \tag{14.92}$$

where

$$\mathbf{x} \geq \mathbf{0} \tag{14.93}$$

Further, it is assumed that the objectives (1 through p) have been preemptively ordered.

The equivalent multiplex representation is given as

$$\text{lexmin } \mathbf{u} = \{\mathbf{c}^{(1)}\mathbf{v},\ \mathbf{c}^{(2)}\mathbf{v},\ \ldots,\ \mathbf{c}^{(K)}\mathbf{v}\} \tag{14.94}$$

subject to:

$$\mathbf{A}\mathbf{v} = \mathbf{b} \tag{14.95}$$

$$\mathbf{v} \geq \mathbf{0} \tag{14.96}$$

$$K = p + 1$$

where

$$\mathbf{A} = [\mathbf{A}' \mathbin{\vdots} \mathbf{I} \mathbin{\vdots} -\mathbf{I}]$$

$$\mathbf{v}^T = [\mathbf{x} \mathbin{\vdots} \boldsymbol{\eta} \mathbin{\vdots} \boldsymbol{\rho}]$$

$$\mathbf{c}^{(1)} = [\mathbf{0} \mathbin{\vdots} \boldsymbol{\mu}^{(1)} \mathbin{\vdots} \boldsymbol{\omega}^{(1)}]$$

$$\mathbf{c}^{(2)} = [\mathbf{d}^{(1)} \mathbin{\vdots} \mathbf{0} \mathbin{\vdots} \mathbf{0}]$$

$$\vdots$$

$$\mathbf{c}^{(K)} = [\mathbf{d}^{(p)} \mathbin{\vdots} \mathbf{0} \mathbin{\vdots} \mathbf{0}]$$

To demonstrate, we shall employ the model from Chapter 14 as represented by functions (14.70–14.75), as repeated in what follows. Here, it is assumed that the first objective (i.e., profit) is the highest ranked.

Find x_1 and x_2 so as to

maximize $z_1 = 3x_1 + x_2$ (profit objective)   (14.97)

maximize $z_2 = 2x_1 + 3x_2$ (market-shares objective)   (14.98)

subject to:

$$2x_1 + x_2 \leq 50 \qquad \text{(raw material limitations)} \tag{14.99}$$

$$x_1 \leq 20 \qquad \text{(market-level goal, product 1)} \tag{14.100}$$

$$x_2 \leq 30 \qquad \text{(market-level goal, product 2)} \tag{14.101}$$

$$\text{and all } x_j \text{ are nonnegative} \tag{14.102}$$

The problem, in multiplex form, is then

$$\text{lexmin } \mathbf{u} = \{(\rho_1 + \rho_2 + \rho_3),\ (-3x_1 - x_2),\ (-2x_1 - 3x_1)\} \tag{14.103}$$

subject to:

$$2x_1 + x_2 + \eta_1 - \underline{\rho}_1 = 50 \tag{14.104}$$

$$x_1 + \eta_1 - \underline{\rho}_1 = 20 \tag{14.105}$$

$$x_2 + \eta_1 - \underline{\rho}_1 = 30 \tag{14.106}$$

and all x_j, η_i, and ρ_i are nonnegative $\tag{14.107}$

where

$$\mathbf{A} = \begin{bmatrix} 2 & 1 & 1 & 0 & 0 & -1 & 0 & 0 \\ 1 & 0 & 0 & 1 & 0 & 0 & -1 & 0 \\ 0 & 1 & 0 & 0 & 1 & 0 & 0 & -1 \end{bmatrix}$$

$$\mathbf{c}^{(1)} = \begin{bmatrix} 0 & 0 & 0 & 0 & 0 & 1 & 1 & 1 \end{bmatrix}$$

$$\mathbf{c}^{(2)} = \begin{bmatrix} -3 & -1 & 0 & 0 & 0 & 0 & 0 & 0 \end{bmatrix}$$

$$\mathbf{c}^{(3)} = \begin{bmatrix} -2 & -3 & 0 & 0 & 0 & 0 & 0 & 0 \end{bmatrix}$$

$$\mathbf{v}^T = \begin{bmatrix} x_1 & x_2 & \eta_1 & \eta_2 & \eta_3 & \rho_1 & \rho_2 & \rho_3 \end{bmatrix}$$

$$\mathbf{b}^T = \begin{bmatrix} 50 & 20 & 30 \end{bmatrix}$$

GOOD AND POOR MODELING PRACTICES

Ultimately, the results derived from mathematical programming are dependent upon the quality of the model that has been developed to represent the problem at hand. Over the years, we have been exposed to a variety of modeling practices in both single and multiobjective optimization—some very good and others not nearly so good. Thus, in this section, we shall briefly touch upon certain modeling practices in multiobjective optimization.

First, let us comment on the form of the achievement function in multiplex and, in particular, in goal programming. The particular form used in this text is that of the lexicographic minimum of an ordered vector, or

$$\text{lexmin } \mathbf{u} = \{\mathbf{c}^{(1)}\mathbf{v}, \ldots, \mathbf{c}^{(K)}\mathbf{v}\}$$

However, there are numerous investigators (in particular, those in the goal-programming sector) who employ an alternative form. In general, this alternative representation takes on the following form:

$$\text{minimize } z = P_1(\cdot) + P_2(\cdot) + \cdots + P_K(\cdot) \tag{14.108}$$

where

$(\cdot)$ = represents the specific term of the achievement vector

P_k = indicates the priority level of the kth achievement vector term

For example, by using this latter approach in linear goal programming, a particular achievement function might be written as

$$\text{minimize } z = P_1(\rho_1 + \rho_2) + P_2(\eta_3) + P_3(\eta_1 + 2\eta_2) \qquad (14.109)$$

Now, as long as it is clearly understood that the summation signs within this particular form of an achievement function are *meaningless*, then the use of such a formulation *might* not present a major problem. However, all too often, those employing such a formulation tend to forget (or, perhaps, never fully appreciate) that we are seeking the *lexicographic minimum of an ordered vector*—and *not* the scalar minimum of a summation. And, when this happens, some very strange procedures and interpretations seem to follow. As just one example, those who employ the summation form have had considerable difficulty in their attempts to develop the dual of linear goal programs (i.e., the multidimensional dual to be presented in Chapter 16). In fact, there are even some in the GP sector who actually have claimed that "the dual of a linear goal program does not exist." However, and as we shall see in Chapter 16, the dual of a linear goal program—or the dual of a multiplex model in general—is easily derived when the proper form of the achievement vector is used.

Another problem that seems to have occurred as a result of the summation form is that of the use of "large" coefficients to replace the priority level (P_k) designation. That is, there are those who would advocate the employment of the following function in place of that one employed in (14.109):

$$\text{minimize } z = M_1(\rho_1 + \rho_2) + M_2(\eta_3) + M_3(\eta_1 + 2\eta_2) \qquad (14.110)$$

Here, M_1 is now some "huge positive number" with respect to M_2, and M_2 is a "huge positive number" with respect to M_3. For example, we might let

$$M_1 = 10^6 \qquad M_2 = 10^3 \qquad M_3 = 1$$

Hillier and Liebermann (1990) have, in fact, given a name to the above concept. They call it the "streamlined procedure" for preemptive linear goal programming. Unfortunately, there are at least four problems with this approach:

- The use of "huge positive numbers" is something rarely advocated in mathematical programming due to the problems such numbers impart to the actual solution procedure.
- The use of the "streamlined" form serves to lend unwarranted credibility to the summation format of the achievement function rather than its proper, ordered vector form.
- The rationale behind the "streamlined" form is evidently based upon the erroneous notion that, with such a form, one may employ an easier and computationally efficient approach: the conventional simplex.
- And, finally, there is always the question as to just how "huge" the huge positive numbers have to be in order to be assured of developing the correct solution.

Another poor modeling practice is that of the use of goals consisting solely of deviation variables. Consider, for example, the following two goals. And note that the second goal consists solely of deviation variables.

$$x_1 + x_2 + \eta_1 - \underline{\rho}_1 = 40 \qquad (14.111)$$

$$\eta_1 + \eta_2 - \underline{\rho}_2 = 10 \qquad (14.112)$$

These two goals represent a portion of a linear goal program that was developed in an attempt to represent the desire, on the part of the analyst, to simultaneously (in the first goal) keep labor hours to less than 40 per week and (in the second goal) also keep idle time (η_1) to less than 10 hours per week. However, there are some major problems with such a modeling practice.

A considerably more efficient form of the previous two goals would be achieved through the use of the following formulations:

$$x_1 + x_2 + \eta_1 - \underline{\rho}_1 = 40 \qquad (14.113)$$

$$x_1 + x_2 + \underline{\eta}_2 - \rho_2 = 30 \qquad (14.114)$$

This latter (equivalent) formulation will require fewer variables than the former. This is so because by using (14.112) we have rather obviously destroyed the dependency that *should* exist between η_1 and ρ_1. Thus, η_1 can now no longer serve as an initial basic variable—and a new variable must be added to (14.111).

Another very perplexing error that is made in model development is that of including *wanted* deviation variables along with *unwanted* deviation variables in the achievement vector. Remember, the only deviation variables that should appear in the achievement vector are those we wish to minimize, that is, those that are unwanted.

Finally, consider yet one more example of a poor modeling practice. This is the case of assigning unrealistically high or low values to the aspiration levels employed in transforming an objective into a goal. For example, let us assume that we wish to maximize a particular objective function and intend to employ a goal-programming model. For example, let us assume that we wish to maximize the number of units of product type 1 that is manufactured—where x_1 denotes this number. Thus, initially, we have an objective function denoted as

$$\text{maximize } x_1$$

To transform this objective into a goal, we must first establish an aspiration level for x_1. What some analysts do is to "simply" establish some huge positive number for this aspiration level (e.g., 99999) and thus the goal function becomes

$$x_1 + \underline{\eta} - \rho = 99999$$

Alternatively, had we wished to minimize x_1, these same analysts would have employed 0 (zero) as the aspiration level. It should be clear to the reader, we hope, that the use of such unrealistic aspiration level values can have a major impact on the solution derived. For example, if the objective/goal discussed earlier were at priority level 2, then its achievement would likely render any lower-ranked goals virtually insignificant.

In summary, there are a number of do's and don'ts when building a multiplex model. These include the following:

- don't limit your choice of model type (e.g., to just non-Archimedean goal programming or just traditional linear programming, etc.)
- don't include anything other than the deviation variables associated with rigid goals in the achievement term of priority level 1
- don't include any *wanted* deviation variables in the achievement function
- don't include an excessive number of priority levels (i.e., any model consisting of more than five or so priority levels is generally suspect)
- don't construct goals consisting solely of deviation variables
- don't employ completely unrealistic aspiration level values
- do use the ordered vector form of the achievement function if you wish to both better appreciate the underlying concepts as well as permit extension to such notions as duality in multiplex

SUMMARY

In this chapter, we have presented the terminology to be used, described certain fundamental components of the modeling process (e.g., weighting and ranking), and indicated just how one represents a number of different model types by means of the multiplex format. Our purpose in all this is to provide a *single* model type to be dealt with, regardless of the particular class of problem addressed. Then, by means of the multiplex algorithm to be presented in the next chapter, one may solve and analyze the problem—using a *single*, *unified* approach.

EXERCISES

14.1. For the problem described in Exercise 13.4, Chapter 13, develop a conventional linear programming representation by means of the assertion of the following assumptions:
- the single most important objective is to maximize profit
- assembly hours and test hours may not, *under any conditions*, use any overtime
- at least 50 units of all types of computers should be assembled each week

14.2. For the problem described in Exercise 13.4, Chapter 13, develop an Archimedean LGP representation by means of the assertion of the following assumptions:
- market visibility is twice as important as profit
- at least $9000 in profit is desired per week

- the production of at least 50 computers (or any type or combination) is desired per week
- assembly and test hours goals are the only rigid goals

14.3. For the problem described in Exercise 13.4, Chapter 13, develop a non-Archimedean representation by means of the assertion of the same assumptions as in Exercise 13.6 except that we no longer wish to make profit and market shares commensurable (i.e., they are to be in separate priority levels with market visibility at a higher priority level).

14.4. For the problem described in Exercise 13.4, develop a Chebyshev representation by means assuming that the limits on assembly and test hours are rigid. Recall that this will require the solution of two LP models.

14.5. For the problem described in Exercise 13.4, Chapter 13, develop both the lexicographic vectormax representation and the generating (vectormax) representation. In the first case, assume that profit is more important than market visibility.

14.6. Establish a matrix, or table, in which the columns are headed by the various model types discussed in both this chapter and Chapter 13 and the rows list the necessary assumptions to convert a baseline model into that particular type of problem—and then fill in the table. For example, one column of this table would be headed "Archimedean LGP" and one row would be headed "Declaration of Aspiration Levels." Then, in the matrix cell at the intersection of this column and row, we would place a "yes."

14.7. For the following exercises, develop the associated multiplex representation (include, in this representation, a statement of all associated vectors and matrices):
(a) Exercise 14.1.
(b) Exercise 14.2.
(c) Exercise 14.3.
(d) Exercise 14.4.
(e) Exercise 14.5.

14.8. A groundwater pumping station to provide potable water is to be constructed in a relatively small country town. The site of the station is fixed because of the availability of well water. Thus, the only questions remaining are as follows.
- Which type of monitoring station (differing in terms of automation) should be used?
- From which firm should the associated machinery be purchased?
- How many workers need to be hired to run the station (using three 8-hour shifts)? The town residents wish to minimize total initial costs as well as future operating costs (with twice as much importance given to initial costs). However, because there is a high level of unemployment in the area, they also want to maximize the number of workers gainfully employed. Assuming that the maximization of employment levels has top priority, and that minimizing costs has a lower priority, formulate the associated goal-programming model—and then represent that model in multiplex form. The data associated with this problem are listed in Table 14.4.

14.9. Formulate the problem described in Exercise 14.8 as a conventional linear programming model. List all of the assumptions that you had to employ in order to formulate the single-objective model—and discuss your impression as to both the validity of these assumptions and their potential impact on the problem solution.

TABLE 14.4

	Monitoring station type		Pumping machinery type		
	A	B	I	II	III
1. Initial costs ($ millions)	2	1.5	5	4	3.5
2. Yearly Operating Costs ($ thousands)	100	250	200	300	800
3. Number of personnel required per shift and associated hourly salary	4 @ $15/h	6 @ $15/h	6 @ $10/h	10 @ $14/h	15 @ $12/h

14.10. By means of the AHP method (i.e., for the weighting of unwanted goal deviations) in conjunction with the extension proposed by Gass (the eigenvalue translation), develop an Archimedean LGP representation for Exercise 14.8. In support of this effort, assume that the personnel goal is very strongly more important than that of initial costs, and initial costs are moderately more important than operating costs.

14.11. Using the model developed in Exercise 14.10, *solve* this model by means of conventional simplex. Note that there is no need to revise the conventional simplex method when the Archimedean LGP model is under consideration. Comment on the results.

MULTIPLEX ALGORITHM

CHAPTER OVERVIEW

In the previous chapter, we demonstrated an approach for constructing a single type of model, the multiplex model, for the representation of a variety of problem classes. In this chapter, we shall address the algorithm to be employed in solving any multiplex model. Actually, we shall present two multiplex algorithms. The first, the "revised form," is theoretically the most efficient. However, the second approach, "sequential multiplex," permits one to employ any existing commercial LP software to solve the multiplex model.* As a consequence, it is this latter approach that has seen the widest application to real-world problems—having been employed, in fact, since the late 1960s (Ignizio, 1967). We conclude the chapter with a brief discussion of computational considerations. However, we wish to emphasize that, despite some additional terminology and notation, the multiplex algorithm is little more than a very simple, very straightforward, and quite transparent extension of the conventional simplex method. In fact, the easiest way to think of the relationship between multiplex and simplex is to simply view the simplex algorithm as a special case of multiplex. As such, if something can be done with simplex (e.g., sensitivity analysis and computational enhancements), it may certainly be accomplished via multiplex.**

---

* Even further, the sequential approach may be easily extended to encompass the solution of either discrete or nonlinear multiplex problems. We will touch on this topic in chapter 17.

** The observations of these last two sentences have, however, seemed to have eluded some of

REDUCED FORM OF THE MULTIPLEX MODEL

Before we introduce the multiplex algorithm, let us consider the fundamental basis for this algorithm. This may be accomplished by transforming the general multiplex model into the so-called reduced form. In fact, the transformation follows virtually the same steps as that used in Part I to develop the reduced form of the LP model (Chapter 4). The primary difference is that we ultimately derive a *matrix* of reduced costs (i.e., the so-called shadow prices, or marginal prices) rather than a *vector*. To begin, let us rewrite the general form of the linear multiplex model, as follows.

Find **v** so as to

$$\text{lexmin } \mathbf{u} = \{\mathbf{c}^{(1)}\mathbf{v}, \mathbf{c}^{(2)}\mathbf{v}, \ldots, \mathbf{c}^{(K)}\mathbf{v}\} \tag{15.1}$$

subject to:

$$\mathbf{Av} = \mathbf{b} \tag{15.2}$$

$$\mathbf{v} \geq \mathbf{0} \tag{15.3}$$

where

$$\mathbf{v} = \begin{pmatrix} \mathbf{x} \\ \boldsymbol{\eta} \\ \boldsymbol{\rho} \end{pmatrix} \tag{15.4}$$

The reader should observe that we have eliminated the use of dashed lines to indicate the partitions within the vector for **v**. From here on, such lines will only be employed when deemed necessary to emphasize the particular partitioning format under consideration.

Let us now note that the technological matrix, **A**, may be partitioned as

$$\mathbf{A} = (\mathbf{B} : \mathbf{N}) \tag{15.5}$$

where

B = $m \times n$ nonsingular matrix, designated as the *basis matrix*

N = $m \times (n - m)$ matrix (the matrix of nonbasic columns)

Further, the variable set, **v**, may be similarly partitioned into

$$\mathbf{v} = \begin{pmatrix} \mathbf{v}_B \\ \mathbf{v}_N \end{pmatrix} \tag{15.6}$$

---

the investigators in the multiobjective sector. As a result, we see a fairly steady flow of papers in which one analyst or another has "discovered" that some approach common to LP may also be utilized in multiobjective optimization. Unfortunately, this is particularly true in the goal-programming sector, where reinvention is almost the order of the day.

where

$\mathbf{v}_B$ = set of *basic* variables (those associated with $\mathbf{B}$, the basic columns of $\mathbf{A}$)

$\mathbf{v}_N$ = set of *nonbasic* variables (those associated with $\mathbf{N}$, the nonbasic columns of $\mathbf{A}$), variables that are typically set to a value of zero

Consequently, we may rewrite (15.2) as

$$\mathbf{Bv}_B + \mathbf{Nv}_N = \mathbf{b} \tag{15.7}$$

Now, because $\mathbf{B}$ is nonsingular (and thus has an inverse), we may premultiply each term in (15.7) by $\mathbf{B}^{-1}$ to obtain:

$$\mathbf{B}^{-1}\mathbf{Bv}_B + \mathbf{B}^{-1}\mathbf{Nv}_N = \mathbf{B}^{-1}\mathbf{b}$$

which leads to

$$\mathbf{v}_B + \mathbf{B}^{-1}\mathbf{Nv}_N = \mathbf{B}^{-1}\mathbf{b}$$

solving for $\mathbf{v}_B$, we have

$$\mathbf{v}_B = \mathbf{B}^{-1}\mathbf{b} - \mathbf{B}^{-1}\mathbf{Nv}_N \tag{15.8}$$

Thus, we may express the set of basic variables ($\mathbf{v}_B$) in terms of a set of constants ($\mathbf{B}^{-1}\mathbf{b}$) minus a function of the nonbasic variable set ($\mathbf{B}^{-1}\mathbf{Nv}_N$). As a consequence, this last relationship will tell us precisely what will happen to the value of the basic variables if any nonbasic variable is given a positive value.

Let us next examine the multiplex achievement vector, as given in (15.1). The kth term of the achievement vector is $\mathbf{c}^{(k)}$. However, using the partition of $\mathbf{v}$ as in (15.6), we may rewrite $\mathbf{c}^{(k)}$ as

$$\mathbf{c}_B^{(k)}\mathbf{v}_B + \mathbf{c}_N^{(k)}\mathbf{v}_N \tag{15.9}$$

where the subscripts for $\mathbf{c}$ are used to indicate whether these coefficients are associated with the basic (B) or nonbasic (N) variables.

Using (15.8), we may substitute for $\mathbf{v}_B$ in (15.9) to obtain

$$\mathbf{c}^{(k)}\mathbf{v} = \mathbf{c}_B^{(k)}\mathbf{B}^{-1}\mathbf{b} - (\mathbf{c}_B^{(k)}\mathbf{B}^{-1}\mathbf{N} - \mathbf{c}_N^{(k)})\mathbf{v}_N \tag{15.10}$$

Then, for sake of clarity, we may let

$$\boldsymbol{\beta} = \mathbf{B}^{-1}\mathbf{b} \tag{15.11}$$

$$\boldsymbol{\pi}^{(k)} = \mathbf{c}_B^{(k)}\mathbf{B}^{-1} \tag{15.12}$$

$$\boldsymbol{\alpha} = \mathbf{B}^{-1}\mathbf{N} \quad \text{and} \quad \boldsymbol{\alpha}_j = \mathbf{B}^{-1}\mathbf{a}_j \quad \text{(where } \mathbf{a}_j \text{ is the } j\text{th column of } \mathbf{A}) \tag{15.13}$$

Using the results just developed, we are now ready to write the multiplex model in reduced form. This follows.

Find $\mathbf{v}$ so as to

$$\text{lexmin } \mathbf{u} = \{[\mathbf{c}_B^{(1)}\boldsymbol{\beta} - (\boldsymbol{\pi}^{(1)}\mathbf{N} - \mathbf{c}_N^{(1)})\mathbf{v}_N], \dots ,$$
$$[\mathbf{c}_B^{(K)}\boldsymbol{\beta} - (\boldsymbol{\pi}^{(K)}\mathbf{N} - \mathbf{c}_N^{(K)})\mathbf{v}_N]\} \tag{15.14}$$

subject to:

$$\mathbf{v}_B = \mathbf{B}^{-1}\mathbf{b} - \mathbf{B}^{-1}\mathbf{N}\mathbf{v}_N = \boldsymbol{\beta} - \boldsymbol{\alpha}\mathbf{v}_N \qquad (15.15)$$

$$\mathbf{v} = \begin{pmatrix} \mathbf{v}_B \\ \mathbf{v}_N \end{pmatrix} \geq \mathbf{0} \qquad (15.16)$$

Despite the additional notation necessary in the multiplex model, if we examine the previously reduced form carefully, we should note that it is essentially nothing more than the reduced form of the LP model, with the exception of K terms in the achievement vector rather than one term in the reduced LP objective function. This is made even more obvious when one places the reduced form into a tableau representation, as depicted in Table 15.1. Again, we may note that this tableau is identical to the LP (reduced) tableau—with the exception of *multiple shadow price* (or reduced cost) rows (i.e., the rows headed by the P_k designator, where P_k indicates the priority level under consideration).

The reader should note that the reduced costs have been placed at the bottom *of this tableau, rather than at the top of the tableau—as was the case in Part 1. We have done this so as to draw a clear distinction between the maximizing approach of Part 1 and the minimizing (i.e., lexicographic minimum) approach of Part 3. Consequently, look for the reduced costs, as well as the pricing matrix, to appear at the* bottom *of all tableaux employed in Part 3.*

TABLE 15.1   MULTIPLEX TABLEAU

	$\mathbf{v}_B$	$\mathbf{v}_N$	β (or the rhs)
$\mathbf{v}_B$	$\mathbf{B}^{-1}\mathbf{B} = \mathbf{I}$	$\mathbf{B}^{-1}\mathbf{N}$	$\mathbf{B}^{-1}\mathbf{b}$
P_1		$\mathbf{c}_B^{(1)}\mathbf{B}^{-1}\mathbf{N} - \mathbf{c}_N^{(1)}$	$\mathbf{c}_B^{(1)}\mathbf{B}^{-1}\mathbf{b}$
$\vdots$	$\mathbf{0}$		
P_K		$\mathbf{c}_B^{(K)}\mathbf{B}^{-1}\mathbf{N} - \mathbf{c}_N^{(K)}$	$\mathbf{c}_B^{(K)}\mathbf{B}^{-1}\mathbf{b}$

Basic Solutions and Basic Feasible Solutions

As in the case of LP, a *basic solution* is one in which all nonbasic variables are set to their lower bound. This bound, for our purposes, shall simply be zero. Thus, if $\mathbf{v}_N = \mathbf{0}$, there is an associated basic solution, $\mathbf{v}_B$. Further, if all the variables in the basic solution are nonnegative, we have a *basic feasible solution*.

This may be defined using matrix notation. Specifically, if $\mathbf{v}_N = \mathbf{0}$, then $\mathbf{B}^{-1}\mathbf{N}\mathbf{v}_N = \mathbf{0}$ and thus a basic solution may be written as

$$\mathbf{v}_B = \mathbf{B}^{-1}\mathbf{b} - \mathbf{B}^{-1}\mathbf{N}\mathbf{v}_N = \mathbf{B}^{-1}\mathbf{b} \qquad (15.17)$$

A basic feasible solution is thus given as

$$\mathbf{v} = \begin{pmatrix} \mathbf{v}_B \geq \mathbf{0} \\ \mathbf{v}_N = \mathbf{0} \end{pmatrix} \geq \mathbf{0} \tag{15.18}$$

In multiplex, as in LP, the optimal solution found by the associated algorithm will always be a basic feasible solution. And this is true even in the case of goal programming, where the optimal solution may form a region (i.e., wherein convergence will be to one of the extreme points that serve to define the region).

Associated Conditions

The four primary conditions associated with the reduced form of the multiplex model are *feasibility*, *implementability*, *optimality*, and *unboundedness*. Given that $\mathbf{v}_N = \mathbf{0}$, these conditions may be defined as follows.

> *Feasibility.* If $\boldsymbol{\beta} = \mathbf{B}^{-1}\mathbf{b} \geq \mathbf{0}$, the resultant solution (i.e., program) is considered to be feasible. Otherwise (i.e., if any element of β is negative), the program is infeasible.
>
> *Implementability.* If $\mathbf{c}_B^{(1)\prime}\boldsymbol{\beta} = \mathbf{c}_B^{(1)\prime}\mathbf{B}^{-1}\mathbf{b} = 0$, then the resultant program (i.e., $\mathbf{v}_B$) is termed an implementable solution. That is, the top-ranked goal set (i.e., the hard goals) have been satisfied.
>
> *Optimality.* If every reduced price vector (designated as $\mathbf{d}_j$) is *lexicographically nonpositive*, the associated basic feasible solution is optimal. This optimal program is typically designated as $\mathbf{v}_B^*$. Note carefully that in the case of goal programming, the "optimal" program simply corresponds to the satisficing solution.
>
> *Unboundedness.* An unbounded solution exists if it is possible to exchange, in a given basic feasible solution, a nonbasic variable for a basic variable and that nonbasic variable can take on an infinite value—and the associated basic solution remains feasible.

Before proceeding further, we should address the new terminology introduced in the definition of optimality. This is the notion of a *lexicographically nonpositive* vector. Quite simply, a lexicographic nonpositive vector is one in which the first nonzero element is negative. Of course, a vector consisting solely of zeros is also considered to be lexicographically nonpositive. Thus, for the vectors listed, $\mathbf{r}$, $\mathbf{s}$, and $\mathbf{q}$ are lexicographically nonpositive, whereas $\mathbf{t}$ is not.

$$\mathbf{r} = \begin{pmatrix} 0 \\ -1 \\ 90 \\ 5 \\ -7 \end{pmatrix} \qquad \mathbf{s} = \begin{pmatrix} 0 \\ 0 \\ 0 \\ 0 \\ 0 \end{pmatrix} \qquad \mathbf{t} = \begin{pmatrix} 0 \\ 1 \\ -999 \\ 587 \\ 667 \end{pmatrix} \qquad \mathbf{q} = \begin{pmatrix} 0 \\ 0 \\ 0 \\ 0 \\ -7 \end{pmatrix}$$

Recall that in the reduced form of the multiplex model, the kth term of the achievement vector is given as $\mathbf{c}_B^{(k)'}\boldsymbol{\beta} - (\boldsymbol{\pi}^{(k)}\mathbf{N} - \mathbf{c}_N^{(k)})\mathbf{v}_N$. The first portion of this term $(\mathbf{c}_B^{(k)'}\boldsymbol{\beta})$ is simply a constant. The second part of the term, $(\boldsymbol{\pi}^{(k)}\mathbf{N} - \mathbf{c}_N^{(k)})\mathbf{v}_N$, is a function of the nonbasic variable set—and this portion of the term indicates just how much the kth term of the achievement vector will change as a function of any change in the value of one or more of the nonbasic variables. Thus, the term

$$\boldsymbol{\pi}^{(k)}\mathbf{N} - \mathbf{c}_N^{(k)} \tag{15.19}$$

is designated as the *vector of reduced costs*. It should be obvious that any improvement (i.e., decrease) in the kth term of the achievement vector is only possible if at least one of the terms within $\boldsymbol{\pi}^{(k)}\mathbf{N} - \mathbf{c}_N^{(k)}$ is positive.

The jth element of (15.19) is denoted as d_j and is given as

$$d_j^{(k)} = z_j^{(k)} - c_j^{(k)} \tag{15.20}$$

wherein

$$z^{(k)} = (\boldsymbol{\pi}^{(k)}\mathbf{N})_j = \boldsymbol{\pi}^{(k)}\mathbf{a}_j \tag{15.21}$$

As may be seen, associated with each nonbasic variable is a column vector of $d_j^{(k)}$ elements. This vector is termed, in multiplex, as the vector of multidimensional shadow prices (or multidimensional reduced costs) or, simply, as the shadow price vector, $\mathbf{d}_j$. The entire matrix of these shadow price vectors is denoted as $\mathbf{D}$. Further, if each column of $\mathbf{D}$ is lexicographically nonpositive, the associated program is optimal.

Recall that with a conventional linear program, we may have *alternative optimal solutions*—and this is denoted by a zero-valued reduced cost for a nonbasic variable in the final (optimal) tableau. Alternative optimal solutions also exist in multiplex models. However, in the case of a multiplex model, alternative optimal solutions are indicated whenever there is a complete *vector* of zero-valued reduced costs associated with a nonbasic variable in the final tableau.

With this material and terminology behind us, we are ready to list the steps of the revised multiplex algorithm. The reader should note both the similarities and differences between this algorithm and the simplex algorithm presented in Part I.

REVISED MULTIPLEX ALGORITHM

Before we list the steps of the revised multiplex algorithm, let us first describe the algorithm in strictly narrative terms. We shall assume that we have developed a basic feasible solution (bfs), where this bfs will serve as a starting point in the search for the optimal program. When we employ a multiplex model, this initial basic feasible solution consists of the set of negative deviation variables. Stated another way, the initial basic feasible solution for a multiplex model is developed by setting all structural variables and all positive deviation variables to zero.

Initially, our attention shall be focused on priority level 1, the satisfaction of the set of hard goals. As soon as the first term of the achievement vector is zero, we know that we have achieved these goals. We move from one bfs to another by exchanging one basic variable for one nonbasic variable—*if* such an exchange will: (1) retain solution feasibility, and (2) result in, at best, an improvement in the achievement vector (i.e., a reduction in the value of the first term) or, at worst, no change in the achievement vector. These exchanges are designated as multiplex pivots, and they are repeated until the first term in the achievement vector has reached its minimal value. If this value is zero, then we will be able to develop an implementable program. Otherwise, the problem has no implementable program (as it is presently modeled).

Having minimized the first term in the achievement vector, $u^{(1)}$, we move to the second term—and attempt to minimize it, *but without degrading the value previously achieved for the first term*. Minimization is achieved through multiplex pivots, the one-for-one exchanges of basic and nonbasic variables. However, in addition to adhering to feasibility and solution-improvement requirements, we must avoid any exchange that would serve to degrade the value of $u^{(1)}$. This process continues until the lexicographic minimum of the achievement vector is found (designated as $\mathbf{u}^*$).

Algorithm Steps

Although the reader should certainly employ the multiplex algorithm to solve a certain number of example problems by hand, solution of anything other than quite small problems should be accomplished on the computer. As such, we shall assume that the algorithm will ultimately be implemented on the computer. Thus, we must store $\mathbf{A}$, $\mathbf{b}$, and all $\mathbf{c}^{(k)}$. We will also assume that $\mathbf{b} \geq \mathbf{0}$.

Further, we shall assume that we have an initial bfs (i.e., as formed by the negative deviation variables) and thus an initial representation of $\mathbf{B}^{-1}$ and $\boldsymbol{\beta} = \mathbf{B}^{-1}\mathbf{b}$. Because the first bfs is associated with $\boldsymbol{\eta}$, $\mathbf{B}^{-1} = \mathbf{I}$ initially.

In actual practice, the following steps may be (and almost always are) augmented by certain refinements so as to increase the computational efficiency. These are, for the most part, the same refinements (e.g., LU decomposition, product form of the inverse, coefficient packing, and multiple pricing) used to enhance the simplex algorithm, and a complete description of such methods may be found in the literature (Murtagh, 1981).

STEP 1. *Initialization.* Let $\mathbf{v}_B = \boldsymbol{\eta}$. Thus, $\mathbf{B}^{-1} = \mathbf{I}$ and $\boldsymbol{\beta} = \mathbf{B}^{-1}\mathbf{b} = \mathbf{b}$. Set $k = 1$. Initially, all variables are *unchecked*.

STEP 2. *Develop the pricing vector for level k.* Determine:

$$\boldsymbol{\pi}^{(k)} = \mathbf{c}_B^{(k)}\mathbf{B}^{-1} \tag{15.22}$$

STEP 3. *Price out all **unchecked**, nonbasic variables at level k.* Compute

$$d_j^{(k)} = \boldsymbol{\pi}^{(k)}\mathbf{a}_j - c_j^{(k)} \qquad \text{for all } j \in \mathcal{N} \tag{15.23}$$

where $\mathcal{N}$ represents the set of nonbasic *and* unchecked variables.

STEP 4. *Selection of the entering nonbasic variables.* Examine those $d_j^{(k)}$ computed in Step 3. If none is positive, proceed to Step 8. Otherwise, select the nonbasic variable with the most positive $d_j^{(k)}$ (ties may be broken arbitrarily) as the entering variable. Designate this entering variable as v_q.

STEP 5. *Update the entering column.* Compute

$$\boldsymbol{\alpha}_q = \mathbf{B}^{-1}\mathbf{a}_q \tag{15.24}$$

STEP 6. *Determine the leaving basic variable.* Designate the row associated with the leaving basic variable as row p. Using the present representation of $\boldsymbol{\beta}$, and the values of $\boldsymbol{\alpha}_q$ as derived in Step 5, determine

$$\beta_p/\alpha_{p,q} = \min_i\{\beta_i/\alpha_{i,q}\}, \qquad \text{for } \alpha_{i,q} > 0 \tag{15.25}$$

Ties may be broken arbitrarily. If no such ratios may be determined (i.e., if $\alpha_{i,q} \leq 0$, for all i), the problem is *unbounded* and the algorithm may be terminated. Otherwise, the basic variable associated with row p is the leaving variable—designated as $v_{B,p}$.

STEP 7. *Pivot.* Replace $\mathbf{a}_p$ in $\mathbf{B}$ by $\mathbf{a}_q$ and compute the new basis inverse, $\mathbf{B}^{-1}$. Return to Step 2.

STEP 8. *Convergence check.* If either (or both) of the following conditions holds, terminate the algorithm as the optimal program has been determined:
(i) if all $d_j^{(k)}$ as computed in Step 3 are negative, or
(ii) if $k = K$.
Otherwise, *check* all nonbasic variables associated with *negative* values of $d_j^{(k)}$, set $k = k + 1$, and return to Step 2.

These steps, with one exception, serve to completely define the revised multiplex algorithm. The only step open to interpretation is Step 7, the pivoting process. Here, the reader may wish to employ whatever method he or she wishes to compute the new basis inverse—and such approaches are detailed in the literature (Murtagh, 1981). However, in order to proceed in a consistent manner, let us utilize one specific approach to the development of the inverse, that is, we shall employ the so-called *explicit form of the inverse*. This particular approach has two advantages. First, it is easy to carry out by hand. Second, it is even quite efficient for a computer-based version of the algorithm. Consequently, let us now describe the establishment and interpretation of the tableau in support of the explicit form of the inverse.

Explicit Form of the Inverse: Tableau Representation

The general form of the tableau for the explicit form of the inverse (EFI) is depicted in Table 15.2. Again note that we have placed the elements of $\boldsymbol{\Pi}$ (the pricing matrix) at the bottom of this tableau so as to call attention to the fact that we are employing lexicographic minimization.

TABLE 15.2 EFI TABLEAU

	Basis inverse	RHS
$v_{B,1}$		β_1
$\vdots$	$\mathbf{B}^{-1}$	$\vdots$
$v_{B,m}$		β_m
$\boldsymbol{\pi}^{(1)'}$		$u^{(1)}$
$\vdots$	Π	$\vdots$
$\boldsymbol{\pi}^{(K)'}$		$u^{(K)}$

Note carefully that this tableau contains only that information absolutely necessary to carry out the multiplex algorithm (where, as we have noted, it is assumed that we have stored, in memory, $\mathbf{A}$, $\mathbf{b}$, and all $\mathbf{c}^{(k)}$). Specifically, the EFI tableau contains

$\mathbf{B}^{-1}$ = inverse of the present basis (which, initially, is simply $\mathbf{I}$)

$\boldsymbol{\beta}$ = present right-hand side, as given by $\mathbf{v}_B = \mathbf{B}^{-1}\mathbf{b}$, where β_m represents the specific value of $v_{B,m}$

$u^{(k)}$ = kth term of the achievement vector, as given by $\mathbf{c}_B^{(k)}\mathbf{B}^{-1}\mathbf{b}$

$\boldsymbol{\pi}^{(k)}$ = pricing vector for priority level k, as given by $\boldsymbol{\pi}^{(k)} = \mathbf{c}_B^{(k)}\mathbf{B}^{-1}$

Π = complete pricing matrix

The substantial advantages, in terms of both storage and computations, of the EFI tableau over the full tableau of Table 15.1 should be obvious.

Actually, in performing the steps of the multiplex algorithm, it is not necessary to include any of the achievement-vector terms or the complete Π matrix. That is, we need only include the pricing vector for the priority level under consideration. The employment of the multiplex algorithm, as well as the EFI tableau, is best illustrated by means of illustration. Consequently, two specific examples follow.

Example 15.1: Solution of an LP via the Multiplex Algorithm

Even though our main interest is in the solution of multiobjective models, the solution of a conventional LP model serves to effectively illustrate the multiplex algorithm. Further, one may easily compare the multiplex process with that of simplex. Thus, for our first example, let us solve the LP model depicted that follows.

Find x so as to

$$\text{maximize } z = 6x_1 + 4x_2$$

subject to:

$$x_1 + x_2 \leq 10$$

$$2x_1 + x_2 \geq 4$$

$$x_1, x_2 \geq 0$$

First, let us transform this LP problem into its multiplex equivalent. The resulting formulation follows.

Find **v** so as to

$$\text{lexmin } u = \{(\rho_1 + \eta_2), (-6x_1 - 4x_2)\}$$

subject to:

$$x_1 + x_2 + \eta_1 - \underline{\rho}_1 = 10$$

$$2x_1 + x_2 + \underline{\eta}_2 - \rho_2 = 4$$

$$\text{all variables} \geq 0$$

This may be placed into the "full" multiplex tableau (see Table 15.1) as shown in Table 15.3. However, this tableau, as has been noted, is terribly inefficient. Consequently, we shall employ the equivalent EFI tableau depicted in Table 15.4.

TABLE 15.3 "FULL" TABLEAU FOR EXAMPLE 15.1

	η_1	η_2	x_1	x_2	ρ_1	ρ_2	$\boldsymbol{\beta}$
η_1	1	0	1	1	-1	0	10
η_2	0	1	2	1	0	-1	4
P_1	0	0	2	1	-1	-1	4
P_2	0	0	6	4	0	0	0

TABLE 15.4 EFI TABLEAU FOR EXAMPLE 15.1

	Basis inverse		$\boldsymbol{\beta}$
η_1	1	0	10
η_2	0	1	4
$\pi^{(1)}$	0	1	4
$\pi^{(2)}$	0	0	0

Note that the elements of Table 15.4 have been computed as follows:

- The basis inverse is simply the identity matrix in the first tableau (i.e., the matrix associated with the negative deviation variables).
- The right-hand side, $\boldsymbol{\beta}$, is given by

$$\boldsymbol{\beta} = \mathbf{B}^{-1}\mathbf{b} = \begin{bmatrix} 1 & 0 \\ 0 & 1 \end{bmatrix} \cdot \begin{pmatrix} 10 \\ 4 \end{pmatrix} = \begin{pmatrix} 10 \\ 4 \end{pmatrix}$$

- The elements of the achievement vector (which, recall, are not really necessary) are found by $u^{(k)} = \mathbf{c}_B^{(k)}\mathbf{B}^{-1}\mathbf{b}$. Thus, $u^{(1)}$ and $u^{(2)}$ are given by

$$u^{(1)} = (0 \quad 1) \cdot \begin{bmatrix} 1 & 0 \\ 0 & 1 \end{bmatrix} \cdot \begin{pmatrix} 10 \\ 4 \end{pmatrix} = 4$$

$$u^{(2)} = (0 \quad 0) \cdot \begin{bmatrix} 1 & 0 \\ 0 & 1 \end{bmatrix} \cdot \begin{pmatrix} 10 \\ 4 \end{pmatrix} = 0$$

- The elements of the pricing vectors are found by $\boldsymbol{\pi}^{(k)} = \mathbf{c}_B^{(k)}\mathbf{B}^{-1}$. Thus, $\boldsymbol{\pi}^{(1)}$ and $\boldsymbol{\pi}^{(2)}$ are given by

$$\boldsymbol{\pi}^{(1)} = (0 \quad 1) \cdot \begin{bmatrix} 1 & 0 \\ 0 & 1 \end{bmatrix} = (0 \quad 1)$$

$$\boldsymbol{\pi}^{(2)} = (0 \quad 0) \cdot \begin{bmatrix} 1 & 0 \\ 0 & 1 \end{bmatrix} = (0 \quad 0)$$

In essence, the establishment of the EFI tableau for the model accomplishes the first two steps of the revised multiplex algorithm. The third step is to compute the shadow prices, for the priority level under consideration (i.e., $k = 1$ at this time), for all nonbasic, unchecked variables. Recall that these shadow prices (reduced costs) may be found by solving

$$d_j^{(k)} = z_j^{(k)} - c_j^{(k)}, \text{ or, more precisely, } d_j^{(k)} = \boldsymbol{\pi}^{(k)}\mathbf{a}_j - c_j^{(k)}$$

Thus, we need to find the reduced prices for x_1, x_2, ρ_1, and ρ_2. These values are computed as follows:

$$d_{x_1}^{(1)} = (0 \quad 1) \cdot \begin{pmatrix} 1 \\ 2 \end{pmatrix} - 0 = 2 \qquad d_{\rho_1}^{(1)} = (0 \quad 1) \cdot \begin{pmatrix} -1 \\ 0 \end{pmatrix} - 1 = -1$$

$$d_{x_2}^{(1)} = (0 \quad 1) \cdot \begin{pmatrix} 1 \\ 1 \end{pmatrix} - 0 = 1 \qquad d_{\rho_2}^{(1)} = (0 \quad 1) \cdot \begin{pmatrix} 0 \\ -1 \end{pmatrix} - 0 = -1$$

Proceeding to Step 4 of the algorithm, we note that there are positive shadow prices. Thus, we select variable x_1 (the variable with the most positive shadow price) as the entering variable. Consequently, $v_q = x_1$. That is, x_1 will now enter the basis, in exchange for a basic variable to be determined.

We now move to Step 5, the update of the entering column. Using (15.24), we may compute the updated column for x_1 as

$$\boldsymbol{\alpha}_{x_1} = \mathbf{B}^{-1}\mathbf{a}_{x_1} = \begin{bmatrix} 1 & 0 \\ 0 & 1 \end{bmatrix} \cdot \begin{pmatrix} 1 \\ 2 \end{pmatrix} = \begin{pmatrix} 1 \\ 2 \end{pmatrix}$$

As strictly a bookkeeping measure, let us add this updated column, and the value of the shadow price for x_1 (i.e., d_{x_1}), to the second tableau for this problem. This is depicted in Table 15.5.

We now must determine, via Step 6, the departing variable (i.e., the basic variable that x_1 is to replace). To accomplish this, we employ (15.25). However, in order to keep track of these computations, let us also add these results to our EFI tableau. The ratios determined via (15.25) have thus been listed to the right of the tableau, under the column designated as θ (this procedure is often termed the θ-rule) in Table 15.5. Thus, by the θ-rule, variable η_2 should depart. Note that the minimum ratio (i.e., $\frac{4}{2}$) has been so designated by means of an asterisk. Furthermore, we have circled the element of $\boldsymbol{\alpha}_{x_1}$ associated with the departing row. This circled element will serve as the *pivoting element*, or $\alpha_{p,q}$, in our pivoting process.

TABLE 15.5 SECOND TABLEAU
FOR EXAMPLE 15.1

	Basis inverse		β	α_{x_1}	θ
η_1	1	0	10	1	$\frac{10}{1}$
η_2	0	1	4	②	$\frac{4}{2}*$
$\pi^{(1)}$	0	1	4	2	

As noted in Table 15.5, the departing variable is η_2 and thus row p = row 2. Consequently, the new basis will consist of the columns of A associated with η_1 and x_1, respectively. Moving to Step 7, we next must compute the inverse for this new basis. Although there are numerous ways of performing this step, let us simply proceed with a fairly mechanical procedure that is effective for hand computation.

First, let us define $\mathbf{B}^{-1}$ as the ''old'' basis inverse and $b_{i,j}$ as the element of $\mathbf{B}^{-1}$ in the ith row and jth column of this matrix. Then let $\underline{\mathbf{B}}^{-1}$ and $\underline{b}_{i,j}$ be the ''new'' basis inverse and basic matrix element, respectively. Then, to derive the new basis inverse, we employ the following formulas:

$$\underline{b}_{p,j} = b_{p,j}/\alpha_{p,q}, \qquad \text{for all } j \tag{15.26}$$

$$\underline{b}_{i,j} = b_{i,j} - [(\alpha_{i,q}) \cdot (b_{p,j})/\alpha_{p,q}], \qquad \text{for all } i \text{ and } j, \text{ where } i \neq p \tag{15.27}$$

Using these formulas, and those for computing $u^{(k)}$, $\boldsymbol{\beta}$, and $\boldsymbol{\pi}^{(k)}$, we may develop the next tableau, as shown in Table 15.6, and return to Step 2 of the algorithm. Actually, with the development of this latest tableau, we have completed the second iteration of Step 2. Thus, we may move on to Step 3, wherein we need to compute the shadow prices for all nonbasic, unchecked variables (i.e., x_2, η_2, ρ_1, and ρ_2).

TABLE 15.6 THIRD TABLEAU FOR EXAMPLE 15.1

	Basis inverse		β
η_1	1	$-\frac{1}{2}$	8
x_1	0	$\frac{1}{2}$	2
$\pi^{(1)}$	0	0	0

The development of these reduced prices follows. As may be noted, none of the shadow prices is positive. Consequently, we must proceed to Step 8 of the algorithm.

$$d_{x_2}^{(1)} = (0 \quad 0) \cdot \begin{pmatrix} 1 \\ 1 \end{pmatrix} - 0 = 0 \qquad d_{\rho_1}^{(1)} = (0 \quad 1) \cdot \begin{pmatrix} -1 \\ 0 \end{pmatrix} - 1 = -1$$

$$d_{\eta_2}^{(1)} = (0 \quad 0) \cdot \begin{pmatrix} 0 \\ 1 \end{pmatrix} - 0 = 0 \qquad d_{\rho_2}^{(1)} = (0 \quad 0) \cdot \begin{pmatrix} 0 \\ -1 \end{pmatrix} - 0 = 0$$

Following the guidelines of Step 8, we may immediately note that convergence has not yet been reached. Thus, we check variable ρ_1 (because it is the only nonbasic variable with a negative shadow price at this step), set $k = 2$, and return to Step 2 of the process. However, before we return to Step 2, observe closely just what has happened. At this point in time, $u^{(1)} = 0$ and thus we have an implementable (although not yet necessarily optimal) solution. Thus, the observant reader may note that the process thus far has been analogous to the first phase of the well-known two-phase simplex process. And, in our next phase, we shall attempt to optimize the original LP objective function—while still remaining implementable.

Returning to Step 2, we must now compute the reduced price vector for priority level 2. This is given as

$$\pi^{(2)} = \mathbf{c}_B^{(2)}\mathbf{B}^{-1} = (0 \quad -6) \cdot \begin{bmatrix} 1 & -\frac{1}{2} \\ 0 & \frac{1}{2} \end{bmatrix} = (0 \quad -3)$$

The associated tableau, for priority level 2, is depicted in Table 15.7. Notice that we have also computed the value of $u^{(2)}$ (i.e., a value of -12) and included yet another

TABLE 15.7 FOURTH TABLEAU FOR EXAMPLE 15.1

	Basis inverse		β	$\sqrt{}$
η_1	1	$-\frac{1}{2}$	8	ρ_1
x_1	0	$\frac{1}{2}$	2	
$\pi^{(2)}$	0	-3	-12	

column to the tableau. Thus last column, designated by the check mark ($\sqrt{}$) simply lists those variables that have thus far been checked (i.e., ρ_1).

We are now ready to proceed with Step 3, and compute the reduced prices for the nonbasic, unchecked variables. These variables are x_2, η_2, and ρ_2. Note carefully that it is not necessary to compute this value for ρ_1 as it is now a checked variable.* The shadow prices for these variables are

$$d_{x_2}^{(2)} = (0 \quad -3) \cdot \begin{pmatrix} 1 \\ 1 \end{pmatrix} - 4 = 1 \qquad d_{\rho_2}^{(12)} = (0 \quad -3) \cdot \begin{pmatrix} 0 \\ -1 \end{pmatrix} - 0 = 3$$

$$d_{\eta_2}^{(2)} = (0 \quad -3) \cdot \begin{pmatrix} 0 \\ 1 \end{pmatrix} - 0 = -3$$

As may be noted, two variables have positive shadow prices. However, proceeding with Step 4, it may be seen that ρ_2 has the largest value and thus it becomes the new entering variable. Thus, $v_q = \rho_2$. We now move to Step 5, the update of the entering column. The updated column for ρ_2 is

$$\boldsymbol{\alpha}_{\rho_2} = \mathbf{B}^{-1}\mathbf{a}_{\rho_2} = \begin{bmatrix} 1 & -\frac{1}{2} \\ 0 & \frac{1}{2} \end{bmatrix} \cdot \begin{pmatrix} 0 \\ -1 \end{pmatrix} = \begin{pmatrix} \frac{1}{2} \\ -\frac{1}{2} \end{pmatrix}$$

We then add this updated column, and the value of the shadow price for ρ_2 (i.e., $d_{\rho_2}^{(1)}$), to the next tableau for this problem. This is depicted in Table 15.8.

TABLE 15.8 FIFTH TABLEAU FOR EXAMPLE 15.1

	Basis inverse		β	$\boldsymbol{\alpha}_{\rho_2}$	θ	$\sqrt{}$
η_1	1	$-\frac{1}{2}$	8	$\textcircled{\frac{1}{2}}$	16*	ρ_1
x_1	0	$\frac{1}{2}$	2	$-\frac{1}{2}$	—	
$\pi^{(2)}$	0	-3	-12	3		

We next determine, via Step 6, the departing variable. The ratios determined via (15.25) have been listed to the right of the tableau, under the θ column. Thus, by the θ-rule, the variable η_1 should depart. Note in particular that there is no ratio computed for x_1 because the value of α in that row is negative. Again, we have circled the element of $\boldsymbol{\alpha}_{\rho_2}$ associated with the departing row.

As noted in Table 15.8, the departing variable is η_1 and thus row p = row 1. Consequently, the new basis will consist of the columns of A associated with ρ_2 and x_1, respectively. Moving to Step 7, we next compute the inverse for this new basis— by means of (15.26–15.27). This new basis, and the complete new tableau, is shown in Table 15.9.

---

* In essence, variable ρ_1 has been dropped from further consideration. Thus, the checking process has also been termed the "column drop rule." Note that the return of ρ_1 to the basis can only serve to degrade the value previously achieved for $u^{(1)}$.

TABLE 15.9 SIXTH TABLEAU FOR
EXAMPLE 15.1

	Basis inverse		β	$\checkmark$
ρ_2	2	-1	16	ρ_1
x_1	1	0	10	
$\pi^{(2)}$	-6	0	-60	

The development of the new tableau of Table 15.9 completes Step 2. Turning to Step 3, we must compute the reduced prices for x_2, η_1, and η_2. These values are

$$d_{x_2}^{(2)} = (-6 \quad 0) \cdot \binom{1}{1} - 4 = -2 \qquad d_{\eta_2}^{(2)} = (-6 \quad 0) \cdot \binom{0}{1} - 0 = 0$$

$$d_{\eta_1}^{(2)} = (-6 \quad 0) \cdot \binom{1}{0} - 0 = -6$$

Because there are no positive shadow prices, we move to Step 8. Following the guidelines of Step 8, we then note that we have converged to the optimal solution, where this solution is given by the final tableau—of Table 15.9. That is

$$x_1^* = 10 \qquad \eta_1^* = 0 \qquad \rho_1^* = 0$$

$$x_2^* = 0 \qquad \eta_1^* = 0 \qquad \rho_2^* = 16$$

and $\mathbf{u}^* = (0, -60)$. Consequently, the solution to the original LP is given by $z^* = 60$ and $\mathbf{x}^* = (10, 0)$.

Example 15.2: Solution of a Non-Archimedean LGP via the Multiplex Algorithm

Let us now examine just how one may solve a multiobjective problem via the multiplex algorithm. Because the procedure is identical regardless of problem type (i.e., for an LP, an Archimedean LGP, a non-Archimedean LGP, a fuzzy LGP, or the lexicographic vectormin problem), we shall demonstrate the approach on the following non-Archimedean LGP model.

$$\text{lexmin } \mathbf{u} = \{(\rho_1 + \rho_2), (\eta_3), (\rho_4), (\eta_1 + 1.5\eta_2)\}$$

subject to:

$$x_1 + \underline{\eta}_1 - \underline{\rho}_1 = 30$$

$$x_2 + \underline{\eta}_2 - \underline{\rho}_2 = 15$$

$$8x_1 + 12x_2 + \underline{\eta}_3 - \rho_3 = 1000$$

$$x_1 + 2x_2 + \eta_4 - \underline{\rho}_4 = 40$$

$$\mathbf{v} \geq \mathbf{0}$$

Because the non-Archimedean LGP model is already in multiplex form, we may proceed to the algorithm. The tableaux associated with the solution of this particular problem are listed in Tables 15.10 through 15.15.

TABLE 15.10 FIRST TABLEAU FOR EXAMPLE 15.2

	Basis inverse				β
η_1	1	0	0	0	30
η_2	0	1	0	0	15
η_3	0	0	1	0	1000
η_4	0	0	0	1	40
$\pi^{(1)}$	0	0	0	0	0

In Table 15.10, we have established the tableau for the initial basic feasible solution. Computing the associated reduced costs for the nonbasic, unchecked variables, we find that the tableau is optimal for priority level 1. Further, variables ρ_1 and ρ_2 may be checked off. We then move to priority level 2, as depicted in Table 15.11. Here, we find that there are positive reduced prices (for x_1 and x_2) and that x_2 is the entering variable. Through the θ-rule, we find that η_2 is the departing variable. The resultant pivoting operations lead to the tableau of Table 15.12.

TABLE 15.11 SECOND TABLEAU FOR EXAMPLE 15.2

	Basis inverse				β	α_{x_2}	θ	$\checkmark$
η_1	1	0	0	0	30	0	—	ρ_1, ρ_2
η_2	0	1	0	0	15	①	$\frac{15}{1}*$	
η_3	0	0	1	0	1000	12	$\frac{1000}{12}$	
η_4	0	0	0	1	40	2	$\frac{40}{2}$	
$\pi^{(2)}$	0	0	1	0	1000	12		

TABLE 15.12 THIRD TABLEAU FOR EXAMPLE 15.2

	Basis inverse				β	α_{x_1}	θ	$\checkmark$
η_1	1	0	0	0	30	1	$\frac{30}{1}$	ρ_1, ρ_2
x_2	0	1	0	0	15	0	—	
η_3	0	−12	1	0	820	8	$\frac{820}{8}$	
η_4	0	−2	0	1	40	①	$\frac{10}{1}*$	
$\pi^{(2)}$	0	−12	1	0	820	8		

For the tableau of Table 15.12, we compute the reduced costs. The single positive reduced cost is that associated with x_1—and thus x_1 is the new entering variable. From the θ-rule, we find that η_4 is the departing variable. The resultant pivoting operations lead to the tableau of Table 15.13.

TABLE 15.13 FOURTH TABLEAU FOR EXAMPLE 15.2

	Basis inverse				β	α_{p_4}	θ	$\checkmark$
η_1	1	2	0	-1	20	①	$\frac{20}{1}*$	p_1, p_2
x_2	0	1	0	0	15	0	—	
η_3	0	4	1	-8	740	8	$\frac{740}{8}$	
x_1	0	-2	0	1	10	-1	—	
$\pi^{(2)}$	0	4	1	-8	740	8		

Computing the shadow prices for the tableau of Table 15.13, we find that p_4 is the new entering variable. The θ-rule concludes that η_1 is the departing variable. The resultant tableau, after pivoting on p_4 and η_1, is given in Table 15.14.

TABLE 15.14 FIFTH TABLEAU FOR EXAMPLE 15.2

	Basis inverse				β	$\checkmark$
p_4	1	2	0	-1	20	p_1, p_2
x_2	0	1	0	0	15	
η_3	-8	-12	1	0	580	
x_1	1	0	0	0	30	
$\pi^{(2)}$	-8	-12	1	0	580	

The tableau of Table 15.14 is optimal for level 2. That is, none of the reduced prices is positive. Further, we may check off variables η_1, η_2, and p_3. Thus, we move to priority level 3, as depicted in Table 15.15. However, the shadow price for the remaining nonbasic, unchecked variable η_4 is negative for this tableau. Thus, the process stops with the solution depicted in Table 15.15. That is:

$$\mathbf{x}^* = (30, 15) \qquad \mathbf{u}^* = (0, 580, 20, 0)$$

Notice that, for this problem, that it was not necessary to proceed to priority level 4.

TABLE 15.15 FINAL TABLEAU FOR EXAMPLE 15.2

	Basis inverse				β	$\checkmark$
p_4	1	2	0	-1	20	p_1, p_2
x_2	0	1	0	0	15	η_1, η_2
η_3	-8	-12	1	0	580	p_3
x_1	1	0	0	0	30	
$\pi^{(3)}$	1	2	0	-1	20	

SEQUENTIAL MULTIPLEX

The revised multiplex algorithm of the previous section is, as has been mentioned, little more than a very transparent extension of the well-known *two-phase simplex process* (see Chapter 4). If properly implemented (i.e., if all the computational enhancements as found in the better commercial LP packages—such as data packing, reinversion, LU decomposition, and so forth—are included), the algorithm is efficient, effective, and robust. However, the development of any truly computationally efficient, computer-based version of this algorithm requires a considerable amount of time and effort. This time and effort can be avoided by means of an alternative approach: *sequential multiplex.*

Sequential multiplex is a straightforward extension of sequential goal programming, a method first developed by Ignizio and Huss (Ignizio, 1967) in the late 1960s (and later discovered independently by Kornbluth [1973] and, even later, by Dauer and Krueger [1977]). In essence, it is a method of solution by problem decomposition, or partitioning—wherein each problem addressed (i.e., in its partitioned format) is a conventional, single-objective model. As such, the approach is appropriate for linear models, integer and discrete models, and nonlinear models. The primary advantage of the sequential approach is that one may utilize any existing commercial single-objective software package (e.g., any commercial LP software in the case of linear multiplex models) as the primary problem-solving component. As a consequence, the amount of effort required to develop extremely efficient software in support of a sequential multiplex approach is minimal.* Further, as a result, *one may solve, with equivalent computational efficiency, problems of sizes equal to those solved by means of the supporting single-objective code.*

The Sequential Algorithm: For Linear Multiplex

The sequential multiplex algorithm proceeds by first addressing the portion of the (linear) multiplex model associated with priority level 1 (i.e., the hard goals and the portion of the achievement vector associated with the first priority). This results in a conventional LP model given as

$$\text{minimize } u^{(1)} = g_1(\mathbf{x}, \boldsymbol{\eta}, \boldsymbol{\rho}) \tag{15.28}$$

subject to:

$$\sum a_{i,j}x_j + \eta_i - \rho_i = b_i, \qquad \text{for } i \in P_1 \tag{15.29}$$

---

* As just one example, students in one of the authors' classes on multiobjective optimization have been required to construct such a software package, using LINDO™ as the supporting LP component, as a take-home assignment. Although the students are given 1 week in which to construct and test their software, many students have reported that the development of a working version of sequential multiplex has been accomplished in less than an hour. And this process can be made even more efficient, and easy, by means of the employment of the EXSYS/LINDO interface—a feature provided in the EXSYS™ expert system package as provided by EXSYS, Inc.

$$\mathbf{x}, \boldsymbol{\eta}, \boldsymbol{\rho} \geq \mathbf{0} \tag{15.30}$$

where $g_1(\mathbf{x}, \boldsymbol{\eta}, \boldsymbol{\rho})$ is the first term of the achievement vector.

Notice carefully that the single term of the achievement function in (15.28) is a single linear function, whereas the goals of (15.29) are the linear goals at priority level 1 (P_1). Consequently, this first model is a conventional LP model that may be solved by any simplex algorithm, or LP software. Once solved, we have found the optimal value for $u^{(1)}$ (i.e., $u^{(1)*}$).

We may then move to the next priority level (i.e., level 2) and construct a new LP model. This model is given as

$$\text{minimize } u^{(2)} = g_2(\mathbf{x}, \boldsymbol{\eta}, \boldsymbol{\rho}) \tag{15.31}$$

subject to:

$$\sum a_{i,j} x_j + \eta_i - \rho_i = b_i, \qquad \text{for } i \in P_1 \cup P_2 \tag{15.32}$$

$$g_1(\boldsymbol{\eta}, \boldsymbol{\rho}) \leq u^{(1)*} \tag{15.33}$$

$$\mathbf{x}, \boldsymbol{\eta}, \boldsymbol{\rho} \geq \mathbf{0} \tag{15.34}$$

where $g_2(\mathbf{x}, \boldsymbol{\eta}, \boldsymbol{\rho})$ is the second term of the achievement function, and $i \in P_1 \cup P_2$ is used to indicate that all goals at priorities 1 *and* 2 are to be included in the "constraint set" for this LP model.

Notice carefully the "constraint" of (15.33). This is called an *augmented constraint* in sequential multiplex. It is included in the model to make sure that the solution previously derived for any higher-level achievement-vector term is not degraded. As such, we may use either (1) $g_1(\mathbf{x}, \boldsymbol{\eta}, \boldsymbol{\rho}) \leq u^{(1)*}$ or (2) $g_1(\mathbf{x}, \boldsymbol{\eta}, \boldsymbol{\rho}) = u^{(1)*}$ as a representation of this function.

In essence, the foregoing process is repeated until convergence is reached. Convergence, in turn, is obtained when either (1) the LP models for all priority levels have been examined or (2) the program is "fixed." The first instance should be self-explanatory. The second will be demonstrated in the example to follow. First, however, let us list the steps of the sequential multiplex algorithm.

Algorithm steps

STEP 1. *Initialization.* Set $k = 1$. Initially, all variables are *unchecked*.

STEP 2. *Establish the first LP model.* That is, set up the model pertaining to priority level 1, as given in (15.28–15.30).

STEP 3. *Solve the LP model for priority level k.* Using any conventional LP algorithm or software, solve the LP model associated with priority level k. Let the optimal solution to this model be given as $u^{(k)*}$, where $u^{(k)*}$ is simply the optimal value of $g_k(\mathbf{x}, \boldsymbol{\eta}, \boldsymbol{\rho})$.

STEP 4. *Variable check-off/column drop.* For the optimal tableau of Step 3, check any nonbasic variable having a negative reduced price. This variable, and its associated column, may then be dropped from consideration in any future tableaux.

STEP 5. *Move to next priority level and check for convergence.* Set $k = k + 1$. If $k > K$, go to Step 8.

STEP **6.** *Establish the equivalent LP model for the new priority level (i.e., level k).* This model is given as

$$\text{minimize } u^{(k)} = g_k(\mathbf{x}, \boldsymbol{\eta}, \boldsymbol{\rho})$$

subject to:

$$\sum a_{i,j} x_j + \eta_i - \rho_i = b_i, \qquad \text{for } i \in P_1 \cup P_2 \cdots \cup P_{k-1}$$

$$g_s(\mathbf{x}, \boldsymbol{\eta}, \boldsymbol{\rho}) \leq u^{(s)*}, \qquad \text{for } s = 1, 2, \ldots, k-1$$

$$\mathbf{x}, \boldsymbol{\eta}, \boldsymbol{\rho} \geq \mathbf{0}$$

where $g_k(\mathbf{x}, \boldsymbol{\eta}, \boldsymbol{\rho})$ is the kth term of the achievement vector, and $i \in P_1 \cup P_2 \cdots \cup P_{k-1}$ is used to indicate that all goals at priorities 1 through $k - 1$ are to be included in the "constraint set" for this LP model.

STEP **7.** *Check for convergence.* If the LP model of Step 6 is fixed (i.e., if the program is fixed), proceed to Step 8. Otherwise, return to Step 3.

STEP **8.** *Determination of the final solution.* The program ($\mathbf{x}^*$) associated with the last LP model solved is the optimal program for the original multiplex model.

Other than for, perhaps, Step 7, the algorithm is quite straightforward. To clarify this particular step, consider the LP model that follows. For sake of discussion, let us assume that this is the LP model to be solved for the second priority level.

$$\text{minimize } u^{(2)} = \rho_3$$

subject to:

$$x_1 = 3$$

$$x_1 + x_2 = 10$$

$$x_2 + \eta_3 - \underline{\rho}_3 = 40$$

all variables nonnegative

Now, one could certainly solve this problem by means of the simplex algorithm. However, this would be a waste of resources, as it should be obvious that the solution is fixed. That is, from the first constraint, $x_1 = 3$. Thus, from the second, x_2 must equal 7. Consequently, no matter how many priority levels may remain, the values of x_1 and x_2 are fixed at 3 and 7, respectively.

Example 15.3: Solution of a Non-Archimedean LGP via the Sequential Multiplex Algorithm

In order to illustrate the employment of the sequential multiplex algorithm, let us employ it to solve the non-Archimedean LGP model of Example 15.2—and compare results. Recall that this model is given as

$$\text{lexmin } \mathbf{u} = \{(\rho_1 + \rho_2), (\eta_3), (\rho_4), (\eta_1 + 1.5\eta_2)\}$$

subject to:

$$x_1 + \underline{\underline{\eta_1}} - \underline{\rho_1} = 30$$

$$x_2 + \underline{\underline{\eta_2}} - \underline{\rho_2} = 15$$

$$8x_1 + 12x_2 + \underline{\underline{\eta_3}} - \rho_3 = 1000$$

$$x + 2x_2 + \eta_4 - \underline{\rho_4} = 40$$

$$\mathbf{v} \geq \mathbf{0}$$

Thus, setting $k = 1$, we may establish the LP model for priority level 1, as listed in what follows. This completes Steps 1 and 2 of the algorithm.

$$\text{minimize } u^{(1)} = \rho_1 + \rho_2$$

subject to:

$$x_1 + \underline{\underline{\eta_1}} - \underline{\rho_1} = 30$$

$$x_2 + \underline{\underline{\eta_2}} - \underline{\rho_2} = 15$$

$$\mathbf{v} \geq \mathbf{0}$$

Proceeding to Step 3, we may solve this most recent model. The initial tableau is, in fact, optimal, and is shown in Table 15.16.

TABLE 15.16   INITIAL *AND* FINAL TABLEAU FOR $k = 1$

	Basis inverse		β
η_1	1	0	30
η_2	0	1	15
$\pi^{(1)}$	0	0	0

For the tableau of Table 15.16, the optimal solution is $u^{(1)*} = 0$. Moving to Step 4, we may note that both ρ_1 and ρ_2 may be checked off (i.e., their reduced prices are both negative for the final tableau for $k = 1$).

Moving to Step 5, we set $k = k + 1 = 2$. We then proceed to Step 6, the development of the LP model for $k = 2$. This model follows:

$$\text{minimize } u^{(2)} = \eta_3$$

subject to:

$$x_1 + \underline{\underline{\eta_1}} - \underline{\rho_1} = 30$$

$$x_2 + \underline{\underline{\eta_2}} - \rho_2 = 15$$

$$8x_1 + 12x_2 + \underline{\underline{\eta_3}} - \rho_3 = 1000$$

$$\rho_1 + \rho_2 = u^{(1)*} = 0$$

$$\mathbf{v} \geq \mathbf{0}$$

Notice that the fourth constraint is the augmented constraint associated with the solution of the model for priority level 1. Further, because ρ_1 and ρ_2 are both checked variables, they are, in essence, set to a value of zero. As such, the model for $k = 2$ may be rewritten simply as

$$\text{minimize } u^{(2)} = \eta_3$$

subject to:

$$x_1 + \underline{\eta_1} = 30$$
$$x_2 + \underline{\eta_2} = 15$$
$$8x_1 + 12x_2 + \underline{\eta_3} - \rho_3 = 1000$$
$$\mathbf{v} \geq \mathbf{0}$$

Moving to Step 7, we may note that this model is not yet fixed (i.e., we have six variables and three equations) and thus we return to Step 3. In Step 3, we solve the LP model for $k = 2$. The final tableau for this model is given in Table 15.17. As may be noted, $u^{(2)*} = \eta_3 = 580$.

TABLE 15.17   FINAL TABLEAU FOR $k = 2$

	Basis inverse			β	$\checkmark$
x_1	1	0	0	30	ρ_1, ρ_2
x_2	0	1	0	15	
η_3	-8	-12	1	580	
$\pi^{(2)}$	-8	-12	1	580	

Proceeding to Step 4, we note that variables η_1, η_2, and ρ_3 all have negative reduced prices for the final solution represented in the tableau of Table 15.17. Thus, these variables may be added to the list of checked variables.

Moving to Step 5, we set $k = k + 1 = 3$ and then go to Step 6, where we establish the LP model for the problem at priority level 3. This model is given as

$$\text{minimize } u^{(3)} = \rho_4$$

subject to:

$$x_1 + \underline{\eta_1} - \underline{\rho_1} = 30$$
$$x_2 + \underline{\eta_2} - \underline{\rho_2} = 15$$
$$8x_1 + 12x_2 + \underline{\eta_3} - \rho_3 = 1000$$
$$x_1 + 2x_2 + \eta_4 - \underline{\rho_4} = 40$$
$$\rho_1 + \rho_2 = u^{(1)*} = 0$$
$$\eta_3 = u^{(2)*} = 580$$
$$\mathbf{v} \geq \mathbf{0}$$

Again, we may simplify this model by removing all checked variables (and substituting, in the third constraint, for the value of η_3). Thus, the revised model for $k = 3$ is given as

$$\text{minimize } u^{(3)} = \rho_4$$

subject to:

$$x_1 = 30$$

$$x_2 = 15$$

$$8x_1 + 12x_2 = 420$$

$$x_1 + 2x_2 + \eta_4 - \underline{\rho}_4 = 40$$

$$\mathbf{v} \geq \mathbf{0}$$

By proceeding to Step 7, it should be obvious that the most recent model is indeed fixed. That is, the values of the structural variables are fixed, regardless of any remaining priority levels. Consequently, we may proceed to Step 8 and list the final, optimal solution for the original multiplex model. That is:

$$\mathbf{x}^* = (30, 15) \qquad \mathbf{u}^* = (0, 580, 20, 0)$$

which is, of course, the same solution arrived at by means of the employment of the revised multiplex algorithm for Example 15.2.*

Observations

It should be noted that because of variable checking, it may not be possible to have an initial basis at each priority level that consists solely of negative deviation variables. That is, the negative deviation variable for a given constraint may have been checked (i.e., dropped), leaving only the structural variables and a positive deviation variable. When this occurs, all one need do is to simply add an artificial variable to the constraint—and then drive this artificial variable from the basis by means of either two-phase simplex or the so-called Big-M method (i.e., include the artificial in the objective function at some large cost).

In anything other than hand calculations, all of these matters will, of course, automatically be taken care of by the LP software package employed to find the solutions of each LP model in the sequence. However, this is an observation that should definitely be considered when solving the exercises at the end of this chapter.

Finally, let us briefly comment on just how one might go about developing a software package for the implementation of the sequential multiplex algorithm. When Ignizio and Huss (Ignizio, 1967) first developed this procedure (which was, at the time limited to just linear goal programming), it was initially implemented by means of a rather crude process. That is, the first LP model was developed by

---

* Note that the values for $u^{(3)}$ and $u^{(4)}$ may be found by simply substituting the optimal values for the structural variables into the original multiplex model.

hand and its data were submitted for solution by an LP package. When the solution was found (remember that this was in the days of batch processing and punched cards), the next LP model was again formed by hand, the data input via punched cards, and the process repeated. This was continued until the algorithm reached convergence.

Today, the complete process is easily automated. This may be accomplished by means of the development of a "buffer program." The buffer program must be developed by the analyst and it serves to accept the initial multiplex model, convert this to the associated LP model for the priority level of interest, submit the data to the simplex software, receive and interpret the results, and reformulate the next LP problem in the sequence. Descriptions with regard to such a process (although focused on non-Archimedean LGP) may be found in the references (Crowder and Sposito, 1991; Ignizio and Perlis, 1977).

COMPUTATIONAL CONSIDERATIONS

Because the algorithms presented in this textbook, or virtually any other text on the topic of mathematical programming, are intended primarily for hand calculation, one should never infer that they represent the latest and/or most computationally efficient versions of the algorithms actually employed in commercial (or proprietary) mathematical programming software. In fact, neither of the multiplex algorithms described in this chapter include features that are found in the most efficient of their commercial versions. Specifically, such software implementations will typically include

- basis reinversion
- automatic model scaling
- data packing (e.g., note that the technological coefficient matrix, along with most other matrices and vectors, are extremely sparse)
- LU decomposition
- improved entering variable rules (i.e., the nonbasic variable with the largest positive reduced cost is not always the best one to select for entry into the basis)
- "crashing" (i.e., starting the process with a specific starting basis)
- bounded variable operations

For those readers who desire further information with regard to such computational enhancements, a particularly outstanding text is that by Murtagh (1981). Although Murtagh focuses exclusively on conventional models, all of the methods covered in his text apply equally as well to the multiplex algorithm.

In addition to the enhancements listed, there is yet one further improvement available for the sequential multiplex algorithm. The way in which the algorithm has been written and presented might tend to imply that, at each priority level, we solve an LP model by starting with the initial basic feasible solution consisting of the negative deviation variables. In actual practice, *this is not done*.

The most efficient way in which to solve the initial LP (for problems of modest to large sizes) *at each level* is to utilize the results achieved for the previous level. That is, one should force those variables that were in the final basis for the previous level into the initial basis of the next level. This may be easily accomplished via postoptimality procedures (and these, in turn, will be addressed in the next chapter). Proceeding in this manner, one may significantly reduce the number of iterations necessary to solve a typical multiplex model.

As a final comment on the sequential multiplex algorithm, the reader may recall that we noted that this approach may be employed to solve integer/discrete and nonlinear multiplex models, as well as linear multiplex problems. However, it should be noted that to extend the sequential approach to these other model types, two things must be done:

- Use the software package appropriate for the model type under consideration (i.e., if dealing with an integer multiplex model, use an integer programming code, and if dealing with a nonlinear multiplex model, use a nonlinear programming code).
- Delete Step 4 of the algorithm. Specifically, if dealing with integer/discrete or nonlinear models, one cannot employ variable checking.

Finally, we once again wish to impress on the reader the fact that one can solve multiplex models that are just as large (and with equivalent computational efficiency) as the largest single-objective models that may be solved. Because either form of the multiplex algorithm is but a simple extension of the simplex method, this should be obvious. However, as has been noted, there are still those who remain oblivious to such results—and, even today, one will see statements lamenting the computational efficiency of algorithms and software for those multiobjective models encompassed by the multiplex approach.

SUMMARY

In this chapter, we have described two forms of the multiplex algorithm. The first, revised multiplex, is simply an extension of the well-known two-phase simplex procedure. The second approach, sequential multiplex, is somewhat less elegant but remains the algorithm of choice for most serious investigators. This is because sequential multiplex may utilize, as its primary computational resource, any commercial LP code—and thus exhibits computational efficiencies on the same order as is available for single-objective models.

Although any real-world problems would be solved on the computer, the authors believe that it is vital to first learn how to solve small problems by hand.

In this way, the reason for each step of the associated algorithm should become more clear—and a better understanding of the overall process should result. Consequently, we strongly recommend that the reader attempt the problems listed in the exercise set that follows.

EXERCISES

15.1. Solve the following non-Archimedean LGP model by means of the revised multiplex algorithm of this chapter.

$$\text{lexmin } \mathbf{u} = \{(\rho_1 + \rho_2), (\eta_3), (\rho_4)\}$$

subject to:

$$4x_1 + 5x_2 + \eta_1 - \rho_1 = 80$$

$$4x_1 + 2x_2 + \eta_2 - \rho_2 = 48$$

$$80x_1 + 100x_2 + \eta_3 - \rho_3 = 800$$

$$x_1 + \eta_4 - \rho_4 = 6$$

$$\mathbf{x}, \, \boldsymbol{\eta}, \, \boldsymbol{\rho} \geq \mathbf{0}$$

15.2. Solve Exercise 15.1 *graphically*, comparing—at each iteration—the graphical results with those obtained by the multiplex algorithm. Comment on the final solution achieved both graphically and by the algorithm. Does the final tableau obtained by the algorithm provide an indication of alternative optimal solutions?

15.3. Add the following goal to the model of Exercise 15.1 and then solve both *graphically* and by means of the revised multiplex algorithm.

new goal: $x_1 + x_2 + \eta_5 - \rho_5 = 7$, where ρ_5 is to be minimized at priority level 4

Comment on the final solution. Does this new final tableau indicate alternative optimal solutions?

15.4. Solve the following non-Archimedean LGP model *graphically*. Note carefully that the third priority level plays no part in the determination of the final solution. Resolve, *graphically*, this problem if the achievement level for the second goal is set to a value of 14 rather than 26.

$$\text{lexmin } \mathbf{u} = \{(\rho_1), (\eta_2), (\eta_3)\}$$

subject to:

$$x_1 + x_2 + \eta_1 - \rho_1 = 810$$

$$2x_1 + x_2 + \eta_2 - \rho_2 = 26$$

$$-x_1 + 2x_2 + \eta_3 - \rho_3 = 6$$

$$\mathbf{x}, \, \boldsymbol{\eta}, \, \boldsymbol{\rho} \geq \mathbf{0}$$

15.5. Formulate the problem in Exercise 15.1 in its *reduced* form.

15.6. Solve the problem in Exercise 15.4 (with the original rhs values) by means of the revised multiplex algorithm. Compare your results with those achieved graphically.

15.7. Solve the following model by means of the *sequential* multiplex algorithm. Notice carefully that the final solution is fixed once the third priority level has been optimized.

$$\text{lexmin } \mathbf{u} = \{(\rho_1 + \rho_2), (\eta_3), (\rho_4)\}$$

subject to:

$$2x_1 + x_2 + \eta_1 - \rho_1 = 12$$

$$x_1 + x_2 + \eta_2 - \rho_2 = 10$$

$$x_1 + \eta_3 - \rho_3 = 7$$

$$x_1 + 4x_2 + \eta_4 - \rho_4 = 4$$

$$\mathbf{x}, \, \boldsymbol{\eta}, \, \boldsymbol{\rho} \geq \mathbf{0}$$

15.8. Solve the following model by means of the *sequential* multiplex. Comment on any problems encountered during the procedure.

$$\text{lexmin } \mathbf{u} = \{(\eta_1 + \rho_1), (2\rho_2 + \rho_3)\}$$

subject to:

$$x_1 - 10x_2 + \eta_1 - \rho_1 = 50$$

$$3x_1 + 5x_2 + \eta_2 - \rho_2 = 20$$

$$8x_1 + 6x_2 + \eta_3 - \rho_3 = 100$$

$$\mathbf{x}, \, \boldsymbol{\eta}, \, \boldsymbol{\rho} \geq \mathbf{0}$$

16

DUALITY AND SENSITIVITY ANALYSIS IN LINEAR MULTIPLEX

CHAPTER OVERVIEW

By now, the reader must surely recognize that solving a problem, be it a conventional single-objective problem or one having multiple objectives, involves far more than simply "plugging" the problem data into a computer program and printing out the "optimal" values of the structural variables. First, one has to develop a "good and faithful" model—and this is, far and away, the most critical facet of mathematical programming. And, ultimately, one needs to determine just how appropriate the solution developed actually is—and just how sensitive it may be to changes in the original system (which are reflected by changes in model parameters or even model structure).

One of the major advantages of the traditional linear programming formulation is that sensitivity–and "what-if"—analysis is straightforward and powerful. But this is due primarily to the existence of the linear program *dual*, and through various exploitations of this property. Fortunately, there is also a dual for the linear multiplex model—and thus *any type of sensitivity analysis/what-if analysis that can be performed on an LP model can also be performed on a linear multiplex model*. This dual is termed, herein, the multidimensional dual (or the MDD) and, in this chapter, we shall describe the development of the MDD and discuss its economic interpretation as well as its use in the support of algorithm development. We then close the chapter with a discussion of just how one performs sensitivity analysis for a linear multiplex model.

THE FORMULATION OF THE MULTIDIMENSIONAL DUAL

The multidimensional dual was developed by Ignizio (Ignizio, 1974a, 1974b, 1976, 1982, 1984, 1985a, 1985b, 1985c; Markowski and Ignizio, 1983a, 1983b) in the early 1970s—and encompassed, at that time, only the dual of linear goal-programming models. Later, in the 1980s (Ignizio, 1985b), the MDD was extended to the more general, linear multiplex model.

For purpose of discussion, we shall call the general form of the linear multiplex model, as depicted in what follows, as the *primal* form. And this is, of course, the model that we have been employing throughout this section of the text.

$$\text{lexmin } \mathbf{u} = \{\mathbf{c}^{(1)}\mathbf{v}, \ \mathbf{c}^{(2)}\mathbf{v}, \ \dots, \ \mathbf{c}^{(K)}\mathbf{v}\} \tag{16.1}$$

subject to:

$$\mathbf{Av} = \mathbf{b} \tag{16.2}$$

$$\mathbf{v} \geq \mathbf{0} \tag{16.3}$$

However, recall—from Chapter 15—that we may write this model in its *reduced* form as follows.

Find $\mathbf{v}$ so as to

$$\text{lexmin } \mathbf{u} = \{[\mathbf{c}_B^{(1)}\boldsymbol{\beta} - (\boldsymbol{\pi}^{(1)}\mathbf{N} - \mathbf{c}_N^{(1)})\mathbf{v}_N], \ \dots, \ [\mathbf{c}_B^{(K)}\boldsymbol{\beta} - (\boldsymbol{\pi}^{(K)}\mathbf{N} - \mathbf{c}_N^{(K)})\mathbf{v}_N]\} \tag{16.4}$$

subject to:

$$\mathbf{v}_B = \mathbf{B}^{-1}\mathbf{b} - \mathbf{B}^{-1}\mathbf{N}\mathbf{v}_N = \boldsymbol{\beta} - \boldsymbol{\alpha}\mathbf{v}_N \tag{16.5}$$

$$\mathbf{v} = \begin{pmatrix} \mathbf{v}_B \\ \mathbf{v}_N \end{pmatrix} \geq \mathbf{0} \tag{16.6}$$

Let us also call this last formulation the multiplex *primal* (or, more properly, the *reduced* form of the linear multiplex primal).

If we note that we may rewrite (16.5) as

$$\mathbf{B}\mathbf{v}_B - \mathbf{N}\mathbf{v}_N = \mathbf{b} \tag{16.6}$$

then the *dual of the reduced form of the linear multiplex model* is simply as follows.

Find $\mathbf{Y}$ so as to

$$\text{lexmax } \mathbf{w} = \mathbf{b}^T\mathbf{Y} + \{\mathbf{c}_B^{(1)}\mathbf{B}^{-1}\mathbf{b}, \ \dots, \ \mathbf{c}_B^{(K)}\mathbf{B}^{-1}\mathbf{b}\} \tag{16.7}$$

subject to:

$$\begin{pmatrix} \mathbf{B}^T \\ \text{----} \\ \mathbf{N}^T \end{pmatrix} \Leftarrow \begin{pmatrix} \mathbf{0} \\ \text{------------------} \\ (\mathbf{c}_N^{(1)} - \mathbf{N}^T(\mathbf{B}^{-1})^T\mathbf{c}_B^{(1)}) \end{pmatrix}, \ \dots, \ \begin{pmatrix} \mathbf{0} \\ \text{------------------} \\ (\mathbf{c}_N^{(K)} - \mathbf{N}^T(\mathbf{B}^{-1})^T\mathbf{c}_B^{(K)}) \end{pmatrix} \tag{16.8}$$

$$\text{and } \mathbf{Y} \textit{ is unrestricted and multidimensional} \tag{16.9}$$

There are certain features of the MDD that are most definitely unusual and deserve explanation. First, note carefully that the set of dual variables exists as a *matrix* ($\mathbf{Y}$) rather than just a vector (i.e., as in single-objective optimization). In addition, each element of $\mathbf{Y}$, designated as $y_i^{(k)}$, is unrestricted in sign. Thus, we may define $y_i^{(k)}$ as

$$y_i^{(k)} = i\text{th dual variable for the } k\text{th right-hand side}$$

Consequently, for *each* right-hand side of the primal, there is a corresponding *vector* of dual variables.

Next, note that the achievement function of the MDD (i.e., relationship (16.7)) is an ordered vector for which we seek a *lexicographic maximum*. Finally, the goals of the MDD (relationship (16.8)) have *multiple and prioritized* right-hand sides. This is reflected by the special symbol used within (16.8): the $\Leftarrow$ operator— an operator that represents the lexicographic nature of the inequalities involved (i.e., as used to indicate that the left-hand side of each goal is *lexicographically* less than or equal to the multiple right-hand sides). Physically, this means that we first seek a solution to the problem subject to the first column of right-hand-side elements. Next, we seek a solution subject to the second column of right-hand-side elements—*but one that cannot degrade the solution obtained for the previous column of right-hand side values.* We continue in this manner until a complete set of solutions has been derived for all right-hand sides.

Those readers who recall the development of the dual of the conventional LP model, as addressed in Part 1 of the text, may recognize that the development of the MDD from the reduced form of the multiplex primal follows a set of rules directly analogous to those used to form the conventional LP dual. Specifically, we may note that

- a "lexmin" (lexicographically minimized) achievement function for the primal translates into a *set* of *lexicographically ordered right-hand sides* for the dual
- if the primal is to be lexicographically minimized (lexmin), then the dual is to be lexicographically maximized (lexmax)
- for every priority level in the primal, there is an associated *vector* of dual variables in the dual
- each element of the (*reduced form* of the) achievement function corresponds to an element in a right-hand side of the dual (i.e., that right-hand side corresponding to the priority level in the achievement function)
- the technological coefficients of the dual are the transpose of the technological coefficients of the primal

Example 16.1

In an attempt to further clarify the MDD concept, let us illustrate just how a specific MDD may be developed from a given multiplex primal. The following is the primal form of a linear multiplex model that we will use in the illustration.

Primal

Find x_1 and x_2 so as to

$$\text{lexmin } \mathbf{u} = \{(\rho_1 + \rho_2), (2\eta_3 + 3\eta_4)\} \tag{16.10}$$

satisfy:

$$x_1 + x_2 + \eta_1 - \underline{\rho_1} = 12 \tag{16.11}$$

$$2x_1 + x_2 + \eta_2 - \underline{\rho_2} = 20 \tag{16.12}$$

$$16x_1 + 10x_2 + \underline{\underline{\eta_3}} - \rho_3 = 160 \tag{16.13}$$

$$3x_1 + 5x_2 + \underline{\underline{\eta_4}} - \rho_4 = 60 \tag{16.14}$$

and all x_j, η_i, and ρ_i are nonnegative $\tag{16.16}$

By using (16.7–16.9), the multidimensional dual may immediately be written as follows.

Multidimensional Dual

Find $\mathbf{Y}$ so as to

$$\text{lexmax } \mathbf{w} = (12 \quad 20 \quad 160 \quad 60)\mathbf{Y} + \{0, 500\} \tag{16.17}$$

satisfy:

$$\begin{vmatrix} 1 & 0 & 0 & 0 \\ 0 & 1 & 0 & 0 \\ 0 & 0 & 1 & 0 \\ 0 & 0 & 0 & 1 \\ \hline 1 & 2 & 16 & 3 \\ 1 & 1 & 10 & 5 \\ -1 & 0 & 0 & 0 \\ 0 & -1 & 0 & 0 \\ 0 & 0 & -1 & 0 \\ 0 & 0 & 0 & -1 \end{vmatrix} \cdot \mathbf{Y} \Leftarrow \begin{vmatrix} 0 \\ 0 \\ 0 \\ 0 \\ \hline 0 \\ 0 \\ 1 \\ 1 \\ 0 \\ 0 \end{vmatrix}, \begin{vmatrix} 0 \\ 0 \\ 0 \\ 0 \\ \hline -41 \\ -35 \\ 0 \\ 0 \\ 2 \\ 3 \end{vmatrix} \tag{16.18}$$

where $\mathbf{Y}$ is multidimensional and unrestricted $\tag{16.19}$

With the mechanics of the formulation of the MDD now behind us, we are prepared to interpret the components of the MDD—and then employ the concept in the development of various algorithms and for sensitivity analysis.

ECONOMIC INTERPRETATION

If we solve the MDD (i.e., via an algorithm for the specific solution of the MDD as to be described later in this chapter) listed in Example 16.1, we will find that the optimal MDD program is given by

$$\mathbf{Y}^* = \begin{bmatrix} 0 & -25 \\ 0 & 0 \\ 0 & -1 \\ 0 & 0 \end{bmatrix}$$

The first column of $\mathbf{Y}$ simply corresponds to the solution of the MDD for the first right-hand side (i.e., first priority). That is:

$$y_1^{(1)*} = 0 \qquad y_2^{(1)*} = 0 \qquad y_3^{(1)*} = 0 \qquad y_4^{(1)*} = 0 \qquad w^{(1)*} = 0$$

Similarly, the second column of $\mathbf{Y}$ serves to indicate the solution to the MDD for the second right-hand side, or second priority, that is,

$$y_1^{(2)*} = -25 \qquad y_2^{(2)*} = 0 \qquad y_3^{(2)*} = -1 \qquad y_4^{(2)*} = 0 \qquad w^{(2)*} = 40$$

The economic interpretation of $y_i^{(k)}$ is that its value represents the per unit contribution of resource i (of the primal) to the kth term of the primal's achievement function. As an example, note the values of $y_1^{(1)}$ and $y_1^{(2)}$. These terms simply represent the dual variables associated with the *first* right-hand-side value of the primal (i.e., $b_1 = 12$ from relationship (16.11)). Because, at optimality

$$y_1^{(1)*} = 0 \qquad \text{and} \qquad y_1^{(2)*} = -25$$

then we can say that an increase of one unit to b_1 will result in

- no impact on u, the first term of the achievement function (and thus no impact on the implementability of the solution)
- an improvement (i.e., reduction) in u_2 of -25 units

Consequently, if the measure being used in the second priority level were dollars, then it would be to our advantage to purchase an extra unit of b_1 as long as it cost something less than \$25. Of course, such observations hold true only as long as the present (and optimal) basis remains unchanged. The range over which such changes are appropriate will be discussed in the material to follow.

MDD ALGORITHMS

A number of algorithms have been developed by Ignizio (1974a, 1974b, 1976, 1985a) and Markowski and Ignizio (1983a, 1983b) to solve the MDD model (it must be noted, however, that the earlier versions of such approaches dealt only with non-Archimedean LGP). However, in this text, we shall only deal with two of these. The first, the *general* MDD algorithm, is—as its name implies—a completely general approach to solving *any* MDD. The second, the *restricted* MDD algorithm, deals only with a special form of the MDD—but is, as we shall see, essential to sensitivity analysis in linear multiplex.

The General MDD Algorithm

The general MDD algorithm is appropriate for any MDD model and, in essence, solves the MDD via the solution of a *sequence* of conventional LP models. Further, each LP in the sequence is generally much smaller than its predecessor. As such, the general MDD algorithm is a very straightforward and computationally efficient approach to the MDD problem (and thus to its corresponding primal). The algorithm for the general MDD algorithm is as follows:

1. Establish the MDD (i.e., in the form as listed in (16.7–16.9)). Set $k = 1$ (i.e., begin with the first priority level).
2. Form the LP model from the MDD that includes only the kth right-hand-side vector (i.e., of 16.8). Solve this LP using any conventional simplex algorithm.
3. If $k = K$, then go to Step 5. Otherwise, proceed to Step 4.
4. For the LP model previously solved (i.e., in Step 2), remove all *nonbinding constraints* (note that this is equivalent to the checking of nonbasic variables in the algorithm used to solve primal multiplex models). If the subsequent model has *no* constraints left, go to Step 5. Otherwise, set $k = k + 1$ and return to Step 2.
5. The present solution is that which is optimal for the MDD and its kth right-hand side. The corresponding optimal solution to the primal model is provided by the reduced prices (i.e., shadow prices) as associated with the *initial* set of basic variables for the kth dual model.

The general MDD algorithm may be illustrated by means of a numerical example. Specifically, we shall use the model as represented previously in primal form by (16.10–16.16), and in MDD form by (16.17–16.19). As such, the first step of the algorithm represents the formulation of the MDD, as given by (16.17–16.19). Consequently, the first LP model to be solved is given by the following.

Find $\mathbf{y}^{(1)}$ so as to

$$\text{maximize } w^{(1)} = 12y_1^{(1)} + 20y_2^{(1)} + 160y_3^{(1)} + 60y_4^{(1)} + \{0\}$$

subject to:

$$
\begin{vmatrix}
1 & 0 & 0 & 0 \\
0 & 1 & 0 & 0 \\
0 & 0 & 1 & 0 \\
0 & 0 & 0 & 1 \\
\hline
1 & 2 & 16 & 3 \\
1 & 1 & 10 & 5 \\
-1 & 0 & 0 & 0 \\
0 & -1 & 0 & 0 \\
0 & 0 & -1 & 0 \\
0 & 0 & 0 & -1
\end{vmatrix}
\cdot \mathbf{y}^{(1)} \le
\begin{vmatrix}
0 \\
0 \\
0 \\
0 \\
\hline
0 \\
0 \\
1 \\
1 \\
0 \\
0
\end{vmatrix}
$$

where $\mathbf{y}^{(1)}$ is unrestricted.

We may solve this LP via any conventional LP algorithm to obtain the following solution:*

$$
\mathbf{y}^{(1)*} = \begin{pmatrix} 0 \\ 0 \\ 0 \\ 0 \end{pmatrix}
\quad \text{and} \quad
w^{(1)*} = 0 + \{0\} = 0
$$

Note that in this solution, both the seventh and eighth constraints are nonbinding and thus may be dropped from the LP model used at the next priority level (i.e., $k = 2$). The next (and final) LP model in the sequence is as follows:

Find $\mathbf{y}^{(2)}$ so as to

$$
\text{maximize } w^{(2)} = 12y_1^{(2)} + 20y_2^{(2)} + 160y_3^{(2)} + 60y_4^{(2)} + \{500\}
$$

subject to:

$$
\begin{vmatrix}
1 & 0 & 0 & 0 \\
0 & 1 & 0 & 0 \\
0 & 0 & 1 & 0 \\
0 & 0 & 0 & 1 \\
\hline
1 & 2 & 16 & 3 \\
1 & 1 & 10 & 5 \\
0 & 0 & -1 & 0 \\
0 & 0 & 0 & -1
\end{vmatrix}
\cdot \mathbf{y}^{(2)} \le
\begin{vmatrix}
0 \\
0 \\
0 \\
0 \\
\hline
-41 \\
-35 \\
2 \\
3
\end{vmatrix}
$$

where $\mathbf{y}^{(2)}$ is unrestricted.

---

*Note carefully that the first four constraints in conjunction with the last constraint simply provide upper and lower bounds on the values of the dual variables. By using this information, the solution to the problem becomes even more simple.

Solving this latest LP, we have

$$\mathbf{y}^{(2)*} = \begin{pmatrix} -25 \\ 0 \\ -1 \\ 0 \end{pmatrix} \qquad \text{and} \qquad w^{(2)*} = -460 + \{500\} = 40$$

All that remains to do is to read off the values of the reduced prices from the final LP tableau (we advise the reader to construct this tableau and read these values). Doing this, we will note that the optimal solution to the primal is simply

$$\mathbf{x}^* = \begin{pmatrix} \frac{20}{3} \\ \frac{16}{3} \end{pmatrix} \qquad \text{and} \qquad \mathbf{u}^* = \{0, 40\}$$

Although this brief introduction to the general multiplex algorithm serves to describe the basic approach, in practice, the algorithm may be enhanced and simplified by means of various straightforward techniques. As a result, our experience with this algorithm has been quite positive. In fact, when comparing the general MDD algorithm with the primal algorithm (as previously discussed), we observed that *the dual version was superior in virtually all computational aspects*.

The Restricted MDD Algorithm

Although the previous dual-based algorithm may be applied to *any* MDD model, the restricted MDD algorithm (Ignizio, 1974a, 1974b, 1976, 1982, 1985a) can only be used when the following conditions hold for the linear multiplex *primal*:

- at least one element in $\mathbf{v}_B$ (i.e., β) must be negative, *and*
- all reduced prices column vectors (i.e., $\mathbf{d}_j$) must be lexicographically nonpositive

In essence, this simply means that the primal must be simultaneously superoptimal *and* infeasible.

Should these two conditions hold, we may use the restricted MDD algorithm to obtain—for the primal problem—a feasible solution (all nonnegative right-hand sides) while maintaining, at every iteration, an indication of optimality. The algorithm for the accomplishment of the restricted MDD method follows:

1. Given that these two conditions hold, examine the right-hand-side values, as they appear in the present *primal* tableau, and select that row with the most negative $v_{B,j}$ element (ties may be broken arbitrarily). The basic variable associated with this row will be the *departing* variable in the primal (and the entering variable in the associated dual). Denote this row as $i = p$.

2. For *all* k, compute

$$\boldsymbol{\pi}^{(k)T} = \mathbf{c}_B^{(k)} \mathbf{B}^{-1}, \qquad \text{for all } k$$

3. Compute $\alpha_{p,j}$, for all nonbasic j:

$$\alpha_{p,j} = \mathbf{b}_p' \mathbf{a}_j$$

where

$$\mathbf{b}_p' = p\text{th row of } \mathbf{B}^{-1}$$

$$\mathbf{a}_j = j\text{th column of } \mathbf{A}$$

4. For all levels of k, and for *only* those j associated with a *negative* $\alpha_{p,j}$ (i.e., as computed in Step 3), determine the associated reduced costs:

$$d_j^{(k)} = \boldsymbol{\pi}^{(k)} \mathbf{a}_j - c_j^{(k)}, \qquad \text{for all } k \text{ and } j \text{ associated with a negative } \alpha_{p,j}$$

5. Determine the nonbasic variable associated with the lexicographically minimum "column ratio," where this column ratio is given by

$$\mathbf{r}_j = \begin{pmatrix} d_j^{(1)}/\alpha_{p,j} \\ \vdots \\ d_j^{(K)}/\alpha_{p,j} \end{pmatrix}, \qquad \text{for all } \alpha_{p,j} < 0$$

6. Designate the nonbasic variable with the lexicographically minimum $\mathbf{r}_j$ as being column $j = q$. This will be the *entering* variable in the primal tableau. (Ties may be arbitrarily broken.)

7. Using the pivoting procedure, exchange the entering variable for the departing variable and develop the new tableau.

8. Repeat steps 1 through 7 until all $v_{B,i}$ are nonnegative.

The restricted MDD algorithm may be illustrated by means of the following example. Consider the problem initially posed by the following model.

Find x_1 and x_2 so as to

$$\text{lexmin } \mathbf{u} = \{(\rho_1 + \rho_2), (\eta_3), (\rho_4)\}$$

satisfy:

$$x_1 + x_2 + \eta_1 - \rho_1 = 10$$
$$x_1 + \eta_2 - \rho_2 = 12$$
$$5x_1 + 3x_2 + \eta_3 - \rho_3 = 56$$
$$x_1 + x_2 + \eta_4 - \rho_4 = 12$$

and all x_j, η_i, and ρ_i are nonnegative

For the sake of discussion, let us assume that we have developed a solution with x_1, x_2, η_3, and η_4 as the basic variables. The tableau for this basis appears in Table 16.1. Notice that the basis is infeasible. Further, if one computes the matrix of reduced costs, it will be seen that all column vectors are lexicographically nonpositive.

TABLE 16.1   INFEASIBLE SOLUTION

	Basis inverse				β
x_2	1	-1	0	0	-2
x_1	0	1	0	0	12
η_3	-3	-2	1	0	2
η_4	-1	0	0	1	2
$\pi^{(1)}$	0	0	0	0	0
$\pi^{(2)}$	-3	-2	1	0	2
$\pi^{(3)}$	0	0	0	0	0

Thus, for Table 16.1, we have satisfied the two conditions necessary in the employment of the restricted MDD algorithm. We may then use the previous algorithm to determine that the departing variable must be x_2. The values of $\pi^{(1)}$, $\pi^{(2)}$, and $\pi^{(3)}$ are listed in the tableau, and thus we may proceed to compute the $\alpha_{p,j}$ values. The results are

$$\alpha_{1,3} = 1 \qquad \alpha_{1,4} = -1 \qquad \alpha_{1,7} = -1$$

$$\alpha_{1,8} = 1 \qquad \alpha_{1,9} = 0 \qquad \alpha_{1,10} = 0$$

Because only $\alpha_{1,4}$ and $\alpha_{1,7}$ are negative, we need consider only the reduced prices for the fourth and seventh variables (i.e., η_2 and ρ_1). Computing these reduced prices, we have

$$\mathbf{d}_4 = \begin{pmatrix} 0 \\ -2 \\ 0 \end{pmatrix} \qquad \mathbf{d}_7 = \begin{pmatrix} -1 \\ 3 \\ 0 \end{pmatrix}$$

Finally, we compute the column ratios associated with η_2 and ρ_1. These are

$$\mathbf{r}_4 = \begin{pmatrix} 0/-1 \\ -2/-1 \\ 0/-1 \end{pmatrix} \qquad \mathbf{r}_7 = \begin{pmatrix} -1/-1 \\ 3/-1 \\ 0/-1 \end{pmatrix}$$

or

$$\mathbf{r}_4 = \begin{pmatrix} 0 \\ 2 \\ 0 \end{pmatrix} \qquad \mathbf{r}_7 = \begin{pmatrix} 1 \\ -3 \\ 0 \end{pmatrix}$$

Thus, the lexicographically minimum column ratio is associated with the fourth variable (i.e., η_2). By letting x_2 depart and η_2 enter, the new tableau is shown in Table 16.2. This new solution is both feasible *and* optimal. We shall now proceed to a discussion of sensitivity analysis in linear multiplex. There, we shall also note just how the restricted MDD algorithm may be employed to assist in these sensitivity computations.

TABLE 16.2 FEASIBLE SOLUTION

	Basis inverse				β
η_2	-1	1	0	0	2
x_1	1	0	0	0	10
η_3	-5	0	1	0	6
η_4	-1	0	0	1	2

SENSITIVITY ANALYSIS IN LINEAR MULTIPLEX

We shall, in this section, present two types of sensitivity analysis. First, we shall describe how to deal with the analysis of a *discrete* change for any model parameter. Next, we shall discuss the manner in which we may consider the sensitivity of a problem *over a complete range* of parameter variations.

Discrete Sensitivity Analysis

Typically, discrete sensitivity analysis is employed *after* one has first developed an "optimal" solution to a given linear multiplex model—and wherein it is then desired to evaluate the impact of a specific change in one or more model parameters. We shall assume this to be the case in the discussions that immediately follow. The parameters that we shall consider for the analysis of discrete changes include

- a change in some $c_j^{(k)}$
- a change in some b_i
- a change in some $a_{i,j}$
- the inclusion of a new structural variable
- the inclusion of a new goal (either rigid or soft)

Dealing with discrete changes in any of these parameters is quite straightforward. However, let us first restate some of the key formulas associated with the linear multiplex model, as it is these formulas that shall be used in the implementation of sensitivity analysis.

$$d_j^{(k)} = \mathbf{c}_B^{(k)} \mathbf{B}^{-1} \mathbf{a}_j - c_j^{(k)} \tag{16.20}$$

$$u_k = \mathbf{c}_B^{(k)} \mathbf{B}^{-1} \mathbf{b} \tag{16.21}$$

$$\boldsymbol{\beta} = \mathbf{B}^{-1} \mathbf{b} \tag{16.22}$$

$$\boldsymbol{\alpha}_j = \mathbf{B}^{-1} \mathbf{a}_j \tag{16.23}$$

The first equation (16.20) serves to compute the reduced cost of the jth nonbasic variable. The second equation (16.21) derives the kth term of the achievement vector for a given basis. The third (16.22) computes the right-hand side associated with a given basis, and the last (16.23) determines the present value of the jth column of the matrix of technological coefficients.

Next, let us adopt the following notation. Whenever the value of a parameter has either been changed, or is indirectly changed due to a discrete change in some other parameter, we shall *underline* those parameters either changed or impacted by the change. For example, the following equation indicates that the value of the achievement function coefficient for variable j, at priority level k, has been changed—and that this change impacts on the value of the reduced cost of that variable (at that priority level).

$$\underline{d}_j^{(k)} = \mathbf{c}_B^{(k)}\mathbf{B}^{-1}\mathbf{a}_j - \underline{c}_j^{(k)}$$

Let us now consider the impact of various discrete changes.

A change in $c_j^{(k)}$, where x_j is a *nonbasic* variable.　To evaluate the result of this change, we simply modify Equation (16.20) as shown:

$$\underline{d}_j^{(k)} = \mathbf{c}_B^{(k)}\mathbf{B}^{-1}\mathbf{a}_j - \underline{c}_j^{(k)} \tag{16.24}$$

Thus, the only impact of such a change will be on a single, reduced cost element. Of course, such a change may also result in a solution that is no longer optimal. If so, the multiplex algorithm must be continued so as to develop a new, optimal result.

A change in $c_j^{(k)}$, where x_j is a *basic* variable.　If a change is made to an achievement-function coefficient that is associated with a *basic* variable, then this can result in change to the entire vector of reduced prices associated with that variable, as well as have an impact on the value of u_k. Consequently, such a change can have an impact on both optimality and/or implementability. Via modifications to our earlier set of formulas, we may list the equations that may be used to evaluate this particular change as follows.

$$\underline{d}_j^{(k)} = \underline{\mathbf{c}}_B^{(k)}\mathbf{B}^{-1}\mathbf{a}_j - c_j^{(k)} \tag{16.25}$$

$$\underline{u}_k = \underline{\mathbf{c}}_B^{(k)}\mathbf{B}^{-1}\mathbf{b} \tag{16.26}$$

A change in some b_i.　The change to a right-hand-side element is felt on both the value of $\boldsymbol{\beta}$ and $\mathbf{u}$. That is, should b_i be changed to $\underline{b}_i$, then

$$\underline{\boldsymbol{\beta}} = \mathbf{B}^{-1}\underline{\mathbf{b}} \tag{16.27}$$

$$\underline{u}_k = \mathbf{c}_B^{(k)}\mathbf{B}^{-1}\underline{\mathbf{b}} \tag{16.28}$$

Note, in the last two equations, that the change in b_i has been denoted as a change to $\mathbf{b}$—even though only a single element of $\mathbf{b}$ has been changed. Further, realize that a change in a right-hand-side element may cause the solution to go infeasible (i.e., one or more elements of β become negative). However, because such a change has no impact on optimality, we may use the restricted MDD algorithm to regain feasibility in such instances. In addition, this particular change may also have an impact on solution implementability as the value of u_k can change.

A change in some $a_{i,j}$. Let us next evaluate the impact of a change to a given technological coefficient. The manner in which we deal with such a change is dependent upon whether or not x_j is a basic or nonbasic variable. If x_j is basic, then it is possible to evaluate this change, but the process is so lengthy and involved that most analysts seem to agree that one might as well simply resolve the problem, from the beginning. Consequently, we shall only describe the process to be used when x_j is *nonbasic*. In that event, we note that a change to such a technological coefficient may have an impact on both the new value of α_j as well as on the entire, associated vector of reduced costs ($\mathbf{d}_j$). Thus, the equations used to evaluate such a change are

$$\underline{\alpha}_j = \mathbf{B}^{-1}\mathbf{a}_j \tag{16.29}$$

$$\underline{d}_j^{(k)} = \mathbf{c}_B^{(k)}\underline{\alpha}_j - c_j^{(k)}, \qquad \text{for all } k \tag{16.30}$$

Note that the change in a single technological coefficient ($a_{i,j}$) is depicted as a change in $\mathbf{a}_j$, the jth vector of technological coefficients. Further, it should be apparent that such a change may have an impact on the optimality of the previous solution.

The addition of a new structural variable. The addition of a new structural variable, x_j, is identical to that of a change in some $\mathbf{a}_j$ vector as associated with a nonbasic variable—and as described directly before. That is, we may think of this new variable as having always existed—but with *zero* technological coefficients. Using the new values of the $a_{i,j}$, and in conjunction with Equations (16.29) and (16.30), we may then compute the new values of α_j and $d_j^{(k)}$, respectively. Thus, the addition of a new structural variable may have an impact on solution optimality. If, after the addition of this new structural variable, the solution is no longer optimal, then we have an indication that this new variable should enter into the basis (i.e., it will improve the present solution). Otherwise, it should not.

The addition of a new goal (rigid or soft). If we add a new rigid goal to the original linear multiplex model, then this goal enters at priority level 1 and thus we need not worry as to whether or not it is in commensurable units with the other goals at level 1. However, if this new goal is at some lower priority level, care must be taken to assure commensurability (i.e., through the use of weights).

A new goal will also change the size of the basis. Further, to determine this new basis, we must first clear the new goal of any nonzero coefficients for any basic variables that appear in the new goal. This is easily accomplished by means of elementary matrix row and column operations. Even further, it should be obvious that the inclusion of a new goal can have an impact on both the feasibility and optimality of the present solution.

Finally, one should note that the addition of a new goal in the primal is the same as the addition of a new structural variable in the MDD (and vice versa). As such, we could—if we so wish—deal with such a change through either the primal or the dual.

Range Analysis

The previous discussion focused on only *discrete* modifications to the original linear multiplex model. However, it is often of considerable interest to examine the sensitivity of a given solution to changes in a parameter, or parameters, over a continuous range. In this section, we discuss how such range, or parametric, analysis may be conducted—but we shall limit our presentation to just changes in a b_i or $c_j^{(k)}$. Further, the easiest way to explain range analysis is through the use of numerical examples—and such examples shall be employed in the discussion to follow.

Range analysis for a parameter in the achievement function $(c_j^{(k)})$.   To demonstrate this analysis, we shall employ the following linear goal-programming model in its multiplex form.

Find x_1 and x_2 so as to

$$\text{lexmin } \mathbf{u} = \{(\rho_1 + \rho_2), (\eta_3 + 2\eta_4), (\eta_1)\} \tag{16.31}$$

satisfy:

$$x_1 + \eta_1 - \rho_1 = 20 \tag{16.32}$$

$$x_2 + \eta_2 - \rho_2 = 35 \tag{16.33}$$

$$-5x_1 + 3x_2 + \eta_3 - \rho_3 = 220 \tag{16.34}$$

$$x_1 + x_2 + \eta_4 - \rho_4 = 60 \tag{16.35}$$

and all x_j, η_i, and ρ_i are nonnegative $\tag{16.36}$

The optimal solution to this model is depicted in the final tableau of Table 16.3.

TABLE 16.3 OPTIMAL SOLUTION
TO ORIGINAL MODEL

	Basis inverse				β
η_1	1	0	0	0	20
x_2	0	1	0	0	35
η_3	0	-3	1	0	115
η_4	0	1	0	1	95
$\pi^{(1)}$	0	0	0	0	0
$\pi^{(2)}$	0	-1	1	2	305
$\pi^{(3)}$	1	0	0	0	20

For sake of discussion, let us assume that we wish to investigate the weight associated with η_4 (i.e., the achievement-vector coefficient for η_4). Specifically, we wish to determine the range over which this value may be changed in which the foregoing program remains basic.

To accomplish this, replace the original weight of η_4 in the achievement vector (which was a value of 2—as indicated in the second term of (16.31)) with a parameter, say, t. Such a change will obviously have an impact on the second row of the $\mathbf{\Pi}$ matrix, or $\pi^{(2)}$. Specifically:

$$\underline{\pi}^{(2)} = \underline{c}_B^{(2)} B^{-1} = (0 \quad 0 \quad 1 \quad t) \begin{bmatrix} 1 & 0 & 0 & 0 \\ 0 & 1 & 0 & 0 \\ 0 & -3 & 1 & 0 \\ 0 & 1 & 0 & 1 \end{bmatrix}$$

or

$$\underline{\pi}^{(2)} = (0 \quad -3 + t \quad 1 \quad t)$$

We may then use this new value of $\pi^{(2)}$ to compute the reduced costs for all nonbasic variables. This results in

$$d(x_1) = \begin{bmatrix} 0 \\ -5 + t \\ 1 \end{bmatrix} \qquad d(\eta_2) = \begin{bmatrix} 0 \\ -3 + t \\ 0 \end{bmatrix} \qquad d(\rho_1) = \begin{bmatrix} -1 \\ 0 \\ -1 \end{bmatrix}$$

$$d(\rho_2) = \begin{bmatrix} -1 \\ 3 - t \\ 0 \end{bmatrix} \qquad d(\rho_3) = \begin{bmatrix} 0 \\ -1 \\ 0 \end{bmatrix} \qquad d(\rho_4) = \begin{bmatrix} 0 \\ -t \\ 0 \end{bmatrix}$$

If the previous program, as depicted in Table 16.3, is to remain optimal, then these reduced costs must each be lexicographically nonpositive. For this to be true, the following conditions must hold:

$$-5 + t < 0$$
$$-3 + t \leq 0$$
$$-t \leq 0$$

And these three relationships are simultaneously satisfied only when t lies between 0 and 3, that is, when

$$0 \leq t \leq 3$$

Consequently, as long as all the other parameters in the problem remain unchanged, the weight on η_4 at priority level 2 may vary from 0 up to as much as 3 and the original basis will remain optimal. We may continue, in this same manner, to examine each of the other achievement-function variable weights.

Range analysis for a parameter in the right-hand side (b_i).   Changes in the coefficients of variables in the achievement function can have an impact of solution optimality. Consequently, as demonstrated earlier, we must investigate the range of values of the parameter (e.g., t) over which the solution remains optimal. However, when one investigates changes in a right-hand-side value, we must consider the impact on solution *feasibility*.

Using the same example as above, let us then examine the range of values over which b_3 may be changed while maintaining the same basis. To do this, we shall replace b_3 by the parameter t. Extending the approach used to investigate a discrete change in a right-hand-side value, we note that

$$\underline{\beta} = \boldsymbol{B}^{-1}\underline{\boldsymbol{b}} = \begin{bmatrix} 1 & 0 & 0 & 0 \\ 0 & 1 & 0 & 0 \\ 0 & -3 & 1 & 0 \\ 0 & 1 & 0 & 1 \end{bmatrix} \begin{bmatrix} 20 \\ 35 \\ t \\ 60 \end{bmatrix} = \begin{bmatrix} 20 \\ 35 \\ t - 105 \\ 95 \end{bmatrix}$$

Thus, for the present basis to remain feasible, $t - 105$ must equal or exceed a value of zero. And this is true as long as t is greater than 105. Thus, the range for t is

$$105 \le t < \infty$$

The range analysis for any other right-hand-side elements may be performed in a similar fashion.

Sensitivity Analysis in Sequential Linear Multiplex

The discussion thus far has assumed the employment of the standard linear multiplex algorithm. However, should one instead use sequential linear multiplex, sensitivity analysis is equally important and should be considered. The easiest way to accomplish sensitivity analysis after one has solved a model by means of the sequential process is to simply use the solution thus developed to construct an equivalent standard multiplex tableau. This is done by noting the basic variables from the final sequential tableau, selecting those that correspond to variables in the original formulation, and then developing the associated standard tableau by forcing these variables into the basis. This is easily done via the process used for reinversion. Once this is done, one may employ the foregoing procedures to perform whatever sensitivity analysis that may be desired.

SUMMARY

We have, in this chapter, presented a rather brief discussion of the dual in linear multiplex—and its use in sensitivity analysis. In essence, the reader should recognize that any type of sensitivity analysis that can be done in conventional linear programming can be achieved, in an analogous fashion, in linear multiplex.

EXERCISES

16.1. Formulate the multidimensional dual (MDD) representation for the following model.

$$\text{lexmin } \mathbf{u} = \{(\eta_1), (\eta_3), (\eta_2), (\rho_1 + \rho_2)\}$$

subject to:

$$2x_1 + x_2 + \eta_1 - \rho_1 = 20$$
$$x_1 + \eta_2 - \rho_2 = 12$$
$$x_2 + \eta_3 - \rho_3 = 10$$
$$\mathbf{x}, \boldsymbol{\eta}, \boldsymbol{\rho} \geq \mathbf{0}$$

16.2. Formulate the MDD representation for the following model.

$$\text{lexmin } \mathbf{u} = \{(\eta_1), (\rho_2), (2\eta_3 + \eta_4), (\rho_1)\}$$

subject to:

$$x_1 + x_2 + \eta_1 - \rho_1 = 100$$
$$x_1 + x_2 + \eta_2 - \rho_2 = 90$$
$$x_1 + \eta_3 - \rho_3 = 80$$
$$x_2 + \eta_4 - \rho_4 = 55$$
$$\mathbf{x}, \boldsymbol{\eta}, \boldsymbol{\rho} \geq \mathbf{0}$$

16.3. For the tableau of Table 16.4, first explain *why* the restricted multiplex dual algorithm may be applied and then use this algorithm to obtain an optimal and feasible solution. Note also that the original achievement vector is given by

$$\text{lexmin } \mathbf{u} = \{(\rho_1 + \rho_2), (\rho_3)\}$$

TABLE 16.4

	Basis inverse			β
η_1	1	0	0	-5
η_2	0	1	0	-4
η_3	0	0	1	0
$\pi^{(1)}$	0	0	0	0
$\pi^{(3)}$	0	0	0	0

16.4. Solve the problem in Exercise 16.1 by means of the revised multiplex algorithm. Then, evaluate the impact of the following changes (where each change is individually considered):

(a) A change in the weighting of ρ_1, at priority 4, from 1 to 3.

(b) A change in the weighting of η_3, at priority 2, from 1 to 100.

(c) A change in $a_{1,2}$ from 1 to 5.

(d) A change in $a_{1,1}$ from 2 to -1.

16.5. Given the following model, determine the set of optimal policies over all reasonable ranges in the value of the parameter d.

$$\text{lexmin } \mathbf{u} = \{(\eta_1 + \rho_1), (2\rho_2 + \rho_3)\}$$

subject to:

$$x_1 + 10x_2 + \eta_1 - \rho_1 = 50 - d$$

$$3x_1 + 5x_2 + \eta_2 - \rho_2 = 20 - 2d$$

$$8x_1 + 6x_2 + \eta_3 - \rho_3 = 100$$

$$\mathbf{x}, \, \boldsymbol{\eta}, \, \boldsymbol{\rho} \geq \mathbf{0}$$

16.6. Refer to Examples 6.13, 6.15, 6.16, 6.17, and 6.18 from Chapter 6. Formulate these problems in *multiplex* format and then repeat the sensitivity analysis as originally performed on the conventional LP formulation. Compare your results with that obtained earlier.

16.7. Assume that we wish to employ sensitivity analysis on Example 15.3, as solved in Chapter 15 via sequential multiplex. Use the final solution to this example to reconstruct the final tableau that would have been derived via the restricted multiplex algorithm (i.e., as derived for Example 15.2)—and compare this tableau with that derived in Example 15.2.

CHAPTER 17

EXTENSIONS OF MULTIPLEX

CHAPTER OVERVIEW

As discussed and demonstrated in the preceding four chapters, the multiplex concept provides a unified approach to the modeling, solution, and analysis of a variety of linear, single, and multiobjective problems. However, the methodology is certainly not limited to just linear models, as we can, through relatively straightforward—if not transparent—means, extend multiplex to models involving integer variables and/or nonlinear relationships. However, because the focus of this text is that of linear models, we shall provide but a brief overview of such extensions in this, our concluding chapter.

INTEGER MODELS

In the early 1970s, Ignizio developed the first algorithms and software for integer goal programming. (Much of this work was summarized in his early text on goal programming and the interested reader is advised to consult that text for details (Ignizio, 1976).) In essence, these were but transparent extensions of methods commonly used for the solution of single-objective integer models. For example,

619

one may extend the well-known branch-and-bound process to integer goal programming by simply replacing the scalar bounds at each node (as derived by linear programming) with bounds represented as ordered vectors (i.e., the achievement-function vector as derived by linear goal programming). In the same manner, we may further extend such a process to integer or discrete linear multiplex models.

Although such extensions are straightforward, transparent, and easily accomplished, it is even easier to simply extend sequential multiplex to integer models. And this is the approach that we recommend.

Recalling the sequential multiplex procedure of Chapter 15, we may modify this process so as to solve integer models by simply

- replacing the term ''linear multiplex model'' with the term ''linear integer multiplex model,'' and
- eliminating step 4 (the column drop procedure—a procedure no longer appropriate when dealing with integer models) of the previous algorithm.

We may then use the revised procedure, along with any commercial linear integer programming code, to solve linear integer multiplex models of sizes comparable to those involving but a single objective.

NONLINEAR MODELS

The first nonlinear goal-programming algorithm and software was developed in the early 1960s by Ignizio (1963)—and actually used to design the antennas that were installed on the Saturn II (the second stage of the Saturn/Apollo rocket that was to ultimately land men on the moon by the late 1960s). This algorithm was based upon a straightforward extension of the so-called pattern-search algorithm of Hooke and Jeeves (Wilde and Beightler, 1967) and a simple version of its code, in FORTRAN, is provided in reference (Ignizio, 1976). In essence, the only significant change to pattern search was in the evaluation process. Specifically, rather than evaluating each trial solution as a scalar, one employs the ordered vector of the goal-programming model. As such, extension of this procedure to nonlinear multiplex models is trivial—and has been accomplished with quite good results. Somewhat more recent versions of this approach have been used to solve nonlinear multiobjective models with hundreds to thousands of variables and goals (Ignizio, 1981).

However, another alternative exists. That is, one may also employ sequential multiplex in the solution of nonlinear multiplex models. Again, as with integer models, we must eliminate step 4 of the algorithm as presented in Chapter 15. Once this is done, the process may be implemented using any commercial single-objective nonlinear programming software.

INTELLIGENT INTERFACES

Although integer and nonlinear multiplex methods may certainly be accomplished by means of the sequential solution process, it does require a certain degree of skill and judgment. For this reason, we have recently utilized the concept of *intelligent interfacing* for dealing with the complete modeling, solution, and analysis effort involved in multiplex (as well as, for that matter, the process involved in conventional single-objective procedures). Such intelligent interfaces involve the use of expert systems (Ignizio, 1991, 1992a) in conjunction with the mathematical programming software. Such a combination permits one to not only easily extend the methodology (e.g., to integer and nonliner models), but to also allow a non-technical user to actually use the procedure to solve and interpret real-world problems.

SUMMARY AND CONCLUSIONS

Although this final chapter may appear remarkably short to the reader wishing to learn more about extensions of the multiplex concept, we believe that the best way in which to learn about multiobjective optimization is to first learn all one can about single-objective optimization, that is, the material covered in the first two parts of this text. With an understanding of single-objective optimization, coupled with an appreciation of the simplicity of the multiplex model, it should require little effort to extend the multiplex approach to whatever model type one wishes.

A REVIEW OF LINEAR ALGEBRA

Techniques for solving linear programming problems and the other related problems covered in this text are based upon the concepts and tools of linear algebra. The material in this appendix provides a brief look at those facets of linear algebra that are believed to be necessary for a full and better understanding of the material presented in this text. Readers interested in further details or rigorous proofs are directed to Hadley (1973) and Noble (1969).

VECTORS

A *vector* is an ordered array of numbers. It can be either a row or a column of elements wherein the position of each element is important. Vectors are printed in lowercase boldface type (e.g., **a**, **b**, **x**, **0**, **1**). Some examples of vectors are

$$\mathbf{a} = \begin{pmatrix} 15 \\ 24 \end{pmatrix} \qquad \mathbf{c} = (4 \quad -2 \quad 3 \quad 1) \qquad \mathbf{0} = \begin{pmatrix} 0 \\ 0 \\ 0 \end{pmatrix}$$

A general *m*-dimensional column vector **b** is designated as

$$\mathbf{b} = \begin{pmatrix} b_1 \\ b_2 \\ \vdots \\ b_m \end{pmatrix}$$

The collection of all *m*-dimensional vectors is called *Euclidean m-space* and is denoted by E^m. Vectors can also be represented geometrically. For example, in E^2, Euclidean 2-space, we could represent vectors as in Figure A.1. A vector can be thought of as either a point or as an arrow directed from the origin to the point.

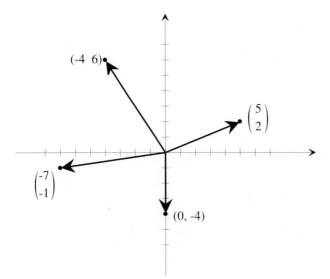

Figure A.1 Geometric representation of vectors.

Vector Addition

Vectors of the same type (i.e., either row or column) may be added together if they have the same number of entries. Given two vectors, **a** and **b**, and given that both are either row or column vectors with the same number of entries, the addition of **a** and **b** is carried out by adding the elements of **a**, position by position, to the elements of **b**. That is, letting **c** = **a** + **b**, and with c_i being the element in the ith position (with a similar designation for a_i and b_i), we have

$$c_i = a_i + b_i \qquad (A.1)$$

Vector addition satisfies both the commutative law,

$$\mathbf{a} + \mathbf{b} = \mathbf{b} + \mathbf{a} \qquad (A.2)$$

and the associative law,

$$\mathbf{a} + (\mathbf{b} + \mathbf{c}) = (\mathbf{a} + \mathbf{b}) + \mathbf{c} = \mathbf{a} + \mathbf{b} + \mathbf{c} \qquad (A.3)$$

Example A.1: Vector Addition

Given that

$$\mathbf{a} = \begin{pmatrix} 4 \\ 0 \\ 7 \end{pmatrix} \qquad \mathbf{b} = \begin{pmatrix} 5 \\ 9 \\ 1 \end{pmatrix} \qquad \mathbf{c} = (6 \quad 8 \quad 0) \qquad \mathbf{d} = \begin{pmatrix} 4 \\ 10 \\ 2 \\ 3 \end{pmatrix}$$

then

$$\mathbf{a} + \mathbf{b} = \begin{pmatrix} 4 \\ 0 \\ 7 \end{pmatrix} + \begin{pmatrix} 5 \\ 9 \\ 1 \end{pmatrix} = \begin{pmatrix} 9 \\ 9 \\ 8 \end{pmatrix}$$

$\mathbf{a} + \mathbf{c}$ is undefined (both vectors must be of the same type)

$\mathbf{a} + \mathbf{d}$ is undefined (both vectors must have the same number of elements)

Multiplication of a Vector by a Scalar

Numbers such as 3, 19, 37.5, and $\frac{2}{3}$ are called scalars, that is, a scalar is simply an element of E^1, Euclidean 1-space. Multiplication of a vector by a scalar is carried out by simply multiplying *each* element in the vector by the scalar. Thus, if α is a scalar, $\mathbf{a}$ is a row vector, and $\mathbf{b}$ is a column vector, then

$$\alpha\mathbf{a} = \alpha(a_1, a_2, \ldots, a_n) = (\alpha a_1, \alpha a_2, \ldots, \alpha a_n) \tag{A.4}$$

$$\alpha\mathbf{b} = \alpha \begin{pmatrix} b_1 \\ b_2 \\ \vdots \\ b_m \end{pmatrix} = \begin{pmatrix} \alpha b_1 \\ \alpha b_2 \\ \vdots \\ \alpha b_m \end{pmatrix} \tag{A.5}$$

Vector Multiplication

Two vectors can be multiplied together only if they both have the same number of entries, one is a row vector, and the other is a column vector. The result, often termed the *dot product (or inner product)*, is a *scalar*. By convention, the row vector is placed first and the column vector second. That is, $\mathbf{a} \cdot \mathbf{b}$, or $\mathbf{ab}$, infers that $\mathbf{a}$ is a row vector and $\mathbf{b}$ is a column vector for the multiplication to be meaningful. To multiply the vectors together, corresponding entries are multiplied and the results are added together. That is, assuming that $\mathbf{a}$ and $\mathbf{b}$ have m entries,

$$\mathbf{a} \cdot \mathbf{b} = \mathbf{ab} = \sum_{i=1}^{m} a_i b_i = \alpha \tag{A.6}$$

where α is a scalar.

Vector multiplication also satisfies the distributive law,

$$\mathbf{a}(\mathbf{b} + \mathbf{c}) = \mathbf{ab} + \mathbf{ac} \tag{A.7}$$

Example A.2: The Dot Product of Vectors
Given that

$$\mathbf{a} = \begin{pmatrix} 3 \\ 0 \\ 7 \end{pmatrix} \qquad \mathbf{b} = \begin{pmatrix} -2 \\ 10 \\ 1 \end{pmatrix} \qquad \mathbf{c} = (4 \quad 9 \quad 2)$$

$$\mathbf{d} = (5 \quad 1 \quad 4 \quad 2) \qquad \mathbf{e} = (3 \quad -2)$$

then

$$\mathbf{ca} = (4 \quad 9 \quad 2) \begin{pmatrix} 3 \\ 0 \\ 7 \end{pmatrix} = 12 + 0 + 14 = 26$$

$$\mathbf{cb} = (4 \quad 9 \quad 2) \begin{pmatrix} -2 \\ 10 \\ 1 \end{pmatrix} = -8 + 90 + 2 = 84$$

$\mathbf{da}$ is undefined

$\mathbf{ea}$ is undefined

Norm of a Vector

The *Euclidean norm* of a vector $\mathbf{a} \in E^n$, denoted by $\|\mathbf{a}\|$, is a measure of the size of $\mathbf{a}$ and is given by

$$\|\mathbf{a}\| = \left(\sum_{i=1}^{n} a_i \right)^{1/2} \tag{A.8}$$

For example, the Euclidean norm of the vector

$$\mathbf{a} = \begin{pmatrix} 3 \\ 2 \\ -1 \end{pmatrix}$$

is

$$\|\mathbf{a}\| = [3^2 + 2^2 + (-1)^2]^{1/2} = (14)^{1/2}$$

The dot product of two vectors can also be defined using the Euclidean norm and is given by

$$\mathbf{a} \cdot \mathbf{b} = \|\mathbf{a}\| \, \|\mathbf{b}\| \cos \theta \tag{A.9}$$

where θ is the angle between the two vectors.

Special Vector Types

There are many vectors that occur so frequently that they have been given their own special names. These include the following.

UNIT VECTOR A unit vector has a 1 in the jth position and 0's elsewhere. A unit vector will generally be denoted by $\mathbf{e}_j$, where the 1 appears in the jth position. For example, if $\mathbf{e}_j \in E^4$,

$$\mathbf{e}_1 = \begin{pmatrix} 1 \\ 0 \\ 0 \\ 0 \end{pmatrix} \qquad \mathbf{e}_2 = \begin{pmatrix} 0 \\ 1 \\ 0 \\ 0 \end{pmatrix} \qquad \mathbf{e}_3 = \begin{pmatrix} 0 \\ 0 \\ 1 \\ 0 \end{pmatrix} \qquad \mathbf{e}_4 = \begin{pmatrix} 0 \\ 0 \\ 0 \\ 1 \end{pmatrix}$$

NULL OR ZERO VECTOR The null vector, written $\mathbf{0}$, is a vector having only 0's as its elements. Graphically, the null vector corresponds to the origin.

SUM VECTOR The sum vector, written $\mathbf{1}$, has only 1's as its elements. It is referred to as the sum vector because forming the dot product of $\mathbf{1}$ and a vector $\mathbf{a}$ results in a scalar that is equal to the sum of the entries of $\mathbf{a}$. That is:

$$\mathbf{1a} = (1, 1, \ldots, 1) \begin{pmatrix} a_1 \\ a_2 \\ \vdots \\ a_n \end{pmatrix} = a_1 + a_2 + \cdots + a_n = \sum_{i=1}^{n} a_i$$

Linear Dependence and Independence

The notion of linear dependence or independence plays an important part in the material that follows. We discuss these concepts next.

A set of vectors, $\mathbf{a}_1, \mathbf{a}_2, \ldots, \mathbf{a}_m$, is termed *linearly dependent* if there exist some scalars, α_i, that are *not all* zero such that

$$\alpha_1 \mathbf{a}_1 + \alpha_2 \mathbf{a}_2 + \cdots + \alpha_m \mathbf{a}_m = \mathbf{0} \tag{A.10}$$

If the only set of scalars, α_i, for which (A.10) holds is $\alpha_1 = \alpha_2 = \cdots = \alpha_m = 0$, the vectors are *linearly independent*.

Example A.3: Linear Dependence

The vectors

$$\mathbf{a}_1 = \begin{pmatrix} 1 \\ 1 \end{pmatrix} \qquad \mathbf{a}_2 = \begin{pmatrix} 2 \\ 3 \end{pmatrix} \qquad \mathbf{a}_3 = \begin{pmatrix} 8 \\ 11 \end{pmatrix}$$

are linearly dependent because

$$2\mathbf{a}_1 + 3\mathbf{a}_2 - 1\mathbf{a}_3 = 2 \begin{pmatrix} 1 \\ 1 \end{pmatrix} + 3 \begin{pmatrix} 2 \\ 3 \end{pmatrix} - 1 \begin{pmatrix} 8 \\ 11 \end{pmatrix} = \begin{pmatrix} 0 \\ 0 \end{pmatrix}$$

That is, $\alpha_1 = 2$, $\alpha_2 = 3$, and $\alpha_3 = -1$.

Example A.4: Linear Independence

Consider the vectors

$$\mathbf{a}_1 = \begin{pmatrix} 1 \\ 1 \end{pmatrix} \qquad \mathbf{a}_2 = \begin{pmatrix} 2 \\ 0 \end{pmatrix}$$

We can investigate linear independence by considering the system

$$\alpha_1 \begin{pmatrix} 1 \\ 1 \end{pmatrix} + \alpha_2 \begin{pmatrix} 2 \\ 0 \end{pmatrix} = \begin{pmatrix} 0 \\ 0 \end{pmatrix} \tag{A.11}$$

Expanding (A.11), we have

$$\alpha_1 + 2\alpha_2 = 0 \tag{A.12}$$

$$\alpha_1 = 0 \tag{A.13}$$

Obviously, the only solution to (A.12–A.13) is $\alpha_1 = \alpha_2 = 0$. Therefore, $\mathbf{a}_1$ and $\mathbf{a}_2$ are linearly independent.

Spanning Sets and Bases

The vectors $\mathbf{b}_1, \mathbf{b}_2, \ldots, \mathbf{b}_p \in E^n$ are called a *spanning set* if every vector in E^n can be written as a linear combination of the $\mathbf{b}_i$. That is, if $\mathbf{v} \in E^n$, then there exist scalars $\alpha_1, \alpha_2, \ldots, \alpha_p$ such that $\mathbf{v} = \alpha_1\mathbf{b}_1 + \alpha_2\mathbf{b}_2 + \cdots + \alpha_p\mathbf{b}_p$.

The vectors $\mathbf{b}_1, \mathbf{b}_2, \ldots, \mathbf{b}_n \in E^n$ are called a *basis* for E^n if they are linearly independent and form a spanning set for E^n. A basis is a minimal spanning set. That is, the addition of a vector to the set results in the loss of linear independence, whereas if any $\mathbf{b}_i$ is removed from the set, the remaining vectors no longer span E^n. For example, the unit vectors

$$\mathbf{e}_1 = \begin{pmatrix} 1 \\ 0 \\ 0 \\ 0 \end{pmatrix} \qquad \mathbf{e}_2 = \begin{pmatrix} 0 \\ 1 \\ 0 \\ 0 \end{pmatrix} \qquad \mathbf{e}_3 = \begin{pmatrix} 0 \\ 0 \\ 1 \\ 0 \end{pmatrix} \qquad \mathbf{e}_4 = \begin{pmatrix} 0 \\ 0 \\ 0 \\ 1 \end{pmatrix}$$

are a basis for E^4.

MATRICES

A *matrix* is simply a rectangular array of numbers. Matrices are typically represented by uppercase boldface type (e.g., $\mathbf{A}$, $\mathbf{B}$, $\mathbf{E}$) and a general matrix with m rows and n columns may be represented as

$$\mathbf{A} = \begin{pmatrix} a_{1,1} & a_{1,2} & \cdots & a_{1,n} \\ a_{2,1} & a_{2,2} & \cdots & a_{2,n} \\ \vdots & \vdots & \ddots & \vdots \\ a_{m,1} & a_{m,2} & \cdots & a_{m,n} \end{pmatrix} \tag{A.14}$$

The position of each element within the matrix is important. The *order* of the matrix refers to the number of *rows and columns* that the matrix contains. The rows are always given first. Consequently, an *m* by *n* (denoted $m \times n$) matrix is one with exactly *m* rows and *n* columns. A *square* matrix has exactly the same number of rows as columns. Thus, $q \times q$, 3×3, 7×7, and so on, all represent orders of square matrices.

As indicated in (A.14), an element within a matrix is usually denoted by a lowercase letter with two subscripts (often *i* and *j*). The first subscript refers to the row in which the element is found, and the second denotes the respective column. Thus, $a_{i,j}$ is the element in row *i*, column *j* of the matrix. Alternatively, we may specify an entire column or row of the matrix by a vector. That is, if $\mathbf{a}_1$, $\mathbf{a}_2$, and $\mathbf{a}_3$ are each column vectors with the same number of elements, we might represent a matrix $\mathbf{A}$ by

$$\mathbf{A} = (\mathbf{a}_1, \mathbf{a}_2, \mathbf{a}_3)$$

Two matrices, say, $\mathbf{A}$ and $\mathbf{B}$, are equal (i.e., $\mathbf{A} = \mathbf{B}$) if and only if their corresponding elements are equal. This, of course, implies that they must also be of the same order. Consequently, if $a_{i,j}$ are the elements of $\mathbf{A}$ and $b_{i,j}$ are the elements of $\mathbf{B}$, then $\mathbf{A} = \mathbf{B}$ if and only if $a_{i,j} = b_{i,j}$, for all *i* and *j*.

Matrix Addition

If two matrices are of the same order, they may be added. Otherwise, addition is undefined. To add matrix $\mathbf{A}$ to matrix $\mathbf{B}$, we simply add the elements in each corresponding position. Thus, if $\mathbf{C} = \mathbf{A} + \mathbf{B}$, then $c_{i,j} = a_{i,j} + b_{i,j}$, for every *i* and *j*. Matrix addition satisfies both the commutative law,

$$\mathbf{A} + \mathbf{B} = \mathbf{B} + \mathbf{A} \qquad (A.15)$$

and the associative law,

$$\mathbf{A} + (\mathbf{B} + \mathbf{C}) = (\mathbf{A} + \mathbf{B}) + \mathbf{C} = \mathbf{A} + \mathbf{B} + \mathbf{C} \qquad (A.16)$$

Example A.5: Matrix Addition

Given that

$$\mathbf{A} = \begin{pmatrix} 7 & 1 & -2 \\ 3 & 3 & 0 \end{pmatrix} \qquad \mathbf{B} = \begin{pmatrix} 2 & -3 & 4 \\ 1 & 5 & 9 \end{pmatrix} \qquad \mathbf{C} = \begin{pmatrix} 2 & 1 \\ 7 & 3 \\ 9 & 2 \end{pmatrix}$$

then

$$\mathbf{A} + \mathbf{B} = \begin{pmatrix} 7 & 1 & -2 \\ 3 & 3 & 0 \end{pmatrix} + \begin{pmatrix} 2 & -3 & 4 \\ 1 & 5 & 9 \end{pmatrix} = \begin{pmatrix} 9 & -2 & 2 \\ 4 & 8 & 9 \end{pmatrix} = \mathbf{B} + \mathbf{A}$$

$\mathbf{A} + \mathbf{C}$ is undefined

$\mathbf{B} + \mathbf{C}$ is undefined

$$\mathbf{A} - \mathbf{B} = \mathbf{A} + (-1)\mathbf{B} = \begin{pmatrix} 7 & 1 & -2 \\ 3 & 3 & 0 \end{pmatrix} + (-1)\begin{pmatrix} 2 & -3 & 4 \\ 1 & 5 & 9 \end{pmatrix} = \begin{pmatrix} 5 & 4 & -6 \\ 2 & -2 & -9 \end{pmatrix}$$

Multiplication by a Scalar

As with vectors, if α is a scalar and $\mathbf{A}$ is a matrix, the product $\alpha\mathbf{A}$ is obtained by multiplying each element $(a_{i,j})$ of $\mathbf{A}$ by α. That is:

$$\alpha\mathbf{A} = \begin{pmatrix} \alpha a_{1,1} & \alpha a_{1,2} & \cdots & \alpha a_{1,n} \\ \alpha a_{2,1} & \alpha a_{2,2} & \cdots & \alpha a_{2,n} \\ \vdots & \vdots & \ddots & \vdots \\ \alpha a_{m,1} & \alpha a_{m,2} & \cdots & \alpha a_{m,n} \end{pmatrix} \tag{A.17}$$

Example A.6: Multiplication of a Matrix by a Scalar

Given that

$$\alpha = 0.7 \qquad \beta = 3$$

$$\mathbf{A} = \begin{pmatrix} 8 & 3 \\ -1 & 2 \\ 7 & 1 \end{pmatrix} \qquad \mathbf{B} = \begin{pmatrix} 10 & 7 \\ -3 & 5 \end{pmatrix}$$

then

$$\alpha\mathbf{B} = 0.7 \begin{pmatrix} 10 & 7 \\ -3 & 5 \end{pmatrix} = \begin{pmatrix} 7 & 4.9 \\ -2.1 & 3.5 \end{pmatrix}$$

$$\beta\mathbf{A} = 3 \begin{pmatrix} 8 & 3 \\ -1 & 2 \\ 7 & 1 \end{pmatrix} = \begin{pmatrix} 24 & 9 \\ -3 & 6 \\ 21 & 3 \end{pmatrix}$$

Matrix Multiplication

Two matrices, $\mathbf{A}$ and $\mathbf{B}$, may be multiplied together (i.e., $\mathbf{AB}$) if and only if they are *conformable*, that is, only if the number of columns in $\mathbf{A}$ is equal to the number of rows of $\mathbf{B}$. Thus, $\mathbf{AB}$ is defined for matrix $\mathbf{A}$ of order $m \times n$ and matrix $\mathbf{B}$ of order $p \times q$, if and only if $n = p$. If the two matrices *are* conformable, their multiplication produces the matrix $\mathbf{C}$ of order $m \times q$. That is, the number of rows in $\mathbf{C}$ equals the number of rows in $\mathbf{A}$, and the number of columns in $\mathbf{C}$ equals the number of columns in $\mathbf{B}$. Each element of $\mathbf{C}$ is given by

$$c_{i,j} = \sum_{k=1}^{n} a_{i,k} b_{k,j} \tag{A.18}$$

where

$$n = \text{number of columns of } \mathbf{A} \text{ or rows of } \mathbf{B}$$

$$i = 1, \ldots, m \ (m = \text{number of rows of } \mathbf{A})$$

$$j = 1, \ldots, q \ (q = \text{number of columns of } \mathbf{B})$$

Matrix multiplication satisfies the associative law

$$(\mathbf{AB})\mathbf{C} = \mathbf{A}(\mathbf{BC}) = \mathbf{ABC} \tag{A.19}$$

(wherein the matrices are conformable) and the distributive law

$$A(B + C) = AB + AC \tag{A.20}$$

but does *not* satisfy the commutative law, that is, in *general*;

$$AB \neq BA \tag{A.21}$$

Consequently, **AB** does not equal **BA** except under special conditions. Note that even if both **AB** and **BA** are defined, the resulting matrices need not be of the same order.

We may also perform matrix multiplication through the use of the multiplication rules previously defined for vectors. Recall that the product of a row and column vector results in a scalar. If we wish to multiply two matrices **A** (order $m \times n$) and **B** (order $n \times q$) together, where

$$\mathbf{A} = \begin{pmatrix} \mathbf{a}_1 \\ \mathbf{a}_2 \\ \vdots \\ \mathbf{a}_m \end{pmatrix} \qquad \mathbf{B} = (\mathbf{b}_1, \mathbf{b}_2, \ldots, \mathbf{b}_q)$$

then $\mathbf{C} = \mathbf{AB}$ is a matrix of order $m \times q$, wherein

$$c_{i,j} = \mathbf{a}_i \mathbf{b}_j$$

Example A.7: Matrix Multiplication

Given that

$$\mathbf{A} = \begin{pmatrix} 7 & 1 \\ 4 & -3 \\ 2 & 0 \end{pmatrix} \qquad \mathbf{B} = \begin{pmatrix} 2 & 1 & 7 \\ 0 & -1 & 4 \end{pmatrix} \qquad \mathbf{C} = \begin{pmatrix} 1 & -1 & 3 \\ 2 & 2 & 3 \\ -1 & 4 & 7 \end{pmatrix}$$

then

$$\mathbf{AB} = \begin{pmatrix} 7 & 1 \\ 4 & -3 \\ 2 & 0 \end{pmatrix} \begin{pmatrix} 2 & 1 & 7 \\ 0 & -1 & 4 \end{pmatrix} = \begin{pmatrix} 14 & 6 & 53 \\ 8 & 7 & 16 \\ 4 & 2 & 14 \end{pmatrix}$$

$$\mathbf{BA} = \begin{pmatrix} 2 & 1 & 7 \\ 0 & -1 & 4 \end{pmatrix} \begin{pmatrix} 7 & 1 \\ 4 & -3 \\ 2 & 0 \end{pmatrix} = \begin{pmatrix} 32 & -1 \\ 4 & 3 \end{pmatrix}$$

AC is undefined

$$\mathbf{CA} = \begin{pmatrix} 1 & -1 & 3 \\ 2 & 2 & 3 \\ -1 & 4 & 7 \end{pmatrix} \begin{pmatrix} 7 & 1 \\ 4 & -3 \\ 2 & 0 \end{pmatrix} = \begin{pmatrix} 9 & 4 \\ 28 & -4 \\ 23 & -13 \end{pmatrix}$$

$$\mathbf{BC} = \begin{pmatrix} 2 & 1 & 7 \\ 0 & -1 & 4 \end{pmatrix} \begin{pmatrix} 1 & -1 & 3 \\ 2 & 2 & 3 \\ -1 & 4 & 7 \end{pmatrix} = \begin{pmatrix} -3 & 28 & 58 \\ -6 & 14 & 25 \end{pmatrix}$$

CB is undefined

(handwritten annotations in left margin:)

$7(2) + 1(0) \qquad 7(1) + 1(-1)$

$4(2) + -3(0) \qquad 4(1) \quad 3(-1)$

$2(2) + 0(0) \qquad 2(1) \quad 0(-1)$

$2(7) + 1(4) + 7(2)$

$0(7) + 1(4) + 4(2)$

Note that although both **AB** and **BA** are defined, they are not even of the same order.

Special Matrices

There are a number of special matrices that are encountered. These include the diagonal matrix, the identity matrix, the null or zero matrix, the matrix transpose, symmetric matrices, and augmented matrices.

DIAGONAL MATRIX   The diagonal matrix is a square matrix (i.e., $m = n$) whose off-diagonal elements (those elements, $a_{i,j}$, where $i \neq j$) are all equal to zero. For example, a diagonal matrix has the form

$$\mathbf{A} = \begin{pmatrix} a_{1,1} & 0 & 0 \\ 0 & a_{2,2} & 0 \\ 0 & 0 & a_{3,3} \end{pmatrix}$$

IDENTITY MATRIX   The identity matrix, also known as the unit matrix, is a special case of the diagonal matrix in which *all* diagonal elements are equal to 1. An identity matrix of order m is designated as either $\mathbf{I}_m$ or just $\mathbf{I}$. The identity matrix is the matrix algebra equivalent of the number 1. For example:

$$\mathbf{I}_2 = \begin{pmatrix} 1 & 0 \\ 0 & 1 \end{pmatrix} \qquad \mathbf{I}_4 = \begin{pmatrix} 1 & 0 & 0 & 0 \\ 0 & 1 & 0 & 0 \\ 0 & 0 & 1 & 0 \\ 0 & 0 & 0 & 1 \end{pmatrix}$$

NULL OR ZERO MATRIX   A null matrix has all of its elements equal to zero. The null matrix does not have to be square. The matrix is the matrix algebra equivalent to the number zero and is usually denoted by **0**. For example:

$$\mathbf{0} = \begin{pmatrix} 0 & 0 & 0 \\ 0 & 0 & 0 \end{pmatrix} \qquad \mathbf{0} = \begin{pmatrix} 0 & 0 & 0 \\ 0 & 0 & 0 \\ 0 & 0 & 0 \end{pmatrix}$$

MATRIX TRANSPOSE   The transpose of a matrix **A**, denoted $\mathbf{A}^t$, is a reordering of the original matrix. To obtain the transpose of a matrix, we simply interchange the rows and columns, in order. That is, row 1 of the original matrix becomes column 1 of the transpose, row 2 of the original becomes column 2 of transpose, and so on. Thus, if

$$\mathbf{A} = \begin{pmatrix} a_{1,1} & a_{1,2} & \cdots & a_{1,n} \\ a_{2,1} & a_{2,2} & \cdots & a_{2,n} \\ \vdots & \vdots & \ddots & \vdots \\ a_{m,1} & a_{m,2} & \cdots & a_{m,n} \end{pmatrix}$$

then

$$
\mathbf{A}^t = \begin{pmatrix} a_{1,1} & a_{2,1} & \cdots & a_{m,1} \\ a_{1,2} & a_{2,2} & \cdots & a_{m,2} \\ \vdots & \vdots & \ddots & \vdots \\ a_{1,n} & a_{2,n} & \cdots & a_{m,n} \end{pmatrix}
$$

Note that the transpose of a diagonal matrix would remain unchanged.

SYMMETRIC MATRIX   A symmetric matrix is one whose transpose and the matrix itself are equal. That is, $\mathbf{A} = \mathbf{A}^t$. Such a matrix must, obviously, be square, and $a_{i,j} = a_{j,i}$, for all i and j. Examples of symmetric matrices are

$$
\mathbf{A} = \begin{pmatrix} 1 & 2 & 3 \\ 2 & 6 & 4 \\ 3 & 4 & 9 \end{pmatrix} \qquad \mathbf{B} = \begin{pmatrix} 1 & 6 \\ 6 & 2 \end{pmatrix}
$$

From the definition, it should be obvious that diagonal matrices and identity matrices are symmetric.

AUGMENTED MATRIX   An augmented matrix is one in which rows or columns of another matrix, of appropriate order, are appended to the original matrix.

If $\mathbf{A}$ is augmented (on the right) with matrix $\mathbf{B}$, the resulting matrix is denoted by either $(\mathbf{A}, \mathbf{B})$ or $(\mathbf{A} \mid \mathbf{B})$. All the rules of matrix algebra apply equally as well to the augmented matrix.

For example, if

$$
\mathbf{A} = \begin{pmatrix} 1 & 4 \\ 5 & 6 \end{pmatrix} \qquad \mathbf{b} = \begin{pmatrix} 3 \\ 1 \end{pmatrix} \qquad \mathbf{I} = \begin{pmatrix} 1 & 0 \\ 0 & 1 \end{pmatrix}
$$

then

$$
(\mathbf{A} \mid \mathbf{b}) = \begin{pmatrix} 1 & 4 & \mid & 3 \\ 5 & 6 & \mid & 1 \end{pmatrix} \qquad \text{or} \qquad \begin{pmatrix} 1 & 4 & 3 \\ 5 & 6 & 1 \end{pmatrix}
$$

$$
(\mathbf{A} \mid \mathbf{I}) = \begin{pmatrix} 1 & 4 & \mid & 1 & 0 \\ 5 & 6 & \mid & 0 & 1 \end{pmatrix} \qquad \text{or} \qquad \begin{pmatrix} 1 & 4 & 1 & 0 \\ 5 & 6 & 0 & 1 \end{pmatrix}
$$

Determinants

Every square matrix (and *only* square matrices) has a number, called the *determinant*, associated with it. Given a square matrix $\mathbf{A}$, the determinant of $\mathbf{A}$ is denoted by $|\mathbf{A}|$. If $\mathbf{A}$ is a 1×1 matrix, then $|\mathbf{A}|$ is called a first-order determinant. Similarly, the determinant of a 2×2 matrix is called a second-order determinant, and so on.

By definition, a first-order determinant is always equal to the value of its single element, that is:

$$|a_{1,1}| = a_{1,1} \qquad \text{or} \qquad |-3| = -3$$

and the value of a second-order determinant is given by

$$\begin{vmatrix} a_{1,1} & a_{1,2} \\ a_{2,1} & a_{2,2} \end{vmatrix} = a_{1,1}a_{2,2} - a_{1,2}a_{2,1}$$

The *principal diagonal* (read from left to right) of a determinant consists of those elements $a_{i,j}$ for which $i = j$. Thus, in a determinant of order 3, the principal diagonal would consist of elements $a_{1,1}$, $a_{2,2}$, and $a_{3,3}$. The other diagonal (i.e., from right to left) is referred to as the *secondary diagonal*. Again, for a third-order determinant $|\mathbf{A}|$, the secondary diagonal would contain elements $a_{1,3}$, $a_{2,2}$, and $a_{3,1}$.

Every element of a determinant, except for the special case of a first-order determinant, has an associated *minor*. To determine the minor of an element, we simply cross out the row and column associated with that element. The remaining elements represent a determinant, of order 1 less than the original, that is the minor for that element. If the element is given as $a_{i,j}$, we omit row i and column j to obtain the minor denoted as $|\mathbf{A}_{i,j}|$.

The *cofactor* of an element is simply its minor with the sign $(-1)^{i+j}$ assigned to it. That is, the cofactor of $a_{i,j}$ is $(-1)^{i+j}|\mathbf{A}_{i,j}|$.

Example A.8: Cofactors

Given the determinant

$$|\mathbf{A}| = \begin{vmatrix} 7 & -1 & 0 \\ 3 & 2 & 1 \\ 8 & 1 & -4 \end{vmatrix}$$

the cofactor for $a_{2,1} = 3$ is

$$(-1)^{2+1}|\mathbf{A}_{2,1}| = (-1)\begin{vmatrix} -1 & 0 \\ 1 & -4 \end{vmatrix} = -4$$

and the cofactor for $a_{3,3} = -4$ is

$$(-1)^{3+3}|\mathbf{A}_{3,3}| = (+1)\begin{vmatrix} 7 & -1 \\ 3 & 2 \end{vmatrix} = 17$$

Thus far, we have only discussed and illustrated how one may find the value for a determinant of order 1 or 2. In general, most determinants will be of considerably higher order. We shall present just one of several ways to obtain the value for the determinant of any order. This involves the use of cofactors and an expansion of the determinant by any row or column.

The *value* of a determinant of order n may be found by adding the products of each element, $a_{i,j}$, of any row or column, by its respective cofactor. This may be written as

$$|\mathbf{A}| = \sum_{j=1}^{n} a_{i,j}(-1)^{i+j}|\mathbf{A}_{i,j}| \tag{A.22}$$

for any row i, or

$$|\mathbf{A}| = \sum_{i=1}^{n} a_{i,j}(-1)^{i+j}|\mathbf{A}_{i,j}| \tag{A.23}$$

for any column j.

Example A.9: Values of Determinants

Evaluate the following determinants

$$|\mathbf{A}| = \begin{vmatrix} a_{1,1} & a_{1,2} & a_{1,3} \\ a_{2,1} & a_{2,2} & a_{2,3} \\ a_{3,1} & a_{3,2} & a_{3,3} \end{vmatrix} \qquad |\mathbf{B}| = \begin{vmatrix} 1 & 4 & 3 \\ 2 & 0 & 2 \\ 1 & 3 & 5 \end{vmatrix}$$

Expanding $|\mathbf{A}|$ by row 2, we obtain

$$|\mathbf{A}| = a_{2,1}(-1)^{2+1}|\mathbf{A}_{2,1}| + a_{2,2}(-1)^{2+2}|\mathbf{A}_{2,2}| + a_{2,3}(-1)^{2+3}|\mathbf{A}_{2,3}|$$

$$= -a_{2,1}\begin{vmatrix} a_{1,2} & a_{1,3} \\ a_{3,2} & a_{3,3} \end{vmatrix} + a_{2,2}\begin{vmatrix} a_{1,1} & a_{1,3} \\ a_{3,1} & a_{3,3} \end{vmatrix} - a_{2,3}\begin{vmatrix} a_{1,1} & a_{1,2} \\ a_{3,1} & a_{3,2} \end{vmatrix}$$

Expanding $|\mathbf{B}|$ by column 3, we get

$$|\mathbf{B}| = 3(-1)^{1+3}\begin{vmatrix} 2 & 0 \\ 1 & 3 \end{vmatrix} + 2(-1)^{2+3}\begin{vmatrix} 1 & 4 \\ 1 & 3 \end{vmatrix} + 5(-1)^{3+3}\begin{vmatrix} 1 & 4 \\ 2 & 0 \end{vmatrix}$$

$$= 3(6) - 2(-1) + 5(-8) = -20$$

Notice that if a determinant of order greater than 3 is to be evaluated, we would have to perform a *series* of expansions. Unfortunately, such an approach is quite unwieldy for large problems.

The expansion of determinants may often be simplified considerably by utilizing the following five properties, where the word "row" may be replaced by the word "column" (and vice versa) without affecting the validity of each statement.

1. If one complete row of a determinant is all zero, the value of the determinant is zero.

2. If two rows have elements that are proportional to one another, the value of the determinant is zero.

3. If two rows of a determinant are interchanged, the value of the new determinant is equal to the negative of the value of the old determinant.

4. Elements of any row may be multiplied by a nonzero constant if, in turn, the entire determinant (i.e., outside the determinant) is multiplied by the reciprocal of the constant.

5. To the elements of any row, you may add a constant times the corresponding element of any other row without changing the value of the determinant.

Using some of our previous definitions, we may now define the *adjoint of a matrix*. If **A** is a square matrix, the adjoint of **A** (designated by $\mathbf{A}^{\alpha}$) may be found by the following process:

1. Replace each element $a_{i,j}$ of **A** by its cofactor (i.e., by $(-1)^{i+j}|\mathbf{A}_{i,j}|$).
2. Take the transpose of the matrix of cofactors, found in step 1.
3. The resulting matrix is $\mathbf{A}^{\alpha}$, the adjoint of **A**. $\mathbf{A}^{\alpha}$ may be described mathematically as follows.

Let $\gamma_{i,j} = (-1)^{i+j}|\mathbf{A}_{i,j}|$ be the cofactor for element $a_{i,j}$, then

$$\mathbf{A}^{\alpha} = \begin{pmatrix} \gamma_{1,1} & \gamma_{2,1} & \cdots & \gamma_{n,1} \\ \gamma_{1,2} & \gamma_{2,2} & \cdots & \gamma_{n,2} \\ \vdots & \vdots & \ddots & \vdots \\ \gamma_{1,n} & \gamma_{2,n} & \cdots & \gamma_{n,n} \end{pmatrix} \tag{A.24}$$

Having set forth the concept and some properties of determinants, we may now present two additional, very important concepts: the inverse of a matrix and the rank of a matrix.

The Inverse of a Matrix

The *inverse* of a square matrix, say, **A**, is denoted by $\mathbf{A}^{-1}$. *Only* square matrices have inverses. Further, for the inverse to exist, the square matrix must be *nonsingular*, wherein a nonsingular matrix is one whose determinant does *not* equal zero. Conversely, the determinant of a *singular* matrix *does* equal zero. Every nonsingular matrix has an inverse, and only nonsingular matrices have inverses. Given that **A** is nonsingular, its inverse may be found by

$$\mathbf{A}^{-1} = \frac{1}{|\mathbf{A}|} \mathbf{A}^{\alpha} \tag{A.25}$$

Example A.10: Computation of Matrix Inverses

Given that

$$\mathbf{A} = \begin{pmatrix} a_{1,1} & a_{1,2} \\ a_{2,1} & a_{2,2} \end{pmatrix} \qquad \mathbf{B} = \begin{pmatrix} 2 & 1 \\ 6 & 5 \end{pmatrix}$$

then $\mathbf{A}^{-1}$ and $\mathbf{B}^{-1}$ are found as follows.

From (A.25), we have that

$$\mathbf{A}^{-1} = \frac{1}{|\mathbf{A}|} \mathbf{A}^{\alpha}$$

But for a 2×2 matrix, we have

$$|\mathbf{A}| = \begin{vmatrix} a_{1,1} & a_{1,2} \\ a_{2,1} & a_{2,2} \end{vmatrix} = a_{1,1}a_{2,2} - a_{1,2}a_{2,1}$$

and

$$\mathbf{A}^{\alpha} = \begin{pmatrix} |a_{2,2}| & -|a_{1,2}| \\ -|a_{2,1}| & |a_{1,1}| \end{pmatrix} = \begin{pmatrix} a_{2,2} & -a_{1,2} \\ -a_{2,1} & a_{1,1} \end{pmatrix}$$

Thus, $\mathbf{A}^{-1}$ is

$$\mathbf{A}^{-1} = \frac{1}{a_{1,1}a_{2,2} - a_{1,2}a_{2,1}} \begin{pmatrix} a_{2,2} & -a_{1,2} \\ -a_{2,1} & a_{1,1} \end{pmatrix}$$

and this is the *general* formula for the inverse of any 2×2 matrix.

Therefore, the inverse of the matrix **B** is simply

$$\mathbf{B}^{-1} = \frac{1}{10 - 6} \begin{pmatrix} 5 & -1 \\ -6 & 2 \end{pmatrix} = \begin{pmatrix} \frac{5}{4} & -\frac{1}{4} \\ -\frac{3}{2} & \frac{1}{2} \end{pmatrix}$$

The inverse of a square nonsingular matrix **A** may also be defined as follows: Given a square matrix **A**, if there exists a square matrix **B** such that

$$\mathbf{AB} = \mathbf{BA} = \mathbf{I}$$

then **B** is the inverse of **A**.

Going back to Example A.10, we may multiply **B** by $\mathbf{B}^{-1}$ to obtain

$$\mathbf{BB}^{-1} = \begin{pmatrix} 2 & 1 \\ 6 & 5 \end{pmatrix} \begin{pmatrix} \frac{5}{4} & -\frac{1}{4} \\ -\frac{3}{2} & \frac{1}{2} \end{pmatrix} = \begin{pmatrix} 1 & 0 \\ 0 & 1 \end{pmatrix} = \mathbf{I}$$

and, similarly,

$$\mathbf{B}^{-1}\mathbf{B} = \begin{pmatrix} \frac{5}{4} & -\frac{1}{4} \\ -\frac{3}{2} & \frac{1}{2} \end{pmatrix} \begin{pmatrix} 2 & 1 \\ 6 & 5 \end{pmatrix} = \begin{pmatrix} 1 & 0 \\ 0 & 1 \end{pmatrix} = \mathbf{I}$$

Some of the more useful properties of the matrix inverse include the following.

1. The inverse of a nonsingular matrix is unique.
2. Given that nonsingular, square matrices **A** and **B** are of the same order, then
   (a) $(\mathbf{AB})^{-1} = \mathbf{B}^{-1}\mathbf{A}^{-1}$
   (b) $(\mathbf{A}^{-1})^{-1} = \mathbf{A}$
   (c) $(\mathbf{A}^{t})^{-1} = (\mathbf{A}^{-1})^{t}$

It should be stressed that (A.25) is not the only way by which the inverse may be computed. Another, rather common, approach is to augment square matrix **A** with identity matrix **I**, and then by performing elementary row operations, we may develop $\mathbf{A}^{-1}$. This method of computation is often referred to as *Gauss-Jordan elimination* and is most easily explained by an example. However, let us first review what is meant by an elementary row operation.

Given matrix **A**, an *elementary row operation* on **A** is one of the following:

1. Interchange a row i with a row j.
2. Multiply a row i by a nonzero scalar α.
3. Replace a row i by row i plus a multiple of some row j.

Example A.11: Finding a Matrix Inverse Using Gauss-Jordan Elimination

Consider the same matrix **B** as in Example A.10, that is:

$$\mathbf{B} = \begin{pmatrix} 2 & 1 \\ 6 & 5 \end{pmatrix}$$

Augmenting a second-order identity to the right of **A**, we have

$$(\mathbf{B}|\mathbf{I}) = \begin{pmatrix} 2 & 1 & | & 1 & 0 \\ 6 & 5 & | & 0 & 1 \end{pmatrix}$$

Our objective then is, by using elementary row operations on $(\mathbf{A}|\mathbf{I})$, to form the identity matrix on the *left*. This will result in $\mathbf{A}^{-1}$ on the right, or $(\mathbf{I}|\mathbf{A}^{-1})$. First, we multiply row 1 by $\frac{1}{2}$ to obtain

$$\begin{pmatrix} 1 & \frac{1}{2} & | & \frac{1}{2} & 0 \\ 6 & 5 & | & 0 & 1 \end{pmatrix}$$

Next, let us multiply row 1 by -6 and add the result to row 2. This results in

$$\begin{pmatrix} 1 & \frac{1}{2} & | & \frac{1}{2} & 0 \\ 0 & 2 & | & -3 & 1 \end{pmatrix}$$

Multiplying row 2 by $\frac{1}{2}$, we get

$$\begin{pmatrix} 1 & \frac{1}{2} & | & \frac{1}{2} & 0 \\ 0 & 1 & | & -\frac{3}{2} & \frac{1}{2} \end{pmatrix}$$

Finally, multiplying row 2 by $-\frac{1}{2}$ and adding the result to row 1 yields

$$(\mathbf{I}|\mathbf{B}^{-1}) = \begin{pmatrix} 1 & 0 & | & \frac{5}{4} & -\frac{1}{4} \\ 0 & 1 & | & -\frac{3}{2} & \frac{1}{2} \end{pmatrix}$$

Notice that the inverse of **B** is now to the right of the identity matrix.

The Rank of a Matrix

The *rank* of an $m \times n$ matrix, **A**, is designated by $r(\mathbf{A})$ and is equal to the number of linearly independent columns (or, alternatively, the number of linearly independent rows) of **A**. **A** does not have to be a square matrix. Note that it follows immediately from the definition that $r(\mathbf{A}) \leq \text{minimum}\{m, n\}$. If $r(\mathbf{A}) = \text{minimum}\{m, n\}$, then $\mathbf{A}$ is said to be of *full rank*.

As with computing the inverse of a matrix, there are a number of ways of determining the rank of a matrix. One such way involves determining the order of the largest nonsingular determinant that may be formed from $\mathbf{A}$. However, a more straightforward approach involves using elementary row operations to reduce the matrix $\mathbf{A}$ to the form

$$\left(\begin{array}{c|c} \mathbf{I}_k & \mathbf{D} \\ \hline \mathbf{0} & \mathbf{0} \end{array} \right)$$

The rank of matrix $\mathbf{A}$ is then $r(\mathbf{A}) = k$. This process is now illustrated via an example.

Example A.12: The Rank of a Matrix

$$\mathbf{A} = \begin{pmatrix} 1 & 1 & 1 & 3 & 1 \\ 2 & 1 & 2 & 3 & 0 \\ 1 & 3 & 1 & 9 & 5 \end{pmatrix}$$

By using a sequence of elementary row operations, $\mathbf{A}$ can be reduced to

$$\mathbf{A} = \left(\begin{array}{cc|ccc} 1 & 0 & 1 & 0 & -1 \\ 0 & 1 & 0 & 3 & 2 \\ \hline 0 & 0 & 0 & 0 & 0 \end{array} \right) = \left(\begin{array}{c|c} \mathbf{I}_2 & \mathbf{D} \\ \hline \mathbf{0} & \mathbf{0} \end{array} \right)$$

Thus, the rank of $\mathbf{A}$ is 2. That is, $r(\mathbf{A}) = 2$.

THE SOLUTION OF SIMULTANEOUS LINEAR EQUATIONS

Probably the best known use of matrices and determinants is their employment in the solution of simultaneous linear equations. The use of matrices and vectors permits a shorthand, concise means of expressing the problem and the use of the matrix inverse and determinants provides the tools for a procedural approach to their solution.

Consider the set of simultaneous linear equations:

$$\begin{aligned} a_{1,1}x_1 + a_{1,2}x_2 + \cdots + a_{1,n}x_n &= b_1 \\ a_{2,1}x_1 + a_{2,2}x_2 + \cdots + a_{2,n}x_n &= b_2 \\ &\vdots \\ a_{m,1}x_1 + a_{m,2}x_2 + \cdots + a_{m,n}x_n &= b_m \end{aligned} \tag{A.26}$$

where the coefficients $a_{i,j}$ are constants, as are the right-hand-side values b_i. The $x_j, j = 1, \ldots, n$, are the variables that must satisfy (if possible) the equations of (A.26).

These equations may be written in concise matrix form as simply $\mathbf{Ax} = \mathbf{b}$, where

$$\mathbf{A} = \begin{pmatrix} a_{1,1} & a_{1,2} & \cdots & a_{1,n} \\ a_{2,1} & a_{2,2} & \cdots & a_{2,n} \\ \vdots & \vdots & \ddots & \\ a_{m,1} & a_{m,2} & \cdots & a_{m,n} \end{pmatrix}$$

$$\mathbf{b} = \begin{pmatrix} b_1 \\ b_2 \\ \vdots \\ b_m \end{pmatrix}$$

$$\mathbf{x} = \begin{pmatrix} x_1 \\ x_2 \\ \vdots \\ x_n \end{pmatrix}$$

Matrix $\mathbf{A}$ is known as the coefficient matrix, column vector $\mathbf{b}$ is the constant or right-hand-side vector, and column vector $\mathbf{x}$ is the vector of variables.

The set of linear equations $\mathbf{Ax} = \mathbf{b}$ has either no solution, a unique solution, or an infinite number of solutions. To determine if a solution exists, let us first rewrite the system $\mathbf{Ax} = \mathbf{b}$ by writing $\mathbf{A}$ in terms of its columns as $\mathbf{A} = (\mathbf{a}_1, \mathbf{a}_2, \ldots, \mathbf{a}_n)$. This results in

$$\mathbf{Ax} = (\mathbf{a}_1, \mathbf{a}_2, \ldots, \mathbf{a}_n) \begin{pmatrix} x_1 \\ x_2 \\ \vdots \\ x_n \end{pmatrix} = x_1 \mathbf{a}_1 + x_2 \mathbf{a}_2 + \cdots + x_n \mathbf{a}_n = \mathbf{b} \qquad (A.27)$$

Note, from (A.27), that when determining a solution of $\mathbf{Ax} = \mathbf{b}$, we are actually trying to find scalars $x_1, x_2, \ldots, x_n$, so that vector $\mathbf{b}$ can be written as a linear combination of the columns of $\mathbf{A}$. Clearly, it will *not* be possible to write $\mathbf{b}$ as a linear combination of the columns of $\mathbf{A}$ if $\mathbf{b}$ is linearly independent of the columns of $\mathbf{A}$. Mathematically, if $\mathbf{b}$ is linearly independent of the columns of $\mathbf{A}$, then the rank of the augmented matrix $(\mathbf{A}|\mathbf{b})$ is 1 greater than the rank of $\mathbf{A}$, that is, $r(\mathbf{A}|\mathbf{b}) = r(\mathbf{A}) + 1$.

Thus, we may summarize the conditions under which a solution exists for the system $\mathbf{Ax} = \mathbf{b}$.

1. If $r(\mathbf{A}|\mathbf{b}) = r(\mathbf{A}) + 1$, then the system of linear equations $\mathbf{Ax} = \mathbf{b}$ is inconsistent and no solution exists. That is, augmenting matrix $\mathbf{A}$ with the additional column vector $\mathbf{b}$ increases the rank by 1. Thus, it is not possible to write $\mathbf{b}$ as a linear combination of the columns of $\mathbf{A}$.

2. If $r(\mathbf{A}|\mathbf{b}) = r(\mathbf{A})$, then the system $\mathbf{Ax} = \mathbf{b}$ is consistent and there exists a solution. In this case, adding column $\mathbf{b}$ to matrix $\mathbf{A}$ does not increase the rank. Therefore, it is possible to write $\mathbf{b}$ as a linear combination of the columns of $\mathbf{A}$.

Example A.13: A System with No Solution

Consider the following system:

$$2x_1 + x_2 + 2x_3 = 6$$

$$x_1 + 3x_2 + x_3 = 9$$

$$x_1 + x_2 + x_3 = 3$$

Then,

$$\mathbf{A} = \begin{pmatrix} 2 & 1 & 2 \\ 1 & 3 & 1 \\ 1 & 1 & 1 \end{pmatrix} \qquad \mathbf{b} = \begin{pmatrix} 6 \\ 9 \\ 3 \end{pmatrix} \qquad \mathbf{x} = \begin{pmatrix} x_1 \\ x_2 \\ x_3 \end{pmatrix}$$

$$(\mathbf{A}|\mathbf{b}) = \begin{pmatrix} 2 & 1 & 2 & | & 6 \\ 1 & 3 & 1 & | & 9 \\ 1 & 1 & 1 & | & 3 \end{pmatrix}$$

By using row reductions, it is easy to determine that the rank of $\mathbf{A}$ is 2. However, the rank of $(\mathbf{A}|\mathbf{b})$ is 3 and thus the system is inconsistent and has no solution. The reader is invited to verify the rank of matrices $\mathbf{A}$ and $(\mathbf{A}|\mathbf{b})$.

A Unique Solution of Ax = b

If a system of simultaneous linear equations is consistent and the coefficient matrix $\mathbf{A}$ is nonsingular (and thus square), a *unique* solution exists. Or, equivalently, if $\mathbf{A}$ is a square matrix and $r(\mathbf{A}) = r(\mathbf{A}|\mathbf{b}) = n$ (where n is the number of variables), then a unique solution exists.

There are numerous methods used to solve for a unique solution, including Cramer's rule and Gaussian elimination (e.g., see Hadley, 1973; Nobel, 1969). We first present Cramer's rule, although it should be noted that it is not a very (computationally) efficient approach.

Let $\mathbf{A}_j$ represent the matrix in which the jth column of $\mathbf{A}$ is replaced by vector $\mathbf{b}$. Then Cramer's rule specifies that the unique solution is given by

$$x_j = \frac{|\mathbf{A}_j|}{|\mathbf{A}|}, \qquad \text{for all } j = 1, \ldots, n \qquad (A.28)$$

Example A.14: Using Cramer's Rule to Find the Unique Solution

We now use Cramer's rule to solve the following set of simultaneous linear equations. Notice that $\mathbf{A}$ is square and its rank is equal to the number of variables, and thus a unique solution does exist.

$$2x_1 + x_2 + 2x_3 = 6$$

$$2x_1 + 3x_2 + x_3 = 9$$

$$x_1 + x_2 + x_3 = 3$$

$$\mathbf{A} = \begin{pmatrix} 2 & 1 & 2 \\ 2 & 3 & 1 \\ 1 & 1 & 1 \end{pmatrix} \qquad \mathbf{b} = \begin{pmatrix} 6 \\ 9 \\ 3 \end{pmatrix} \qquad \mathbf{x} = \begin{pmatrix} x_1 \\ x_2 \\ x_3 \end{pmatrix}$$

Using (A.28), we obtain x_1, x_2, and x_3:

$$x_1 = \frac{\begin{vmatrix} 6 & 1 & 2 \\ 9 & 3 & 1 \\ 3 & 1 & 1 \end{vmatrix}}{\begin{vmatrix} 2 & 1 & 2 \\ 2 & 3 & 1 \\ 1 & 1 & 1 \end{vmatrix}} = \frac{6}{1} = 6$$

$$x_2 = \frac{\begin{vmatrix} 2 & 6 & 2 \\ 2 & 9 & 1 \\ 1 & 3 & 1 \end{vmatrix}}{\begin{vmatrix} 2 & 1 & 2 \\ 2 & 3 & 1 \\ 1 & 1 & 1 \end{vmatrix}} = \frac{0}{1} = 0$$

$$x_3 = \frac{\begin{vmatrix} 2 & 1 & 6 \\ 2 & 3 & 9 \\ 1 & 1 & 3 \end{vmatrix}}{\begin{vmatrix} 2 & 1 & 2 \\ 2 & 3 & 1 \\ 1 & 1 & 1 \end{vmatrix}} = \frac{-3}{1} = -3$$

Another approach to the solution of simultaneous linear equations for a unique solution is by means of the inverse. Notice that

$$\mathbf{A}\mathbf{x} = \mathbf{b}$$

and that

$$\mathbf{A}^{-1}\mathbf{A} = \mathbf{I}$$

Thus, if we premultiply $\mathbf{A}\mathbf{x} = \mathbf{b}$ by $\mathbf{A}^{-1}$, we obtain

$$\mathbf{A}^{-1}\mathbf{A}\mathbf{x} = \mathbf{A}^{-1}\mathbf{b}$$

or

$$\mathbf{I}\mathbf{x} = \mathbf{A}^{-1}\mathbf{b}$$

And because $\mathbf{I}\mathbf{x} = \mathbf{x}$, we have

$$\mathbf{x} = \mathbf{A}^{-1}\mathbf{b}$$

Example A.15: Using the Inverse to Find the Unique Solution

We now solve the problem given in Example A.14 by the use of the matrix inverse and (A.29). $\mathbf{A}^{-1}$ may be found as in Example A.11 and is given by

$$\mathbf{A}^{-1} = \begin{pmatrix} 2 & 1 & -5 \\ -1 & 0 & 2 \\ -1 & -1 & 4 \end{pmatrix}$$

Thus,

$$\mathbf{x} = \mathbf{A}^{-1}\mathbf{b} = \begin{pmatrix} 2 & 1 & -5 \\ -1 & 0 & 2 \\ -1 & -1 & 4 \end{pmatrix} \begin{pmatrix} 6 \\ 9 \\ 3 \end{pmatrix} = \begin{pmatrix} 6 \\ 0 \\ -3 \end{pmatrix}$$

which is the same solution as obtained through the use of Cramer's rule, as it should be.

The solution can also be found directly by the use of Gauss-Jordan elimination. The solution strategy is essentially the same as that used to find the matrix inverse in Example A.13. The basic idea is to use elementary row operations to reduce the augmented matrix $(\mathbf{A}|\mathbf{b})$ to the form $(\mathbf{I}|\mathbf{A}^{-1}\mathbf{b})$. The process is now illustrated by an example.

Example A.16: Using Gauss-Jordan Elimination to Find the Unique Solution

Consider again the linear system given in Example A.14. We begin the solution process by forming the augmented matrix $(\mathbf{A}|\mathbf{b})$ as follows:

$$(\mathbf{A}|\mathbf{b}) = \left(\begin{array}{ccc|c} 2 & 1 & 2 & 6 \\ 2 & 3 & 1 & 9 \\ 1 & 1 & 1 & 3 \end{array} \right)$$

First, we multiply row 1 by $\frac{1}{2}$ to obtain

$$\left(\begin{array}{ccc|c} 1 & \frac{1}{2} & 1 & 3 \\ 2 & 3 & 1 & 9 \\ 1 & 1 & 1 & 3 \end{array} \right)$$

Next, we multiply row 1 by -2 and add the result to row 2. Also, we multiply row 1 by -1 and add the result to row 3. These operations result in

$$\left(\begin{array}{ccc|c} 1 & \frac{1}{2} & 1 & 3 \\ 0 & 2 & -1 & 3 \\ 0 & \frac{1}{2} & 0 & 0 \end{array} \right)$$

Multiplying row 2 by $\frac{1}{2}$ yields

$$\left(\begin{array}{ccc|c} 1 & \frac{1}{2} & 1 & 3 \\ 0 & 1 & -\frac{1}{2} & \frac{3}{2} \\ 0 & \frac{1}{2} & 0 & 0 \end{array} \right)$$

Now, multiplying row 2 by $-\frac{1}{2}$ and adding the result to both row 1 and row 3, we have

$$\left(\begin{array}{ccc|c} 1 & 0 & \frac{5}{4} & \frac{9}{4} \\ 0 & 1 & -\frac{1}{2} & \frac{3}{2} \\ 0 & 0 & \frac{1}{4} & -\frac{3}{4} \end{array} \right)$$

Multiplying row 3 by 4, we got

$$\begin{pmatrix} 1 & 0 & \frac{5}{4} & \bigm| & \frac{9}{4} \\ 0 & 1 & -\frac{1}{2} & \bigm| & \frac{3}{2} \\ 0 & 0 & 1 & \bigm| & -3 \end{pmatrix}$$

Finally, we multiply row 3 by $-\frac{5}{4}$, adding the result to row 1, and multiply row 3 by $\frac{1}{2}$, adding the result to row 2. This results in

$$(\mathbf{I} \mid \mathbf{A}^{-1}\mathbf{b}) = \begin{pmatrix} 1 & 0 & 0 & \bigm| & 6 \\ 0 & 1 & 0 & \bigm| & 0 \\ 0 & 0 & 1 & \bigm| & -3 \end{pmatrix}$$

and, thus, the solution is

$$\mathbf{x} = \mathbf{A}^{-1}\mathbf{b} = \begin{pmatrix} 6 \\ 0 \\ -3 \end{pmatrix}$$

which is exactly the same solution found in Examples A.14 and A.15.

An Infinite Number of Solutions of Ax = b

The final case to be considered is actually the one of most interest because it reflects the problem class that is typically encountered in linear programming. This is the case of a set of simultaneous linear equations for which an *infinite* number of solutions exist. An infinite number of solutions will exist for a set of simultaneous linear equations if $r(\mathbf{A}) = r(\mathbf{A} \mid \mathbf{b}) < n$, where n is the number of variables.

Example A.17: An Infinite Number of Solutions

Consider the following system:

$$3x_1 + x_2 - x_3 = 8$$

$$x_1 + x_2 + x_3 = 4$$

This system is consistent because $r(\mathbf{A}) = r(\mathbf{A} \mid \mathbf{b}) = 2$, where

$$\mathbf{A} = \begin{pmatrix} 3 & 1 & -1 \\ 1 & 1 & 1 \end{pmatrix} \qquad \mathbf{b} = \begin{pmatrix} 8 \\ 4 \end{pmatrix}$$

Also, the number of variables ($n = 3$) is greater than the rank of the system ($r(\mathbf{A}) = 2$). For such a system, we may choose r equations (where r denotes the rank) and find r of the variables in terms of the remaining $n - r$ variables. That is, for the example,

$$3x_1 + x_2 = 8 + x_3$$

$$x_1 + x_2 = 4 - x_3$$

Solving these two equations for x_1 and x_2 only, we obtain

$$x_1 = 2 + x_3$$

$$x_2 = 2 - 2x_3$$

Because we can assign *any* real value to x_3, the number of solutions to the system of equations is infinite, and, in fact, the set of solutions is given by

$$\mathbf{x} = \begin{pmatrix} x_1 \\ x_2 \\ x_3 \end{pmatrix} = \begin{pmatrix} 2 + \alpha \\ 2 - 2\alpha \\ \alpha \end{pmatrix}$$

where α is any real number.

As mentioned earlier, the type of problem usually encountered in linear programming is one in which an infinite number of solutions (i.e., decision variable values) do exist. However, only a finite number of these infinite many solutions will really be of interest. This finite set of solutions is called the *basic solution set* and is discussed in detail in Chapter 3.

REFERENCES

AARTS, E., AND J. KORST. 1989. *Simulated Annealing and Boltzmann Machines*. New York: John Wiley.

AARTS, E., AND P. VAN LAARHOVEN. 1985. "A New Polynomial Time Cooling Schedule." In *Proceedings of the IEEE International Conference on Computer-Aided Design*, pp. 206–208. Santa Clara, CA.

ACKOFF, R. L., AND M. W. SASIENI. 1968. *Fundamentals of Operations Research*. New York: John Wiley.

AGIN, N. 1965. "Optimum Seeking with Branch and Bound." *Management Science*, Vol. 13, No. 4: 400–412.

ALDER, I., N. KARMARKAR, M. G. C. RESENDE, AND G. VEIGA. 1986. "An Implementation of Karmarkar's Algorithm for Linear Programming," Operations Research Center, Report 86-8. Berkeley: University of California.

ANDERSON, D., AND C. ORTIZ. 1987. "AALPS: A Knowledge-Based System for Aircraft Loading." *IEEE Expert* (Winter): 71–79.

ASPVALL, B., AND R. E. STONE. 1980. "Khachiyan's Linear Programming Algorithm." *Journal of Algorithms*, Vol. 1: 1–13.

BARNES, E. R. 1986. "A Variation of Karmarkar's Algorithm for Solving Linear Programming Problems." *Mathematical Programming*, Vol. 36: 123–134.

———. 1987. "A Polynomial Time Version of the Affine Scaling Algorithm." Paper presented at the Joint National ORSA/TIMS Meeting, St. Louis, October.

BAZARAA, M. S., J. J. JARVIS, AND H. D. SHERALI. 1990. *Linear Programming and Network Flows*, 2d ed. New York: John Wiley.

BAZARAA, M. S., H. D. SHERALI, AND C. M. SHETTY. 1993. *Nonlinear Programming: Theory and Algorithms*, 2d ed. New York: John Wiley.

BEALE, E. M. L. 1955. "Cycling in the Dual Simplex Method." *Naval Research Logistics Quarterly,* Vol. 2, No. 4: 269–276.

BENNETT, K. P., AND O. L. MANGASARIAN. 1990. "Neural Network Training via Linear Programming." Technical Report. Madison: University of Wisconsin, Center for Parallel Optimization.

BLAND, R. G. 1977. "New Finite Pivoting Rules for the Simplex Method." *Mathematics of Operations Research,* Vol. 2: 103–107.

BLAND, R. G., D. GOLDFARB, AND M. J. TODD. 1981. "The Ellipsoid Method: A Survey." *Operations Research,* Vol. 26, No. 6: 1039–1091.

BROWN, D. E., AND C. L. HUNTLEY. 1992. "A Practical Application of Simulated Annealing to Clustering." *Pattern Recognition,* Vol. 14, No. 4 (April): 401–412.

CAMPBELL, H., AND J. P. IGNIZIO. 1972. "Using Linear Programming for Predicting Student Performance." *Journal of Educational and Psychological Measurement,* Vol. 32: 397–401.

CAVALIER, T. M., AND K. C. SCHALL. 1987. "Implementing an Affine Scaling Algorithm for Linear Programming." *Computers and Operations Research,* Vol. 14, No. 5: 341–347.

CAVALIER, T. M., AND A. L. SOYSTER. 1985. "Some Computational Experience and a Modification of the Karmarkar Algorithm." Paper presented at the 12th International Symposium on Mathematical Programming, Massachusetts Institute of Technology, Cambridge, MA, August.

CAYLEY, A. 1874. "On the Mathematical Theory of Isomers." *Philosophical Magazine,* Vol. 67: 444.

CHARNES, A., AND W. W. COOPER. 1961. *Management Models and Industrial Applications of Linear Programming,* Vols. 1 and 2. New York: John Wiley.

———. 1975. "Goal Programming and Constrained Regression: A Comment." *OMEGA,* Vol. 3: 403–409.

CHARNES, A., W. W. COOPER, AND A. HENDERSON. 1953. *An Introduction to Linear Programming.* New York: John Wiley.

CHARNES, A., AND K. O. KORTANEK. 1963. "An Opposite Sign Algorithm for Purification to an Extreme Point Solution," Office of Naval Research, Memorandum No. 84. Evanston, IL: Northwestern University.

CHRISTOFIDES, N. 1975. *Graph Theory: An Algorithmic Approach.* New York: Academic Press.

CHRISTOPHERSON, D., AND E. C. BAUGHAN. 1992. "Reminiscences of Operational Research in World War II by Some of Its Practitioners: II." *Journal of the Operational Research Society,* Vol. 43, No. 6 (June): 569–577.

CHURCHMAN, C. W., R. L. ACKOFF, AND E. L. ARNOFF. 1957. *Introduction to Operations Research.* New York: John Wiley.

CHVATAL, V. 1980. *Linear Programming.* New York and San Francisco: W. H. Freeman.

COCHARD, D. D., AND K. A. YOST. 1985. "Improving Utilization of Air Force Cargo Aircraft." *Interfaces,* Vol. 15, No. 1: 53–68.

COHON, J. L. 1978. *Multiobjective Programming and Planning.* New York: Academic Press.

CROWDER, H., E. L. JOHNSON, AND M. PADBERG. 1983. "Solving Large-Scale Zero-One Linear Programming Problems." *Operations Research,* Vol. 31, No. 5: 803–834.

CROWDER, L. J., AND V. A. SPOSITO. 1991. "Sequential Linear Goal Programming: Implementation via MPSX/370E." *Computers and Operations Research,* Vol. 18, No. 3: 291–296.

CUNNINGHAM, W. H. 1979. "Theoretical Properties of the Network Simplex Method." *Mathematics of Operations Research,* Vol. 4, No. 2: 196–208.

DANTZIG, G. B. 1948. *Programming in a Linear Structure.* Washington, D.C.: Comptroller, U.S. Air Force.

———. 1963. *Linear Programming and Extensions.* Princeton: Princeton University Press.

DANTZIG, G. B., A. ORDEN, AND P. WOLFE. 1955. "The Generalized Simplex Method for Minimizing a Linear Form under Linear Inequality Restraints." *Pacific Journal of Mathematics,* Vol. 5, No. 2: 183–195.

DANTZIG, G. B., AND P. WOLFE. 1960. "Decomposition Principle for Linear Programs." *Operations Research,* Vol. 8, No. 1: 101–111.

———. 1961. "The Decomposition Algorithm for Linear Programs." *Econometrica,* Vol. 29, No. 4: 767–778.

DAUER, J. P., AND R. J. KRUEGER. 1977. "An Iterative Approach to Goal Programming." *Operational Research Quarterly,* Vol. 28, No. 3: 671–681.

DIKIN, I. I. 1967. "Iterative Solution of Problems of Linear and Quadratic Programming." *Soviet Mathematics Doklady,* Vol. 8: 674–675.

DIKIN, I. I. 1974. "On the Speed of an Iterative Process," *Upravlyaemye Sistemi,* Vol. 12, pp. 54–60.

ECKER, J. G., AND M. KUPFERSCHMID. 1988. *Introduction to Operations Research.* New York: John Wiley.

EVANS, J. R., AND E. MINIEKA. 1992. *Optimization Algorithms for Networks and Graphs,* 2d ed. New York: Marcel Dekker.

FIACCO, A. V., AND G. P. McCORMICK. 1968. *Nonlinear Programming.* New York: John Wiley.

FLEISHER, H. J., M. A. TAVEL, AND D. B. MARTIN. 1985. "Simulated Annealing as a Tool for Logic Optimization in a CAD Environment." In *Proceedings of the IEEE International Conference on Computer-Aided Design,* pp. 203–205. Santa Clara, CA.

FRAZER, R. J. 1968. *Applied Linear Programming.* Englewood Cliffs, NJ: Prentice Hall.

FREED, N., AND F. GLOVER. 1981. "A Linear Programming Approach to the Discriminant Problem." *Decision Sciences,* Vol. 12: 68–74.

GAL, T. 1977. "A General Method for Determining the Set of All Efficient Solutions to a Linear Vectormaximum Problem." *European Journal of Operational Research,* Vol. 1, No. 5: 307–322.

GAREY, M. R., AND D. S. JOHNSON. 1979. *Computers and Intractability: A Guide to the Theory of NP-Completeness.* San Francisco: W. H. Freeman.

GARFINKEL, R. S., AND G. L. NEMHAUSER. 1972. *Integer Programming.* New York: John Wiley.

GASS, S. I. 1986. "A Process for Determining Priorities and Weights for Large-Scale Linear Goal Programming Models." *Journal of the Operational Research Society,* Vol. 37, No. 8: 779–784.

———. 1987. "The Setting of Weights in Linear Goal Programming Problems." *Computers and Operations Research,* Vol. 14, No. 3: 227–229.

GEMAN, S., AND D. GEMAN. 1985. "Stochastic Relaxation, Gibbs Distributions, and the Restoration of Images." *IEEE Transactions on Pattern Analysis and Machine Intelligence,* PAMI-6: 721–741.

GLOVER, F. 1986. "Future Paths for Integer Programming and Links to Artificial Intelligence." *Computers and Operations Research,* Vol. 13: 533–549.

GLOVER, F., E. TAILLARD, AND D. DE WERRA. 1993. "A User's Guide to Tabu Search," in F. Glover, D. de Werra, eds. *Annals of Operations Research: Tabu Search,* J. C. Baltzer AG: Switzerland, Vol. 41.

GOLDBERG, D. E. 1989. *Genetic Algorithms.* Reading, MA: Addison-Wesley.

GOMORY, R. E. 1960. "An Algorithm for the Mixed Integer Problem," Research Memorandum RM-2597. Santa Monica, CA: The RAND Corporation.

———. 1963. "An Algorithm for Integer Solutions to Linear Programs." In R. L. Graves and P. Wolfe, eds. *Recent Advances in Mathematical Programming,* pp. 269–302. New York: McGraw-Hill.

HADLEY, G. 1963. *Linear Programming.* Reading, MA: Addison-Wesley.

———. 1973. *Linear Algebra.* Reading, MA: Addison-Wesley.

HALL, A. D. 1962. *A Methodology for Systems Engineering.* New York: Van Nostrand Reinhold.

HILLIER, F. S., AND G. J. LIEBERMANN. 1990. *Introduction to Operations Research,* 5th ed. New York: McGraw-Hill.

HOOKER, J. N. 1986. "Karmarkar's Linear Programming Algorithm." *Interfaces,* Vol. 16, No. 4: 75–90.

HU, T. C. 1970. *Integer Programming and Network Flows.* Reading, MA: Addison-Wesley.

HWANG, C. L., AND A. S. MASUD. 1979. *Multiple Objective Decision Making—Methods and Applications.* New York: Springer Verlag.

IGNIZIO, J. P. 1963. "S-II Trajectory Study and Optimum Antenna Placement," Report SID-63. Downey, CA: North American Aviation Corporation.

———. 1967. "A FORTRAN Code for Multiple Objective LP," Internal Report. Downey, CA: North American Aviation Corporation.

———. 1968. "A Method to Achieve Optimum Air Defense Sensor Allocation." M. S. thesis, University of Alabama, Alabama.

———. 1971. "A Heuristic Solution to Generalized Covering Problems," Ph.D. dissertation, Virginia Polytechnic Institute, Blacksburg.

———. 1974a. "The Development of the Multidimensional Dual in Linear Goal Programming," Technical Paper. Pennsylvania State University, University Park.

———. 1974b. "A Primal-Dual Algorithm for Linear Goal Programming," Technical Paper. Pennsylvania State University, University Park.

———. 1976. *Goal Programming and Extensions.* Lexington, MA: D. C. Heath.

———. 1978. "Solving Large Scheduling Problems by Minimizing Conflict," *Simulation* (March): 75–79.

———. 1980. "Solving Large Scale Problems: A Venture Into a New Dimension." *Journal of the Operational Research Society,* Vol. 31: 217–225.

———. 1981. "Antenna Array Beam Pattern Synthesis via Goal Programming." *European Journal of Operational Research,* Vol. 6: 286–290.

———. 1982. *Linear Programming in Single and Multiple Objective Systems.* Englewood Cliffs, NJ: Prentice Hall.

——— (guest editor and contributor). 1983. *Generalized Goal Programming,* Special Issue of *Computers and Operations Research,* Vol. 10, No. 4.

———. 1984. "A Note on the Multidimensional Dual." *European Journal of Operational Research,* Vol. 17, No. 1: 116–122.

———. 1985a. *Introduction to Linear Goal Programming.* Beverly Hills: Sage.

———. 1985b. "Multiobjective Mathematical Programming via the MULTIPLEX Model and Algorithm." *European Journal of Operational Research,* Vol. 22, No. 3: 338–346.

———. 1985c. "An Algorithm for Solving the Linear Goal Programming Problem by Solving Its Dual." *Journal of the Operational Research Society,* Vol. 36, No. 6: 507–515.

———. 1986. "Discriminant Analysis via Goal Programming: On the Elimination of Certain Problems." In *Proceedings of Southwestern Institute of Decision Sciences,* Houston, TX: Institution for Decision Sciences, pp. 125–126.

———. 1991. *Introduction to Expert Systems.* New York: McGraw-Hill.

———. 1992a. "Knowledge Programming Meets Linear Programming I," *PC AI,* Vol. 6, No. 5, pp. 25–29.

———. 1992b. "Modifications to the Baek and Ignizio Method for Pattern Classification," Technical Paper. Charlottesville: University of Virginia.

IGNIZIO, J. P., AND W. BAEK. 1993. "Knowledge Programming Meets Linear Programming II," *PC AI,* Vol. 7, No. 1, pp. 45–49.

IGNIZIO, J. P., AND L. IGNIZIO-BURKE (guest editors). 1992. *Neural Networks and Operations Research,* Special Issue of *Computers and Operations Research,* Vol. 19, No. 3/4 (April–May).

IGNIZIO, J. P., AND K. E. CASE. 1972. "A Method for Validating Certain Set-Covering Heuristics." In Paul Brock, ed., *The Mathematics of Large-Scale Simulation,* Chap. 13. La Jolla, CA: Simulation Councils.

IGNIZIO, J. P., AND T. M. CAVALIER. 1986. "Classification Analysis via Goal Programming and Implications for Supercomputer Implementation." In *Proceedings of Conference on Supercomputing,* pp. 446–450. Santa Clara, CA: International Supercomputing Institute, Inc.

IGNIZIO, J. P., AND R. M. HARNETT. 1974. "Heuristically Aided Set Covering Algorithms." *International Journal of Computer and Information Systems,* Vol. 3, No. 1: 59–70.

IGNIZIO, J. P., AND J. H. PERLIS. 1977. "Sequential Linear Goal Programming: Implementation via MPSX." *Computers and Operations Research,* Vol. 3: 141–145.

IGNIZIO, J. P., K. W. WIEMANN, AND W. J. HUGHES. 1987. "Sonar Array Element Location: A Hybrid Expert Systems Application." *European Journal of Operational Research,* Vol. 32: 76–85.

IGNIZIO-BURKE, LAURA. 1985. "A Multicriteria Approach to the Interceptor Scheduling Problem." Masters thesis, Pennsylvania State University, University Park.

JAMES, MIKE. 1985. *Classification Algorithms.* London: John Wiley.

JENSEN, P. A., AND J. W. BARNES. 1980. *Network Flow Programming.* New York: John Wiley.

JOHNSON, D. E., AND J. R. JOHNSON. 1972. *Graph Theory.* New York: Ronald Press.

JOHNSON, D. S., C. R. ARAGON, AND L. A. MCGEOCH. 1989. "Optimization by Simulated Annealing: An Experimental Evaluation; Graph Partitioning." *Operations Research,* Vol. 37: 865–892.

KARMARKAR, N. 1984. "A New Polynomial-Time Algorithm for Linear Programming." *Combinatorica,* Vol. 4: 373–395.

KARP, R. 1975. "On the Computational Complexity of Combinatorial Problems." *Networks,* Vol. 5: 45–68.

KARUSH, W. 1939. "Minima of Functions of Several Variables with Inequalities as Side Constraints." M. S. thesis, Department of Mathematics, University of Chicago.

KEENEY, R. L., AND H. RAIFFA. 1976. *Decisions with Multiple Objectives.* New York: John Wiley.

KHACHIAN, L. G. 1979. "A Polynomial Algorithm in Linear Programming." *Soviet Mathematics Doklady,* Vol. 20: 191–194.

KIRKPATRICK, S., C. GELATT, AND M. VECCHI. 1983. "Optimization by Simulated Annealing." *Science,* Vol. 220: 671–680.

KLEE, V., AND G. MINTY. 1972. "How Good Is the Simplex Algorithm?" In O. Shisha, ed., *Inequalities III,* pp. 159–175. New York: Academic Press.

KORNBLUTH, J. H. S. 1973. "A Survey of Goal Programming." *OMEGA,* Vol. 1: 193–205.

KORTANEK, K. O., AND Z. JISHAN. 1988. "New Purification Algorithms for Linear Programming." *Naval Research Logistics Quarterly,* Vol. 35: 571–583.

KORTANEK, K. O., AND M. SHI. 1987. "Convergence Results and Numerical Experiments on a Linear Programming Hybrid Algorithm." *European Journal of Operational Research,* Vol. 32, No. 1: 47–61.

KORTANEK, K. O., AND H. M. STROJWAS. 1984. "An Application of the Charnes-Kortanek-Raike Purification Algorithm to Extreme Points and Extreme Directions." Pittsburgh: Carnegie-Mellon University, Department of Mathematics.

KUHN, H. W. 1955. "The Hungarian Method for the Assignment Problem." *Naval Research Logistics Quarterly,* Vol. 2, No. 1–2: 83–97.

KUHN, H. W., AND A. W. TUCKER. 1950. "Nonlinear Programming." In J. Neyman, ed., *Proceedings of the Second Berkeley Symposium on Mathematical Statistics and Probability,* pp. 481–492. Berkeley: University of California Press.

LASDON, L. S. 1970. *Optimization Theory for Large Systems.* New York: Macmillan.

LAWLER, E. L., J. K. LENTRA, A. H. G. RINNOOY KAN, AND D. B. SHMOYS. 1985. *The Traveling Salesman.* New York: John Wiley.

LAWLER, E. L., AND D. E. WOOD. 1966. "Branch and Bound Methods: A Survey." *Operations Research,* Vol. 14: 699–719.

LEMKE, C. E. 1954. "The Dual Method of Solving the Linear Programming Problem." *Naval Research Logistics Quarterly,* Vol. 1: 48–54.

LEONTIEF, W. W. 1951. *The Structure of the American Economy, 1919–1939.* New York: Oxford University Press.

LEWIS, H. R., AND C. H. PARADIMITRION. 1978. "The Efficiency of Algorithms." *Scientific American* (January): 96–109.

LLEWELLYN, R. W. 1964. *Linear Programming.* New York: Holt, Rinehart, and Winston.

LUENBERGER, D. G. 1984. *Introduction to Linear and Non-Linear Programming,* 2d ed. Reading, MA: Addison-Wesley.

LUNDY, M., AND A. MEES. 1986. "Convergence of an Annealing Algorithm." *Mathematical Programming,* Vol. 34: pp. 111–124.

LUSTIG, I. J. 1985. "A Practical Approach to Karmarkar's Algorithm," Technical Report Sol 85-5. Stanford, CA: Stanford University, Department of Operations Research.

MANGASARIAN, O. L. 1965. "Linear and Nonlinear Separation of Patterns by Linear Programming." *Operations Research,* Vol. 13: 444–452.

———. 1969. *Nonlinear Programming.* New York: McGraw-Hill.

MARCH, J. G., AND H. A. SIMON. 1958. *Organizations.* New York: John Wiley.

MARKOWSKI, C. A., AND J. P. IGNIZIO. 1983a. "Theory and Properties of the Lexicographic LGP Dual." *Large Scale Systems,* Vol. 5, No. 2: 115–121.

———. 1983b. "Duality and Transformations in Multiphase and Sequential LGP." *Computers and Operations Research,* Vol. 10, No. 4 (October): 321–334.

MARSHALL, K. T., AND J. W. SUURBALLE. 1969. "A Note on Cycling in the Simplex Method." *Naval Research Logistics Quarterly,* Vol. 16: 121–137.

MARSTEN, R. E., M. J. SALTZMAN, D. F. SHANNO, G. S. PIERCE, AND J. F. BALLINTIJN. 1988. "Implementation of a Dual Affine Interior Point Algorithm for Linear Programming," Working Paper CMI-WPS-88-06. Tucson: University of Arizona Center for the Management of Information.

MARTELLO, S., AND P. TOTH. 1990. *Knapsack Problems: Algorithms and Computer Implementations.* New York: John Wiley.

MEGIDDO, N. 1986. "Pathways to the Optimal Set in Linear Programming," Research Report RJ 5295. San Jose, CA: IBM Almaden Research Center.

METROPOLIS, N., A. ROSENBLUTH, M. A. TELLER, AND E. TELLER. 1953. "Equations of State Calculations by Fast Computing Machines." *Journal of Chemical Physics,* Vol. 21: 1087–1092.

MILLER, C. E., A. W. TUCKER, AND R. A. ZEMLIN. 1960. "Integer Programming Formulations and Traveling Salesman Problems." *Journal of the Association for Computing Machinery,* Vol. 7: 326–329.

MIN, H., AND J. STORBECK. 1991. "On the Origin and Persistence of Misconceptions in Goal Programming." *Journal of the Operational Research Society*, Vol. 42, No. 4: 301–312.

MURTAGH, B. A. 1981. *Advanced Linear Programming*. New York: McGraw Hill.

MURTY, K. G. 1976. *Linear and Combinatorial Programming*. New York: John Wiley.

———. 1983. *Linear Programming*. New York: John Wiley.

NEMHAUSER, G. L., AND L. A. WOLSEY. 1988. *Integer and Combinatorial Optimization*. New York: John Wiley.

NICHOLSON, T. A. J. 1971. *Optimization in Industry*. Chicago: Aldine-Atherton.

NOBLE, B. 1969. *Applied Linear Algebra*. Englewood Cliffs, NJ: Prentice-Hall.

PADBERG, M. 1986. "A Different Convergence Proof of the Projective Method for Linear Programming." *Operations Research Letters*, Vol. 4: 253–257.

PHILLIPS, D. T., AND A. GARCIA-DIAZ. 1990. *Fundamentals of Network Analysis*. Prospect Heights, IL: Waveland Press.

PHILLIPS, D. T., A. RAVINDRAN, AND J. SOLBERG. 1976. *Operations Research: Principles and Practice*. New York: John Wiley.

ROY, A., AND S. MUKHOIPADHYAY. 1991. "Pattern Classification using Linear Programming." *ORSA Journal on Computing*, Vol. 3, No. 1 (Winter): 66–80.

SAATY, T. L. 1980. *The Analytic Hierarchy Process*. New York: McGraw-Hill.

SALKIN, H. M., AND K. MATHUR. 1989. *Foundations of Integer Programming*. New York: North-Holland.

SHERALI, H. D. 1987. "Algorithmic Insights and a Convergence Analysis for a Karmarkar-Type Algorithm for Linear Programs." *Naval Research Logistics Quarterly*, Vol. 34: 399–416.

SIEGEL, S. 1956. *Nonparametric Statistics for the Behavioral Sciences*. New York: McGraw-Hill.

SILVANO, M., AND P. TOTH. 1990. *Knapsack Problems: Algorithms and Computer Implementations*. Chicester, England: John Wiley.

SIMON, H. A. 1957. *Administrative Behavior*. New York: Macmillan.

SIMON, H. A., AND A. NEWELL. 1958. "Heuristic Problem Solving: The Next Advance in Operations Research." *Operations Research* (January–February): 1–10.

SMYTHE, W. R., AND L. A. JOHNSON. 1966. *Introduction to Linear Programming, with Applications*. Englewood Cliffs, NJ: Prentice Hall.

TAHA, H. A. 1975. *Integer Programming: Theory, Applications, and Computations*. New York: Academic Press.

TEN DYKE, R. P. 1990. "A Taxonomy of AI Applications." *AI Week*, February 1, pp. 6–7.

THERRIEN, C. W. 1989. *Decision Estimation and Classification: An Introduction to Pattern Recognition and Related Topics*. New York: John Wiley.

TODD, M. J., AND B. P. BURRELL. 1986. "An Extension of Karmarkar's Algorithm for Linear Programming Using Dual Variables." *Algorithmica*, Vol. 1: 409–424.

VAN LAARHOVEN, P. J. M., AND E. H. L. AARTS. 1987. *Simulated Annealing: Theory and Applications*. Dordrecht: D. Reidel.

VANDERBEI, R. J., M. S. MEKETON, AND B. A. FREEDMAN. 1986. "A Modification of Karmarkar's Linear Programming Algorithm." *Algorithmica,* Vol. 1, No. 4: 395–408.

WADDINGTON, C. H. 1973. *OR in World War II: Operational Research Against the U-Boat.* London: Elek Science.

WILDE, D. J., AND C. S. BEIGHTLER. 1967. *Foundations of Optimization.* Englewood Cliffs, NJ: Prentice Hall.

WINSTON, W. L. 1991. *Introduction to Mathematical Programming: Applications and Algorithms.* Boston: PWS-Kent.

WOLFE, C. S. 1973. *Linear Programming with FORTRAN.* Glenview, IL: Scott, Foresman.

WU, N., AND R. COPPINS. 1981. *Linear Programming and Extensions.* New York: McGraw-Hill.

YOUNG, T. Y., AND T. W. CALVERT. 1974. *Classification, Estimation and Pattern Recognition.* New York: American Elsevier.

ZIMMERMANN, H. J. 1978. "Fuzzy Programming and Linear Programming with Several Objectives." *Fuzzy Sets and Systems,* Vol. 1: 45–55.

———. 1985. *Fuzzy Set Theory—and Its Applications.* Boston: Kluwer-Nijhoff.

INDEX